AN INTRODUCTION TO COMMUNITY DEVELOPMENT

Beginning with the foundations of community development, *An Introduction to Community Development* offers a comprehensive and practical approach to planning for communities. Road-tested in the authors' own teaching, and through the training they provide for practicing planners and community developers, it enables students to begin making connections between academic study and practical know-how from both private and public sector contexts.

An Introduction to Community Development shows how planners and community developers can utilize local economic interests and integrate finance and marketing considerations into their strategy. Most importantly, the book is strongly focused on outcomes, encouraging students to ask: what is best practice when it comes to planning for communities, and how do we accurately measure the results of planning and development practice?

This newly revised and updated edition includes:

- increased coverage of sustainability issues;
- discussion of localism and its relation to community development;
- quality of life, community well-being, and public health considerations; and
- content on local food systems.

Each chapter provides a range of reading materials for the student, supplemented with text boxes, a chapter outline, keywords, reference lists, and new skills-based exercises at the end of each chapter to help students turn their learning into action, making this the most user-friendly text for community development now available.

Rhonda Phillips, Ph.D., AICP, is Dean of Purdue University's Honors College, USA, and a Professor in the Agricultural Economics Department. Previously, at Arizona State University, USA, she served as Professor in the School of Community Resources and Development; Senior Sustainability Scientist at the Global Institute of Sustainability; and Affiliate Faculty in the School of Geographical Sciences and Urban Planning. Rhonda is author or editor of 18 books, including *Sustainable Communities: Creating a Durable Local Economy* (Routledge Earthscan, 2013, co-editor). She is editor of the book series *Community Development Research and Practice* (Community Development Society and Routledge) and serves as the President of the International Society for Quality-of-Life Studies (www.isqols.org).

Robert H. Pittman, Ph.D., PCED, is Founder and Executive Director of the Janus Institute and Janus Forum, USA, whose mission is to advance the community and economic development profession and help communities in new and innovative ways. Attendees include leading community and economic developers from around the country and overseas, state directors, corporate executives, and other stakeholders in community and economic development. Pittman has held a number of executive positions in the field of community and economic development, including Director of the Global Business Location and Economic Development Consulting Group at Lockwood Greene, USA, and Deputy Director of the International Development Research Council (now Corenet). He also served as Associate Professor of Economics and Director of the Community Development Institute at the University of Central Arkansas, USA. Pittman is a widely published author and frequent speaker in the field of community and economic development, and an Honorary Life Member of the Southern Economic Development Council.

"While there are plenty of books on community development, this book is very comprehensive, covers most of the aspects of community development, and is perfect for undergraduate courses on community development. The authors are well respected in the area of community development with extensive experience and expertise."

Muthusami Kumaran, Assistant Professor,
Nonprofit Management & Community Organizations, University of Florida, USA

"The topics are excellent and cover almost every facet of community development."

Joan Wesley, Associate Professor, Urban and Regional Planning,
Jackson State University, USA

"The chapters are eloquent and accessible and provide a good introductory insight into the key considerations associated with this subject."

Adam Sheppard, JDL Course Director and Senior Lecturer,
Department of Geography and Environmental Management,
University of the West of England, UK

"I am attracted first by the sustentative content of the book. I especially like the first two parts that cover the theoretical foundations of community development and the how-to's of preparation and planning. The numerous case studies are also essential to illuminate the theoretical points and show that they work in practice."

Martin Saiz, Professor, Political Science,
California State University, Northridge, USA

AN INTRODUCTION TO COMMUNITY DEVELOPMENT

Second Edition

Edited by
Rhonda Phillips and Robert H. Pittman

NEW YORK AND LONDON

First published 2015
by Routledge
711 Third Avenue, New York, NY 10017

and by Routledge
2 Park Square, Milton Park, Abingdon, Oxon, OX14 4RN

Routledge is an imprint of the Taylor & Francis Group, an informa business

Library of Congress Cataloging in Publication Data
An introduction to community development / edited by Rhonda Phillips, Robert H. Pittman. – 2 Edition.
pages cm
Includes bibliographical references and index.
1. Community development. 2. Economic development. I. Phillips, Rhonda, editor of compilation. II. Pittman, Robert H., editor of compilation.
HN49.C6I554 2014
307.1'4–dc23

2014015459

ISBN: 978-0-415-70356-7 (hbk)
ISBN: 978-0-415-70355-0 (pbk)
ISBN: 978-0-203-76263-9 (ebk)

Typeset in Futura
by Saxon Graphics Ltd, Derby

Contents

Figures

Tables

Boxes

Contributors

Jeni Burnell studied architecture at the University of New South Wales in Sydney, Australia, before pursuing a career in participatory arts, community development, and design. She has a Master's in Development and is Research Associate at the Centre for Development and Emergency Practice (CENDEP) based at Oxford Brookes University, UK. Jeni is Chair of the development initiative, the Small Change Forum, and Director of the architecture practice, Space Program Ltd.

Timothy R. Dahlstrom, Ph.D. is a faculty associate in the School of Public Affairs at Arizona State University, USA. His research interests include entrepreneurial development programs, the nexus of business and government, and organizational leadership. He has practice experience in public entrepreneurial financing programs, small organization consulting, and public–private partnerships.

Jessica LeVeen Farr is Regional Community Development Manager for Tennessee for the Federal Reserve Bank of Atlanta, Nashville Branch, USA. She works with banks, nonprofit organizations, and government agencies to develop programs that promote asset building, affordable housing, job creation, and other related community development initiatives. Previously, Farr was Assistant Vice President at Bank of America in Nashville in the Community Development Corporation (CDC), where she oversaw the single family housing development program. Farr received her Master's in City Planning from UNC-Chapel Hill with an emphasis on community economic development in 1999. She graduated from University of California, San Diego, USA, in 1993 with a BA in urban studies.

Hamilton Galloway is an economist and consultant with extensive experience working in the US as well as the UK on economic research and labor market analysis. Mr. Galloway's primary focus is on helping organizations understand economic development opportunities, workforce needs and education linkages through economic modeling techniques. He has led numerous strategic planning and development projects, ranging from workforce development to comprehensive economic development strategies. Galloway has also provided research and assistance in areas of international business finance and large-scale adjustments, including NASA's shuttle program shutdown and Maytag's closure. He has authored numerous academic papers and several book chapters. Galloway is a frequent speaker at national conferences about knowledge economies, data-driven analysis for regional decision making and new methods of estimating the impacts of further and higher education.

Gary Paul Green is a professor in the Department of Community and Environmental Sociology at the University of Wisconsin–Madison, USA, and a community development specialist at the University of Wisconsin–Extension. Green's teaching and research interests are primarily in the areas of community and economic development. He received his Ph.D. at the University of Missouri-Columbia, USA, and has taught at the University of Wisconsin–Madison, USA, for the past 20 years. His recent books include *Mobilizing Communities: Asset Building as a Community Development Strategy* (Temple University Press, 2010); *Introduction to Community Development: Theory, Practice, and Service-Learning* (Sage, 2011); *Asset Building and Community Development*, 3rd ed. (Sage, 2012); and *Handbook of Rural Development* (Edward Elgar, 2014). In addition to his work in the US, he has been involved in research and teaching in China, New Zealand, South Korea, Uganda, and Ukraine.

John Gruidl is a professor in the Illinois Institute for Rural Affairs at Western Illinois University, USA, where he teaches, conducts research, and creates new outreach programs in community and economic development. Gruidl earned a Ph.D. in Agricultural and Applied Economics from the University of Wisconsin-Madison in 1989, with a major in the field of community economics. Gruidl has created and directed several successful programs in community development. From 1994–2005, he directed the award-winning Peace Corps Fellows Program, a community-based internship program for returned Peace Corps volunteers. Gruidl also helped to create the MAPPING the Future of Your Community, a strategic visioning program for Illinois communities. He currently serves as Director of the Midwest Community Development Institute, a training program for community leaders.

Anna Haines is a professor in the College of Natural Resources at the University of Wisconsin–Stevens Point, USA, and a land use and community development specialist with the University of Wisconsin–Extension. Haines received her Ph.D. from the University of Wisconsin–Madison, USA, in the Department of Urban and Regional Planning. Her research and teaching focuses on planning and community development from a natural resources or environmental perspective. Her research has focused on factors that influenced land division in amenity-rich areas of Wisconsin, understanding how zoning codes affect future build out scenarios, examining sustainable zoning, and understanding the origination and resilience of local food clusters. Her extension work has focused on comprehensive planning and planning implementation tools and techniques, sustainable communities, and property rights issues.

Janet R. Hamer is Senior Community Development Manager with the Federal Reserve Bank of Atlanta, Jacksonville Branch, USA. Her primary geographic areas of responsibility are north, central, and southwest Florida. Hamer has over 20 years of experience in housing, community and economic development, and urban planning. Prior to joining the Federal Reserve, she served as Chief of Housing Services for the Planning and Development Department of the City of Jacksonville for three years. For the previous 18 years, she served as Deputy Director of the Community Development Department of the City of Daytona Beach. Before moving to Florida, Hamer was employed as a regional planner in Illinois. She has a BA from Judson College, Elgin, Illinois and an MA in Public Affairs from Northern Illinois University, DeKalb, Illinois.

Ronald J. Hustedde is a professor in the department of community and leadership development at the University of Kentucky, USA. He teaches graduate courses in community development and has an Extension (public outreach) appointment. He is a past president of the Community Development Society and has served on the board of directors of the International Association for Community Development (based in Scotland). His community development work has focused on public issues deliberation, public conflict analysis and resolution, leadership development, and rural entrepreneurship. He has a Ph.D. in sociology from the University of Wisconsin–Madison, USA, with three other graduate degrees in community development, agricultural economics, and rural sociology.

Patsy Kraeger received her Ph.D. in Public Administration from the School of Public Affairs at Arizona State University (ASU), USA. She has worked in the public, nonprofit, and private sectors. She is also a faculty associate at the ASU School of Public Affairs. She has published in academic and practitioner journals and presented at scholarly and professional conferences; her area of focus is on strategic nonprofit management and social enterprise development.

Paul R. Lachapelle is Associate Professor in the Department of Political Science at Montana State University–Bozeman (MSU), USA, and serves as the Extension Community Development Specialist. Working in partnership with the MSU Local Government Center, his responsibilities involve providing research, technical assistance and training on various community development topics including community strategic visioning, local governance, and leadership development. Paul earned his Ph.D. at the University of Montana's College of Forestry and Conservation with a focus on natural resource policy and governance. He has conducted research and taught in both Montana and internationally, doing applied research in Nepal as a U.S. Fulbright Scholar, working with Inuit in several arctic national parks with Parks Canada in the Canadian Territory of Nunavut, and working in South Africa and Guinea, West Africa. Paul is author of many peer-reviewed journal articles on local governance and strategic planning.

Michael McGrath is Chief Information Officer of NCL and Senior Editor of the National Civic League. He edits the award-winning *National Civic Review*, a 99-year-old quarterly civic affairs journal, and is the organization's chief blogger. Before joining NCL, he was a reporter and feature writer for *Westword* newspaper in Colorado and an associate editor of the *East Bay Express* in Berkeley/Oakland, California, USA, covering environmental policy and local and state government. His freelance articles have appeared in magazines, newspapers, and online publications. Recently, he researched and authored a report on local government and civic engagement for PACE: Philanthropy for Active Civic Engagement. He has a Master's degree from UC Berkeley's Graduate School of Journalism and a Bachelor's degree (general and comparative studies) from the University of Texas, Austin, USA.

Deborah Markley is Managing Director and Director of Research for the Rural Policy Research Institute's Center for Rural Entrepreneurship, a national research and policy center in Lincoln, Nebraska, USA. Her research has included case studies of entrepreneurial support organizations, evaluation of state industrial extension programs, and consideration of the impacts of changing banking markets on small business finance. She has extensive experience conducting field-based survey research projects and has conducted focus groups and interviews with rural bankers, entrepreneurs, business service providers, venture capitalists, small manufacturers, and others. Her research has been presented in academic journals, as well as to national public policy organizations and congressional committees.

Paul Mattessich is the Executive Director of Wilder Research, one of the largest applied social research organizations in the United States, dedicated to improving the lives of individuals, families, and communities. Dr. Mattessich has done research, lecturing, and consulting with nonprofit organizations, foundations, and government in North America,

Europe, and Africa since 1973. His book on effective partnerships, *Collaboration: What Makes It Work*, (Wilder Publishing 2001, 2nd ed.) is used worldwide, along with the online Wilder Collaboration Factors Inventory. His 1997 book, *Community Building: What Makes It Work* (Wilder Publishing) is widely recognized and used by leaders and practitioners in community and neighborhood development. He received his Ph.D. in sociology from the University of Minnesota, USA.

Rhonda Phillips, Ph.D., AICP, is Dean of Purdue University's Honors College, USA, and a Professor in the Agricultural Economics Department. Previously, at Arizona State University, USA, she served as Professor in the School of Community Resources and Development; Senior Sustainability Scientist at the Global Institute of Sustainability; and Affiliate Faculty in the School of Geographical Sciences and Urban Planning. Rhonda is author or editor of 18 books, including *Sustainable Communities: Creating a Durable Local Economy* (Routledge Earthscan, 2013, co-editor). She is editor of the book series *Community Development Research and Practice* (Community Development Society and Routledge) and serves as the President of the International Society for Quality-of-Life Studies (www.isqols.org).

Robert H. Pittman, Ph.D., PCED, is Founder and Executive Director of the Janus Institute and Janus Forum, USA, whose mission is to advance the community and economic development profession and help communities in new and innovative ways. Attendees include leading community and economic developers from around the country and overseas, state directors, corporate executives, and other stakeholders in community and economic development. Pittman has held a number of executive positions in the field of community and economic development, including Director of the Global Business Location and Economic Development Consulting Group at Lockwood Greene, USA, and Deputy Director of the International Development Research Council (now Corenet). He also served as Associate Professor of Economics and Director of the Community Development Institute at the University of Central Arkansas, USA. Pittman is a widely published author and frequent speaker in the field of community and economic development, and an Honorary Life Member of the Southern Economic Development Council.

Alan Price is the founder of Falcon Leadership, LLC. Since his retirement from Delta Air Lines in 2004, Captain Price has worked with clients across the country and overseas in areas relating to leadership development/crisis leadership, strategic planning, safety systems, communications, and team development. As Chief Pilot for Delta's Atlanta base, Capt Price led the recovery efforts for the Atlanta base following 9/11. His particular passion, which grew out of these experiences, is leadership during a time of crisis, a topic upon which he has spoken extensively. Price was a first-in-class facilitator for Delta Air Lines CRM program, and Pilot Manager of the Human Factors department. He developed and led Delta's new captain's leadership program for five years while conducting classes for over 2,000 new captains. He was for many years a regular contributor to Delta's flight operations safety magazine, *Up Front*, and has written extensively in many publications on the topics of leadership, safety, and CRM. Capt. Price is a frequent speaker, workshop developer, and master facilitator for a range of topics dealing with leadership/crisis leadership, teamwork, communications, and strategic planning. He holds a BS degree in engineering science from the USAF Academy and a Master's in Decision Science from Georgia State University, USA, and he has done postgraduate work at the

Georgia Institute of Technology in Man-Machine Systems Engineering.

Kenneth M. Reardon is Professor of City and Regional Planning at the University of Memphis, USA, and engages in research, teaching, and outreach activities focused on issues related to neighborhood planning, community development, community/university partnerships, and municipal government reform. Prior to joining the Memphis faculty, Ken was Associate Professor and Chair of Cornell University's Department of City and Regional Planning. He began his academic career as Assistant Professor in Urban and Regional Planning at the University of Illinois at Urbana-Champaign where he helped establish the East St. Louis Action Research Project.

Richard T. Roberts is a graduate of the University of Alabama, who has practiced economic development in three Southeastern states and worked for a variety of communities in Alabama, Mississippi, and Northwest Florida. During the past 30 years, he has managed both rural and metro chambers of commerce, economic development authorities, and a convention and tourism organization. He has written marketing plans and existing industry programs and has established and organized multi-county as well as local economic development programs. Mr. Roberts currently resides in Dothan, Alabama and is employed by the Covington County Economic Development Commission.

Brock Stout is an Associate Professor at the Solbridge International Business School in Daejeon, South Korea and has experience in entrepreneurial community-building in the United States.

John W. Vincent II is currently President of Performance Development Plus of Metairie, LA, a consulting company that he co-founded in 1993. The firm specializes in community development, economic development and organizational development consulting. Mr. Vincent has over 35 years' experience in various professional and technical positions, and he is currently an owner and board member of a homeland security company. He holds a Bachelor's degree in government from Southeastern Louisiana University, USA, and a Master's degree from the University of New Orleans, USA, in curriculum and instruction (adult learning and development). He is a graduate of the University of Central Arkansas' Community Development Institute in Conway, Arkansas, USA, and is a certified Professional Community and Economic Developer.

Monieca West is an experienced economic and community development professional. In 2001, she retired from SBC-Arkansas where she had served as Director of Economic Development since 1992. She is a past chair of the International Community Development Society and the Community Development Council and has received numerous awards for services to community organizations and public education. In 2004, West returned to full-time work for the Arkansas Department of Higher Education where she is a program manager for the Carl Perkins federal program which funds career and technical education projects at the post-secondary level. She also does freelance work in the areas of facilitation, leadership development, technical writing, association management, and webmaster services.

Stephen M. Wheeler joined the Landscape Architecture faculty at the University of California at Davis, USA, in January 2007. He is also a member of UCD's Community Development and Geography Graduate Groups. He previously taught in the Community and Regional Planning Program at the University of New Mexico, USA, where he initiated a Physical Planning and Design concentration, and the University of California at Berkeley, USA, where he received his Ph.D. and Master of City Planning degrees. He received his Bachelor's degree from Dartmouth College, USA. Professor Wheeler's areas of academic and professional expertise include urban design, physical planning, regional planning, climate change, and sustainable development. His current research looks at state and local planning for climate change, and evolving built landscape patterns in metropolitan regions.

Acknowledgements

Like community development itself, creating a book on the subject is a team effort. We are thankful first of all to the chapter authors who have graciously consented to share their professional knowledge and experience with the readers. While the editors' and authors' names appear in the book, so many other community and economic development scholars and practitioners have played important roles in bringing this book to fruition. One of the most rewarding aspects of studying and practicing the discipline is learning from others along the way. This volume reflects the collective input over decades of countless community developers who may not be cited by name in the book but have nonetheless had a profound impact on it.

We also wish to express our thanks to Routledge and the Taylor & Francis Publishing Group. Their expert assistance throughout the process of updating this book is acknowledged and appreciated. And finally we owe our families much gratitude, for without their support and encouragement this book would not exist.

Rhonda Phillips and Robert H. Pittman

Editors' Introduction

In this edition, we continue to explore community development as a complex and interdisciplinary field of study—one that is boundary spanning in its scope and multidimensional in its applications. Why is this? It's because community development concerns not only the physical realm of community, but also the social, cultural, economic, political, and environmental aspects as well. Evolving from an original needs-based emphasis to one that is more inclusive and asset-based, community development is a now a distinct and recognized field of study. Today, scholars and practitioners of community development are better equipped to respond to the challenges facing communities and regions. Because its applications are wide-ranging yet always aimed at improving quality of life and overall community well-being, it is important to understand the underlying foundations and theory of community development as well as the variety of strategies and tools used to achieve desired outcomes.

This text seeks to address the challenging and exciting facets of community development by presenting a variety of essential and important topics to help students understand its complexities. The chapter authors represent perspectives from both academe and practice, reflecting the applied nature of the discipline. Importantly, this book emphasizes the strong link between community development and economic development which is all too often overlooked in the literature. We believe a discussion of one is incomplete without a discussion of the other. We hope that this book will serve to more closely align the study and practice of these two inextricably related disciplines.

This text is presented in the spirit of community development as planned efforts to improve quality of life. With this goal in mind, 25 chapters covering a range of issues have been selected and organized into four major categories: (1) foundations; (2) preparation and planning; (3) programming techniques and strategies; and (4) issues impacting community development.

Part I: Foundations, provides an introduction and overview of the discipline as well as its underlying premises. In Chapter 1 we present the basic concepts and definitions of community development and how it relates to economic development, a central theme of this book. Additionally, we explore the history and foundations of community development, including a rich legacy of social action and advocacy. Chapter 2, Seven Theories for Seven Community Developers, distills a variety of ideas from different fields into a theoretical underpinning for community development. Hustedde offers seven contextual perspectives that provide this theoretical core: organizations, power relationships, shared meanings, relationship building, choice making, conflicts, and integration of paradoxes. Chapter 3, Asset-Based Community Development, focuses on the concept of capacity building, both inside and outside the community. Haines explains the value of adopting an asset-based approach, and how it is dramatically different from the needs-based approaches of the past. Mattessich explains in Chapter 4, Social Capital and Community Building, how social capital (or capacity) lies at the heart of community development. Analogous to other forms of capital, social capital constitutes a resource that can be used by communities to guide outcomes. Chapter 5, Sustainability in Community Development by Wheeler, provides background on the concept of sustainability and how it applies to both the theory and practice of community development and its relations to planning. It also gives examples of strategies that can be implemented to help increase sustainable approaches. Chapter 6, The "New" Local by Burnell and Phillips addresses newer ideas of the local economy movement, along with small change approaches. It includes a discussion of local foods and community development. Chapter 7, Community Development Practice, is the final chapter in this section and outlines the foundation of processes and applications introducing students to community development as a practice.

Part II: Preparation and Planning covers the variety of ways in which communities organize, assess, and plan for community development. McGrath's Chapter 8, Community Visioning and Strategic Planning, takes the reader through the process of establishing goals and a vision for the future—essential activities for success in community development. Without this foundation, it is difficult to accomplish the desired outcomes. Chapter 9, Establishing Community-Based Organizations, by West, Kraeger, and Dahlstrom addresses the all-important question, "How should we be organized?" They outline different types of community-based organizations and their structures, including social enterprise, and show examples in practice. Chapter 10, Leadership and Community Development by Price and Pittman, discusses the need for communities to effectively integrate skill development into their activities. The premise is that great leadership leads to the most desirable community development outcomes. Vincent's second contribution, Chapter 11, Community Development Assessments, provides a broad perspective on the total community assessment process. It discuses comprehensive assessments and the areas that should be considered, including a community's physical, social, and human infrastructure and capital. Chapter 12, Community Asset Mapping and Surveys by Green provides information on techniques such as asset inventories, identifying potential partners and collaborators, various survey instruments, and data collection methods. The final chapter in this section, Chapter 13 Understanding Community Economies by Galloway, discusses how to assess the underlying strengths and weaknesses of the local economy. It provides an overview of economic impact analysis and how it can be used to allocate scarce community financial resources.

Part III: Programming Techniques and Strategies gives several specific application areas for community development and how these areas can be approached. Chapter 14, Human Capital and Workforce Development, addresses the vital question of how to develop a quality workforce and educational components in the community. It provides examples of initiatives that communities have used to address this need. Pittman's Chapter 15, Marketing the Community, provides an overview of how to attract new businesses into a community to strengthen the local economy. Creating recognition for the community, identifying the appropriate target audience and the most effective marketing message are discussed. In Chapter 16, Retaining and Expanding Existing Community Businesses, Pittman and Roberts explain the importance of focusing on businesses already present in communities. An existing business program can help communities in many direct and indirect ways and is often more effective in job creation than other approaches. Gruidl and Markley present Entrepreneurship as a Community Development Strategy in Chapter 17 as a vital component driving economic growth and job creation. The fundamentals for implementing a strategy of supporting entrepreneurs and creating a nurturing environment are outlined. Phillips' Chapter 18, Arts, Culture and Community Development, explores ways communities can build on cultural assets to develop approaches around historic preservation, arts and culture. A variety of models and approaches are reviewed. Chapter

19, Housing and Community Development by Phillips, provides a basic understanding of how housing typology, density, and affordability affect housing and community development. It also includes innovative approaches to housing, such as ecovillages and cohousing. Reardon's Chapter 20, Neighborhood Planning for Community Development and Revitalization, discusses the model of participatory neighborhood planning. This model seeks to improve quality of life with comprehensive revitalization strategies grounded in an asset-based approach. Our final contribution to this part, Chapter 21, Measuring Progress, by Phillips and Pittman, begins with the premise that progress evaluation is not only challenging but vital and organizations must be able to asses and demonstrate the value and outcome of their activities. Community indicators are presented, as well as a discussion of quality of life and happiness considerations for communities.

Part IV: Issues Impacting Community Development focuses on a few of the many and diverse issues relevant to community development theory and practice. Chapter 22, Perspectives on Current Issues by Lachapelle, provides a summary of responses from community development scholars and practitioners from across the globe. They discuss challenges and opportunities, and provide a wealth of insight into both theory and practice. Chapter 23 by Phillips, Community-Based Energy, addresses issues emerging in alternative energy sources from a community economic development perspective. Chapter 24, Community and Economic Development Finance by Hamer and Farr, gives an overview and explanation of the different types of community development financing from public and private sources. It includes definitions of key terms as well as ideas on structuring funding partnerships. The final chapter offers some concluding observations on issues and challenges facing our communities, as well as discussing the important role of community development in helping shape the future of our society.

As stated at the outset, community development is indeed a complex and interdisciplinary field, as evidenced by the breadth and scope of the chapters presented. We encourage students of community development to embrace the "ethos" of the community development discipline as one that focuses on creating better places to live and work and increasing the quality of life for all.

Rhonda Phillips
Purdue University, USA
Robert H. Pittman
Janus Institute, USA

PART I
Foundations

1 A Framework for Community and Economic Development

Rhonda Phillips and Robert H. Pittman

Overview

Community development has evolved into a recognized discipline of interest to both practitioners and academicians. Community development is defined in many different ways; some think of community development as an outcome—physical, social, and economic improvement in a community—while others think of community development as a process—the ability of communities to act collectively and enhancing the ability to do so. This chapter defines community development as both a process and an outcome and explains the relationship between the two. A related area, economic development, is also defined in different ways. This chapter offers a holistic definition that includes not only economic dimensions but vital other dimensions impacting community quality of life and well-being. The model of the community and economic development chain shows the links, causal relationships, and feedback loops between community and economic development.

Introduction

Community development has many different definitions. Unlike mathematics or physics where terms are scientifically derived and rigorously defined, community development has evolved with many different connotations. Community development has probably been practiced for as long as there have been communities. Many scholars trace the origin of community development as a discipline and known profession to post-World War II reconstruction efforts to improve less-developed countries (Wise 1998). In the US, some cite the "war on poverty" of the 1960s with its emphasis on solving neighborhood housing and social problems as a significant influence on contemporary community development (Green and Haines 2011). A major contribution of community development has been the recognition that a city or neighborhood is not just a collection of buildings but a "community" of people facing common problems with untapped capacities for self-improvement. Today community is defined in many different ways: in geographic terms, such as a neighborhood or town (place-based or communities of place definitions); or in social terms, such as a group of people sharing common chat rooms on the internet, a national professional association, or a labor union (communities of interest definitions).

BOX 1.1 EVOLUTION OF COMMUNITY DEVELOPMENT

Community development as a profession has deep roots, tracing its origins to social movements (it is, after all, about collective action) of earlier times throughout the globe. Activism such as the Sanitary Reform movement in the 1840s and housing reforms a bit later helped push forward positive changes at community levels. For example, in the 1880s, Jane Addams was among the first to respond to deplorable housing conditions in tenements by establishing Hull House in Chicago, a community center for poor immigrant workers. Beyond North America, community development may be called "civil society," or "community regeneration," and activities are conducted by both government and nongovernmental organizations (NGOs). There may or may not be regulation of organizations, depending on different countries' policy frameworks (for a review of community development, see Hautekeur 2005). The Progressive movement of the 1890s through the first few decades of the twentieth century was all about community development (von Hoffman 2012), although the term itself did not arise until mid-century.

During the 1950s and 1960s, social change and collective action again garnered much attention due to the need to rectify dismal conditions within poverty-stricken rural and areas of urban decline. The civil rights and anti-poverty movements led to the recognition of community development as a practice and emerging profession, taking form as a means to elicit change in social, economic, political, and environmental aspects of communities. During the 1960s, literally thousands of community development corporations (CDCs) were formed, including many focusing on housing needs as prompted by U.S. Federal legislation providing funding for nonprofit community organizations. This reclaiming of citizen-based governing was also prompted in response to urban renewal approaches by government beginning with the U.S. Housing Act of 1949. The richness of the CDC experience is chronicled in the Community Development Corporation Oral History Project by the Pratt Center for Community Development. This includes one of the first CDCs in the US, the Bedford Stuyvesant Restoration Corporation in New York City.

Today, there are about 4,000 CDCs in the US, with most focusing on housing development as well as other related activities for improving community quality of life. However, many also include a full range of community development activities, with about 25% providing a comprehensive array of housing development, home ownership programs, commercial and business development, community facilities, open space/environmental, workforce, and youth programs, and planning and organizing activities (Walker 2002). Throughout the world, many organizations practice community development, including the public sector as well as the private, for-profit and other nonprofit groups. As the variety of topics in this book attests, community development continues to be built on social activism and housing to encompass a broad spectrum of processes and activities dealing with multiple dimensions of community including physical, environmental, social, and economic factors.

Sources

G. Hautekeur (2005) Community Development in Europe. *Community Development Journal* 40(4): 385–398.

A. von Hoffman (2012) The Past, Present, and Future of Community Development in the United States. In Andrews, N.O. & Erickson, D.J. (eds), *Investing in What Works for America's Communities*. San Francisco: Federal Reserve Bank.

C. Walker (2002) *Community Development Corporations and Their Changing Support Systems*. Washington, D.C.: Urban Institute.

For more details on the CDC Oral History Project, see www.prattcenter.net/cdcoralhistory.php

To learn more details on the history of community development in the US, see Alexander von Hoffman's, *The Past, Present, and Future of Community Development in the United States*, published by the Joint Center for Housing Studies, Harvard University, 2012. www.jchs.harvard.edu/sites/jchs.harvard.edu/files/w12-6_von_hoffman.pdf

Civil and social activism throughout the world has served as a basis for community development approaches, including those built upon works by Paulo Freire, a noted activist educator inspiring many with his classic 1970 *Pedagogy of the Oppressed*. See Margaret Ledwith's 2005 *Community Development: A Critical Approach* for discussion of Freire's ideas in community development theory and practice.

For a history of community development in the UK, see *The Community Development Reader, History, Theory, and Issues*, by Gary Craig, Marjorie Mayo, Keith Popple, Mae Shaw, and Marilyn Taylor (2011).

The rich and diverse history of community development has roots in many areas, evolving into a recognized discipline drawing from a wide variety of academic fields including urban and regional planning, sociology, economics, political science, geography, and many others. Today there is a variety of academic and professional journals focusing on community development. The interest of researchers and practitioners from many different disciplines has contributed greatly to the growth and development of the field. However, community development's growth and interdisciplinary nature have led to the current situation where it is defined and approached in a great variety of ways. Given community development's origins in social advocacy and calls to social action, the emphasis on equity is a common thread. The need for social reform and social justice has always been present, and continue to be critical across communities globally. See Box 1.2 for more information about some of the history of social equity.

This chapter takes a broad approach to community development. While it is impossible in one chapter (or book) to completely cover such a large field, many different aspects of community development are included. In particular, we have observed that the strong interrelationship between community and economic development is often overlooked in research and practice. This interrelationship is one focus of this chapter and book. Economic development focuses predominately on the monetary aspects of the processes and outcomes—generating wealth in a community. It is, by default, often centered on issues of efficiency. This is counter to community development's emphasis on equity, yet balance can be sought between these two elements. It is our intent to explore this balance, bringing together both community and economic development to explore ways to foster better outcomes or community well-being and improved quality of life for our towns, cities, regions, and beyond.

BOX 1.2 THE UNDERREPRESENTED

More than a century prior to the Occupy Wall Street movement, Upton Sinclair, in *The Jungle* (1906), described the horrible conditions of the meatpacking factories and the poverty and hopelessness prevalent in the Back of the Yards neighborhood in Chicago. Inspired by the work of Sinclair, the other Muckrakers, and the labor organizer John Lewis, Saul Alinsky became a community organizer and took it to the streets, convinced that the interests of workers and the poor were not being represented and protected by traditional political processes and institutions. Alinsky "walked the talk," served many jail terms for nonviolent demonstrations and, shortly before his death, published his seminal work, *Rules for Radicals* in 1971. The book is a primer for community organizing and extols the importance and justification of using non-conventional means to close the gap between the "haves" and "have-nots."

Martin Luther King, Jr. and the early civil rights movement adopted Alinsky's and other nonviolent protesters' methods. In his eloquent "Letter from a Birmingham Jail," King wrote:

> You may well ask: "Why direct action? Why sit ins, marches and so forth? Isn't negotiation a better path?" You are quite right in calling for negotiation. Indeed, this is the very purpose of direct action. Nonviolent direct action seeks to create such a crisis and foster such a tension that a community which has constantly refused to negotiate is forced to confront the issue.

The Mavis Staples video (www.youtube.com/watch?v=0ZWdDI_fkns) dramatically captures the bravery of the civil rights protesters.

Why has there been a long tradition of community activism in the United States? The reason is that traditional institutions and political processes have not always given sufficient voice and a "seat at the table," for the nation's poor and minorities. As the urban planning theorist and activist for social justice and equity Paul Davidoff has written:

> the "great issues" in economic organization, those resolving around the central issue of the nature of distributive justice, have yet to be settled. The world is still in turmoil over the way in which the resources of the nations are to be distributed.
>
> (1965: 50)

These issues are as relevant today as when first written by Davidoff 50 years ago. Although economic growth and development can bring many benefits, those benefits are not evenly distributed and the costs are often paid by the poorest members of our society. One noteworthy attempt at the local level to address these issues is the Cleveland experiment from late 1969–1979 in the practice of equity planning. Norman Krumholz and his staff of professional planners instituted progressive programs and policies that resulted in property law changes, improvements in public service delivery, protection of transit services for the most transit-dependent, and the rescue of city parklands and beaches to serve the poor. The Cleveland experience is well worth reviewing for a politically effective and savvy approach to progressive economic development and growth.

Jay M. Stein, Ph.D., FAICP
Arizona State University, USA

Sources

Saul Alinsky (1971) *Rules for Radicals.* New York: Random House.
Paul Davidoff (1965) Advocacy and Pluralism in Planning, *Journal of the American Institute of Planners,* 31(4): 331–338.
Dr. Martin Luther King, Jr. (1963) Letter from a Birmingham Jail. Accessed December 21, 2013 at: www.africa.upenn.edu/Articles_Gen/Letter_Birmingham.html
Norman Krumholz and John Forester (1990) *Making Equity Planning Work.* Philadelphia: Temple University Press.
Upton Sinclair (1906) *The Jungle.* Chicago: Doubleday, Jabber and Company.

The terms community development and economic development are widely used by academicians, professionals, and citizens from all walks of life and have almost as many definitions as users. Economic development is perhaps more familiar to laypersons. If random individuals on the street were asked what economic development is, some might define it in physical terms such as new homes, office buildings, retail shops, and "growth" in general. Others might define it as new businesses and jobs coming into the community. A few thoughtful individuals might even define it in socioeconomic terms such as an increase in per capita income, enhanced quality of life, or issues around equity such as reduction in poverty.

Ask the same individuals what community development is, and they would probably think a while before answering. Some might say it is physical growth—new homes and commercial buildings—just like economic development. Others might say it is community improvement such as new infrastructure, roads, schools, or improving quality of life in areas such as reducing poverty or improving housing conditions. Most respondents would probably define community and economic development in terms of an *outcome*—physical growth, new infrastructure, better housing, or jobs. Probably no one would define them in terms of a *process* and many would not understand how they are interrelated. This is unfortunate because some of these passers-by are probably involved in community and economic development efforts serving as volunteers or board members for community-oriented nonprofit organizations, economic development agencies, or charitable organizations.

The purpose of this introductory chapter is to provide meaningful descriptions of community development and economic development as both processes and outcomes, explore what they entail, and understand them as distinct but closely related disciplines. First the focus will be on community development, followed by a discussion of economic development and, finally, an examination of the relationship between the two.

Community Development

The first step in defining community development is to define "community." As mentioned previously, community can refer to a location (communities of place) or a collection of individuals with a common interest or tie whether in close proximity or widely separated (communities of interest). A review of the literature conducted by Mattessich and Monsey (2004) found many definitions of community such as:

> People who live within a geographically defined area and who have social and psychological ties with each other and with the place where they live.
>
> (Mattessich and Monsey 2004: 56)

> A grouping of people who live close to one another and are united by common interests and mutual aid.
>
> (National Research Council 1975, cited in Mattessich and Monsey 2004: 56)

> A combination of social units and systems which perform the major social functions … the organization of social activities.
>
> (Warren 1963, cited in Mattessich and Monsey 2004: 57)

These definitions refer first to people and the ties that bind them and second to geographic locations. They remind us that without people and the connections among them, a community is just a collection of buildings and streets.

Given these definitions, the term community development takes on the context of developing stronger "communities" of people and the social and psychological ties that they share. Indeed this is how community development is defined in much of the literature. Again Mattessich and Monsey found many such definitions in their literature review:

> Community development is an educational process designed to help adults in a community solve their problems by group decision making and group action.
>
> (Long 1975, cited in Mattessich and Monsey 2004: 58)

> Community development is the "active voluntary involvement in a process to improve some identifiable aspect of community life; normally such action leads to the strengthening of the community's pattern of human and institutional relationships."
>
> (Ploch 1976, cited in Mattessich and Monsey 2004: 59)

These definitions imply that community development is the *process* of teaching people how to work together to solve common problems.

Other authors define community development more in terms of an action, result, or *outcome*:

> Local decision making and the development of programs designed to make the community a better place to live and work.
>
> (Huie 1976, cited in Mattessich and Monsey 2004: 58)

> A group of people in a locality initiating a social action process to change their economic, social, cultural and/or environmental situation.
>
> (Christenson and Robinson 1989, cited in Mattessich and Monsey 2004: 57)

> A series of community improvements which take place over time as a result of the common efforts of various groups of people.
>
> (Dunbar 1972, cited in Mattessich and Monsey 2004: 59)

All of these definitions are valid in that community development should be considered both a process and an outcome. However, as mentioned previously, the person on the street would probably think of community development in terms of a physical result (improved infrastructure, better health care, etc.). Therefore, a definition of community development in simple but broad terms is:

> A *process*: developing the ability to act collectively, and
> An *outcome*: (1) taking collective action and (2) the result of that action for improvement in a community in any or all realms: physical, environmental, cultural, social, political, economic, etc.

Having arrived at a comprehensive definition of community development, the focus can now be on what facilitates or leads to community development. The community development literature generally refers to this as *social capital* or *social capacity*:

> The abilities of residents to organize and mobilize their resources for the accomplishment of consensual defined goals.
> (Christenson and Robinson 1989, cited in Mattessich and Monsey 2004: 61)

> The resources embedded in social relationships among persons and organizations that facilitate cooperation and collaboration in communities.
> (Committee for Economic Development 1995, cited in Mattessich and Monsey 2004: 62)

Simply put, social capital or capacity is the extent to which members of a community can work together effectively to develop and sustain strong relationships; solve problems and make group decisions; and collaborate effectively to plan, set goals, and get things done. There is a broad literature on social capital, with some scholars making the distinction between *bonding capital* and *bridging capital* (Agnitsch et al. 2006). Bonding capital refers to ties within homogeneous groups (e.g. races, ethnicities, social action committees, or people of similar socioeconomic status) while bridging capital refers to ties among different groups.

There are four other forms of "community capital" often mentioned in the community development literature (Green and Haines 2002: viii):

- Human capital: labor supply, skills, capabilities and experience, etc.;
- Physical capital: buildings, streets, infrastructure, etc.;
- Financial capital: community financial institutions, microloan funds, community development banks, etc.;
- Environmental capital: natural resources, weather, recreational opportunities, etc.

All five types of community capital are important. However, it is difficult to imagine a community making much progress without some degree of social capital or capacity. The more social capital a community has, the more likely it can adapt to and work around deficiencies in the other types of community capital. When conducting community assessments (see Chapter 11), it is useful to think in terms of these five types of community capital.

So far, working definitions of community, community development, and social capital have been provided. To complete the community development equation, it is necessary to identify how to create or increase social capital or capacity. This process is generally referred to as *social capital building* or *capacity building*. Here is one definition from the literature:

> An on-going comprehensive effort that strengthens the norms, and supports the problem solving resources of the community.
> (Committee for Economic Development 1995, cited in Mattessich and Monsey 2004: 60)

Notice that this sounds exactly like the definitions of the process of community development given above. We have come full circle. The *process* of community development *is* social capital/capacity building which leads to social capital which in turn leads to the *outcome* of community development.

Figure 1.1 depicts the community development chain. The solid lines show the primary flow of causality. However, there is a feedback loop shown by the dotted lines. Progress in the outcome of community development (taking positive action resulting in physical and social improvements in the community) contributes to capacity building (the process of community development) and social capital. For example, better infrastructure (e.g. public transportation, internet access, etc.) facilitates public interaction, communications, and group meetings. Individuals who are materially, socially, and psychologically better off are likely to have more time to spend on community issues because they have to devote less time to meeting basic human and family needs. Success begets success in community development. When local citizens see positive results (outcome), they generally get more enthused and plow more energy into the process because they see the payoff. Research has shown there are certain characteristics of communities that influence their ability to do capacity building and create social capital (Mattessich and Monsey 2004). Chapter 4 of this book discusses some of these community characteristics.

Now that the components of community development and their relationships have been identified, we can return to the random individuals on the street and ask them what the difference is between *growth* and *development*. From the definitions above, it would seem that development is a more encompassing term than growth. Green and Haines (2002: 5) define growth as:

> Increased quantities of specific phenomena such as jobs, population and income, whereas development involves structural change in a community including: 1) how resources are used; and 2) the functioning of institutions.

By these definitions, a community can have growth without development and vice versa. The important point to note, however, is that development not only facilitates growth but also influences the kind and amount of growth a community experiences. Successful communities control their own destiny through the successful practice of community development. Community development empowers communities to change.

Economic Development

As with community development, economic development has its origins in efforts to improve less developed countries and the American war on poverty (Malizia and Feser 1999). Immediately after World War II and certainly before, parts of the US were not unlike developing countries with rampant poverty and unemployment due to the decline of agricultural jobs. Many southern states developed programs to recruit industries from the northern US with cheap labor and government incentives (e.g. tax breaks) as bait. In the 1960s, the emphasis on economic development was at the federal level with Great Society programs aimed at eliminating pockets of poverty such as

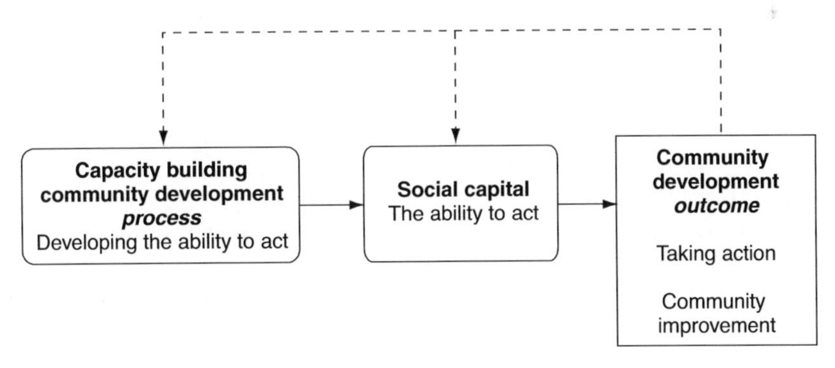

Figure 1.1 *Community Development Chain*

BOX 1.3 COMMUNITY WELL-BEING: THE NEXT GENERATION FOR COMMUNITY DEVELOPMENT?

Aristotle's term *eudaemonia* consists of the words "*eu*" and "*daimon*," meaning "good" and "spirit." Modern scholars have seized on this concept in the context of communities and use it to explain renewed interest in promoting improvements in well-being (Lee and Kim 2014). Common frameworks of concepts and measures can assist communities in prioritizing goals and values (Phillips 2014); community well-being can then be thought of as encompassing "the broad range of economic, social, environmental, cultural and governance goals and priorities identified as of greatest importance by a particular community, population group or society" (Cox et al. 2010: 72) or, "the concept of community well-being is focused on understanding the contribution of a community in maintaining itself and fulfilling the various needs of local residents" (Haworth and Hart 2007: 95). It is further described as a way to bring together social relationship and social organization with the idea that community "*well-being is something that we do together, not something that we each possess*" (Haworth and Hart 2007: 128). The underlying idea is that this broader concept is an umbrella concept, encompassing ideas of sustainability, happiness, quality of life, and other social domains. Community and economic development, along with urban and regional planning are seen as approaches to helping foster improved community well-being in this context. See for example Chicago's efforts toward "Quality of Life Planning," in conjunction with the Local Initiatives Support Corporation (LISC).

Sources

Cox, D., Frere, M., West, S., and Wiseman, J. (2010) Developing and Using Local Community Wellbeing Indicators: Learning from the Experiences of Community Indicators Victoria. *Australian Journal of Social Issues* 45(1): 71–89.

Haworth, J. and Hart, G. (2007) *Well-Being, Individual, Community and Social Perspectives*. Hampshire, UK: Palgrave.

Lee, S.J. and Kim, Y. (2014). *Community Wellbeing: Concepts and Measurements*. In S.J. Lee, Y. Kim, and R. Phillips (eds.), *Community Wellbeing: Conceptions and Applications*. Dordrecht: Springer, forthcoming.

LISC New Communities Program, Chicago. *Quality of Life Planning Handbook*, www.newcommunities.org/tools/qofl.asp

Phillips, R. (2014) Building Community Well-Being across Sectors: "For Benefit" Community Business. In S.J. Lee, Y. Kim and R. Phillips, *Community Wellbeing: Conceptions and Applications*. Dordrecht: Springer.

the southern Appalachian Mountains region. In the 1970s and 1980s, the emphasis shifted to states and localities. If they did not already have them, many communities created economic development organizations with public as well as private funding. In many communities, economic development was part of city or country government, while in other communities it was under the auspices of the private nonprofit sector such as local chambers of commerce.

Initially, most economic development agencies focused on industrial recruitment or on enticing new companies to locate in their communities. State and local departments of economic development aggressively courted industry and increased the amount and variety of incentives. Soon it became apparent that tens of thousands of economic development organizations were chasing a limited number of corporate relocations/expansions and playing a highly competitive game of incentives and marketing efforts. This highly competitive situation became known as a "zero sum game" after Lester Thurow's description of macroeconomics in the now classic *The Zero-Sum Society Distribution and the Possibilities for Economic Change* (1980). In other words, there are no true winners, given a tendency toward a "beggar thy neighbor approach" to enticing industry. The benefit to communities or regions of giving expansive tax incentives to keep or lure industry into the area is not often clear. Timothy Bartik addressed this question in *Who Benefits from State and Local Economic Development Policies?* in 1991; policy makers, scholars, and practitioner alike are still debating.

As communities realized that there were other ways to create jobs, they began to focus on internal opportunities such as small business development and ensuring that businesses already located in the community stayed and expanded there. Many communities also realized that by improving education, aesthetics and amenities to make life better, environmental conditions, government services, local labor skills and supply, and the business climate in general, they could make themselves more attractive to both internal and external investments. These wider approaches help encourage better quality-of-life conditions.

Like community development, economic development has evolved into a broad and multidisciplinary field. In earlier decades, economic development was synonymous with growth, not necessarily a positive correlation. A more holistic definition of economic development includes an emphasis on increasing the *quality of life* (the general well-being of individuals and collectively as community) through opportunities and improvement. From this perspective, economic development is undertaken for the purposes of increasing standards of living through attaining greater per capita income, quality and quantity of employment opportunities, and enhanced quality of life for the persons represented within the development district.

BOX 1.4 GROWTH VS. DEVELOPMENT

Growth by itself could be either an improvement or detriment. For instance, a facility that paid very low wages might open in an area. As a result, the population and overall size of the economy might increase, but per capita incomes might fall, and the quality of life might suffer (Blair 1995: 14). Of course many communities would be happy to get any new facility regardless of the wage rate. A minimum wage job may be better than no job in some situations, and there is a portion of the labor force in most communities that is a good match for minimum wage jobs (e.g. teenagers and adult low-skilled workers).

Since growth does not always equate with a better standard of living, Blair suggests a higher-order definition of economic development: Economic development implies that the welfare of residents is improving. Increases in per capita income (adjusted for inflation) is one important indicator of welfare improvements (1995: 14).

There are many other indicators of welfare and quality of life for community residents such as poverty rates, health statistics, and income distribution, but per capita income or income per household is a common measurement of wealth. Whether wealth equates to a good quality of life depends on the individual, although there are basic levels needed to have a decent standard of living. Moreover, per capita income is not necessarily a measure of purchasing power. The cost of living varies from place to place, and a dollar goes farther where prices are low. Like community development, these descriptions portray economic development as both a process and an outcome—the process of mobilizing resources to create the outcome of more jobs, higher incomes, and an increased standard of living, however it is measured. From this discussion it should be apparent that development has a very different connotation than growth. Development implies structural change and improvements within community systems encompassing both economic change and the functioning of institutions and organizations (Boothroyd and Davis 1993; Green and Haines 2011). Development is a deliberate action taken to elicit desired structural changes. Growth, on the other hand, focuses on the quantitative aspects of more jobs, facilities construction, and so on—within the context that more is better. One should carefully distinguish then between indicators that measure growth versus development. By these definitions, a community can have growth without development and vice versa. The important point to note, however, is that development not only facilitates growth but also influences the kind and amount of growth a community experiences. Development guides and directs growth outcomes.

Economic development centers on creating wealth and other economic outcomes. One way to view the relationship between these two areas is that community development can be thought of as *producing* assets and capacities while economic development *mobilizes* these assets for economic benefits. As noted, traditional or mainstream economic development has focused on economic outcomes, sometimes to the detriment of ecological or social dimensions of community.

Now that we have defined economic development, we can discuss what it entails in practice. Most economic developers concentrate on creating new jobs. This is generally the key to wealth creation and higher living standards. Many economic developers refer to the "three-legged stool" of job creation: recruiting new businesses, retaining and expanding businesses already in the community, and facilitating new business start-ups. As discussed previously, economic developers originally concentrated on recruiting new businesses. Stiff competition for a limited number of new or expanded facilities in a given year led some communities to realize that another way to create jobs is to work with companies already in the area to maximize the likelihood that if they need to expand existing operations or start new ones, they would do so in the community and not elsewhere. Even if an expansion is not involved, some businesses may relocate their operations to other areas for "pull" or "push" reasons (Pittman 2007). They may relocate to be closer to their customers, closer to natural resources, or for any number of strategic business reasons (pull). Businesses may also relocate because of problems with their current location such as an inadequate labor force, high taxes, or simply lack of community support (push). Although communities cannot influence most pull factors, they can act to mitigate many push factors. If the problem is labor, they can establish labor-training programs. If the problem is high taxes, they can grant tax abatements in return for creating new jobs. Business retention and expansion has become a recognized sub-field of economic development, and there are many guides on the subject such as those by Entergy and other utility companies, for example. Chapter 16 of this book addresses business retention and expansion in more detail.

There is also much communities can do to facilitate the start up of new local businesses. Some communities create business incubators where fledgling companies share support services, benefit from reduced rent, or even get free consulting assistance. Financial assistance, such as revolving loan funds at reduced interest rates, is also a common tool used to encourage small business start-ups. Making a community "entrepreneurial friendly" is the subject of Chapter 17 of this book.

While the majority of new jobs in most regions are created by business retention and expansion and new business start-ups (Roberts 2006), many communities continue to define economic development in terms of recruiting new facilities. Some communities are stuck in the old paradigm of "smokestack chasing"—relying solely on recruiting a new factory when manufacturing employment is declining nationally and many companies are moving offshore. These communities may have historically relied on traditional manufacturing companies for the bulk of their employment, and when these operations shut down, the only thing they knew to do was to pursue more of the same. Elected officials, civic board members, and citizens in these communities need to be educated and enlightened to the fact that the paradigm has long ago shifted. They should be recruiting other types of businesses such as call centers or service industries while practicing business retention and expansion and creating new business start-up programs. Growing businesses and enterprise locally will result in better outcomes for communities.

While economic development today is often defined by the "three-legged stool," there is much more to the profession. As shown in a survey by a large regional professional association (SEDC 2006), economic developers get involved in workforce development, permitting assistance, and many other issues, some of which are better defined as community development.

Many communities are beginning to realize that it is often better, especially when recruiting new companies, to practice economic development on a regional basis and combine resources with nearby communities. Rather than create an economic development agency providing the same services for every community, it is usually more efficient to combine resources and market a region collectively through one larger organization. Regardless of where a new facility locates or expands, all communities in the region will benefit as employees live in different areas and commute to work.

The Relationship between Community and Economic Development

Community development and economic development are sometimes used interchangeably, and the term "community economic development" is often used as well. While they have distinct meanings, in practice they can be inextricably linked on many levels and can be highly synergistic. Striving to achieving balance between community and economic development is essential. Economic development, and economic considerations in general, tend to receive more attention yet the attributes that community development approaches bring are just as vital, particularly as related to community well-being. We call your attention to the following declaration by the United Nations:

> Everyone has the right to a standard of living adequate for the health and well-being of himself and his family, including food, clothing, housing and medical care and necessary social services, and the right to security in the event of unemployment, sickness, disability, widowhood, old age or other loss of livelihood in circumstances beyond his control.
>
> (United Nations Universal Declaration of Human Rights (Article 25), 1948)

As seen, community development is intertwined with the idea of economic security and standard of living. Through the years, most economic analyses and approaches do not adequately address the human element; this is required if we are to build community capacity and resiliency for more durable economies (Phillips et al. 2013). In his now classic *Small Is Beautiful: Economics as If People Mattered* (1973), British economist E.F. Schumacher pointed out the need to have the human element—people—as the main concern of economic analysis and development strategies. His work has subsequently inspired a new generation to explore options to traditional approaches.

To understand the synergies more fully between community and economic development, consider this definition of community development (Green and Haines 2002: vii):

> Community development is ... a planned effort to produce assets that increase the capacity of residents to improve their quality of life. These assets may include several forms of community capital: physical, human, social, financial and environmental.

Now consider a classic definition of economic development focusing on economic outcomes, (AEDC 1984: 18):

> The process of creating wealth through the mobilization of human, financial, capital, physical and natural resources to generate marketable goods and services. The economic developer's role is to influence the process for the benefit of the community through expanding job opportunities and the tax base.

These two definitions are parallel yet not entirely alike. The purpose of community development is to *produce* assets that can be used to improve the community, and the purpose of economic development is to *mobilize* these assets to benefit the community. Both definitions refer to the same community capital assets: human, financial, and physical (environmental or natural resources). A more modern holistic definition of economic development, as Blair (1995) pointed out, would include not only wealth and job creation but increasing standard of living for all citizens. This would bring in the equity dimensions of community development and the desire for fostering community well-being and improved quality of life, resulting in a more sustainable approach. This expanded definition is certainly compatible with community development. The definition of economic development does not include social capital per se, but while economic developers might not have used this term when this definition was created, it will be seen that social capital is important for economic development as well as community development.

As most economic developers will attest, a key to success in economic development—new business recruitment, retention and expansion of existing businesses, and new business start-up—is to have a "development-ready" community. Most businesses operate in competitive markets, and one of the major factors influencing business profitability is their location. When making location decisions, businesses weigh a host of factors that affect their costs and profits such as:

- available sites and buildings;
- transportation services and costs (ground, water, air);
- labor costs, quality, and availability;
- utility costs (electricity, natural gas);
- suitability of infrastructure (roads, water/sewer);
- telecommunications (internet bandwidth); and
- public services (police and fire protection).

Quality-of-life factors (education, health care, climate, recreation, etc.) are also important in many location decisions (Pittman 2006), whether from external or internal investors. If a community scores poorly in these factors, many companies would not consider it development ready. A weakness in even one important location factor can eliminate a community from a company's search list. Aesthetics and the ability of a community to provide a high quality of life are increasingly important in investment decisions.

Whether a community is considered development ready depends on the type of business looking for a location. For example, important location criteria for a microchip-manufacturing facility include a good supply of skilled production labor, availability of scientists and engineers, a good water supply, and a vibration-free site. A call center seeking a location would focus more on labor suitable for telephone work (including students and part-time workers), a good non-interruptible telecommunications network, and, perhaps, a time zone convenient to its customers. While location needs differ, a community lacking in the location factors listed above would be at a disadvantage in attracting or retaining businesses. Shortcomings in these or other location factors would increase a firm's costs and make it less competitive. In addition, companies are risk averse when making location decisions.

There are many location factors that are subjective and not easily quantified. For example, it is not feasible to objectively measure factors such as work ethic, ease of obtaining permits, or a community's general attitude toward business (these and other factors are sometimes collectively referred to as the "business climate"). Yet in the final decision process after as many factors as possible have been quantified, these intangibles often determine the outcome. Most company executives prefer to live in desirable communities with good arts, cultural amenities, recreational facilities and opportunities, low crime, and a neighborly, cordial atmosphere.

It should be apparent now how important community development is to economic development. If a community is not development ready in the physical sense of available sites, good infrastructure, quality-of-life factors, and public services, it will be more difficult to attract new businesses and retain and expand existing ones. New businesses starting up in the community would be at a competitive disadvantage. Being development ready in this physical sense is an *outcome* of community development (taking action and implementing community improvements).

The *process* of community development also contributes to success in economic development. First, as has been discussed, the process of community development (developing the ability to act in a positive manner for community improvement) leads to the outcome of community development and a development-ready community. In addition, some of the intangible but important location factors can be influenced through the process of community development. Companies and enterprises do not like to locate in divided communities where factions are openly fighting with one another, city councils are deadlocked and ineffective, and residents disagree on the types of businesses they want to attract (or even if they want to attract any businesses). As a company grows, it will need the support of the community for infrastructure improvements, good public education, labor training, and many other factors. Communities that are not adept (or, worse, are totally dysfunctional) at the process of community development are less likely to win the location competition. Furthermore, others (potential new residents, investors, etc.) would probably prefer not to live in such a place.

As it has been shown, economic development, like community development, is also a process. Establishing and maintaining good community and economic development programs is not easy. Significant resources must be devoted to accomplishing desired outcomes. Communities that are successful in community and economic development devote the appropriate resources to the effort, design good programs, and stay with them for the long haul. Osceola, Arkansas, is an inspiring example of good community development leading to success (see Box 1.5).

BOX 1.5 OSCEOLA, ARKANSAS: COMMUNITY DEVELOPMENT TURNS A DECLINING COMMUNITY AROUND

Osceola is a town of about 9,000 in the Mississippi Delta in northeast Arkansas. Like many rural communities, Osceola and Mississippi County grew up around the agriculture industry. As farm employment declined, the city attracted relatively low-skilled textile manufacturing jobs which ultimately disappeared as the industry moved offshore. In 2001 a major employer, Fruit of the Loom, shut down its Osceola plant, leaving the community in a crisis. Not only was unemployment rampant, the local schools were classified as academically distressed and were facing a state takeover. Osceola's mayor stated "we almost hit the point of no return" (Shirouzu 2006).

The remaining businesses were having difficulty finding labor with basic math skills for production work. To address this problem, executives from these businesses began to work with the city administration to find a solution. They decided to ask the state for permission to establish a charter school that could produce well-educated students with good work skills. After many community meetings and differences of opinion, the community united behind the effort and opened a charter school.

Shortly after the school was established, Denso, a Japanese-owned auto parts company, was looking for a manufacturing site in the southern states. City representatives showed Denso the charter school and repeatedly told them how the community had come together to solve its labor problem. Denso decided to locate the plant in Osceola, creating 400 jobs with the potential to grow to almost 4,000. While Denso cited the improved labor force as a reason for selecting Osceola, they were also impressed with how the community came together and solved its problem. An executive from Denso was quoted as saying "It was their aggressiveness that really impressed us."

Denso was also concerned about the community's commitment to continuous improvement in the schools. Regarding this issue, the Denso executive was quoted as saying "Is there a future here? Are they doing things that are going to drive them forward? Do they have that commitment to do it? We saw that continuously in Osceola." Denso located in Osceola not only because of its labor force, but because they were convinced the community would continue to practice good community development principles and move forward. Osceola's success story was featured on the front page of the national edition of the *Wall Street Journal* (Shirouzu 2006).

Shaffer et al. (2006) describe the relationship and synergy between community development and economic development as follows:

> We maintain that community economic development occurs when people in a community analyze the economic conditions of that community, determine its economic needs and unfulfilled opportunities, decide what can be done to improve economic conditions in that community, and then move to achieve agreed-upon economic goals and objectives. (p. 61)

They also point out that the link between community development and economic development is sometimes not understood or appreciated:

> Economic development theory and policy have tended to focus narrowly on the traditional factors of production and how they are best allocated in a spatial world. We argue that community economic development must be broader than simply worrying about land, labor and capital. This broader dimension includes public capital, technology and innovation, society and culture, institutions, and the decision-making capacity of the community. (p. 64)

These authors make it clear that community development and economic development are inextricably linked and if scholars and practitioners of economic development do not address community development, they are missing an important part of the overall equation.

Fortunately, there has been a shift to a more holistic approach in community economic development, one that includes more consideration of all dimensions of community well-being as a result of increased awareness of ecological needs and the push for sustainable approaches and longer-term sustainable development. Anglin (2011: 3) provides a newer definition reflecting this shift:

> Community economic development, and its values and practices, are indeed important strategies to help forge a stronger base for addressing key challenges going forward such as (1) development that protects the environment while opening opportunities for the poor to build wealth and opportunity, and (2) assisting in the larger project of strengthening the economic competiveness of cities and regions.

We agree that a definition of community economic development is "a merging of aspects of both the fields of community development and economic development, implying practice aimed at community betterment and economic improvement at the local level, preferably encompassing sustainable development approaches" (Phillips and Besser, 2013: 2). It should be focused on

BOX 1.6 DIFFERENT DEFINITIONS OF COMMUNITY ECONOMIC DEVELOPMENT

Community development and economic development are frequently used interchangeably, and the term "community economic development" is often seen as well. Anglin (2011), Phillips and Besser (2013), and Shaffer and colleagues (2006) use it to describe the integration of community and economic development processes. Some authors, however, also use it to refer to "local economic development," encompassing growth (economics), structural change (development), and relationships (community capital). This term, along with community regeneration is often seen in the UK or Canada (see e.g. Haughton (1999) or Boothroyd and Davis (1993)).

Source:

P. Boothroyd and H.C. Davis (1991) Community Economic Development: Three Approaches. *Journal of Planning, Education and Research* 12: 230–240.

G. Haughton (1999) *Community Economic Development*. London: The Stationery Office.

sustainable approaches that help communities build durable economies with resiliency.

Now the diagram begun in Figure 1.1 can be completed to show the holistic relationship between the process and outcome of community development and economic development (see Figure 1.2).

The community development chain is as depicted in Figure 1.1: capacity building (the process of community development) leads to social capital which in turn leads to the outcome community development. In addition, communities with social capacity (the ability to act) are inherently more capable of creating good economic development programs should they chose to do so. When these communities take action (community development outcome), they create and maintain effective economic development programs that mobilize the community's resources. They also improve their physical and social nature and become more development ready, which leads to success in business attraction, retention and expansion, and start-up.

Citizens should understand the community and economic development chain in order to move their communities forward efficiently and effectively. While community developers might not believe they are practicing economic development and vice versa, in reality, they are all practicing community economic development.

Conclusion

Some communities do not believe they can influence their own destiny and improve their situation. They have a fatalistic attitude and feel they are victims of circumstances beyond their control—a closed factory, a downsized military base, or a natural disaster. In reality, they can build a better future. For every community that fails to act, there is another that is proactively applying the tools of community and economic development to better itself. Successful communities realize that community development is a group effort involving all.

Sometimes it takes a negative event to shock a community into action. On the other hand, many communities realize that change is inevitable and choose to be prepared by practicing good community and economic development. In today's changing global economy, maintaining the status quo is rarely an option—either a community moves itself forward or by default moves backward. Achieving desirable quality of life and community well-being requires informed action.

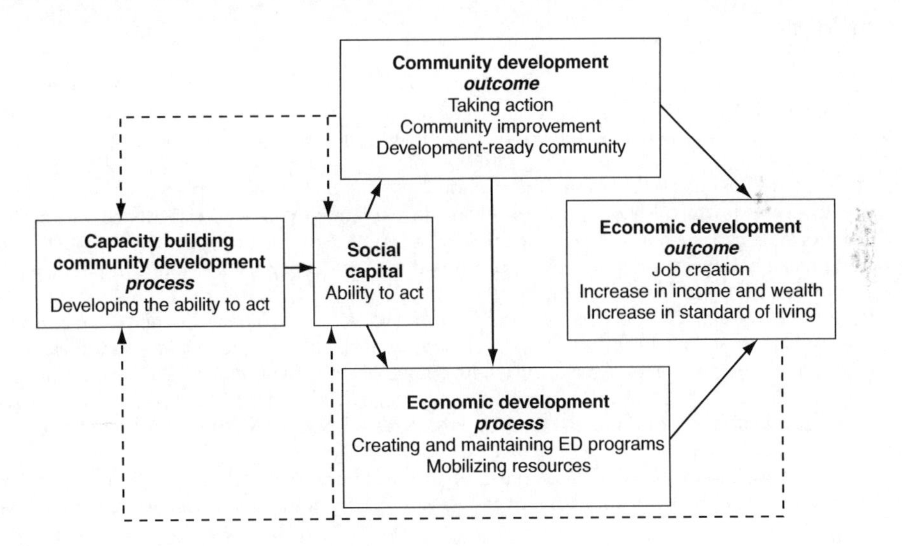

Figure 1.2 *Community and Economic Development Chain*

CASE STUDY: VERMONT SUSTAINABLE JOBS FUND[1]

This case highlights newer ways of thinking about community and economic development, with an emphasis on fostering sustainable enterprises. It incorporates ideas of sustainability, socially responsible businesses, and building competitive advantage via cooperation.

In the summer of 1994, a group of approximately 15 business leaders convened for a day-long retreat in a rural meadow to discuss their long-term vision for the state of Vermont. All were members of an organization called Vermont Businesses for Social Responsibility (VBSR), an organization whose mission is to support and encourage socially responsible business practices and public policy initiatives. The leaders committed to a vision that included the idea of creating a state entity to support the development and creation of "sustainable" jobs, which were defined as jobs consistent with VBSR core values—protection of the environment, social justice, and economic equity. (See below "What Sustainable Means to Vermont.") These leaders formed the Sustainable Jobs Coalition to pursue legislation that would translate their vision into reality.

Since the 1960s, Vermont has had a reputation as a national leader on conservation, thanks to laws protecting the environment, including a landmark land use law, the billboard ban, the bottle deposit law, Green-Up Day, and the Scenic Preservation Council. Most companies that rely on Vermont's green image and fertile land to grow and sell their products approve, and many company leaders are in VBSR. They see an opportunity for growth through cooperation and networking, particularly in the area of marketing and advertising, where pooling resources can really pay off.

Consider one industry example of collaboration. Vermont's artisan and farmstead cheese makers rode a wave of national popularity in the 1990s. But as Cabot Creamery's director of marketing, Jed Davis, recalls, they were "really pretty darn insignificant compared to big states like California. We knew that we weren't going to gain as much by being competitors as we are by cooperating." To grow their businesses—and create new jobs for Vermonters—participants in the state's cheese industry knew they would need to work together to build their reputation as the premier source for small-batch, farmstead cheeses.

So VBSR developed the idea of a sustainable jobs fund. The fund would support growing enterprises and business networks that demonstrated commitment to a dual bottom line—making profits while pursuing social responsibility for the environment, social justice, economic equity, and an increased number of jobs. The purpose of the Vermont Sustainable Jobs Fund (VSJF) was to create a nonprofit arm of the state that, with an initial state investment, would be able to attract funding from the federal government and private foundations to support the development of sustainable jobs. VBSR, in concert with allies such as those in the Sustainable Jobs Coalition, crafted and passed the legislation in 1995 with the modest appropriation of $250,000.

Grant making

The VSJF's initial approach was twofold: plant as many seeds as possible by providing grants, and build networks to support existing or emerging businesses. VSJF has acted as a catalyst, leveraging good ideas, technical know-how, and financial resources to propel innovation in sustainable development, especially in the realms of organic agriculture and local food systems, sustainable forestry, and biofuels (locally grown for local use).

Since 1997, the VSJF has made grants of more than $2.7 million to 150 recipients. Grantees have utilized these funds to leverage an additional $11.8 million to implement projects, test ideas, and assemble the building blocks of a green economy. Their combined efforts have created approximately 800 local jobs, supported community development initiatives, preserved resilient ecosystems, filled vital needs in Vermont's economy, and provided new models for moving forward.

VSJF funding in 1998 and 2003 was used to help form the Vermont Cheese Council (VCC) and to help its first 12 members market their products, share technical assistance, develop quality standards for Vermont cheeses using the Council's label, develop a fundraising plan, complete the VCC website, and produce a logo. Today, two-thirds of VCC's now 36 members have won awards. At the 2006 American Cheese Society Conference, the Clothbound Cheddar collaboration between Jasper Hill Farm and Cabot Creamery won best in show.

"People were scratching their heads," laughs Davis. "Cabot makes the cheese? And Jasper Hill ages the cheese? And you don't fight about this? For that cheese to win best in show provided validation for everything that Vermont is all about. Not that there's just great cheese coming out of here, but that our whole approach to it is innovative and unique."

The VCC represents a transition from commodity to value-added agricultural production, which yields a higher rate of return for farmers and a more diverse range of local products for consumers.

Network Building

With limited funds, VSJF was forced to become innovative in supporting businesses. VSJF saw that every successful business is embedded in a network of relationships, and the stronger the network, the more sustainable, flexible, and resilient the business. The power of networks was a notion derived in part from Robert Putnam, author of *Bowling Alone* (2000), who refers to social capital as "the connections among individuals (social networks) and the norms of reciprocity and trustworthiness that arise from them" (p. 19).

Social capital can be linked positively to innovation, to sales growth, return on investment, international expansion success, and the like. Given Vermont's size and connectedness, it's not surprising that the state is ranked third in the country according to Putnam's social capital index. And VSJF saw that creating a supportive environment to nurture and sustain these kinds of business networks and organizations would be a prudent way to use limited development resources.

VSJF grants helped enable more than a dozen networks, representing 1,600-plus businesses, to do the kind of strategic planning, capacity building, information sharing, market research, joint marketing, and policy development that are crucial to developing a unified voice and competitive advantage in a sustainable economy.

Market Building

VSJF also recognizes the importance of developing markets for existing businesses. They take the perspective that markets are made through interactions among businesses, government, nonprofit organizations, communities, and other resources. For example, four years ago biodiesel was not available in Vermont. VSJF and its partners conducted pilot projects and educational activities that successfully introduced biodiesel to large-scale institutional and commercial diesel users, in addition to residential heating oil customers.

To realize this success, VSJF, in cooperation with the University of Vermont Extension Service, conducted on-farm oilseed production and feasibility studies to help farmers familiarize themselves with these new crops. They helped a small nascent biodiesel producer expand his capacity with new equipment, enabled several farmers to construct on-farm production facilities, and assisted the installation of biodiesel pumps at a fueling station. With a little over $2 million invested over the past four years, more than 30 locations in the state now carry biodiesel, and many farmers are in the process of developing farm-scale biodiesel production capacity.

VSJF also has committed to extending its influence outside Vermont's borders. It is now in the process of codifying Vermont's model of local production for local use into a set of sustainable biofuel principles, policies, and practices that could be applicable in other rural states.

VSJF's experimentation has affirmed that a little goes a long way. With investments and technical assistance targeted at the development of markets for sustainably produced goods and services, the building blocks of Vermont's green economy are ready to go mainstream.

What "Sustainable" Means to Vermont

The Vermont Sustainable Jobs Fund was established by the Vermont Legislature in 1995 to build markets within the following natural-resource-based economic sectors:

- environmental technologies;
- environmental equipment and services;
- energy efficiency;
- renewable energy;
- pollution abatement;
- specialty foods;
- water and wastewater systems;

- solid waste and recycling technologies;
- wood products and other natural-resource-based or value-added industries;
- sustainable agriculture;
- existing businesses, including larger manufacturing companies striving to minimize their impact; and
- waste through environmentally sound products and processes.

The VSJF works with entrepreneurs and consumers to develop both the supply of and demand for goods and services that provide sustainable alternatives to economic practices that could cause negative impact over time. Success has come through a combination of targeted, early-stage funding, technical assistance such as business coaching and the Peer to Peer Collaborative, and a focus on the long term.[2]

Rhonda Phillips, Bruce Seifer, and Ed Antczak, 2013

Keywords

Community, community development, economic development, growth, social capital, capacity building, business development, community/economic development chain.

Review Questions

1 What types of community capital exist in your community?
2 How is community development both a process and an outcome?
3 What is the difference between growth and development?
4 How is community development related to economic development?
5 What issues exist in your community that could be understood in the context of the community development, and community economic development chain?

Notes

1 This case is reprinted with permission of the authors from *Sustainable Communities, Creating a Durable Local Economy*, by R. Phillips, B. Seifer, and E. Antczak, London: Routledge, 2013, pp. 171–175.
2 See www.vsjf.org/peer_collaborative/purpose.shtml

Bibliography

Agnitsch, K., Flora, J., and Ryan, V. (2006) "Bonding and Bridging Capital: The Interactive Effects on Community Action," *Journal of the Community Development Society*, 37(1): 6–52.

American Economic Development Council (AEDC) (1984) *Economic Development Today: A Report to the Profession*, Schiller Park, IL: AEDC.

Anglin, R. (2011) *Promoting Sustainable Local and Community Economic Development*, London: CRC Press.

Blair, J.P. (1995) *Local Economic Development: Analysis and Practice*, Thousand Oaks, CA: Sage.

Christenson, J.A. and Robinson, J.N. (1989) *Community Development in Perspective*, Ames, IA: Iowa University Press.

Committee for Economic Development, Research and Policy Committee (1995) *Inner-City Communities: A New Approach to the Nation's Urban Crisis*, New York: Committee for Economic Development.

Dunbar, J. (1972) "Community Development in North America," *Journal of the Community Development Society*, 7(1): 10–40.

Green, G.P. and Haines, A. (2002) *Asset Building and Community Development*, Thousand Oaks, CA: Sage.

Green, P.G. and Haines, A. (2011) *Asset Building and Community Development*, 2nd ed., Thousand Oaks, CA: Sage.

Huie, J. (1976) "What Do We Do about it?—A Challenge to the Community Development Profession," *Journal of the Community Development Society*, 6(2): 14–21.

Lee, S.J. and Kim, Y. (2014) Searching for the Meaning of Community Well-being. In S.J. Lee, Y. Kim, and R. Phillips (eds.), *Community Well-being and Community Development: Conceptions and Applications*, Dordrecht, Netherlands: Springer.

Long, H. (1975) "State Government: A Challenge for Community Developers," *Journal of the Community Development Society*, 6(1): 27–36.

Malizia, E.E. and Feser, E.J. (1999) *Understanding Local Economic Development*, New Brunswick, NJ: Rutgers University Center for Urban Policy Research.

Martin, H.A. (1996) "There's a 'Magic' in Tupelo-Lee County, Northeast Mississippi," *Practicing Economic Development*, Schiller Park, IL: American Economic Development Foundation.

Mattessich, P. and Monsey, M. (2004) *Community Building: What Makes It Work*, St. Paul, MN: Wilder Foundation.

National Research Council (1975) *Toward an Understanding of Metropolitan America*, San Francisco: Canfield Press.

Phillips, R. and Besser, T. (eds.) (2013) *Introduction to Community Economic Development*, Abingdon, UK: Routledge.

Phillips, R., Seifer, B., and Antczak, E. (2013) *Sustainable Communities, Creating a Durable Local Economy*, Abingdon, UK: Routledge.

Pittman, R.H. (2006) "Location, Location, Location: Winning Site Selection Proposals," *Management Quarterly*, 47(1): 2–26.

Pittman, R.H. (2007) "Business Retention and Expansion: An Important Activity for Power Suppliers," *Management Quarterly*, 48(1): 14–29.

Ploch, L. (1976) "Community Development in Action: A Case Study," *Journal of the Community Development Society*, 7(1): 5–16.

Putnam, R.D. (2000) *Bowling Alone: The Collapse and Revival of American Community*, New York: Simon & Schuster.

Roberts, R.T. (2006) "Retention and Expansion of Existing Businesses," *Community Development Handbook*, Atlanta, GA: Community Development Council.

Schumacher, E.F. (1973) *Small Is Beautiful: Economics as If People Mattered.* London: Harper & Row.

Shaffer, R., Deller, S., and Marcouiller, D. (2006) "Rethinking Community Economic Development," *Economic Development Quarterly*, 20(1): 59–74.

Shirouzu, N. (2006) "As Detroit slashes car jobs, southern towns pick up slack," *Wall Street Journal*, Feb. 1, p. A1.

Southern Economic Development Council (SEDC) (2006) *Member Profile Survey*, Atlanta, GA: SEDC.

Thurow, L. (1980) *The Zero-Sum Society: Distribution and the Possibilities for Economic Change*, New York: Basic Books.

Warren, R.L. (1963) *The Community In America*, Chicago: Rand McNally Press.

Wise, G. (1998) *Definitions: Community Development and Community-Based Education*, Madison, WI: University of Wisconsin Extension Service.

Connections

Kick-start your inner social entrepreneur/changemaker leader with this list, "Top 30 Social Entrepreneurship Reads for Aspiring Changemakers," at www.ashoka.org/bookfutureforum

Explore links to community development at the following websites: www.comm-dev.org, www.iactglobal.org, www.planning.org (link through to the Economic and Community Development Divisions).

2 Seven Theories for Seven Community Developers[1]

Ronald J. Hustedde

Overview

Community developers need theories to help guide and frame the complexity of their work. However, the field is girded with so many theories from various disciplines that it is difficult for practitioners to sort through them. Although many undergraduate and graduate community development programs have emerged in North America and throughout the world, there is no fixed theoretical canon in the discipline. This chapter focuses on the purpose of theory and the seven theories essential to community development practice. Why seven theories? In Western cultures, seven implies a sense of near completeness. There are seven days in a week, seven seas, seven climate zones, and seven ancient and modern wonders of the world. Rome was built on seven hills. While seven may or may not be a lucky number, seven theories are offered as a theoretical core for those who approach community development from at least seven contextual perspectives: organizations; power relationships; shared meanings; relationship building; choice making; conflicts; and integration of the paradoxes that pervade the field. Hence, the chapter's title: "Seven Theories for Seven Community Developers." It is a potential canon for practitioners.

Introduction: Why Theory?

Theories are explanations that can provide help in understanding people's behavior and a framework from which community developers can explain and comprehend events. A good theory can be stated in abstract terms and help create strategies and tools for effective practice. Whether community developers want others to conduct relevant research or they want to participate in the research themselves, it is important that they have theoretical grounding. Theory is the major guide to understanding the complexity of community life and social and economic change (Collins 1988; Ritzer 1996).

The starting point is to offer a definition of community development that is both distinctive and universal and can be applied to all types of societies from postindustrial to preindustrial. Bhattacharyya (2004) met these conditions when he defined community development as the process of creating or increasing solidarity and agency. He asserts that solidarity is about building a deep sense of shared identity and a code of conduct for community developers. The developers need that solidarity as they sort through conflicting visions and definitions of problems among ethnically and ideologically plural populations. It can occur in the context of a "community of place" such as a neighborhood, city, or town. It can also occur in the context of a "community of interest" such as a breast cancer survivors' group, an environmental organization, or any group that wants to address a particular issue. Bhattacharyya contends that creating agency gives people the capacity to order their world. According to Giddens, agency is "the capacity to intervene in the world, or to refrain from intervention, with the effect of influencing a process or the state of affairs" (1984: 14). There are complex forces that work against agency. However, community development is intended to build capacity, which makes it different from other helping professions. Community developers build the capacity of a people when they encourage or teach others to create their own dreams, and to learn new skills and knowledge. Agency or capacity building occurs when practitioners assist or initiate community reflection on the lessons its members have learned from their actions. Agency is about building the capacity to understand, create and act, and reflect.

BOX 2.1 THEORY TO INFORM ACTION

As Green (2008: 50) explains:

> If it is to be used in informing practice, theory must help us to interpret and understand the world. Community theory may benefit from directing more attention to the work that people collectively do to shape community life. As one step in this direction, models of community development practice should be revisited to bring in and synthesize theory describing, interpreting, and understanding people's strategic action.

The Editors

Seven Key Concerns in the Community Development Field

Following this definition of community development, there are seven major concerns involving solidarity and agency building: (1) relationships, (2) structure, (3) power, (4) shared meaning, (5) communication for change, (6) motivations for decision making, and (7) integration of these disparate concerns and paradoxes within the field. Horton (1992) shared similar concerns about African-American approaches to community development. He emphasized historic power differences and the influence of culture and black community institutions in his black community development model. Chaskin et al. (2001) focused on neighborhood and other structures and networks in their work on capacity building. Littrell and Littrell (2006), Green and Haines (2002), and Pigg (2002) all wove concerns about relationships, communicating for change, full participation, rational decision making, and integrating micro and macro forces into their community development insights.

Relationships are linked to a sense of solidarity. How critical are trust and reciprocity in the community development process? What is essential to know about relationship building? Structure refers to social practices, organizations, or groups that play a role in solidarity and capacity building. It also refers to the relationships among them. Some of these social practices and organizations may have a limited role. Therefore, to establish solidarity, new organizations may need to be built and/or existing ones could expand their missions.

Power refers to relationships with those who control resources, such as land, labor, capital, and knowledge, or those who have greater access to those resources than others. Since community development is about building the capacity for social and economic change, the concept of power is essential. Shared meaning refers to social meaning, especially symbols, that people give to a place, physical things, behavior, events, or action. In essence, solidarity must be built within a cultural context. Individuals and groups give different meanings to objects, deeds, and matters. For example, one community might see the construction of an industrial plant as an excellent way to bring prosperity to their town, while another community might see a similar construction as the destruction of their quality of life. Community developers need to pay attention to these meanings if they wish to build a sense of solidarity in a particular community or between communities. It is important to understand the conceptual underpinnings and models of the context of power in communities, especially to increase the likelihood of desirable community-level outcomes (Brennan et al. 2013).

Communication for change is linked to the concept of full participation, a consistent value in the community development literature. Within a framework often dominated by technicians, the corporate sector, or national political constraints, practitioners raise questions about how the voice of citizens can be heard at all. Motivation can influence many aspects of community development. It helps us understand whether people will or will not become involved in a community initiative. It also affects making difficult public choices, a process which usually involves thinking through all the policies to decide which will maximize individual and collective needs. Who is more likely to win or lose if a public policy is implemented? What are the potential consequences on other aspects of life if the policy is carried out? Essentially, the process of making rational choices can be nurtured as a form of capacity building. The integration of paradox and disparate macro and micro concerns are part of community development practice. How does one reconcile concerns about relationships, power, structure, shared meaning, communication for change, and motivational decision making? Is there a theory that ties some of these economic, political and sociological concerns together?

These seven concerns form the basis for essential community development theory: social capital theory, functionalism, conflict theory, symbolic interactionism, communicative action theory, rational choice theory, and Giddens' structuration theory. Table 2.1 lists these concerns and theories. Each of these seven theoretical perspectives will be examined and considered as to how they can be applied to community development practice.

1 Concerns about Relationships: Social Capital Theory

Community developers inherently know that the quality of social relationships is essential for solidarity building and successful community initiatives. Friendships, trust, and the willingness to

Table 2.1 *Concerns and Related Theories*

Concern	*Related theory*
1 Relationships	Social capital theory
2 Structure	Functionalism
3 Power	Conflict theory
4 Shared meaning	Symbolic interactionism
5 Communication for change	Communicative action
6 Motivations for decision making	Rational choice theory
7 Integration of disparate concerns/paradoxes	Giddens' structuration

share some resources are integral to collective action. Community developers intuitively build on these relationships. Social scientists view these relationships as a form of capital. *Social capital* is that set of resources intrinsic to social relations and includes trust, norms, and networks. It is often correlated with confidence in public institutions, civic engagement, self-reliant economic development, and overall community well-being and happiness.

Trust is part of everyday relationships. Most people trust that banks will not steal their accounts or that when they purchase a pound of meat from the grocer, it will not actually weigh less. Life can be richer if there is trust among neighbors and others in the public and private sectors. Think of settings where corruption, indifference, and open distrust might inhibit common transactions and the sense of the common good. Equality is considered to be an important cultural norm that is high in social capital because it reaches across political, economic, and cultural divisions. Reciprocity is another cultural norm that is viewed as part of social capital. It should not be confused with a *quid pro quo* economic transaction; it is much broader than the concept of "I'll scratch your back if you'll scratch mine." When individuals, organizations, or communities provide food banks, scholarship funds, low-cost homes—or other forms of self-help, mutual aid, or emotional support—it stimulates a climate of reciprocity in which the recipients are more likely to give back to the community in some form. A culture with high levels of reciprocity encourages more pluralistic politics and compromise which can make it easier for community development initiatives to emerge.

Putnam (1993, 2000) has argued that social capital has declined in the United States since the 1990s. Social capital indicators have included voter turnout, participation in local organizations, concert attendance, or hosting others for dinner at one's home. Suburban sprawl, increased mobility, increased participation of women in the labor force, and television are among the reasons given for this decline. Some critics claim the indicators are linked too closely with "communities of place" because memberships in organizations such as the Sierra Club and other groups have increased significantly. They have also asserted that communities with strong social capital can also breed intolerance and smugness. They have distinguished between "bonding social capital" and "bridging social capital." They contend a mafia group or the Klu Klux Klan may have strong bonding social capital, but it does not build any new bridges that can expand horizons, provide new ideas, or generate wealth. They suggest focusing more on "bridging social capital"—the formation of new social ties and relationships to expand networks and to provide a broader set of new leaders with fresh ideas and information. For example, some communities have created stronger links between African-American and Caucasian faith-based communities or established leadership programs that nurture emerging and diverse groups of leaders. These activities both create new community linkages to broader resource bases and build new levels of trust, reciprocity, and other shared norms.

How Can Social Capital Theory Serve as a Guide for Community Development Practice?

Community developers can integrate social capital theory into their initiatives. In some cases, they will find communities which have relatively low levels of social capital. In such cases, they may have to begin by nurturing "bonding social capital" through sharing food and drink, celebrations, storytelling, dance, or public art. They will have to create opportunities for people to get to know each other and build new levels of trust through shared interests including music, book clubs, games, or other pursuits.

In other cases, communities may have strong bonding social capital but really need "bridging social capital" if they are going to prosper and increase their quality of life. Take the case of tobacco-dependent counties in rural Kentucky that have limited communications with sister counties to build new regional initiatives such as agricultural and ecological tourism. The Kentucky Entrepreneurial Coaches Institute was created to build a new team of entrepreneurial leaders through a mutually supportive network and linkages with the "best and brightest in rural entrepreneurship" from around the world, nation, and region (Hustedde 2006). Social capital was built through the mutual support of multi-county mini-grant ventures consisting of international and domestic travel seminars in which participants shared rooms, buses, seminars, and programs. These activities led to new forms of bonding and bridging social capital which stimulated not only entrepreneurship but an entrepreneurial culture.

2 Concerns about Structure: Functionalism

Second, it is important to look at structure, which underlies organizational and group capacity to bring about or stop change. In essence, structure is related to Giddens' concept of agency or capacity building. The theoretical concept concerned with structure is known as *structural functionalism*. It is also called *systems theory*, *equilibrium theory*, or simply *functionalism*. According to this theoretical framework, societies contain certain interdependent structures, each of which performs certain functions for societal maintenance. Structures refer to organizations and institutions such as health care, educational entities, business and nonprofits, or informal groups. Functions refer to their purposes, missions, and what they do in society. These structures form the basis of a social system. Talcott Parsons and Robert K. Merton are the specialists most often associated with this theory. According to Merton (1968), social systems have manifest and latent functions. Manifest functions are intentional and recognized. In contrast, latent functions may be unintentional and unrecognized. For example, it could be argued that the manifest function of urban planning is to assure well-organized and efficiently functioning cities, whereas the latent function is to allocate advantages to certain interests such as those involved with the growth machine or real estate developers.

Functionalists such as Parsons argue that structures often contribute to their own maintenance, not particularly to a greater societal good. Concern for order and stability also leads functionalists to focus on social change and its sources. They view conflict and stability as two sides of the same coin. If a community development practitioner wants to build community capacity, he or she will have to pay attention to the organizational capacity for stimulating or inhibiting change. Structural functionalism helps one to understand how the status quo is maintained. Some critics claim that the theory fails to offer much insight into change, social dynamics, or existing structures (Collins 1988; Ritzer 1996; Turner 1998).

How Can Structural Functionalism Guide Community Development Practice?

Structural functionalism is a useful tool for practitioners. Looking at the case of an inner city neighborhood that is struggling to create a microenterprise business that will benefit local people, if one applied structural functionalism to community development practice, one would help the community analyze which organizations are committed to training, nurturing, and financing microenterprise development and what their latent or hidden functions might be. A functionalist-oriented practitioner is more likely to notice dysfunctions in organizations. If existing organizations are not meeting local needs in this area, the functionalist would build community capacity by transforming an existing organization to meet the same concerns. A functionalist would also want to build links with broader social systems, such as external organizations, that could help the community's micro-entrepreneurs to flourish. In essence, a functionalist would see structures as important components of capacity building. While structural functionalism is an important tool for community development, it is limited because it does not fully explore the issue of power that can be found in other theories.

BOX 2.2 SOCIAL JUSTICE AND COMMUNITY DEVELOPMENT IN A CHANGING WORLD

There are many ways of thinking about social justice in the contemporary world. For philosophers, social justice represents a normative ideal concerning the society and polity we should strive to achieve. In both eastern and western traditions, philosophers as varied as the Chinese confucian Mencius (1966) and classical Greek scholar Plato (1894/2000) have associated justice with ideal properties that should be sought in the relationship between the governed and those who govern. In more modern times, the concept of social justice has been linked to liberal notions of fairness and thus a desired future rooted in principles of desert, need, and equality (Miller 1999) in which the procedure for distributing benefits and burdens throughout society is based on the arrangement of social and economic inequalities so that it provides the greatest benefit to the least advantaged (Rawls 1971).

For social scientists, the study of social justice focuses less on aspirational ideals than on legacies of past injustices. Sociologist Avery Gordon (2008), for example, reminds us that the pursuit of social justice requires the recognition of shadowy historical forces of injustice, such as the experience of slavery, and the complex ways in which these experiences continue to impact contemporary life. By examining the cultural, economic, and political forces that have contributed to these injustices, we are better able to understand how and why they occurred and thus take steps to minimize if not prevent unjust actions, processes, and outcomes yet to come.

For activists engaged in organizations and movements pursuing social change, social justice represents the intertwining of theory and praxis as a foundation for transforming the status quo through the active pursuit of a more fully representative society. Grounded in the notion of fundamental human rights, mutual interdependence, and a commitment to the power of human agency, social justice provides a basis for action whose legitimacy is derived in large part from its universal appeal (Shaw 2001). Tapping a set of values that transcend political ideology, economic class, and cultural cleavages created by gender, ethnicity, and race, it provides a framework for uniting and mobilizing disparate interests into a common cause.

The Intersection of Social and Economic Justice with Community Development

Interdisciplinary research focusing on the intersection of social and economic justice and community development bridge knowledge domains across multiple levels of analysis, from global governance and national policies and programs to local community and neighborhood initiatives and projects. Though often focused on processes of change, analytically this research is rooted in a broader critique of the economic, political, and/or cultural systems and conditions in which social interactions are embedded. Consequently, it raises questions and exposes dilemmas often ignored and/or overlooked by mainstream disciplinary approaches, such as why and how income inequality and racial hierarchy persist and how they impact community development in specific contexts around the globe.

Rooted in a concern for communities' most vulnerable populations, much of the theoretical work related to social justice and community development seeks to illuminate the ways in which social stratification and disparities in economic and political power become engrained in cultural and institutional arrangements, thus accounting for the persistence of social and economic injustice over time. Antonio Gramsci (1971), for example, underscores the critical role that cultural hegemony plays in enabling those with power in a society to maintain their dominance by normalizing and naturalizing their values, interests, norms, and ideologies thus making them appear commonsensical even to those that may be harmed by them. In her exploration of the tension between universal ideals and individual and group interest, Iris Marion Young (1990), a prominent theorist of justice, exposes the myriad ways in which the "five faces of oppression"—exploitation, marginalization, powerlessness, cultural imperialism, and violence/harassment—block effective group representation in the democratic process. Focusing to how oppression becomes embedded in the U.S. criminal justice system, scholar-activist Angela Davis (2003) highlights the ways in which low-income and minority communities are adversely and unjustly impacted by practices embedded in the penal system.

Connecting local issues and problems to a broader range of experiences and perspectives regarding the causes and consequences of social and economic (in)justice, brings academics, practitioners, and community activists together to tackle a wide variety of social justice issues that impact community development. These include voter suppression, predatory lending, racial profiling, and residential segregation to name but a few.

As social justice workers, practitioners, academics, and activists function as public intellectuals who expose and directly challenge the types of oppressive ideologies and practices that perpetuate injustice. They do this in several different ways.

First, they raise awareness, educating students, policy makers, and community members about the causes and consequences of (in)justice both formally through courses, papers, and reports and more informally through participation in community and workplace forums, rallies, and events. In so doing they bring otherwise disparate individuals and organizations together in a potentially transformative dialogue focused on finding ways to minimize if not overcome barriers to greater equity and inclusiveness.

Second, they help to organize and mobilize political opposition to the legal and social institutions, practices and processes, which undermine social injustice. This is exemplified in both the Civil Rights and Welfare Rights movements of the 1960s, which challenged segregation, racial discrimination, and inequity by means of nonviolent protest and political and social pressure to extend basic human and social rights. More recently, the immigrants' rights movement and the Human Rights Campaign have worked to organize broad-based coalitions to pursue changes in the law in support of the rights of undocumented workers and immigrant and LGBT families.

Third, social justice workers seek to give both individuals and groups that have been excluded and/or marginalized greater voice, thereby helping to empower them as potential agents of change within their own communities. This is demonstrated in the contemporary era through an explosion of movements and projects aimed at improving social and economic conditions by fostering social innovation and participatory governance. Long engaged in the process of catalyzing solidaristic enterprises rooted in the needs of local communities, the cooperative movement is at the forefront of promoting economic self-sufficiency, empowerment, and cultural integration—themes of particular interest to members of marginalized communities (Gonzales and Phillips 2013). By sharing costs, spreading out risk, and promoting internal and external mutuality, cooperatives allow small producers and socially and environmentally conscious consumers to respond to the needs of the community while competing effectively in the capitalist marketplace (Briscoe and Ward, 2005; Fairbairn, 2003). Guided by a clear set of values such as democratic control and concern for community, which are enshrined in cooperative principles established by the International Cooperative Association (ICA), both ideationally and pragmatically, cooperatives offer a promising structure for poverty reduction (Bendick and Egan, 1995), the development of sustainable food systems (Ward, 2005) and refugee resettlement (Gonzales et al., 2013). Examples of other types of models and projects linked to the so-called social solidarity economy (Fonteneau et al., 2011) include tenant-managed public housing, community-owned alternative energy production, participatory budgeting and a variety of other programs that bring together elected officials, public agencies, and local residents in partnerships that democratize the ownership and management of community assets.

In the areas of community and cooperative development, social justice organizations and movements draw from various approaches and traditions to emphasize social inclusion, democratic participation, and a more equitable distribution of both material and non-material goods. While various groups define social justice in distinctive ways depending on the particular political and social traditions they draw from, promoting social justice within the process of community development entails a recognition and support for human interconnection and social solidarity in the pursuit of a better world (see Adams et al., 2007).

Vanna Gonzales, Ph.D.
Director, Certificate in Economic Justice, Faculty of Justice and Social Inquiry-School of Social Transformation, Arizona State University, USA.

Sources

Adams, M., Bell, L.A., & Griffin, P. (Eds.). (2007). *Teaching for diversity and social justice* (2nd ed.). New York: Routledge.

Bendick, Marc & Egan, Mary Lou. (1995). Worker ownership and participation enhances economic development in low-opportunity communities. *Journal of Community Practice*, 2(1): 61–85.

Briscoe, R. & Ward, M. (2005). What co-ops have in common. In R. Briscoe & M. Ward (Eds.), *Helping ourselves: Success stories in co-operative business and social enterprise.* Cork, Ireland: Oak Tree Press.

Davis, A. (2003). *Are prisons obsolete?* New York: Seven Stories Press.
Fairbairn, B. (2003). History of cooperatives. In C.D. Merrett & N. Walzer (Eds.), *Cooperatives and local development: Theory and applications for the 21st century* (pp. 23–51). Armonk, NY: M.E. Sharpe.
Fonteneau, B., Neamtan, N., Wanyama, F., Morais, L.P., de Poorter, M., Borzaga, C., Galera, G., Fox, T., & Ojong, N. (2011). Social and Solidarity Economy: Our common road towards Decent Work (2nd ed.). Turin, Italy: International Training Center of the ILO.
Gonzales, Vanna, Forrest, N., & Balos, N. (2013). Refugee farmers and the social enterprise model in the American Southwest. *Journal of Community Positive Practices*, XIII (4): 32–54. http: //jppc.ro/?page=current&lang=en
Gonzales, Vanna & Phillips, Rhonda (Eds.). (2013). *Cooperatives and community development.* Boca Raton, FL: Routledge.
Gordon, A. (2004). Theory and justice. In A. Gordon (Ed.), *Keeping good time: Reflections on knowledge, power, and people* (pp. 99–105). Boulder, CO: Paradigm Publishers. [speech in 1994]
Gordon, A. (2008). *Ghostly matters: Haunting and the sociological imagination.* Minneapolis, MN: University of Minnesota Press.
Gramsci, Antonio. (1971). *Selections from the Prison Notebooks.* Translated and edited by Quintin Hoare and Geoffrey Nowell Smith. New York: International Publishers.
Mencius. (1966). Justice and humanity. In Mencius, *On the mind.* W.A.C.H. Dobson (Trans.), *Mencius.* University of Toronto Press.
Miller, D. (1999). *Principles of social justice.* Cambridge, MA: Harvard University Press.
Neamtan, Fredrick Wanyama, Leandro Pereira Morais, Mathieu de Poorter. International Training Centre of the International Labour Organization. Turin: Italy.
Plato. (2000). *The republic.* (B. Lovett, Trans.). Mineola, NY: Dover Publications Inc. [original work published in 1894]
Rawls, J. (1971). *A theory of justice.* Cambridge, MA: Harvard University Press.
Shaw, R. (2001). *The activists' handbook.* Berkeley, CA: University of California Press.
Young, I.M. (1990). *Justice and the politics of difference.* Princeton, NJ: Princeton University Press.
Ward, M. (2005). Feeding ourselves II: Farmers' co-ops and food. In R. Briscoe & M. Ward (Eds.), *Helping ourselves: Success stories in co-operative business and social enterprise* (pp.64–91). Cork, Ireland: Oak Tree Press.

3 Concerns about Power: Conflict Theory

Power is the third key issue for community development. Power is control or access to resources (land, labor, capital, and knowledge). Since community development builds capacity, concerns about power are pivotal. Insights into power tend to be found in political science or political sociology. More contemporary theorists have added to the richness of the literature. In his later writings, Foucault (1985) argued that where there is power there is resistance. He examines the struggles against the power of men over women, administration over the ways people live, and of psychiatry over the mentally ill. He sees power as a feature of all human relations (Foucault, 1965, 1975, 1979, 1980, 1985; Nash, 2000). Power has fluidity in the sense that it can be reversed and exists in different degrees. Beyond conventional politics at the state level, Foucault's focus extends to the organizations and institutions of civil society and to interpersonal relations.

Wallerstein (1984) applied Marxist theory to understand the expansion of capitalism to a globalized system which needs to continually expand its boundaries. "Political states," such as Japan, the UK, the European Union, and the US, are among the core developed states based on higher-level skills and capitalization. These states dominate the peripheral areas such that weak states are economically dependent on the "core." The low-technology states form a buffer zone to prevent outright conflict between the core and the periphery. Some have applied Wallerstein's world system theory to regional economics, with places like Appalachia serving as a "periphery" to global market forces. Mills (1959), one of the earliest American conflict theorists, examined some of the key themes in post-World War II American politics. He argued that a small handful of individuals from major corporations, federal government, and the military were influencing major decisions. He believed this triumvirate shared similar interests and often acted in unison. Mills' research on power and authority still influences theories on power and politics today. However, Mills also had critics such as Dahl (1971), who believed that power was more diffused among contending interest groups. Galbraith (1971) asserted that technical bureaucrats behind the scenes had more power than those in official positions. Neo-Marxists argued that Mills and Dahl focused too much on the role of individual actors. They believed that institutions permit the exploitation of one class by another. They also posited that the state intervenes to correct the flaws of capitalism and preserve the status quo, both of which are in the institutions' interests.

In summary, conflict theory suggests that conflict is an integral part of social life. There are conflicts between economic classes, ethnic groups, young and old, male and female, or among races. There are conflicts among developed "core" countries and regions and those that are less developed. It is argued these conflicts result because power, wealth, and prestige are not available to everyone. Some groups are excluded from dominant discourse. It is assumed that those who hold or control desirable goods and services or who dominate culture will protect their own interests at the expense of others. Conflict theorists such as Coser (1956), Dahrendorf (1959), and Simmel (cited in Schellenberg, 1996) have looked at the integrative aspects of conflict and its value as a contributing force to order and stability. Conflict can be constructive when it forces people with common interests to make gains to benefit them all. Racial inequalities or other social problems would never be resolved to any degree without conflict to disturb the status quo. Simmel discusses how conflict can be resolved in a variety of ways including disappearance of the conflict, victory for one of the parties, compromise, conciliation, and irreconcilability (Schellenberg 1996).

This theoretical framework that underlies both the power of one party over another and the potential for conflict is not intended to be exhaustive. Instead, it points to some of the major concerns that can guide community development practice.

How Can Conflict Theory Serve as a Guide for Community Development Practice?

Community organizers tend to more readily embrace conflict theory as a pivotal component of their work. However, it can be argued that community developers also need conflict theory if their goal is to build capacity. Power differences are a reality of community life and need to be considered as development occurs. Take the case of an Appalachian community near a major state forest. The state Department of Transportation (DOT) wanted to build a highway through the state forest. They claimed it would lead to more jobs and economic development. A group of local citizens questioned this assumption. They believed the highway would pull businesses away from the prosperous downtown area to the edge of town, lead to sprawling development that would detract from the quality of life, destroy a popular fishing hole, and harm the integrity of the forest. The DOT refused to converse with the community; they claimed the proposed highway's economic benefits were irrefutable.

Conflict theory served as a reference point for moving the community's interests further. At first glance, it appeared that the DOT was in charge of making the major decisions about the highway. However, the community developer put conflict theory into practice. Community residents were encouraged to analyze the power of the DOT as well as its being their own political, technical, economic, and social power. Through its analysis, the group was expanded to include downtown businesspeople, hunters, and environmental and religious groups. In this particular case, the community decided it needed more technical power. They were able to secure the services of university researchers, such as economists, foresters, sociologists, and planners, who had the credentials to write an alternative impact assessment of the proposed highway. This report was widely circulated by the community to the media and prominent state legislators. Gradually, external support (power) emerged to help the community and the DOT decided to postpone the project.

In a similar situation, the use of conflict theory took another twist. The opponents of a DOT-proposed road sought a mediator/facilitator to help them negotiate with the DOT and other stakeholders. They believed a neutral third party could create a safe climate for discussion, and that during such discussions power differences would be minimized. In this particular case, their use of conflict theory paid off because the dispute was settled to everyone's satisfaction.

In summary, community developers need conflict theory because it helps them gain insight into why specific differences and competition have developed among groups and organizations in a community. It can help them to understand why some people are silent or have internalized the values of elites even to their own disadvantage. Practitioners and researchers can use Simmel's theory to see how people resolve their differences. Alternately, they can borrow from Marx and the neo-Marxists to consider the sharp differences between and among class economic interests, gender, race, and other concerns.

Conflict theory can help communities understand the kind and extent of competing interests among groups. It also can shed light on the distribution of power, whether concentrated in the hands of a few or more broadly distributed. Communities can also explore the use of conflict to upset the status quo—whether through protests, economic boycotts, peaceful resistance, or other ranges of possibilities—especially if competing groups or institutions refuse to change positions or negotiate.

While conflict theory is an essential tool for capacity building, it should be noted that critics claim it is limited because it ignores the less controversial and more orderly parts of society and does not help in understanding the role of symbols in building solidarity (Collins 1988; Ritzer 1996; Turner 1998). This leads to another theoretical framework about shared meaning.

4 Concerns about Shared Meaning: Symbolic Interactionism

Shared meaning is the fourth key concern in community development. If the field is committed to building or strengthening solidarity, then practitioners must be concerned about the meaning that people give to places, people, and events. Herbert Blumer (1969) named the theory "symbolic interactionism" because it emphasizes the symbolic nature of human interaction rather than a mechanical pattern of stimulus and interaction. For symbolic interactionists, the meaning of a situation is not fixed but is constructed by participants as they anticipate the responses of others. Mead (1982) explored the importance of symbols, especially language, in shaping the meaning of the one who makes the gesture as well as the one who receives it.

Goffman (1959) argued that individuals "give" and "give off" signs that provide information to others on how to respond. There may be a "front" such as social status, clothing, gestures, or a physical setting. Individuals may conceal elements of themselves that contradict general social values and present themselves to exemplify accredited values. Such encounters can be viewed as a form of drama in which the "audience" and "team players" interact. In his last work, Goffman (1986) examined how individuals frame or interpret events. His premise involves group or individual rules about what should be "pictured in the frame" and what should be excluded. For example, a community developer's framework of a community event might exclude ideas such as "citizens are apathetic." It will probably include shared "rules" such as "participation is important." The emphasis is on the active, interpretive, and constructive capacities of individuals in the creation of social reality. It assumes that social life is possible because people communicate through symbols. For example, when the traffic light is red, it means stop; when the thumb is up, it means everything is fine. Flora et al. (2000) investigated how two opposing community narratives moved through the stages of frustration, confrontation, negotiation, and reconciliation. Their case study could be viewed as the employment of social interactionism. They concluded that, among the symbols that humans use, language seems to be the most important because it allows people to communicate and construct their version of reality. Symbolic interactionists contend that people interpret the world through symbols but stand back and think of themselves as objects.

For example, a group of Native Americans view a mountain as a sacred place for prayer and healing and react negatively when someone tries to develop or alter access to it. Developers, foresters, tourism leaders, and others are likely to have other meanings for the mountain. Different individuals or groups attach a different meaning to a particular event. These interpretations are likely to be viewed by others as a form of deviance which may be accepted, rejected, or fought over. Social interactionists argue that one way people build meaning is by observing what other people do, by imitating them, and following their guidance.

How Can Symbolic Interactionism Serve as a Tool for Community Development Practice?

Symbolic interactionism is essential for community development because it provides insight into the ways people develop a sense of shared meaning, an essential ingredient for solidarity. When a community developer helps a community develop a shared vision of their future, she is helping them build a sense of unity. A community-owned vision comes about through the interaction of people and is related through pictorial, verbal, or musical symbols. A symbolic interactionist would be keen on bringing people together to develop a shared understanding.

For example, take a case where some citizens have expressed an interest in preserving the farmland adjacent to the city, and have asked a community developer for assistance. If one employed a symbolic interactionist perspective, one would ask them what the presence of farmland means to them. One would link them with farmers and others to see if there were different or competing meanings. Participants would be asked how they developed their meaning of farmland. A symbolic interactionist wouldn't ignore the concept of power. Participants would be asked questions as to whose concept of farmland dominates public policy. Through the employment of symbolic interaction theory, a sense of solidarity could gradually be established in a community.

A symbolic interactionist would identify groups that deviate from the dominant meaning of something and would engage them with other groups in order to move the community toward solidarity. Symbolic interactionists would also use symbols to build capacity. For example, a community might choose to preserve a historic structure because they believed it was beautiful, or explain its importance in a labor, class, racial, or gender struggle or some other interests. A community developer could augment their meaning with data about the historical and architectural significance that external agents see in the structure. Community capacity could be built in other ways such as providing information about tax credits for historic structures or how to locate grants for preservation. Increasingly, community development researchers and practitioners are asked to help citizens reflect and understand the meaning of their work. The symbolic interactionist concepts can be used to aid in collective evaluations. Essentially, it all boils down to what it means and who gives it meaning.

Symbolic interactionists probe into the factors that help people understand what they say and do by looking at the origins of symbolic meanings and how meanings persist. Symbolic interactionists are interested in the circumstances in which people question, challenge, criticize, or reconstruct meanings. Critics argue that symbolic interactionists do not have an established systematic framework for predicting *which* meanings will be generated, for determining *how* meanings persist, or for understanding how they change. For example, say a group of Mexican workers and a poultry-processing firm move into a poor rural community that was historically dominated by Anglo-Saxon Protestants. The events may trigger cooperation, goodwill, ambivalence, anger, fear, or defensiveness. The cast of characters involved in these events may be endless. What has really happened and whose interpretation captures the reality of the situation? Symbolic interactionists have limited methodologies for answering such questions. In spite of these limitations, it is hoped that a strong case has been made as to why symbolic interactionism is an essential theory for community development practice.

5 Communication for Change: Communicative Action Theory

It is safe to assume that community development occurs within the context of democracy that is deliberative and participatory. Public talk is not simply talk; it is essential for democratic participation. It is about thinking through public policy choices. Deliberation occurs when the public examines the impacts of potential choices and tries them on, just as one might try on clothing in a department store before making a choice. In such settings, public talk involves rich discussions among a variety of networks. From the community development perspective, participation occurs in a setting where a diversity of voices are heard in order to explore problems, test solutions, and make changes to policies when the community finds flaws. Communities with robust democratic networks can be viewed as *communicatively integrated* (Friedland, 2001). This type of integration involves the communicative activities that link individuals, networks, and institutions into a community of place or interest.

Habermas argues that communicative action is shaped at the seam of a system and *lifeworld*. Systems involve macroeconomic and political forces that shape housing, employment, racial, and class divisions in a particular community. Local politics are also influenced by federal and state laws, national party politics, and regulations. Although the system is embedded in language, it is self-producing. Power and markets can be relatively detached from community, family, and group values. At the same time, there is the world of everyday life or the *lifeworld*. Habermas views the lifeworld as constituted of language and culture:

> The lifeworld, is, so to speak, the transcendental site where speaker and hearer meet, where they reciprocally raise claim that their utterances fit the world . . . and where they can criticize and confirm those validity claims, settle their disagreements and arrive at agreements.
>
> (1987: 126)

Habermas is concerned about the domination and rationalization of the lifeworld, in which science and technology are the *modi operandi* to address complex public issues. He believes that science and technology maintain the illusion of being value-free and inherently rational. In practical terms, citizens find it difficult to engage in dialogue with "more rational" scientists, engineers, or political and corporate elites. The problem is compounded when there is technical arrogance or limited receptivity to local voices. For example, many local newspapers and television stations are corporately owned. It is therefore difficult to hear local voices, for they are filtered through more dominant perspectives. Habermas is concerned about the colonization of the lifeworld of culture and language, a colonization that reduces people to the status of things. He also argues that technical knowledge is not sufficient for democratic settings in which community developers work. It must be balanced by hermeneutic knowledge which he calls "practical interests." Hermeneutics deals with the interpretation of technical knowledge and what it means for an individual, his or her family, or community. It is action oriented and involves mutual self-understanding.

The third dimension of knowledge is emancipatory. It regards the liberation of the self-conscious and transcends and synthesizes the other two dimensions of knowledge. While science and technology may help liberation, they can also suffocate it. Emancipatory knowledge incorporates both technical and hermeneutic knowledge into a fresh perspective and outlook that leads to action.

In essence, Habermas' theory of communicative action is that it builds a linkage between the "rational" system and the lifeworld. His communicative action theory and political objective are based on free, open, and unlimited communication. It should be noted that Habermas grew up in Nazi Germany and his focus on reason could be viewed as a response to the unreason of the Holocaust. At the same time, unlimited public talk could be seen as reaction to the curtailment of intellectual freedom and public dialogue during the Hitler years. Habermas' insights about communicative action theory, and his emphasis on reason and unrestrained public talk are viewed by some critics as utopian liberal ideals in which people talk their ideas to death. Others assert that universal principles of justice and democracy have been replaced by relativistic and egocentric perspectives. They assert that "reason" is a rationale for the powerful to suppress others. While Habermas emphasizes the potential to reach common ground, his detractors claim that common ground is not possible and that there is nothing wrong with competition between groups. They say he is merely moralizing and that communicative action theory is a hodgepodge of ideas gathered from the Enlightenment, Karl Marx, Max Weber, and others.

On the other hand, it should also be asserted that Habermas is continually expanding his perceptions and that, in spite of these criticisms, he is one of the world's leading public intellectuals. He and has been a powerful influence on the formation of social democracies in Germany and the rest of Europe. "Communicative action" describes the seam where monetary and bureaucratic structures meet the lifeworld. This emphasis on reason, unfettered public discussion, and the potential for common ground provide an essential theory for community development practice in its concern for process.

How Can Communicative Action Theory Guide Community Development Practice?

By its very nature, community development involves the participation of networks, groups, and individuals whose voices are part of the lifeworld. While this lifeworld operates within the context of technical, political, and market realities, it should be noted that the principles of community development entail participation of citizens in defining their own problems and dreams. If technicians or political and corporate interests dominate discussions, citizen involvement and participation becomes a mere afterthought. If technical knowledge is discarded or minimized, community development efforts may not be successful. Habermas' communicative action theory is guided by the intersection of technical and corporate knowledge with local and practical knowledge. Combined, they can lead to a new kind of "emancipatory knowledge" that offers fresh ideas and action.

There are many ways for community developers to carry out Habermas' communicative action theory. For example, the National Issues Forums are held in many communities wherein individuals, networks, and groups explore public issues through the perspective of several public policy choices. Rather than choose sides, these forums are designed for the participants to examine the applicability, strengths, limitations, and values of each choice. National Issues Forums are conscious acts of deliberation that make it easier for the system and the lifeworld to interact.

In another community development case, an Appalachian Cancer Network was developed by homemakers and health-care professionals to deal with high rates of breast and cervical cancer in that region. The health-care leaders were tempted to tell the homemakers what to do. However, the community developers who guided this initiative did not begin with technical knowledge. They started with storytelling in which technical and lay participants responded to the questions: *Have you or a family member ever been touched by cancer or another serious illness? If so, what happened?* The stories that emerged told of triumph, heartache, loss, and anger. The next set of questions was: *What do our stories have in common? What should we do, if anything, about our common issues?*

Eventually, the community development principles of full participation were carried out. The network acted in ways that brought out technical, practical, and emancipatory knowledge. That is, new ideas and action emerged from this initiative that would have been impossible if technicians or lay leaders had acted independently.

6 Motivation for Decision Making: Rational Choice Theory

The rational economic man model was proposed by Alfred Marshall (1895). He believed that humans were interested in maximizing their utility, happiness, or profits. The rational man would investigate each alternative and choose that which would best suit his individual needs. While Marshall recognized that irrational decisions were made, he believed that the overwhelming number of decision makers would operate in a maximizing fashion and cancel out irrational actions. Marshall assumed all the relevant information was available to the economic man and that he could understand the consequences of his choices. The focus was on the individual rather than the collective. Rational choice theory has several embellishments and spinoffs from various social scientists. For example, Mancur Olson (1965) explored whether rational calculation would lead a few individuals to pursue collective action as a way to obtain public goods because they could pursue these goods whether they were active or not. He believed that collective behavior could be expected under two conditions: (1) selective incentives—such as increased stature in the community, tax breaks, or other benefits—could increase the rewards of those engaging in collective action, and (2) the threat of sanctions against those who fail to participate.

In recent years, social scientists have explored how four structural factors relate to individual participation in collective activities. One is prior contact with a group member because it is easier to recruit through interpersonal channels. A second is prior membership in organizations due to the likelihood that those who are already active may join other groups and, conversely, isolated individuals may perceive joining as a type of risk. The second is a history of prior activism because those with previous experience are more likely to reinforce their identity through new forms of activism. The fourth factor is biographical availability, which pulls people toward and away from social movements. For example, full-time employment, marriage, and family responsibilities may increase the risks and costs of becoming involved. Conversely, those who are free of personal constraints may be more likely to join. There is some empirical evidence that students and autonomous professionals may be more likely to join social movements (McAdam et al., 1988).

Critics of rational choice theory have argued that actors do not have equal access to information or that information is distorted. Others assert that many people's choices are limited by social, political, and economic interests and values, which limits their participation in rational choice making.

How Can Rational Choice Theory Serve as a Guide for Community Development Practice?

Community developers know that while people may have altruistic concerns, they also have their own needs and make choices about how to invest their time. There have been many creative responses to rational choice theory. For example, the Cooperative Extension Service Master Gardener Program offers free horticultural training but participants must volunteer hours back to the community in order to receive the training. Leadership programs have popped up in many communities where participants gain the advantage of expanding their network and knowledge bases. Their positive experience in meeting and working with others in collective settings leads to a greater openness and involvement.

When applied to community development, rational choice theory is concerned with finding appropriate rewards and minimizing risks to individuals who become involved in community initiatives. Such rewards might be as simple as free babysitting services or an awards and recognition banquet. Both examples would facilitate people's choices to invest their time or money in community development efforts. In other situations, there is a tendency toward misinformation, misunderstanding, competing sets of data, or different interpretations of the same data. Any or all of these make it difficult to reach common ground and establish solidarity. In such cases, community developers can find new ways to gather data, interpret information, or glean new information from mutually respected third party sources. It should be asserted that in many settings universities are no longer viewed as neutral or objective. They may be perceived as instruments of the state, the corporate sector, or a particular political or economic interest. One of the limitations of rational choice theory is that it can be implemented by technicians, the corporate sector, and bureaucracies in ways that can overwhelm and silence citizens who may not understand such knowledge. Habermas' theory of communicative action can provide a counterbalance to such shortcomings.

7 Integration of Disparate Concerns and Paradigms: Giddens' Structuration Theory

The classical theories of structural functionalism, conflict theory, and rational choice theory are essential concepts for building community capacity. The fluid contemporary theories of social capital, communicative action, and the classical theory of symbolic interactionism are important for creating or strengthening solidarity. There are obvious tensions inherent in these theories. The dualism of macro versus micro characterizes much of the theoretical thinking in sociology. Sharing the same goal of picturing social reality, these schools choose to proceed from opposite directions. The macro-thinkers attempt to draw a holistic picture and lay down the works of society, whereas the micro-theorists hope to arrive at the same results by scrutinizing what happens "in" and "between" individual people. Neither approach is entirely successful in producing a complete and exhaustive picture for community development practice. In a more recent development, efforts have been made at a "microtranslation," which seeks to visualize social reality as composed of individuals interacting with one another to form "larger interaction ritual chains" (Collins 1988).

However, recent theory also recognizes that social agency itself, pointed out above as a key concern for community development, needs to be theoretically addressed. This must be done in a way that transcends both the established orientations in modern social theory and the whole macro–micro split. In his structuration theory, Anthony Giddens (1984, 1989) offers a perspective that is more fluid and process oriented. He introduces a third dimension, or an "in-between" level of analysis, which is neither macro nor micro. It has to do with cultural traditions, beliefs, and societal norms, and how actors draw upon those in their behavior (Collins 1988: 399). For Giddens, those normative patterns of society exist "outside of time and space" (Collins, 1988: 398–399), meaning they are neither properties of the empirical social system nor of the individual actors. Their actuality consists in the moments when individuals' behaviors rise to that level of society's traditions and norms. People also draw and act upon thought patterns or cultural "molds"; for example, the classical notion of reciprocity—

getting one thing in return for something else. Cultural traditions and patterns become modalities by virtue of placing them on Giddens' analytical scheme. They represent a third level, that between individualistic behavior and the macro-structures. Even though the reality of modalities may be only momentary, when people actually rise to them in their behavior, then the social process and the role of culture and normative patterns can be better visualized. "Actors draw upon the modalities of structuration in reproduction of systems of interaction" (Giddens, 1984: 28). Social structure is upheld and existing divisions of society carry on through these "mental molds."

The laying out of society on the six above-mentioned levels—social capital theory, functionalism, conflict, symbolic interactionism, communicative action theory, and rational choice theory—reflects a fluid process in which all levels interact. Individuals represent the agency whereby interaction among different levels take place. Coming back to the community development profession and its key concerns, Giddens' model is perhaps best suited to grasp how social agency is exercised and solidarity established amid and often against the existing structural divisions of society. Behavior is neither haphazard nor merely a reflection of the existing social structure and its divisions. Modalities represent the levels in which people establish solidarity by following the symbolic norms and patterns of their cultures and traditions.

Similarly, new rules of behavior also occur through the medium of modalities, in this instance their creative redefinition. This is how the existing divisions can be overcome and new bonds between people forged. For this to take place, genuine social creativity is necessary. This means that people come up with solutions and ideas that simultaneously draw on the common reference point of their cultural traditions and transcend those traditions to establish new bonds and patterns of solidarity. Modalities serve not only as the rules for the reproduction of the social system, but for its transformation (Turner, 1998: 494). Giddens' concept of modalities is the link between macro- and micro-theories. Modalities are part of the analytical scheme in a particular place. For example, individualism in the United States is a strong modality and can keep citizens from united action. The notion of the common good is another American modality which can be used to transform a divided community into one with a greater sense of solidarity. Modalities can be used to influence the macro or micro level of social change. There are several substantive analyses looking at cultural patterns and systems of ideas and how they mediate the social process. In these analyses, social processing and the dynamics of social transformation are at least partly carried out on the level of modalities. Gaventa (1980) examines the modalities of Appalachia with a focus on rebellion and quiescence. He analyzes how power is used in the region to prevent or implement decisions. The use of force and threat of sanctions are discussed along with less intrusive aspects such as attitudes that are infused into the dominant culture by elites and internalized by non-elites. For example, there are perspectives such as "you can't change anything around here" or "you don't have to be poor if you want to really work." Gaventa argues that there are other modalities in which Appalachian culture has resisted the penetration of dominant social values. Those with less power can develop their own resources for analyzing issues and can explore their grievances openly. He views the "myth of American democracy" as another modality that can set the stage for greater openness and transparency in local government.

Staniszkis (1984) provides further insights about modalities through her ideas about how workers' solidarity emerged in Poland. She saw the working class under the communist regime as a unified bloc, both in a positive hegemonic way and negatively, as subject to the party's control and manipulation. Solidarity and its charismatic leader Lech Walesa transformed these modalities with references to workers' common identity, as opposed to their identity with the Communist party apparatus. To further create a sense of solidarity and unity in opposition to the Communist party and the system, Walesa incorporated Polish workers' strong Christian identification into helping define their new self-understanding and self-image. In her work on the change in workers' collective identity, Staniszkis' consistent attention to symbolic meanings and their interplay with the social structure aptly demonstrates how modalities can be transformed.

BOX 2.3 POWER

Ah, the age-old struggle, trials, and tribulations of those with power and those without, those that should have it and those who create it or take it. Within communities, power struggles and the lack of power confound efforts. Paulo Freire, an advocate of critical pedagogy and author of *Pedagogy of the Oppressed* (1970) had major influence on those working in community development, education and related fields. He proposed that people can either be passive or engage as active participants. In his words:

> No pedagogy which is truly liberating can remain distant from the oppressed by treating them as unfortunates and by presenting for their emulation models from among the oppressors. The oppressed must be their own example in the struggle for their redemption.
>
> (Freire, 1970: 54)

The power to link knowledge to action so people can actively engage to change their societies is essential for community development.

More recent scholars such as Margaret Ledwith have continued to approach issues of power, community activism and theory within the context of community development approaches. She encourages community developers to reengage with the critical pedagogy of Freire to help address challenging issues in today's societies. See Ledwith's *Community Development: A Critical Approach*, 2nd ed. (2011) Bristol, UK: Policy Press.

The Editors

Analytically, Giddens' structuration theory stands as a middle ground between the micro- and the macro-theories as well as the issue of agency and solidarity. Giddens' structuration theory suggests that the micro-theories associated with symbolic interactionism can influence cultural and traditional norms and patterns (modalities) and vice versa. While the symbolic interactionists tend to ignore structure, Giddens' mid-level theory about modalities is a crucial link among symbolic interactionism, rational choice theory, social capital, the micro–macro conflict, communicative action, and structural functionalist theories (Giddens, 1984).

Max Weber's social action theory was originally cast at an "in-between level." If his theory was not explicit, his intentions were at least implicit. Weber attempted to view society as a fluid process, dissecting it into various components for analytical purposes (Turner, 1998: 17) much like Giddens did. Although Weber never attempted an analytical model of society along micro-theoretical lines, some observers have categorized Weber as a micro-theorist because of his subjective interpretation of behavior and its meaning to the actor. Others argue that Weber is a strong macro-theorist, for his intentions may lie closer to Giddens' perspective. This was especially obvious in his attempts to explain the rise of modern capitalism through the interplay of social structural conditions and the religious beliefs of Protestantism. He followed similar analyses for non-Western societies in his volumes on the sociology of religion. What Giddens delineated in theory Weber actually performed in his works, bridging the macro and the micro dimensions in his attention to society's traditions and norms. He observed how people, independent of the macro-structural forces of society, transform these traditions and norms by interpreting and reinterpreting them. Similarly, Gaventa and Staniszkis demonstrated how one can connect communities or groups to structure them in a way that is not fixed or mechanical.

In contrast to debates on whether structure shapes action to determine social phenomena or the reverse, Giddens believes that structure exists in and through the activities of human agents. He views it as a form of "dualism" in which neither can exist without the other. When humans express themselves as actors and monitor the ongoing flow of activities, they contribute to structure and their own agency. He contends that social systems are often the results of human action's unanticipated outcome. Giddens views time and space as crucial variables. Many interactions are face to face, and hence are rooted in the same space and time. However, with the advent of new technologies, there can be interaction across different times and spaces. Community developers are likely to feel some kinship with Giddens because he has a dynamic rather than static concept of the world. He recognizes the interplay of humans and structure in shaping and being shaped. Critics are likely to argue that he has oversubscribed to the concept of the power of human agency. The space of this chapter limits a response to those critiques and a fuller exploration of Giddens' theoretical insights.

How Can Giddens' Structuration Theory Guide Community Development Practice?

Structuration theory provides many theoretical insights (Ritzer, 1996: 433) for those engaged in community development because it links disparate macro-theories about structure and conflict with micro-theories about individual and group behavior such as social capital, rational choice, and symbols or symbolic interactionism. Giddens' concept of modalities is essential for community development practice.

Revisiting the case of the Appalachian community group that opposed the construction of a road through a nearby state forest (see above: "How Can Conflict Theory Serve as a Guide for Community Development Practice?"), the group believed they were overpowered by the Department of Transportation (DOT) that wanted to build the road. The community found it difficult to argue against the DOT report, which contained sophisticated economic, social, and natural resource information. Here is what the community development practitioner did. First, the practitioner asked community residents to identify the strengths of their local traditions—particularly storytelling and the arts—as a venue for building solidarity regarding the integrity of the forest. Together, the community and the practitioner examined the modalities of storytelling and the arts to see if they could use the media to make an impact on the public and local legislators. The community's strong respect for the local Cooperative Extension Service was identified as another modality to mobilize the broader information resources of the land grant university. Without spending much money, the community developer was able to draw upon the services of professional economists, sociologists, foresters, and others. These professionals developed an alternative to the DOT report that was widely disseminated. Storytelling, the local arts, and links with the local Extension Service influenced broader structures and led to fewer power imbalances. Eventually, the DOT decided to permanently "postpone" the development of the road. Because the community developer understood the power of modalities (local cultural traditions and patterns), the community was able to develop a sense of shared meaning. This led to greater influence on structure and resolved the conflict.

How do Giddens' structuration theory and the concept of modalities relate to some of the theories discussed earlier, particularly the classical theories of structural functionalism, conflict theory, rational choice theory, and symbolic interactionism?

When one looks at functionalism through a Giddens lens, one sees how structures shape and can be shaped by modalities. From a Giddens perspective, community change agents are not powerless when faced with powerful structures. Cultural patterns can be transformed to influence or break down structural constraints that inhibit solidarity and capacity building. Giddens' structuration theory illuminates conflict theory because it suggests that communities can influence power imbalances through cultural norms and patterns. It also suggests that external power can shape behavior.

Based on a Giddens perspective, the micro-theories associated with symbolic interactionism and making rational choices can influence cultural and traditional norms and patterns (modalities) and vice versa. While the symbolic interactions and rational choice theorists tend to ignore structure, Giddens' mid-level theory about modalities is a crucial link among symbolic interactionism, rational choice making, the macro "conflict" theory, and structural functionalism. The fluid theories associated with Habermas' communicative action and social capital can be viewed as mid-level theories, as part of structuration theory. They also address the intersection of modalities and structure.

However, there are several limitations to Giddens' theories. His writing is analytical and abstract to the point of being vague and imprecise. He rarely gives concrete examples, which can be frustrating to those community developers who are more empirically grounded. Giddens' analysis is also difficult because it involves constant movement among the levels of modalities, societal institutions, and the actions of individuals. In spite of these limitations, structuration theory is especially useful for community developers because of the potent role of symbolic norms and cultural patterns (modalities) in creating new structures, influencing power differences, and infusing individual behavior with a sense of solidarity.

Conclusion

Community development is often thought of as intention to build solidarity and agency (capacity building). Theory is essential for community development practice because it provides explanations of individual and group behavior. It also provides frameworks so that community developers may comprehend and explain events. There are seven theories that should be part of a community development canon, or knowledge: (1) social capital; (2) structural functionalism; (3) conflict; (4) symbolic interactionism; (5) communicative action; (6) rational choice; and (7) structuration theory. Each theory should be explored along with its limitations and applicability for community development practice.

This chapter is about reaching across the conceptual divide between theory and action. It should stimulate dialogue and further discussion on essential theory for community development practice. The classical theories of structural functionalism, conflict, symbolic interactionism, and rational choice can be balanced by the more fluid and synthesizing theories of social capital, communicative action, and structuration. These theoretical camps may be linked in novel ways to help community developers become more effective.

CASE STUDY: COMMUNITY DEVELOPMENT AND INTERNATIONAL CONFLICT RESOLUTION

Arguably, one of the most pressing international issues of this and future generations is the relationship between Islamic and Western countries, as evidenced by the wars in Iraq and Afghanistan and the ongoing conflict between the Israelis and Palestinians. One scholar believes that community development could serve as a valuable tool to improve Islamic–Western relations and help ease conflicts across the globe. In a series of articles, Jason Ben-Meir states his belief that participatory, grass-roots community development in conflict areas will empower local residents and encourage them to reject religious extremism, engage in community and nation building and appreciate the foreign aid efforts of Western countries. Ben-Meir is President of the High Atlas Foundation, a U.S. nonprofit organization that assists community development in Morocco.

According to Ben-Meir, the billions spent in foreign aid reconstruction in Iraq and Afghanistan typically channeled through third-party contractors and national governments often fosters resentment toward Western countries because input from the communities where the projects take place is not obtained and local residents feel they are not in control of rebuilding their own economic and social life. Ben-Meir argues that sustained development and genuine reconstruction require funding local projects designed by the entire community. The community's priorities would be established by facilitated interactive dialogue where all local residents have a right to express their opinions and collective priorities are developed in a true inclusive and participatory community development process. He believes this will encourage community residents to actively support local rebuilding and economic development efforts. As they feel empowered, develop hope for the future, and see tangible signs of progress of their own design, they will be less likely to embrace extremism born of frustration and alienation. Ben-Meir also believes that successful community and economic development outcomes fostered by this approach will engender goodwill toward Western countries funding these local projects and helping with the community development capacity-building process. Furthermore, progress will be sustainable, since citizens in communities throughout turbulent regions will have learned community- and nation-building skills and local infrastructure will be improved.

In the case of Iraq, Ben-Meir believes the national government should:

- Train local schoolteachers and other community members in group facilitation methods and begin the community development process in all communities with inclusive, participatory meetings to establish local priorities.
- Create community reconstruction planning and training centers in all communities to help implement local priorities and redevelopment projects. The centers would also provide further training in facilitation, conflict management, modern agricultural techniques, health care, and other development topics.

Encouraging community development and funding local priority projects will also help alleviate the Israeli–Palestinian conflict, according to Ben-Meir. He points out that the Palestinian economy is almost totally dependent on Israel's, and when political tensions rise, economic links and flows of people and goods are severely restricted causing a huge hardship on the Palestinians. He argues that Israel and the West can generate tremendous goodwill and help make the Palestinian people economically self-reliant by promoting the community development process and investing in projects designed and managed by local residents.

Whether or not the community development can help achieve these lofty goals is an open question, but there is no doubt that its principles of conflict resolution, group decision making, inclusiveness, and fairness are certainly relevant to international affairs and foreign policy. Community development is germane to countries all over the world and its principles transcend geopolitical boundaries.

The Editors

Sources

Ben-Meir, J. (2004) "Create a new era of Islamic–Western relations by supporting community development," *International Journal of Sociology and Social Policy*, 24(12): 25–41.

Ben-Meir, J. (2005) "Iraq's Reconstruction: A community responsibility," *The Humanist*, 65(3): 6.

Keywords

Solidarity, agency building, structure, power, shared meaning, social capital theory, structural functionalism, conflict theory, social action.

Review Questions

1. What are the seven concerns of community development discussed?
2. How are the seven theories of community development related to the concerns?
3. What can be learned from theory for community development practice? Give an example of an application.
4. How can theory serve to support a "call to action" for communities, and can you think of an example where this could help address a community issue?
5. Look around your own community. What theory could you identify to help explain a situation impacting community development process or outcomes that you observe?

Note

1 This book chapter is an expansion of the article: Hustedde, R.J. & Ganowicz, J. (2002). "The basics: What's essential about theory for community development practice?" *Journal of the Community Development Society*, 33(1): 1–19. The editor of the journal granted permission to duplicate and integrate parts of the article into this chapter.

Bibliography

Bhattacharyya, J. (2004) "Theorizing community development," *Journal of the Community Development Society*, 34(2): 5–34.

Biddle, W. and Biddle, L. (1965) *The Community Development Process*, New York: Holt, Rinehart & Winston.

Blumer, H. (1969) *Symbolic Interactionism: Perspective and Method*, New York: Prentice Hall.

Brennan, M., Bridger, J.C., and Alter, T.R. (2013) *Theory, Practice, and Community Development*, Abingdon, UK: Routledge.

Chaskin, R.J., Brown, P., Venkatesh, S., and Vidal, A. (2001) *Building Community Capacity*, Hawthorne, NY: Aldine de Gruyter.

Christenson, J. and Robinson, J. (eds.) (1989) *Community Development in Perspective*, Iowa City, IA: University of Iowa Press.

Collins, R. (1988) *Theoretical Sociology*, New York: Harcourt Brace Jovanovich.

Coser, L. (1956) *The Functions of Social Conflict*, New York: The Free Press.

Dahl, R.A. (1971) *Polyarchy: Participation and Opposition*, New Haven, CT: Yale University Press.

Dahrendorf, R. (1959) *Class and Class Conflict in Industrial Society*, Stanford, CA: Stanford University Press.

Flora, C.B., Flora, J.L., and Tapp, R.J. (2000) "Meat, meth and Mexicans: Community responses to increasing ethnic diversity," *Journal of the Community Development Society*, 31(2): 277–299.

Foucault, M. (1965) *Madness and Civilization: A History of Insanity in the Age of Reason*, New York: Vintage.

Foucault, M. (1975) *The Birth of the Clinic: An Archeology of Medical Perception*, New York: Vintage.

Foucault, M. (1979) *Discipline and Punish: The Birth of Prison*, New York: Vintage.

Foucault, M. (1980) *The History of Sexuality, Volume 1, An Introduction*, New York: Vintage.

Foucault, M. (1985) *The History of Sexuality, Volume 2, The Use of Pleasure*, New York: Pantheon.

Friedland, L.A. (2001) "Communication, community and democracy," *Communication Research*, 28(4): 358–391.

Fussell, W. (1996) "The value of local knowledge and the importance of shifting beliefs in the process of social change," *Community Development Journal*, 31(1): 44–53.

Galbraith, J.K. (1971) *The New Industrial State*, Boston, MA: Houghton Mifflin.

Gaventa, J.L. (1980) *Power and Politics: Quiescence and Rebellion in an Appalachian Valley*, Champaign, IL: University of Illinois Press.

Giddens, A. (1984) *The Constitution of Society*, Berkeley, CA: University of California Press.

Giddens, A. (1989) "A Reply to My Critics," in D. Held and J.B. Thompson (eds.) *Social Theory of Modern Societies: Anthony Giddens and His Critics* (pp. 249–301), Cambridge: Cambridge University Press.

Goffman, E. (1959) *The Presentation of Self in Everyday Life*, Garden City, NY: Anchor.

Goffman, E. (1986) *Frame Analysis: An Essay on the Organization of Experience*, Boston, MA: Northeastern University Press.

Green, G.P. and Haines, A. (2002) *Asset Building and Community Development*, Thousand Oaks, CA: Sage.

Green, J.J. (2008) "Community Development as Social Movement: A Contribution to Models of Practice," *Community Development* 39(1): 50–62.

Habermas, J. (1987) *The Theory of Communicative Action, Vol. 2, Lifeworld and System: A Critique of Functionalist Reason*, Boston, MA: Beacon Press.

Horton, J.D. (1992) "A Sociological Approach to Black Community Development: Presentation of the Black Organizational Autonomy Model," *Journal of the Community Development Society*, 23(1): 1–19.

Hustedde, R.J. (2006) "Kentucky Leadership Program Coaches Entrepreneurs," *Economic Development America*, winter: 28–29.

Hustedde, R.J. and Ganowicz, J. (2002) "The Basics: What's Essential about Theory for Community Development Practice?" *Journal of the Community Development Society*, 33(1): 1–19.

Jeffries, A. (2000) "Promoting Participation: A Conceptual Framework for Strategic Practice, with Case Studies from Plymouth, UK and Ottawa, Canada," *Scottish Journal of Community Work and Development*, Special Issue 6 (Autumn): 5–14.

Littrell, D.W. and Littrell, D.P. (2006) *Practicing Community Development*, Columbia, MO: University of Missouri-Extension.

McAdam, D., McCarthy, J., and Zald, M. (1988) "Social Movements," in N.J. Smelser (ed.) *Handbook of Sociology* (pp. 695–738), Newbury Park, CA: Sage.

Marshall, A. (1895) *Principles of Economics, Third Edition*, London: Macmillan and Co.

Mead, G.H. (1982) *The Individual and the Social Self: Unpublished Work of George Herbert Mead*, Chicago, IL: University of Chicago Press.

Merton, R.K. (1968) *Social Theory and Social Structure*, revised ed., New York: Free Press.

Mills, C.W. (1959) *The Sociological Imagination*, New York: Oxford University Press.

Nash, K. (2000) *Contemporary Political Sociology: Globalization, Politics, and Power.* Malden, MA: Blackwell.

Olson, M., Jr. (1965) *The Logic of Collective Action*, Cambridge, MA: Harvard University Press.

Parsons, T. (ed.) (1960) "Some Reflections on the Institutional Framework of Economic Development," in *Structure and Process in Modern Societies* (pp. 98–131), Glencoe, IL: Free Press.

Parsons, T. and Shils, E.A. (eds.) (1951) *Toward a General Theory of Action*, New York: Harper & Row.

Perkins, D.D. (1995) "Speaking Truth to Power: Empowerment Ideology as Social Intervention and Policy," *American Journal of Community Psychology*, 23(5): 569–579.

Pigg, K.E. (2002) "Three Faces of Empowerment: Expanding the Theory of Empowerment in Community Development," *Journal of Community Development Society*, 33(1): 107–123.

Putnam, R.D. (1993) *Making Democracy Work: Civic Traditions in Modern Italy*, Princeton, NJ: Princeton University Press.

Putnam, R.D. (2000) *Bowling Alone: The Collapse and Revival of American Community*, New York: Simon & Schuster.

Ritzer, G. (1996) *Sociological Theory*, 4th ed., New York: McGraw-Hill.

Rothman, J. and Gant, L.M. (1987) "Approaches and Models of Community Intervention," in D.E. Johnson, L.R. Meiller, L.C. Miller, and G.F. Summers (eds.), *Needs Assessment: Theory and Methods*, Ames (pp. 35–44), IA: Iowa State University Press.

Schellenberg, J.A. (1996) *Conflict Resolution: Theory, Research and Practice.* Albany, NY: State University of New York Press.

Shaffer, R.E. (1989) *Community Economics: Economic Structure and Change in Smaller Communities*, Ames, IA: Iowa State University Press.

Staniszkis, J. (1984) *Poland's Self-Limiting Revolution*, Princeton, NJ: Princeton University Press.

Turner, J.H. (1998) *The Structure of Sociological Theory*, 6th ed., Belmont, CA: Wadsworth.

Wallerstein, I. (1984) "The Development of the Concept of Development," *Sociological Theory*, 2: 102–116.

Weber, M. (1947) *The Theory of Social and Economic Organization*, trans. A.M. Henderson and T. Parsons, New York: Oxford University Press.

Connections

Explore change in communities in the context of theory and practice. See the Aspen Institute's *Community Change: Theories, Practice, and Evidence*, available at: www.aspeninstitute.org/sites/default/files/content/docs/rcc/COMMUNITYCHANGE-FINAL.PDF

Watch a classic video—Robert Redford's *The Milagro Beanfield War* (1988)—and see if you can discern which theories could help both understand and guide action in this case.

Explore how "political" a community development issue can become. When the City Market project (a nonprofit member-owned food cooperative) was being developed in Burlington, Vermont, things got a little crazy. See the Public Broadcasting Service documentary, *People Like US: Social Class in America*, at www.pbs.org/peoplelikeus.film/index.html. Happily, issues were resolved and the cooperative is now a valued community partner and place.

3 Asset-Based Community Development

Anna Haines

Overview

Building on a community's assets rather than focusing on its needs for future development is the basic approach of asset-based community development. By focusing on successes and small triumphs instead of looking to what is missing or negative about a place, a positive community outlook and vision for the future can be fostered. This approach also focuses on a sustainable approach to development. This chapter outlines the process and the major steps in identifying individual, organizational, and community asset development.

Introduction

Chapter 1 focused on the philosophical underpinnings for community development, explained the context within which community development operates, and made the argument that communities need to diversify their local economies because of the knowledge-based information-age economy in which we live.

This chapter discusses community development from the perspective of concentrating and building on community assets rather than focusing on needs and problems. This approach leads to a more sustainable approach to development. The term "community" is used throughout this chapter to refer a place. A place can be a governmental entity, such as a city, or it can be a neighborhood that has no specific or official boundaries. Finally, this chapter outlines the major steps in planning for an asset-based community development strategy.

Definitions of Community Development

There are many definitions of community development, including the following:

- "Community development is asset building that improves the quality of life among residents of low- to moderate-income communities, where communities are defined as neighborhoods or multi-neighborhood areas" (Ferguson and Dickens 1999: 5).
- "Community building in all of these efforts consists of actions to strengthen the capacity of communities to identify priorities and opportunities and to foster and sustain positive neighborhood change" (Chaskin 2001: 291).
- "Community development is defined as a planned effort to produce assets that increase the capacity of residents to improve their quality of life" (Green and Haines 2007: vii).

- "Community development is a place-based approach: it concentrates on creating assets that benefit people in poor neighborhoods, largely by building and tapping links to external resources" (Vidal and Keating 2004: 126).

Critical components of these definitions include:

- *A place-based focus*
  Communities can be thought of as the neighborhoods, towns, villages, suburbs, or cities in which people live. These are places that are rooted in a physical environment. In contrast, communities also can be interest based. Many people identify with groups of people that share similar interests, for example, professional associations, sports teams, religious affiliations, service clubs, etc.
- *The building up or creation of assets*
  The next section of this chapter will spend time discussing asset-based community development. For now, the definition of an asset is: a resource or advantage within a community (of place).
- *The improvement of quality of life*
  Quality of life is a vague notion, and, therefore, each community must define indicators in order to be able to monitor whether or not improvement is occurring. Quality of life can refer to economic, social, psychological, physical, and political aspects of a community. Examples of indicators include: number of violent crimes within a neighborhood; hours of work at the median wage required to support basic needs; percentage of employment concentrated in the top 10 employers; percentage of the population that gardens; and tons of solid waste generated and recycled per person.
- *A focus on low- to moderate-income communities*
  Community development is primarily focused on lower-income communities.

Unstated in the above definitions of community development are the following aspects:

- *Financial, economic, environmental, and social sustainability*
  More and more the idea behind community development is to build up resources and advantages in a community so that the community and the individuals within it can be sustained over time.
- *The approach is not focused on wealthy communities*
  This is the notion that unlike wealthy communities which not only have assets but recognize these assets and use them in the formal economy, many lower-income communities do not.

Needs-Based Community Development

There are two primary methods of approaching community development. The conventional or traditional approach is to identify the issues, problems, and needs of a community. In many low-income neighborhoods, it is easy to point to problems—vacant and abandoned houses, boarded-up store fronts, empty lots filled with trash, and countless others. By focusing on problems, community residents tend to concentrate only on what is missing in a community. For example, a neighborhood may point to problems such as high unemployment rates or lack of shopping opportunities and identify the need for more jobs and businesses. If community residents focus only on trying to fix the problems that they see, they may miss or ignore the causes of these problems.

Many of the problems identified, like poverty or unemployment, are issues too large for one community to solve by itself. By focusing on the causes of problems, community residents may end up wringing their hands or giving up because of the overwhelming nature of the causes. This approach can create unreasonable expectations that can lead to disappointment and failure over time. In addition, this approach can point to so many problems and needs that people feel overwhelmed, and, therefore, nothing is done. Figure 3.1 provides an example of a community needs

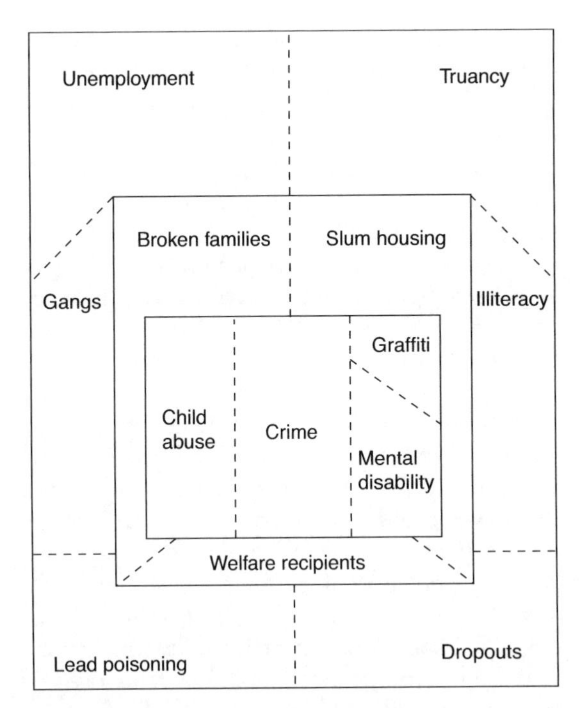

Figure 3.1 *Community Needs Map*

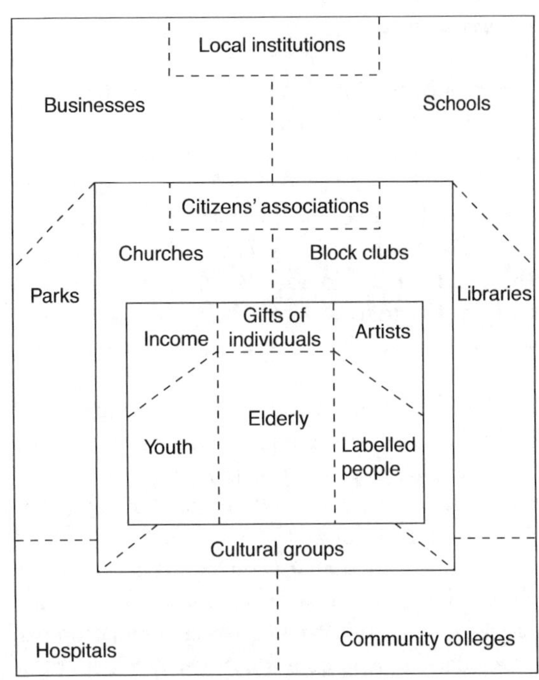

Figure 3.2 *Community Assets Map*

map which outlines problems within a community. This map illustrates numerous problems many of which are difficult to resolve by any one community, neighborhood, or organization.

Asset-Based Community Development

An alternative approach is asset-based community development. One could argue that this approach is the reverse of the conventional approach. The idea is to build capacity within a community—to build and strengthen a community's assets. In contrast to focusing on problems and needs, this alternative approach focuses on a community's strengths and assets.

This asset-based approach is focused on a community's capacity rather than on its deficits. For instance, rather than focusing on missing small businesses, this approach would focus on existing small businesses and their success. Further, by zooming in on its assets, the community as a whole will see its positive aspects (such as community gardens, a mentoring program, and the many skills of its residents) and can then work on developing these assets even more. By implication, concentrating on community assets will create a snowball effect that will influence other areas within a community such as its needs and problems. This alternative approach does not ignore the problems within a community, but focuses first on its strengths and small triumphs in order to provide a positive perspective of the community rather than a discouraging one. Figure 3.2 shows an example of the "mapped" assets within a community and the capacities of its individuals, associations, and institutions. The assets map underscores the potential for community development because it is starting from a positive base rather than from a base rooted in problems.

Assets Defined

Before moving forward in this discussion, it is critical to define the term assets. Assets are the stock of wealth in a household or other unit (Sherraden 1991: 96). Another definition is that assets are "a useful or valuable quality, person or thing; an advantage or resource" (Dictionary.com). Thus, individuals, associations, local institutions and organizations are useful and valuable within the asset-based community development framework. Kretzmann and McKnight (1993) defined assets as the "gifts, skills and capacities" of "individuals, associations and institutions" (p. 25). The idea that individuals within a community are assets is important. What would a community be without its residents?

Within an economic context, assets can be forms of capital such as property, stocks and bonds, and cash. Within a community context, assets can be seen as various forms of capital as well. Assets take a variety of forms within a community. Ferguson and Dickens (1999) talk about five forms of community capital: physical, human, social, financial, and political. Green and Haines (2007) identify seven forms of community capital: physical, human, social, financial, environmental, cultural, and political. Rainey et al. (2003) present three forms of capital that they see as essential: human, public (physical), and social. While there can be debate about the forms of community capital and which forms are more essential than others, the important point here is that a community can identify its own assets—their own capital.

For the purpose of this chapter, three types of capital—physical, human, and social—will be defined and discussed. Why only these three forms of capital? These forms of capital can be subdivided into other forms of capital. For example, physical capital may comprise the built environment and natural resources. Thus, environmental capital, which comprises the natural resources within a community, is part of a community's physical characteristics and, thus, its assets. Natural resources may have shaped the community in the past either through its physical shape—constraining where and how a place grew—or by influencing its economy. The built environment—buildings, infrastructure, etc.—is a critical component of a community and clearly is part of its physical capital.

Physical capital comprises the roads, buildings, infrastructure, and natural resources within a community. When thinking about a specific community, what is often thought about are its physical attributes and key features. These attributes and features can include roads, rail, water and sewer, a downtown, residential neighborhoods, parks, a riverfront, industrial areas, strip development, schools, government buildings, a university or college, a museum, a prison, and many others. In contrast to the other forms of capital, physical capital is largely immobile. Although redevelopment of buildings and infrastructure occurs, physical capital endures over a long time period and is rooted in place. Another quality of physical capital is the degree of both public and private investment—public investment into infrastructure (roads, sewer, water) and private investment into structures (residential, commercial, and industrial)—with the expectation of a return on that investment.

Human capital is defined as the skills, talents, and knowledge of community members. It is important to recognize that not only are adults part of the human capital equation, but children and youth also contribute. It can include labor market skills, leadership skills, general education background, artistic development and appreciation, health, and other skills and experience (Green and Haines 2007: 81). In contrast to physical capital, human capital is mobile. People move in and out of communities, and thus over time, human capital can change. In addition, skills, talents, and knowledge change due to many kinds of cultural, societal, and institutional mechanisms.

Social capital often refers to the social relationships within a community and can refer to the trust, norms, and social networks that are established (Green and Haines 2007). "Social capital consists of the stock of active connections among people: the trust, mutual understanding, and shared values and behaviors that bind the members of human networks and communities and make cooperative action possible" (Cohen and Prusak 2001: 4).

In the community development context, the importance of social relationships is critical to mobilizing residents and is often a critical component for the success of a project or program. Social capital comprises the formal and informal institutions and organizations, networks, and ties that bind community members together. There are many forms of social capital—formal and informal, strong and weak, bonding and bridging—to name the more well-defined types. Formal ties or networks are those ties that are established through organizations, such as service clubs, and are seen as weak ties. Informal ties are those established through personal relationships. Often these ties are strong, and time and energy are involved in maintaining them. Bonding capital refers to bringing together people who already have established relationships or ties. In contrast, bridging capital refers to the idea of widening individuals' networks and ties. By establishing new networks or ties, people will have access to new information and more networks for sharing and using information.

In addition, social capital can be subdivided into various forms such as financial, political, and cultural. While these three forms can easily be separated from social capital and each other, social capital is central to these forms of community capital. Financial capital refers to access to credit markets and other sources of funds. Poor and minority communities often lack access to credit markets. Without sources of financing for home ownership, business start-ups and expansions, etc., these communities are unable to put underused resources to work (Green and Haines 2007). Political capital is the capacity of a community to "exert political influence" (Ferguson and Dickens 1999: 5). Weir (1988) discusses three categories of the relationship between community-based organizations (CBOs) and local political systems. These categories include elite domination with weak ties to CBOs, a political patronage system, and a more inclusionary system where community and political leaders overlap. Based on Weir's categories, the exertion of political influence would differ between communities and the organizations within those communities.

One important concept to take from this discussion is that all forms of community capital are intricately linked together and are necessary for sustaining communities and achieving a better quality of life.

The Process of Asset-Based Community Development

Many community development professionals and others have, or are moving toward, an asset-based approach to community development. This next section introduces a general outline of this approach and describes its major steps. Figure 3.3 illustrates a community development process with four main steps. It is shown moving from community organizing to visioning to planning to implementation and evaluation and back to organizing. While the illustration moves from one step to another and creates a feedback loop, community development is far more messy and nonlinear in practice. Many of these steps continue throughout the process. In addition, one step may be given more emphasis than the others in specific time periods. Every community is different, and the actual process and the time it takes for each step will differ as well. Some communities may be fairly organized and cohesive and can move through organizing, visioning, and planning in a short amount of time and spend the bulk of their time and effort on implementation. However, other communities may find they are spending a great deal of time on organizing. Another aspect to consider in this illustration is the absence of a time frame. All the steps in the process appear to take a similar amount of time, but, as indicated above,

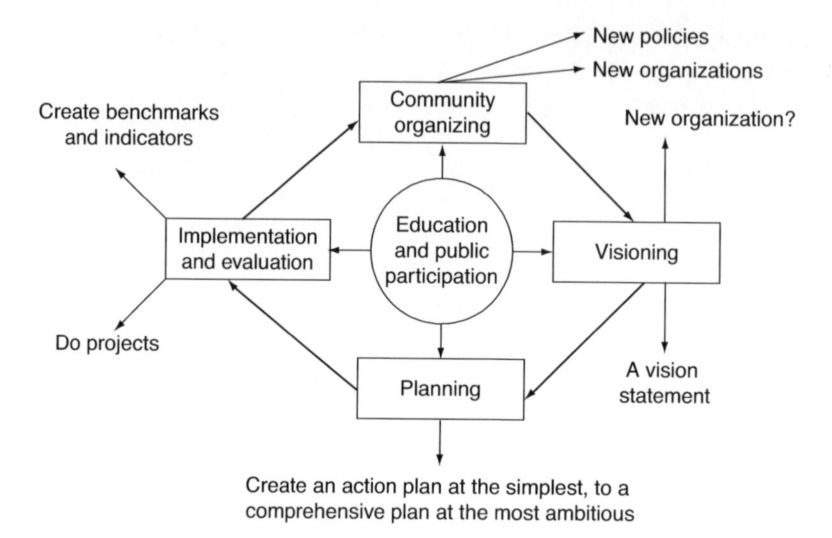

Figure 3.3 *A Community Development Process*

the amount of time spent on any one step will depend on the community's residents and what they are trying to accomplish. The illustration includes a step for implementation—the action phase from which outcomes will be felt and measured. This step is a crucial part of the process—it is not separate from it.

Community Organizing

Community organizing focuses on mobilizing people within a specific neighborhood or community. It is distinct from other forms of organizing because of its focus on communities of place rather than communities of interest. Community organizing does not need to be conceived of as a task for getting everyone in a community mobilized for doing something. In fact, community organizing can be thought of as a way to mobilize small groups of people to accomplish a particular task. Mobilizing community residents involves direct action ranging from writing letters to the editor to organizing a protest outside the offices of the school district. Often community organizing uses a problem-oriented approach rather than an asset-based approach. Community residents are mobilized to "solve" a particular problem recognized in their neighborhood. There are two strategies for mobilizing residents: social action campaigns and the development model. Social action campaigns are those direct actions, like the examples above, that aim to change decisions, societal structures, and cultural beliefs.

Another form of organizing occurs through the development model and is more prevalent at the community level. The development model is a way to organize communities of place to accomplish a variety of community goals. There are several different community organizing models (Rubin and Rubin 1992). The Alinsky model is probably the most popular and involves a professional organizer. The organizer works with existing organizations in a community to identify common issues. In contrast, the Boston model contacts welfare clients individually at their residences and relies heavily on appeals to the self-interest of each person. The Association of Community Organization for Reform Now (ACORN) has mixed the Alinsky and Boston models. The Industrial Areas Foundation (IAF) model emphasizes the importance of intensive training of organizers. This model is a direct descendent of the Alinsky model but emphasizes the importance of maintaining close ties with existing community organizations. Community development corporations (CDCs) use these development models to achieve community development goals. CDCs often represent the type of organization that provides economic and social services in low-income neighborhoods and communities (Rubin and Rubin 1992).

Visioning

Visioning is one method among many, such as future search, to establish a long-range view of a community. Box 3.1 provides a vision statement from one community. The term became popular in the 1990s, and many communities have used the technique to guide their future. While it is often used in the context of community planning, it also has been used to focus on specific topical areas, such as housing, transportation, and education. Many communities have found it useful to create multiple topical visions that can be more detailed and focused rather than creating one broad vision, which many people view as too vague and broad to bring meaning to the necessary actions. The basic idea is to bring together a wide range of individuals, associations, and institutions within a community to arrive at, often through some form of consensus, a written statement—the vision—of the future and to prepare an action plan to move that community toward the vision.

There are at least three critical components of a visioning exercise. The first component is inviting a broad spectrum of the community so that many opinions and perspectives are represented. The second component is preparing a process that is meaningful, effective, and efficient. The process must be meaningful to the participants so that the time they are volunteering will have appropriate and useful results. The process must be effective in that it fulfills the purpose defined for it. Finally, the process must be efficient in terms of people's time, energy, and funds expended. This third component, which is closely related to the second one, involves choosing public participation techniques to accomplish a vision or multiple visions for a community. While visioning is thought of as a public participation technique, it must use techniques such as brainstorming, SWOT analysis, and charrettes, to accomplish its purpose.

Planning

During the planning phase there are at least three tasks in preparing an action plan: data collection and analysis, asset mapping, and a community survey. Data collection and analysis is important

BOX 3.1 VISION STATEMENT FROM RUIDOSO, NEW MEXICO

A Vision Statement for Ruidoso
We treasure…

- The *serene natural environment*—cool pines, high mountains, the Rio Ruidoso, comfortable weather, and clear skies.
- A *sense of community*. People are friendly; we prize the easy lifestyle where people know each other and where kids are safe riding their bikes.
- A *small-town atmosphere*, even during the summer and winter when the village serves an influx of part-time residents and tourists.

We like where we live, take pride in "our place" and we are willing to volunteer our time for community betterment.

Source: www.ruidoso.net/local/vision_statement.htm (accessed June 16, 2004).

to understand current circumstances, changes occurring within a community over time, and the implications of the data collected.

Asset mapping is a process of learning what resources are available in a community. Kretzmann and McKnight (1993) provide the most hands-on and thorough asset mapping process. Their process "maps" or inventories the assets or capacities of:

- individuals including youth, seniors, people with disabilities, local artists, and others;
- local associations and organizations including business organizations, charitable groups, ethnic associations, political organizations, service clubs, sports leagues, veterans groups, religious institutions, cultural organizations, and many others;
- local institutions for community building including parks, libraries, schools, community colleges, police, hospitals, and any other institution that is part of the fabric of a community.

Asset mapping is an ongoing exercise. The purpose is to recognize the skills, knowledge, and resources within a community. It is a good first step in beginning to understand the assets of a community.

Community surveys can be useful in identifying issues at the beginning stages of a planning process and/or to refine particular ideas or policies as a community begins to think about its goals or its action plans. A community survey will allow various organizations in a community to:

- Gather information about public attitudes and opinions regarding precisely defined issues, problems, or opportunities.
- Determine how the public ranks issues, problems, and opportunities in order of importance and urgency.
- Give the public a voice in determining policy, goals, and priorities.
- Determine public support for initiatives.
- Evaluate current programs and policies.
- End speculation about "what people are thinking" or "what people really want."

(Laboratory of Community and Economic Development (LCED) 2002)

While community surveys can communicate important information about public attitudes and opinions, this public participation technique is focused only on input, not on shared decision making. Nevertheless, if carried out well, this technique allows for a much broader range of residents to participate than many other public participation techniques that call for face-to-face interaction.

Public Participation

Determining the future of a community and how that community will get from Point A to Point B are important endeavors. Often residents leave community goals and decisions to others—a consultant, a local government, a state or federal agency, a private developer, a business owner, or corporation. However, residents of a community need to participate in and get actively involved in determining the future course of their community. If they do not, others will determine their future for them. Thus, public participation is critical to the entire community development process. Figure 3.3 illustrates its importance by situating it at the center of the process.

Using public participation effectively and meaningfully is a difficult task for individuals, groups, or organizations that are trying to determine which techniques are most appropriate to use, when to use them, and who should be involved. Effective public participation needs to be both functional for the specific goal and meaningful to the public. Participation is functional when it helps to create better decisions and a more thoughtful community plan or some other document that can help organizations, institutions, and individuals understand how their community is moving forward. Participation is meaningful when it creates opportunities for the public to exercise influence over decisions and feel a sense of ownership toward the product.

BOX 3.2 ONLINE SUPPORT FOR PLACEMAKING

Numerous technologies and techniques have emerged to provide opportunity for participation in community planning, information sharing, development, and decision-making processes—wherever anyone may be in the world. New applications and approaches emerge constantly—a few of these include Ning, an online scalable hosted platform with an affordable pricing structure, used by some communities to gain input on projects, ideas, and programs and SeeClickFix, a web-based tool for residents to report non-emergency issues to local government. Beyond locally focused projects, there are ways for people to communicate widely, and effect change. A mobile publishing platform started in the UK, nOtice.org, in their words, "*enables mobile participation, unlock local knowledge, fuel social actions, supporting* the social–local–mobile ambitions of publishers, broadcasters, brands, communities and developers around the world" (nOtice.org, 2013). Online voting and polling for visual idea-sharing platforms are in use in many areas, across the globe. Crowdmapping is becoming more popular, and provides insights into trends, and ideas for solving issues. Being online and connected is definitely in the future for community development practice!

The Editors

Implementation and Evaluation

Actions in community development are where change occurs and where people can see tangible results. This phase in the community development process is the point at which the rubber meets the road. It is the phase where individuals, groups, and organizations are active rather than passive participants in their community. Up to the implementation phase of a community development process, individuals and organizations have made a concerted effort to understand their assets, community attitudes, and opinions; have arrived at a shared understanding of the future; and have agreed upon initial actions, and possibly broader strategies, to take that will lead to specific actions in the future. Action plans generally identify specific projects, deadlines, responsible parties, funding mechanisms, and other tasks that will accomplish specific goals. An action plan describes a set of activities that need to be accomplished to move the community toward its future vision and/or goals.

Another part of implementation is to consider the regulatory context within which development occurs, in particular the physical capital of a community. It is likely the action plan has identified areas that need changes through local government, such as zoning changes. It is important for the regulations to integrate with the plan.

An often overlooked but important component to community development is monitoring and evaluation. Monitoring is the act of assessing the community development process as it is taking place. Monitoring functions as a way to take the pulse of a community effort. It allows for adjustments to be made in the process rather than letting issues or situations get beyond the control of facilitators, a steering committee, and/or a project team to manage the process (Green et al. 2001).

Evaluation, in contrast, usually occurs after a project or a plan is considered completed. At least two types of accomplishments can be measured: outputs—the direct and short-term results of a project or plan such as the number of people trained, the number of affordable houses built, or the number of jobs created; and outcomes—the long-term results of a project or plan. Outcomes are much more difficult to measure and to link directly to a specific action. Outcomes that many community development plans aim for are long-term goals such as decreased levels of poverty, a better quality of life in a community, or increased levels of personal income (ibid.).

By establishing benchmarks and indicators to more easily track the accomplishments of specific actions, communities are able to more usefully conduct monitoring and evaluation. "Sustainable

Seattle" offers a good example of a nonprofit organization that has used regional and neighborhood indicators to monitor the environmental health of the Seattle region in addition to using public participation to accomplish their goals (Sustainable Seattle 2004).

Challenges of the Community Development Process

An important caveat to any community development process is that they can be difficult, time-consuming, and costly. The difficulty can occur, for example, when many diverse interests cannot or will not find common ground about either specific actions or even the general direction the community should take. Thus, finding consensus and making compromises is not only difficult but can be time-consuming as well. Nevertheless, creating a forum where diverse interests can discuss issues is critical for improvement of a community. In addition to stumbling blocks and obstacles, the process itself can take time, which can be frustrating for those individuals and organizations that like taking action. It is difficult to maintain interest and commitment to the effort if participants cannot point to successes. It is important for motivation and trust to create a balance between initial actions and completing the process that was set out so that residents can see that something is occurring.

The cost of the process depends on the amount and types of public participation that are used within a process in addition to the amount and types of data collected and analyzed. While some forms of public participation are inexpensive, such as developing a website, they may only act to inform residents rather than provide a way to gather input and create partnerships. Forms of public participation that call for facilitators may be more expensive; however, residents can be trained as facilitators to reduce costs. Creating a process that is manageable, fits the community, and fits the budget is imperative; the process of community development can be as important and as valuable as its products. While there continues to be a debate over the importance of process versus outcomes in community development, it is clear that the ultimate goal is to improve the quality of life for the residents in the community. In the long run, both process and outcomes are essential parts of community development.

Conclusion

Asset-based community development is a promising approach to achieving a better quality of life and sustaining communities not only over time or in an economic sense, but through the development of all forms of capital that are necessary for a community to thrive. Kretzmann and McKnight (1993) believe that the key to community revitalization is "to locate all of the available local assets, to begin connecting them with one another in ways that multiply their power and effectiveness, and to begin harnessing those local institutions that are not yet available for local development purposes" (p. 5).

This chapter has defined asset-based community development in contrast to a needs-based approach and has touched upon the steps in a community development process.

CASE STUDY: ASSET BUILDING ON THE SHORES OF LAKE SUPERIOR

Place-based communities across the US are making changes toward a more sustainable future—reducing greenhouse gases, becoming less reliant on fossil fuels, promoting local food systems, encouraging green buildings, higher density and more active lifestyles. In the past two years, approximately 11 communities in Wisconsin have declared themselves "eco-municipalities" to indicate their commitment to move toward a sustainable future. In a small region of Wisconsin comprised of about 32,000 people living along the southern shore of Lake Superior, residents and community leaders joined together in creating a strategic plan for sustainable development. Its three small cities and one township have passed "eco-municipality" resolutions, and a series of groups called "green teams" are taking a variety of actions initially identified in the strategic plan as well as other actions that were not.

One of the region's strongest regional assets is its natural capital. The region is rich in natural resources which underlies the current economy that is based in tourism, wood products, and farming. Within the two counties are over 1,000 lakes, many miles of rivers and streams (including over 100 hundred miles of Lake Superior shoreline), and thousands of acres of wetlands and forests. The major tourist draw is the Apostle Island National Lakeshore comprised of 22 islands in Lake Superior within a short ferry and boating distance to the City of Bayfield. Other key environmental assets are the Kakagon and Bad River sloughs—16,000 acres on and around the Bad River Band of the Lake Superior Tribe of Chippewa Indians Reservation—that represent the largest undeveloped wetland complex in the upper Great Lakes. These sloughs have been called Wisconsin's Everglades and were designated a National Natural Landmark by the U.S. Department of the Interior in 1983 (Nature Conservancy 2000). This area is considered one of the healthiest wetland ecosystems in the area and produces wild rice for members of the Bad River Tribe.

In the first half of 2007, the area cities and several organizations have each created a "Green Team" to review each organization's operations to determine specifics ways in which sustainable practices may continue to occur. For example, one of the first initiatives the City of Ashland Green Team will be completing is an energy audit to determine where its energy consumption is occurring and where it can make changes. The City of Washburn has installed energy-efficient compact fluorescent bulbs and tubes in the civic center and library; studied how lighting and heating can be improved for the city garage; replaced a hot water boiler at a local park's shower building with a tankless coil system that operates on demand; and installed geothermal heating and cooling in public housing designed for low-income and elderly citizens (Boyd 2007).

While the environmental capital is the motivation for the current efforts to create sustainable communities in this small region, it is the social, financial, and political forms of capital that have functioned to bring about the results described above in the first paragraph. Local politicians, wealthy residents, faculty, and staff at Northland College (an environmental liberal arts college), local environmental groups, and the League of Women Voters worked together in multiple ways over many years to get to where they are today.

Keywords

Asset-based community development, asset identification, community organizing, community participation, community revitalization, visioning, sustainable development, physical capital, social capital.

Review Questions

1 What is asset-based development, and why is it different from past approaches?
2 Define the types of assets that a community may have.
3 Name some of the steps in the process of community development and discuss their importance.
4 What is the importance of social relationships in community development?
5 What are some challenges to community development?

Bibliography

Boyd, D. (2007) "Chequamegon Bay Communities Support Sustainability Framework," MSA Professional Services update, 18: 3.

Chaskin, R. (2001) "Building Community Capacity: A Definitional Framework and Case Studies from a Comprehensive Community Initiative," *Urban Affairs Review*, 36(3): 291–323.

Cohen, D. and Prusak, L. (2001) *In Good Company: How Social Capital Makes Organizations Work*, Boston, MA: Harvard Business School Press.

Ferguson, R.F. and Dickens, W.T. (eds.) (1999) "Introduction," in *Urban Problems and Community Development*, Washington: Brookings Institution Press, pp. 1–31.

Future Search Network. Online. Available at: <www.futuresearch.net> (accessed 16 June, 2004).

Green, G.P. and Haines, A. (2007) *Asset Building and Community Development*, 2nd ed., Thousand Oak: Sage.

Green, G.P., Borich, T.O., Cole, R.D., Darling, D.L., Hancock, C., Huntington, S.H., Leuci, M.S., McMaster, B., Patton, D.B., Schmidt, F., Silvis, A.H., Steinberg, R., Teel, D., Wade, J., Walzer, N., and Stewart, J. (2001) *RRD182 Vision to Action: Take Charge Too*, Ames, IA: North Central Regional Center for Rural Development. Online. Available at: <www.ag.iastate.edu/centers/rdev/pubs/contents/182.htm> (accessed August 9, 2004).

Kretzmann, J.P. and McKnight, J.L. (1993) *Building Communities from the Inside Out: A Path Toward Finding and Mobilizing a Community's Assets*, Chicago, IL: ACTA Publications.

Kretzmann, J.P. and McKnight, J.L. (1996) "Assets-Based Community Development," *National Civic Review*, 85(4): 23.

Laboratory of Community and Economic Development (LCED) (2002) *Assessing and Developing: Your Community Resources*, University of Illinois—Extension, Winter 2002 Issue 1(3). Online. Available at: <www.communitydevelopment.uiuc.edu/toolbox/> (accessed August 9, 2004).

Nature Conservancy (2000) "Chequamegon Bay watershed site conservation plan." October 2000, unpublished manuscript.

Rainey, D.V., Robinson, K.L., Allen, I., and Christy, R.D. (2003) "Essential Forms of Capital for Sustainable Community Development," *American Journal of Agricultural Economics*, 85(3): 708–715.

Rubin, H.J. and Rubin, I.S. (1992) *Community Organizing and Development*, 2nd ed., Boston: Allyn & Bacon.

Sherraden, M. (1991) *Assets and the Poor: A New American Welfare Policy*, Armonk, NY: M.E. Sharpe.

Shuman, M.H. (1998) *Going Local: Creating Self-Reliant Communities in a Global Age*, New York: Free Press.

Sustainable Seattle (2004) Online. Available at: <www.sustainableseattle.org/nd/programs/default.htm> (accessed August 27, 2004).

Vidal, A.C. and Keating, W.D. (2004) "Community Development: Current Issues and Emerging Challenges," *Journal of Urban Affairs*, 26(2): 125–137.

Weir, M. (1988) "The federal government and unemployment: The frustration of policy innovation from the New Deal to the Great Society," in M. Weir, A.S. Orloff, and T. Skocpol (eds.) *The Politics of Social Policy in the United States*, Princeton, NJ: Princeton University Press, pp. 149–197.

Connections

Watch a short YouTube video about the power of asset-based community development in action: www.youtube.com/watch?v=leN7_TICGXM

Explore the Asset-Based Community Development Organization at: www.abcdinstitute.org. You'll find links and applications to John McKnight's classic work, *Building Communities from the Inside Out*.

4 Social Capital and Community Building

Paul Mattessich

Overview

Social capital or capacity lies at the heart of community development. If citizens cannot plan and work together effectively and inclusively, then substantial proactive community progress will be limited. This chapter discusses community social capacity and how it relates to community development. Based on an extensive review of previous studies, the chapter lists and categorizes factors that affect the likelihood that community-building efforts can succeed to increase the social capacity of geographically defined communities.

Introduction

Chapter 1 distinguished social capital (sometimes called social capacity) from human, physical, financial, and environmental capital. All of these constitute resources that communities need to function. The extent to which communities have these forms of capital influences their ability to accomplish tasks and to develop themselves.

Development of a community includes, in part, the building of its social capacity. Conversely, the level of a community's social capacity both influences the way development evolves for that specific community and the pace at which its development efforts can occur. The opening chapter of this book defined social capacity and social capital. That chapter explained how the social capital of communities relates to the process of community development, both as an antecedent that predisposes a community to further development in the "community development chain" and, as a consequence, a feature of the community that increases as a result of community development. The discussion in this chapter places emphasis on social relationships embedded within, *and* distributed throughout, a community. Individuals can amass social capital, as well as other forms of capital. They can have strong ties to social networks that enable them to do things for themselves. However, unless the social capital that they possess becomes a resource available to and used by the entire community, it has little or no direct effect upon that community's development.

Social Capital: What Is It?

Chapter 1 provided some definitions of social capital. In the simplest sense, what comprises the core of these definitions are "social networks and the associated norms of reciprocity" (Clarke 2004). Analogous to other forms of capital (financial, human, etc.), social capital constitutes a resource. It provides value to individuals and can also benefit communities. It has effects, often called "externalities" in economics literature.

No single definition has achieved universal acceptance though eventually a common standard may emerge. Most definitions stress interconnections among people or social networks. Early definitions tended to place emphasis on how individuals could use social relationships as a resource to accomplish goals; they did not add a community dimension to the definition. More recent definitions tend to recognize the distinction between social capital at the individual level and social capital at the community level.

Bourdieu's (1986: 242) early definition, for example, focused on individuals:

> Social capital is an attribute of an individual in a social context. One can acquire social capital through purposeful actions and can transform that capital into conventional economic gains. The ability to do so, however, depends on the nature of the social obligations, connections, and networks available to you.

In this tradition, Sobel (2002: 139) stated that social capital "describes circumstances in which individuals can use membership in groups and networks to secure benefits."

Eventually, however, social scientists began to realize that social capital in the control of individuals, but not shared with others in the same community, did not fit into the "community development chain" illustrated in Chapter 1. It does not necessarily produce community social capacity that affects the ability of the community to develop. In fact, individuals who have such capital can even use it to the detriment of their own or other communities.

Consider, for example:

- a gang or a network of organized criminals that lives and operates within a community but uses its close interconnections to commit crimes and to engage in acts that detract from the livability of the community;
- wealthy elite residents of a small city in a developing country who separate themselves from others in the population, maintain social and business connections with individuals and organizations in other locations, and channel their financial wealth to places outside the country;
- in a financially struggling United States city, a gated cluster of town homes for high-wealth residents who don't engage in ongoing relationships with the local population and who invest most of their money elsewhere.

In all the above cases, social capital, not to mention financial capital, exists within a geographically defined area. However, that capital produces few, if any, benefits for the community as a whole. In fact, those who have such capital might actually use it to exploit others in the same locality. A "map" of the social interconnections within any of the communities in the above examples would reveal one or both of the following:

- Individuals with no ties to other individuals in the community but to those outside the community. For example, many of the wealthy members of the developing society, such as residents of the gated community, may have strong ties with business and social associates in locations throughout the world.
- Highly cohesive social networks among one or more subsets of the community, with no strong ties between any of these networks. For example, members of criminal gangs can maintain strong ties with one another or wealthy elite residents in an area may do the same.

Recognition of the distinction between individual social capital and community social capital (or community social capacity) does not negate the importance of individual social capital. An individual's social capital provides an important resource for the individual; it has real effects. However, unless amalgamated with the social capital of others in the same community, it does not necessarily produce benefits for that community. As noted in Chapter 1, "bridging" capital ties individuals to others like themselves (race, economic status, nationality, etc.); "bonding" capital ties individuals to a diverse set of others, some like themselves, some not.

Community Social Capacity: What Is It?

Mattessich and Monsey (1997) define community social capacity as: "The extent to which members of a community can work together effectively." This includes the abilities to: develop and sustain strong relationships; solve problems and make group decisions; and collaborate effectively to identify goals and get work done. Communities with high community social capacity can identify their needs; establish priorities and goals; develop plans, of which the members of that community consider themselves "owners"; allocate resources to carry out those plans; and carry out the joint work necessary to achieve goals.

The term "community social capacity" applies holistically to an entire community. It is an attribute of a community, not of any specific members. The level of community social capacity depends upon the number and strength of ties or bonds that community members have with one another. Thus, it is a form of social capital—since it involves "social networks and norms of reciprocity"—but it distinctly involves interconnections among people who reside in the same community (defined by geographic location).

If people who live within the same geographic area do not know one another and have little contact with one another, the likelihood is low that they can get together to define community goals or respond productively with one voice to a community issue. Their community social capacity is low. On the other hand, if people who live within the same geographic area do know one another, share a large number of social ties, and feel a commitment to the place where they live, then community social capacity is high.

How Does Community Social Capacity Influence Development?

Jane Jacobs recognized the importance of community social capacity for community vitality. She observed that deep and heterogeneous social relationships seemed to enable communities to thrive; barriers imposed upon these relationships seemed to lead to community deterioration (Jacobs 1961).

The level of community social capacity (or community social capital) influences community development in two broad ways, structural and cognitive (for example, see Uphoff 2000). Structurally, interconnections among people within a community create a web of social networks. These networks facilitate community development by enabling the flow of information, ideas, products,

BOX 4.1 CAPACITY FACTORS

Capacity factors in community development are essential and serve as ways to connect diverse interests, drawing in commonalities while remaining cognizant of differences that exist between groups, organizations, neighborhoods, and individuals. These factors are often subjective in nature (as opposed to more concrete, measurable objective factors typically gauged in development). Subjectivity is better suited for gauging activities and orientations such as involvement and strength of relationships, for example. These capacity factors can include such aspects as: positive attitude toward development and activities; participation in community activities; effective implementation of plans; unified vision and mission; and ability to work together and avoid factionalism (Pittman et al. 2009: 84). All of these are highly important in community development, and reflect a community's social capital situation.

The Editors

and services among residents. Cognitively, interconnections create a shared sense of purpose, increase commitment, promote mutual trust, and strengthen norms of reciprocity among community residents.

Intentional Action to Increase Social Capacity

Communities often recognize the need to increase their social capacity and take steps to do so in a community-building process. Mattessich and Monsey (1997: 60) offer a brief definition of community building: "Any identifiable set of activities pursued by a community in order to increase community social capacity."

The elements of community building come across in a longer definition from a review of comprehensive community initiatives by Kubisch et al. (1995: 16):

> Fundamentally, community building concerns strengthening the capacity of neighborhood residents, associations, and organizations to work, individually and collectively, to foster and sustain positive neighborhood change. For individuals, community building focuses on both the capacity and "empowerment" of neighborhood residents to identify and access opportunities and effect change, as well as on the development of individual leadership. For associations, community building focuses on the nature, strength, and scope of relationships (both affective and instrumental) among individuals within the neighborhood and through them, connections to networks of association beyond the neighborhood. These are ties through kinship, acquaintance or other more formal means through which information, resources, and assistance can be received and delivered. Finally, for organizations, community building centers on developing the capacity of formal and informal institutions within the neighborhood to provide goods and services effectively, and on the relationships among organizations both within and beyond the neighborhood to maximize resources and coordinate strategies.

Community organizing can support community building. It refers to the process of bringing community members together and providing them with the tools to help themselves. The process can include:

> identification of key local resources, the gathering of information about the community context, the development and training of local leaders to prepare them to serve effectively as representatives of the community and as full partners in an initiative, and the strengthening of the network of the various interests both internal and external to a community.
>
> Joseph and Ogletree 1996: 94

BOX 4.2 BUILDING COMMUNITY CAPACITY, CAPITAL, AND DEVELOPMENT VIA PHILANTHROPY

Philanthropic support for community development dates back to the Progressive Era and has evolved over the past century but one constant remains: the engagement of private foundations (Lubove, 1962; Vale, 2000; Ylvisaker, 1987). While social justice motivated early efforts, by the middle of the twentieth century, the Civil Rights movement and the War on Poverty spurred private foundations to increase their support for comprehensive community development initiatives or CCIs (Green & Haines, 2012; O'Connor, 1996). CCIs are partnerships between foundations, the public sector, and the private sector, and they are based on the principles of fostering comprehensive community change that improves the physical, social, and economic conditions of specific, geographically defined communities (Fulbright-Anderson & Auspos, 2006, Pitcoff, 1997).

Since the 1980s, CCIs have offered a promising vehicle for philanthropic partnerships by allowing for holistic approaches that aim to reverse the long-term disinvestment in low-income urban neighborhoods. Martinez-Cosio & Bussell (2013) identified over 60 place-based CCIs that have been in existence, for varying lengths of time, since 1980. CCIs can be found across the country from Chicago and Detroit to Los Angeles and San Diego. Many of the country's largest foundations, such as Ford, Rockefeller, and Annie E. Casey, have invested heavily in CCIs. CCIs vary in terms of their organizational and operational logic, sources of funding, governing principles, programming, leadership structures, and outcomes. The unique conditions of each community, and the community development challenges present in these communities, dictate the shape and structure of each individual CCI (Martinez-Cosio & Bussell, 2013).

A comparison of two CCIs in San Diego, located in adjacent communities with many of the same socioeconomic and physical attributes, offers a picture of the complexity of CCIs. San Diego is unusual in that it hosts two CCIs each receiving the majority of its funding from single private family foundations. The two foundations independently decided to embed themselves in each specific geographically defined community in order to bring about comprehensive community change.

Sol Price, the founder of the Price Club, which eventually became Costco, sought out opportunities to focus all his philanthropic resources into one location in an effort to achieve greater outcomes. In 1994, with his foundation, Price Charities, he selected City Heights, a community comprising 74,000 residents spread over four square miles in which more than 40% of residents are foreign born, 27% live in poverty, and only 25% are homeowners. Price's approach to community change has involved strategic partnership with the City of San Diego, the local school district, as well as the thick web of nonprofit organizations located in the community.

Physical revitalization is at the core of Price Charities' philanthropic philosophy, realized in the eight-block Urban Village in the 1990s. Working with its public, private, and nonprofit partners, Price Charities contributed to the construction of a new public library, a new police substation, a community recreational facility, retail services (including a large national chain grocery store), and market rate as well as affordable housing. Price Charities' approach has been characterized as pragmatic and efficient, but also opaque. Price Charities does not promote deep levels of resident engagement and instead engages actively with its nonprofit and public partners. Sol Price passed away in 2009 and his son Robert assumed the helm of the foundation. Because of the insular nature of the foundation, Robert's goals and objectives are unclear but the foundation is expected to continue its work in City Heights for the foreseeable future (Martinez-Cosio & Bussell, 2013).

The second CCI in San Diego was launched by Joseph Jacobs, a Lebanese immigrant, who took his engineering consulting firm public there in the 1980s, and established the Jacobs Family Foundation in 1988. Within a few years of distributing grants to a variety of nonprofits, he and his family realized that this type of philanthropy was neither rewarding nor impactful, and thus they sought out a community in which to embed themselves and target their philanthropic investment. They selected the Diamond Neighborhoods in Southeastern San Diego, a group of ten underserved neighborhoods with a population of just under 94,000 residents. Home to a large number of middle-class black homeowners, the community saw a significant influx of Latinos, along with a growing population of Somalis, Laotians, Samoans, Chamorro, and Filipinos. Located adjacent to City Heights, Southeastern San Diego was notable for its lack of public investment in physical infrastructure as well as a thin network of nonprofits. In 1995, the foundation created its own operating arm, the Jacobs Center for Neighborhood Innovation, to realize its objectives. The Jacobs Family Foundation was motivated less by physical revitalization and more by resident engagement and community empowerment. Its motto, "resident ownership of community change," is manifested by dozens of working teams comprising residents and foundation staff. Initially the foundation eschewed partnerships with the public sector and only selectively worked with community-based nonprofits.

The Jacobs Family Foundation's cornerstone project is Market Creek Plaza, a ten-acre retail center and adjacent multipurpose community center completed in the mid-2000s. This project brought the first chain grocery store into the neighborhood in 30 years. The plaza was also a tool for economic development. Not only were close to 70% of the construction contracts awarded to local minority-owned businesses, but 91% of the initial hires at the grocery store were neighborhood residents (Robinson, 2005). The community designed the plaza itself with over 2,000 adults and 1,000 youths participating. In 2006 it offered shares in Market Creek Plaza through an innovative, first of its kind, initial public offering (IPO).

The Jacobs Family Foundation's approach to comprehensive community development emphasizes resident engagement, but its early inability to partner with the public sector stalled some of its efforts, including its goal to build affordable housing. Furthermore, its emphasis on process has led to lengthy timetables and community impatience. The foundation was intended to be a one-generation foundation with an initial sunset date of 2020. Family members eventually acknowledged that comprehensive change takes a long time, and the sunset date has been pushed back to 2030. In 2013 Joe Jacobs' 30-year-old grandson became chairman of the foundation's board of directors. It is anticipated that he will continue to promote his grandfather's goals and objectives for the foundation (Martinez-Cosio & Bussell, 2013).

Mirle Rabinowitz Bussell, Ph.D.
Urban Studies and Planning Program, University of California San Diego, USA
Maria Martinez-Cosio, Ph.D.
School of Urban and Public Affairs, University of Texas at Arlington, USA

Sources

Fulbright-Anderson, K., & Auspos, P. (2006). *Community change: Theories, practice and evidence*. Washington, DC: Aspen Institute.

Green, G.P., & Haines, A. (2012). *Asset building and community development*. New York, NY: Sage.

Lubove, R. (1962). *The progressives and the slums: Tenement house reform in New York City, 1890–1917*. Pittsburgh, PA: University of Pittsburgh Press.

Martinez-Cosio, M., & Bussell, M.R. (2013). *Catalysts for change: Twenty-first century philanthropy and community change*. New York, NY: Routledge.

O'Connor, A. (1996). Community action, urban reform, and the fight against poverty: The Ford Foundation's gray areas program. *Journal of Urban History, 22*(5), 586–625.

Pitcoff, W. (1997). Comprehensive community initiatives: Redefining community development. *ShelterForce Online*. Retrieved from www.nhi.org/online/issues/96/ccis.html

Robinson, L. (2005). *Market Creek Plaza: Toward resident ownership and neighborhood change*. Oakland, CA: PolicyLink.

Vale, L. (2000). *From the puritans to the projects: Public housing and public neighbors*. Cambridge, MA: Harvard University Press.

Ylvisaker, P.N. (1987). The nonprofit sector. In W.W. Powell (Ed.), *Foundations and nonprofit organizations* (pp. 360–379). New Haven, CT: Yale University Press.

Factors that Influence the Success of Community-Building Efforts

Communities often wish to improve themselves. They want to attract new businesses, improve housing stock, reduce crime, improve the education of their children, or accomplish any number of tasks that will better the quality of life for community residents. All of these goals, whether adopted individually or together by a community, constitute goals that fall under the umbrella of "community development."

After adopting goals, communities often attempt collaborative action involving individuals and organizations in order to improve themselves. Mattessich and Monsey (1997) synthesized the research literature on what makes one aspect of such efforts successful; that is, the aspect of "community building," a term used purposefully to distinguish it from the more inclusive concept of community development.

What Is Community Building?

Community building refers to activities pursued by a community in order to increase the social capacity of its members (the term capacity building is often used interchangeably; see Chapter 1). In the words of Gardner (1993: 5), community building involves "the practice of building connections among residents, and establishing positive patterns of individual and community behavior based on mutual responsibility and ownership."

BOX 4.3 APPRECIATIVE INQUIRY (AI): A BRIEF OVERVIEW[1]

AI began as an approach to helping corporations become more effective and has since grown into a worldwide movement. The focus on "appreciative" means that we think of the world as a glass that is half full as opposed to a glass that is half empty. Our appreciative eye focuses on the things that are working and leading us to explore why and how they work. "Inquiry" refers to the quest for new knowledge and understanding. We rely on the stories people tell about what is working in their lives and their communities and what they wish for the future as both the content and inspiration in our quest for new knowledge and positive social change.

In the traditional approach to AI developed by David Cooperrider (2005), there are four stages in the AI process (Discover, Dream, Design, Deliver). Another approach which we will use here has six Ds (Define, Discover, Dream, Design, Deliver, Debrief). The Define stage allows us to focus our AI process around what we want to see more of. The Debrief, or Drumming and Dancing stage as some practitioners prefer to call it, provides an opportunity to reflect on what we are doing and what we can learn from those experiences at the same time it offers reason to celebrate—with dancing and drumming or whatever the culture supports.

The theoretical basis of the approach emerges from the field of social constructionism, that examines how what we think and talk about determines what we care about and do. Thus, awareness of how we tell the stories about what is working well in communities is critical to learning how to make those communities even better. One way to think about this approach is that if we focus on problems, we create more problems. If we focus on solutions, we can create more solutions. If we talk about our dreams, we can create a vibrant new future. AI is a process that encourages us to think and talk about what is working and how it could work better.

AI involves several key components including:

- the power of storytelling;
- recognizing the wisdom of others;
- the importance of curiosity in our quest for doing better;
- the value of hearing stories; and
- the primacy of conversations and dialogue.

The Power of Storytelling

Each culture has its own tradition of storytelling. Thus, we can easily adapt the AI approach in many different settings. AI asks people to discover the best of what is by sharing their wisdom in a story about a time when things were working very well in regard to the topic that emerged from the Define stage of Appreciative Inquiry. For example, in the Agricultural Workers' Health Initiative (a project of the California Endowment to help agricultural workers identify and build on local assets that support wellness strategies), we asked community members to share their stories of when they and their families felt the healthiest. In an Alaskan Village, we energized people by engaging in a discussion about what was working well in the village-based medical system. In other settings, we have asked people to tell stories about a time when their community was working very well together. What did they value most about living in their community? These stories ground the process in the reality of people's everyday lives. The common themes among these stories constitute the positive core of that community. As we co-construct the positive core of what is working well, we create the foundation for effectively Dreaming for the future.

In AI we also ask people to construct a story of what the best could be; what life would look like if things were even better than the ideas generated in the Discovery stories. For example, we have asked community members to tell us a story of what their community would look like if the village medical system was working even better, or the community was even better at supporting family wellness. We might follow with what three wishes do they have for their community? Based on their awareness of what makes their community work well, they can envision how it could become even better. These stories connect people to their passions and values. Sharing our wisdom about the best that is and that can be through stories creates a unique sense of community. It also generates positive energy and allows people to find a way to act on their passion.

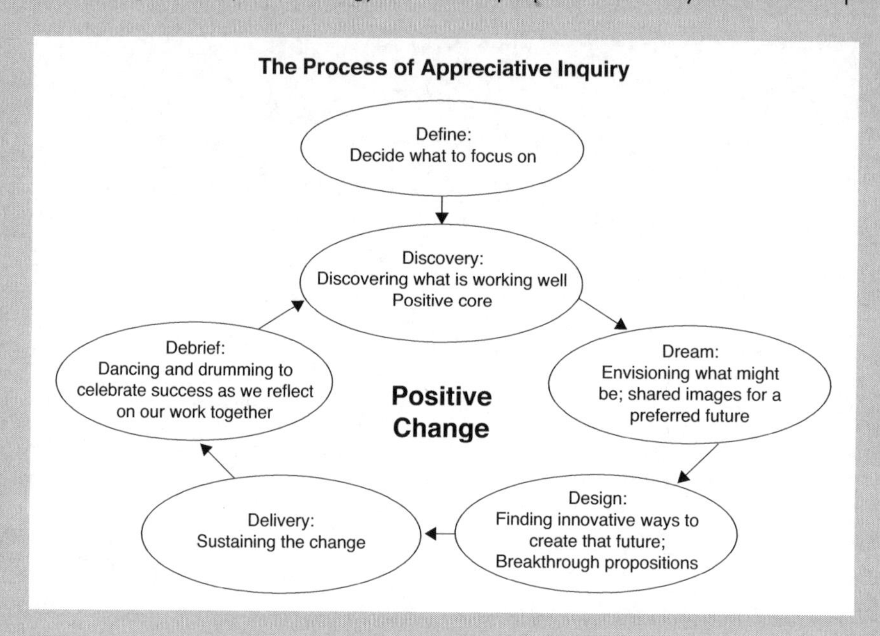

AI approaches typically involve participants by encouraging them to use the AI process to interview one another or to listen to stories about the best that is. The interview process asks people to share their stories around personal and community successes. Furthermore, it asks people to consider what conditions lead to those best experiences and what might come from building on our strengths. Listening to the stories and hearing what people care about creates the opportunity to develop unique relationships. These relationships, based on hearing what people are passionate about, build the foundation for mobilizing people to make change toward a healthier community.

Recognizing the Wisdom of Others

AI works best when we approach our process as co-learners. We must apply active listening skills to fully hear the stories of others; we must eagerly seek the wisdom that emerges from their experience. Active listening means giving others our full attention demonstrated in our responses to their points and our alert and receptive body language. We do not evaluate or judge their stories; we honor their wisdom and strive to learn from their experience. As co-learners, we explore the best of what is, so that we can then co-create with fellow participants a design for the future.

The Importance of Curiosity in Our Quest for Doing Better

AI requires us to be curious about what we hear. Our curiosity allows us to be open to new ideas and perspectives. The AI approach continually searches for new ways of looking at what is by turning an appreciative eye to seek what is working well and why. The AI quest also leads us to consider what might be possible, pushing the possibilities as far as we can. In AI we seek to open doors as well as to look for new doors that need opening. In the Design stage, we ask participants to think outside the box about what strategies can take us from the best of what is now to a future where the best can be even better. Finally, we explore those possibilities and generate strategies to make them real.

The Primacy of Conversations and Dialogue

Traditional planning processes use a lot of brainstorming, post-its, and sticky dots. We suggest that providing the space for people to have real conversations with each other about what they value and their hopes for the community is vitally important to co-creating a future in which we all have a stake. Then in the Design stage, we urge people to construct "provocative propositions" and create "strategy declarations" that can bridge between the positive core of what is to the future we aspire to. The AI process invites people to begin a conversation about what they care about and what their dreams of the future are. In the dialogue that emerges from these conversations, people can share their experiences and insights about their community and the opportunities for positive change. Such conversations help develop the trust and commitment people need to work together to construct a vibrant future. They also lead to new ways of thinking and talking about community and change. Thus, the discovery process leads to a new language with which to describe our work toward a new future.

Using Appreciative Inquiry in Reflection, Evaluation, and Solution-Seeking

1 Define: What do you want more of in your community, team, or project?
2 Discover: Ask those involved in your program to interview each other to learn more about what is working well. You can also use these questions to encourage asset identification and solution-seeking and to uncover tacit wisdom on what is working well within the community.
   a Ask for a story about what is working well: Who was involved, when, where, etc.?
   b What was it about that situation that made it work so well?
   c Share this information and discuss it to come up with those things that need to be in place for the program to operate well (readiness factors) and the positive core of what makes the program work (success factors).
   d Discuss how to use this information to improve program delivery.
3 Dream: Include questions about the future in the interview, such as:
   a What would our program look like if we were even better at doing the things you mentioned in your story? Who would be involved, how, why, when, where?
   b If you have three wishes for the program, what would they be?
   c Share this information within your group and develop a picture or poster of that future.
4 Design: In conversations with those involved in your program:
   a Develop strategies for moving the program forward by asking what would our evaluation look like if we successfully measure change.
   b Discuss what would be needed to successfully implement these strategies. What provocative propositions can your group generate?
5 Deliver: Work on strategies for implementation by:
   a Deciding what needs to happen to be successful;
   b Identifying who will do what to make it happen;
   c Doing something immediately to take a step closer to that future.
6 Dancing and Drumming: Celebrating successes; creating opportunities for reflection and group learning can lead you back to step two: discover what is working well by:
   a Cataloging successes;
   b Storytelling about peak experiences and dialogue to discover core drivers;
   c Determining how to build on success;
   d Strategizing for sustainability;
   e Celebrating the success of all.

Mary Emery
Department Head, Sociology and Rural Studies
South Dakota State University, USA

Source

Cooperrider, D.L. & Whitney, D. (2005). *Appreciative Inquiry: A positive revolution to change.* San Francisco, CA: Berrett-Koehler.

What Influences the Success of Community-Building Efforts?

The review, *Community Building: What Makes It Work* (Mattessich and Monsey 1997) synthesized research on community building to identify factors that influence its success. The factors fall into three categories:

1 Characteristics of community—social, psychological, and geographic attributes of a community and its residents that contribute to the success of a community-building effort.
2 Characteristics of a community-building process—components of the process by which people attempt to build community.
3 Characteristics of community-building organizers—qualities of the people who organize and lead a community-building effort such as commitment, trust, understanding, and experience.

No community-building effort has a 100% likelihood of success. Whether an effort will succeed depends upon many circumstances, some within and others outside the control of community residents. Based on the synthesis of research, it is reasonable to assume that the higher a community stands on the factors influencing the success of community building, the greater the likelihood that it will succeed in a community-building effort.

Building community is much like improving or maintaining one's health. If someone eats nutritious food, exercises, and has regular checkup, for example, that person maximizes the likelihood of good health. Good health is not guaranteed but it's much more likely. It is similar to the factors that influence the success of community building. If present, they maximize the likelihood that an effort will succeed but there are no guarantees. Box 4.4 lists the success factors in community building as identified by the Mattessich and Monsey literature review.

BOX 4.4 28 FACTORS THAT INFLUENCE THE SUCCESS OF COMMUNITY BUILDING

1 Characteristics of the Community

a. Community awareness of an issue
Successful efforts more likely occur in communities where residents recognize the need for some type of initiative. A community-building effort must address an issue which is important enough to warrant attention, and which affects enough residents of a community to spark self-interest in participation. Residents must know that the problem or issue exists.

b. Motivation from within the community
Successful efforts more likely occur in communities where the motivation to begin a community-building process is self-imposed rather than encouraged from the outside.

c. Small geographic area
Successful efforts are more likely to occur in communities with smaller geographic areas, where planning and implementing activities are more manageable. Interaction is harder to achieve if individuals are separated from one another by a great distance.

d. Flexibility and adaptability
Successful efforts more likely occur in communities where organized groups and individuals exhibit flexibility and adaptability in problem solving and task accomplishment.

e. Preexisting social cohesion
The higher the existing level of social cohesion (that is, the strength of interrelationships among community residents), the more likely that a community-building effort will be successful.

f. Ability to discuss, reach consensus, and cooperate
Successful efforts tend to occur more easily in communities that have a spirit of cooperation and the ability to discuss their problems and needs openly.

g. Existing identifiable leadership
Successful efforts more likely occur in communities with existing, identifiable leadership—that is, communities with at least some residents that most community members will follow and listen to—who can motivate, act as spokespersons, and assume leadership roles in a community-building initiative.

h. Prior success with community building
Communities with prior positive experience with community-building efforts are more likely to succeed with new ones.

2 Characteristics of the Community-Building Process

a. Widespread participation
Successful efforts occur more often in communities that promote widespread participation, which is:

- Representative—it includes members of all, or most, segments of the community at any specific point in time.
- Continuous—it recruits new members over time, as some members leave for one reason or another.

b. Good system of communication
Successful efforts have well-developed systems of communication within the community itself and between the community and the rest of the world.

c. Minimal competition in pursuit of goals
Successful efforts tend to occur in communities where existing community organizations do not perceive other organizations or the leaders of a community-building initiative as competitors.

d. Development of self-understanding
Successful efforts are more likely to occur when the process includes developing a group identity, clarifying priorities, and agreeing on how to achieve goals.

e. Benefits to many residents
Successful community-building efforts occur more often when community goals, tasks, and activities have visible benefits to many people in the community.

f. Concurrent focus on product and process
Community-building initiatives are more likely to succeed when efforts to build relationships (the process focus) include tangible events and accomplishments (the product focus).

g. Linkage to organizations outside the community
Successful efforts are more likely to occur when members have ties to organizations outside the community, producing at least the following benefits: financial input; political support; source of knowledge; source of technical support.

h. Progression from simple to complex activities
Successful community-building efforts are more likely to occur when the process moves community members from simple to progressively more complex activities.

i. Systematic gathering of information and analysis of community issues
Successful community-building efforts more likely occur when the process includes taking careful steps to measure and analyze the needs and problems of the community.

j. Training to gain community-building skills
Successful community-building efforts are more likely to occur when participants receive training to increase their community-building skills. Examples include group facilitation, organizational skills, human relations skills, and skills in how to analyze complex community issues.

k. Early involvement and support from existing indigenous organizations
Successful community-building efforts tend to occur most often when community organizations of long tenure and solid reputation become involved early, bringing established contacts, legitimization, and access to resources.

l. Use of technical assistance
Successful community-building efforts are more likely to occur when residents use technical assistance to gain necessary skills.

m. Continual emergence of leaders, as needed
Successful community-building efforts more likely occur when the processes produce new leaders over time.

n. Community control over decision making
Successful community-building efforts more likely occur when residents have control over decisions, particularly over how funds are used.

o. The right mix of resources
Successful community-building efforts occur when the process is not overwhelmed by too many resources or stifled by too few, and when there is a balance between internal and external resources.

3 Characteristics of Community-Building Organizers
a. An understanding of the community
Successful community-building efforts more likely occur when organizers understand the community they serve. This includes an understanding of the community's culture, social structure, demographics, political structures, and issues.

b. Sincerity of commitment
Successful community-building efforts are more likely to occur when organized by individuals who convey a sincere commitment to the community's well-being. They are interested in the community's long-term well-being; have a sustained attachment to community members; are honest; and act primarily to serve the interests of the community, not of an external group.

c. A relationship of trust
Successful community-building efforts are more likely to occur when the organizers develop trusting relationships with community residents.

d. A high level of organizing experience
Successful community-building efforts are more likely when the organizers are experienced.

e. Flexibility and adaptability
Successful community-building efforts are more likely when organizers are flexible and able to adapt to constantly changing situations and environments.

Source: Mattessich and Monsey (1997: 14–17)

Conclusion

As stated in the opening of this chapter, the development of a community in part includes the building of the community's social capacity. Conversely, the level of a community's social capacity influences the way community development evolves for that specific community; it also influences the pace at which community development efforts can occur.

It should be apparent that communities are never "built" in the finite sense of that word. Community building is a continual process, not a set of steps to a permanent conclusion. In the process of building community, a set of individuals who live in the same geographic region can develop social networks and community social capacity; however, sustaining those networks and the social capacity requires ongoing effort.

In this chapter, factors were presented that research has demonstrated as affecting the success of community building, both in initial efforts to build community and ongoing ones to sustain community social capacity. The greater the extent to which those factors are in place, the greater the likelihood that community building can be successful.

CASE STUDY: THE GOOD NEWS GARAGE[2]

This case provides the story of a nonprofit organization working with local government to serve a pressing need in the community. The Good News Garage provides an effective illustration of building social capital for addressing community issues. It is based in Burlington, Vermont.

In 1996, CEDO provided assistance to help launch the Good News Garage, now a nationally acclaimed model for its success at meeting two critical community needs: skilled job training and affordable, reliable transportation. Using $10,000 of CDBG (Community Development Block Grant) funding from the City of Burlington and a $35,000 grant from Lutheran Social Services, the organization started with the concept of donating cars to needy single mothers (at the time, 80% of people in poverty were single mothers). Transportation equity was the primary focus behind its creation and helping those who would benefit greatly from having their own car.

Hal Colston, the founder, explains the success of the program:

> It really comes down to relationships—how do we get to understand and get to know someone who is different from me? Asking people what they need, not what you think they need. Focus on what you can control and let go of what you cannot. At that point, stakeholders and others will appear! How did being in Burlington help? It's a human scale city, and this made a difference. If you show up and want to be involved, you're in. It's also a place where there is a preponderance of nonprofits and people who value social capital, and this is instrumental to forging relationships that can benefit all.

The program, which has now spread to other New England states, solicits donated vehicles, trains people to recondition them, and provides the vehicles to people in need at the cost of the repairs.[3] Their mission is to "promote economic opportunity through our training program while providing dependable private transportation that in turn helps people succeed in the job market." Founded in 1996 as part of Burlington's Old North End Enterprise Community, the program has awarded approximately 4,000 reliable vehicles to individuals and families in need. Working with trainees referred by the Vermont Departments of Labor and Social Welfare, the Good News Garage has a 100% job placement record to date. Half of the program's trainees have been women. To be eligible to receive a Good News Garage vehicle, individuals must possess a valid driver's license and earn less than 150% of the federal poverty level. "When we have to prioritize, we focus on the person who is already employed, whose car has died, and who needs a reliable vehicle to get to work," Colston points out. "We also look at whether the person is about to start a job and needs transportation or needs a car in order to take part in a job training program." Seventy-five percent of the program's vehicle recipients find work and move off the welfare roles." Having reliable, affordable transportation makes all the difference," Colston notes. "It blows away the whole notion that people don't want to work."

Good News Garage has received a multitude of media attention, including an appearance on *Oprah*. All the attention began with a piece about moving people from welfare to work on Vermont Public Radio. This was picked up by National Public Radio's *All Things Considered*, then others followed suit, from NBC Nightly News with Tom Brokaw, and CBS Evening News with Dan Rather, to the *Smithsonian* magazine.

It's more than media, however. Its measurable impacts are impressive, and the program is making a difference.

Lack of a vehicle consumes significant time for parents and their children—time that could be spent working, studying, participating in after-school activities, or investing in family time. Parents are not well positioned to care and provide for children without a vehicle in today's American society.

More than 80% of clients served are single parents who are struggling to achieve financial independence and a better quality of life for themselves and their children. It was found that a total of 61% reported a decrease in their reliance on public assistance (Temporary Assistance for Needy Families, or TANF) due to the vehicle; 58% of the population reported an increase in some sort of community participation due to the vehicle; 48% of the population attributed an increase in education to the car; 60% of the population attributed an increase in training to the car; and 90% of the population reported an improvement in hope for the future of themselves and their family members within Vermont due to the car. Impressive results!

The Editors

Keywords

Individual social capital (capacity), social capacity, social networks, community social capital, community capacity, community building, social cohesion, linkages, indigenous organizations.

Review Questions

1. What is social capital (capacity)?
2. What is the difference between individual social capital and community social capital?
3. What is the interrelationship between community social capacity and development?
4. What is community (capacity) building?
5. What are some key factors that influence the success of community building?

Notes

1 To learn more about AI, visit: http: //appreciativeinquiry.case.edu

2 This case has been excerpted with permission from the authors from *Sustainable Communities, Creating a Local Durable Economy*, London: Routledge, 2013 by R. Phillips, B. Seifer, & E. Antczak.

3 See more details and information at www.goodnewsgarage.org

Bibliography

Bourdieu, P. (1986) "The forms of capital," in Richardson, J.G. (ed.) *Handbook for Theory and Research for the Sociology of Education*, New York: Greenwood Press, pp. 241–260.

Clarke, R. (2004) "Bowling Together," *OECD Observer*, 242 (March) Online. Available at: <http: www.oecdobserver.org> (accessed April 14, 2007).

Gardner J. (1993) *Community Building: An Overview Report and Case Profiles*, Washington, DC: Teamworks.

Jacobs, J. (1961) *The Death and Life of Great American Cities*, New York: Random House.

Joseph, M. and Ogletree, R. (1996) "Community Organizing and Comprehensive Community Initiatives," in R. Stone (ed.) *Core Issues in Comprehensive Community Building Initiatives*, Chicago, IL: Chapin Hall Center for Children, pp. 93–97.

Kubisch, A., Brown, P., Chaskin, R., Hirota, J., Joseph, M., Richman, H., and Roberts, M. (1995) *Voices from the Field: Learning from Comprehensive Community Initiatives*, New York: Aspen Institute.

Mattessich, P. and Monsey, B. (1997) *Community Building: What Makes It Work: A Review of Factors Influencing Successful Community Building*, St. Paul, MN: Wilder Foundation.

Pittman, R., Pittman, E., Phillips, R. and Cangelosi, J. 2009 "The Community and Economic Development Chain: Validating the Links between CED Processes and Outcomes," *Community Development*, 40, 1: 80–93.

Sobel, J. (2002) "Can We Trust Social Capital?" *Journal of Economic Literature*, 40, 1 (March): 139–154.

Uphoff, N. (2000) "Understanding Social Capital: Learning from the Analysis and Experience of Participation," in P. Dasguptaand I. Serageldin (eds.) *Social Capital: A Multifaceted Perspective*, Washington, DC: World Bank, pp. 215–252.

Connections

Watch Robert Putnam speak about social capital at: www.youtube.com/watch?v=XsvD6ud_150

Estimate your social capital network with a toolkit found at: www.kellogg.northwestern.edu/parttime/emails/files/SocialCapitalToolkit.pdf

Become a design thinker for social innovation! Explore the Standford Social Innovation Review article on thinking for social innovation: www.ssireview.org/articles/entry/design_thinking_for_social_innovation

5 Sustainability in Community Development

Stephen M. Wheeler

Overview

The aim of this chapter is to provide basic background on the concept of sustainability and how it may apply to both the practice and content of community development. It starts with a brief overview of the history and theory of this term, then examines its implications for a number of areas within the context of community development. There is substantial agreement in the international literature on many of these implications; however, there is no single ideal of "the sustainable community," nor any examples of such places. Rather, there are many strategies that can potentially improve the long-term health and welfare of communities by working with local geography, history, culture, economy, and ecology. Every existing community has some features that others can learn from as well as many challenges to be addressed. For any given place, the task for professionals is to develop creative strategies and processes that will work within the local context and with its constituencies to improve long-term human and ecological welfare.

History of the Sustainability Concept

The reasons why sustainability has become a leading theme worldwide are well known. Concerns such as climate change, resource depletion, pollution, loss of species and ecosystems, poverty, inequality, traffic congestion, inadequate housing, and loss of community and social capital are ubiquitous. These problems are interrelated; for example, global warming emissions are caused in part by inefficient transportation systems and land use patterns, poorly designed and energy-intensive housing, and economic systems that do not internalize the costs of resource depletion and pollution.

As far as anyone has been able to tell, the term "sustainable development" was used for the first time in two books that appeared in 1972: *The Limits to Growth*, written by a team of MIT researchers led by Donella Meadows, and *Blueprint for Survival*, written by the staff of *The Ecologist* magazine (Goldsmith et al. 1972; Meadows et al. 1972). The Meadows report in particular was significant in that it used newly available computer technology to develop a "systems dynamics" model predicting future levels of global resources, consumption, pollution, and population. Every scenario that the team fed into the model showed the global human system crashing midway through the twenty-first century, and so the researchers concluded that human civilization was approaching the limits to growth on a small planet. This prediction was highly controversial. But revisiting the model in 2002, with three additional decades of actual data, the team concluded that its initial projections had been relatively accurate and that humanity had entered into a period of "overshoot" in which it is well beyond the planet's ability to sustain human society (Meadows et al. 2004).

Other events in the 1970s also helped catalyze concern about the sustainability of human development patterns. The first United Nations Conference on Environment and Development, held in Stockholm in September 1972, brought together researchers and policy makers from around the world to explore humanity's future on the planet. The energy crises of 1973 and 1979 raised global concerns about resource depletion and brought these concerns home to millions of Americans at the gas pump. Public attention to the need for sustainable development received further boosts in the early 1990s as a result of United Nations conferences, such as the "Earth Summit" held in Rio de Janeiro in 1992, and in the early 2000s as knowledge spread about the threat of global warming. In the 2010s, concerns about the sustainability of financial systems and food systems have spread within many countries. Although for many years "sustainability" was dismissed as a faddish or overly idealistic term, in the early twenty-first century it has become well established as a priority worldwide.

Perspectives

Several perspectives on sustainable development emerged early on that have characterized debates ever since (Wheeler 2004). One of these viewpoints is that of global environmentalism, which has focused on resource depletion, pollution, species and habitat loss, and climate change (Brown 1981; Blowers 1993). Some, such as the "deep ecologists," have even argued that other species should be given the same rights as humans and that human population overall is too large and should be substantially reduced, presumably through wise family planning in the long run (Devall and Sessions 1985). Counter to these environmental perspectives—in fact directly opposing the limits-to-growth viewpoint—has been the approach known as technological optimism which holds that human ingenuity and technology will be able to conquer environmental problems. Though clearly this does happen sometimes, technology has not yet addressed many of the concerns described above.

A somewhat different set of perspectives, also originating in the 1970s, focus on the role of economics in addressing environmental and social problems. Economists within the newly emerging disciplines of environmental economics and ecological economics set to work to better incorporate environmental factors into economic models (Repetto 1985; Pearce et al. 1989; Costanza 1991). Some began to question on a much more fundamental level the desirability of endless economic growth on a planet with finite resources. Herman Daly, in particular, advocated a "steady-state society" with qualitative rather than quantitative economic growth (1973, 1980, 1996). Some recent economic thinkers have also advocated new forms of capitalism that better incorporate environmental and social concerns (Hawken et al. 1999; Barnes 2006).

A third main set of perspectives is that of social justice advocates, many of them in the developing world. These critics point out global inequities that have led the United States, for example, with about 4% of the world's population to consume some 25% of its resources (Goldsmith et al. 1992; Barlow and Clarke 2001; Shiva 2005). Such critics have argued that sustainable development first needs to address global disparities and that wealthier countries need to substantially reduce their consumption.

Finally, spiritually and ethically oriented observers have argued that the global crises facing humanity are due to misplaced values, a cognitive perspective that does not adequately recognize interdependency, and/or the lack of an ethical perspective that takes the needs of other societies and the planet into account (Daly and Cobb 1989; Goldsmith 1993; Capra 1996). These writers often build on precedents such as the "land ethic" of

Aldo Leopold (1949) to argue that a new relationship between humans and the earth and between humans and each other is necessary.

These different perspectives on sustainable development have led to different arguments, analyses, and proposals ever since. For example, economists tend to assume that market mechanisms, such as emissions trading systems or steps to set the proper prices on natural resources and pollution, will be able to address sustainability problems. Many environmentalists, on the other hand, argue for strong regulation by the public sector and public investment in areas such as alternative energy and land conservation. Equity activists call for radical rethinking of global capitalism and tend to be highly critical of institutions such as the World Bank and the World Trade Organization. Meanwhile, ethically or spiritually oriented thinkers seek leadership and education toward a different set of societal values and, in some cases, seek guidance within organized religious traditions. Elements of all of these approaches seem useful at different times, and an awareness of all these perspectives is important to form an understanding for pragmatic application of sustainability ideas within communities.

BOX 5.1 THE UNITED NATIONS' (UN) MILLENNIUM DEVELOPMENT GOALS—STRIVING FOR SUSTAINABILITY ACROSS DIMENSIONS OF WELL-BEING

The eight Millennium Development Goals (MDGs)—which range from halving extreme poverty rates to halting the spread of HIV/AIDS and providing universal primary education, all by the target date of 2015—form a blueprint agreed to by all the world's countries and all the world's leading development institutions. They have galvanized unprecedented efforts to meet the needs of the world's poorest. The UN is also working with governments, civil society and other partners to build on the momentum generated by the MDGs and carry on with an ambitious post-2015 development agenda. From this site, explore the efforts of the UN and its partners for building a better world.

(United Nations (2014: 1) UN Millennium Goals; see: www.un.org/millenniumgoals)

Sustainability Definitions and Themes

Despite the extraordinary influence of the sustainable development concept, no perfect definition of the term has emerged. The most widely used formulation is that issued by the United Nations Commission on Environment and Development (the "Brundtland Commission") in 1987, which defines sustainable development as "development that meets the needs of the present without compromising the ability of future generations to meet their own needs" (World Commission on Environment and Development 1987: 43). However, this definition is problematic since it raises the difficult-to-define concept of "needs" and is anthropocentric, discussing the needs of humans rather than those of ecosystems or the planet as a whole.

Other definitions have similar problems. For example, relying on the notion of the "carrying capacity" (the inherent ability of a community or region to support human life and maintain environmental well-being) is difficult since this is hard to determine for human communities. Relying on concepts such as maintaining natural and social capital is problematic since these entities are difficult to measure and would require a complex economic calculus. My own preference is simply to define sustainable development as development that improves the long-term welfare of human and ecological communities and then move on to a more specific discussion of particular strategies.

Approaches

Sustainable development tends to require certain approaches on the part of community development leaders and professionals. One obvious starting point is to emphasize the long-term future. Rather than thinking about the next economic quarter, the next election cycle, or even the next 10 or 20 years (as is common in local planning documents), it becomes important to think what current development trends mean if continued for 50, 100, or 200 years. Often short-term trends that seem acceptable become disastrous when viewed in the longer term. One essential starting place is getting the public and decision makers to understand the long-term implications of current trends in addition to their near-term impacts.

Another main approach within sustainability planning is to emphasize interconnections between community development issues. Land use, transportation, housing, economic development, environmental protection, and social equity are all related. Historically, one main problem in planning has been that these issues have been treated in isolation; for example, highways have been planned without considering the sprawling land use patterns that they will stimulate, and suburban malls and big box stores have been encouraged without realizing that they may lead to disinvestment and poverty in traditional downtowns. Viewing any given topic from a broad and holistic perspective, each individual decision can be better tied into sustainable community development as a whole. Likewise, developing an understanding of how actions at different scales interrelate is important as well. Building, site, neighborhood, city, region, state, national, and global scales fit together; actions at each scale must consider and reinforce actions at other scales. Recent movements such as the New Urbanism have emphasized a similar coordination of action at different scales (Congress for the New Urbanism 1999).

BOX 5.2 GAUGING SUSTAINABILITY IN THE REGIONAL VANCOUVER URBAN OBSERVATORY

In late 2004, the Regional Vancouver Urban Observatory (RVu) was created in response to the twin challenges of launching an innovative new professional masters program in Urban Studies at Simon Fraser University in downtown Vancouver, Canada; and launching a significant and meaningful forum with which to engage questions of measuring, monitoring, and improving outcomes of sustainable development at home in Vancouver and in cities around the world. With regard to the second challenge, the international push at that time came from Vancouver's invitation to the worldwide community of urban development professionals to gather in Vancouver for the 2006 World Urban Forum 3, and this on the 30th anniversary of Vancouver's first major foray into hosting international community development events, Habitat '76. The initial building blocks of RVu were thus laid with considerable excitement in the local community and at UN Habitat headquarters in Nairobi, plus throughout the extended network of cities participating in the World Urban Forum.

RVu (pronounced "Our View") advisers and citizen participants crafted a mission captured by the phrase "measures to match our values." We aligned RVu activities with both the urban and community development and the measurement and monitoring agendas of the UN Global Urban Observatory. The GUO is a group within UN Habitat that aims to support "better information for better cities," initially through the design of a Global Urban Indicators Database, and subsequently through more localized capacity-building efforts. The RVu flagship work was launched in June 2006, a community-built and systems-based sustainability indicator system for the Metro Vancouver region called *Counting on Vancouver: Our view of the region* (Holden, 2006).

What happened next at both local and global scales is, in some ways, disappointing and, at others, indicative of larger trends. RVu has never been able to implement the set of measures that several hundred citizens spent significant time and passion creating to guide their city toward greater sustainability (Holden et al., 2009; Holden, 2009). Neither, on the global front, has the GUO been able to see its global indicator set populated with data, or make headlines with the data it did manage to gather. The appeal of meeting this ideal of measuring and comparing cities in this way according to a central database has not been lost, however, as judged by the new institutions like the Global City Indicators Facility and STAR Communities (see weblinks below) that have popped up to take other approaches to this work.

Nor has the work of RVu ceased. Indeed, the growth in attention to measures of urban, community and sustainable development has been exponential since 2006; and data is taken as the new raw material of the postindustrial economy. RVu has a role in all of this, as advisor to and reviewer of new urban indicator systems in municipalities across Canada and around the world. We also pursue work in the more strategic development of keystone indicators such as those in the more specific realm of housing, transportation (see "walkscore" weblink below), and neighborhood design (see "LEED ND" weblink below).

Meg Holden
Associate Professor, Urban Studies and Geography
Simon Fraser University, Canada

Sources

Holden, M. (ed.) 2006. Special issue: Urban Indicators. *Cities* 23(3).
Holden, M. 2009. Community interests and indicator system success. *Social Indicators Research* 92: 429–448.
Holden, M., C. Owens, and C. Mochrie. 2009. Lessons from a community-based process in regional sustainability indicator selection. In M.J. Sirgy, D. Rahtz, and R. Phillips (eds.) *Community Quality-of-Life Indicators: Best Cases IV*. Springer, pp. 59–80.

Further Reading

Global City Indicators Facility, an institution based at the University of Toronto promoting a common set of indicators and membership network for urban development: http: //cityindicators.org
Holden, M. and R. Phillips (eds.) 2010. Special issue: Best papers from the 2009 Community Indicators Consortium. *Applied Research in Quality of Life* 5(4).
LEED ND, Leadership in Energy and Environmental Design for Neighborhoods, is a rating system for neighborhood scale developments, designed collaboratively by the Natural Resources Defense Council, U.S. Green Building Council, and Congress for the New Urbanism: www.cnu.org/leednd
Regional Vancouver Urban Observatory: www.rvu.ca
STAR Communities is a rating system for local municipalities, designed to report on a broad swath of sustainability goals: http: //starcommunities.org
Walkscore is an organization and a calculation methodology that provides a comparative understanding of the walkability of North American neighborhoods: www.walkscore.com
Wikiprogress, an online forum for sharing information about measuring social, economic, and environmental progress: http: //wikiprogress.org

Another theme within sustainable community development is attention to place. Local history, culture, climate, resources, architecture, building materials, businesses, and ecosystems provide a rich and valuable context for local sustainability efforts. Working with these resources is also a way to build community pride and identity. For example, restoring a stream or riverfront can create an attractive new amenity for a community, as well as helping to support nearby businesses, creating new civic gathering spaces, and bringing people together for stewardship activities and celebratory events.

Tied to an emphasis on place is an acknowledgment of limits. Any given place can only handle so much change before it becomes something different (which is of course sometimes desirable). There are limits to the number of people or the amount of traffic that can be accommodated easily in any given community without undermining those place-based attributes that community members value. Likewise, there are limits to the quantities of resources that our society as a whole can use without damaging either local or global ecosystems. "Growth" itself must be reconsidered within a sustainable development paradigm, following Daly's notion, moving from quantitative expansion of goods consumed to qualitative improvement in community welfare. An organization named Redefining Progress has in fact developed an "Genuine Progress Indicator" that it believes can measure such a shift at the national level instead of the gross domestic product which, as is often pointed out, rises significantly during environmental disasters, such as the BP Deepwater Horizon oil spill, since large sums are spent on clean-up and public relations (Talberth et al. 2006). At the local level, efforts to rethink growth should not take the form of exclusionary growth controls designed to keep out lower-income residents by restricting the amount of multifamily housing, but should be a more comprehensive rethinking of how the community will coexist with local, regional, and global resource limits in the long run.

A final theme implicit within sustainable development is the need for active leadership by planners, politicians, and other community development professionals. In the past, these players have sometimes facilitated unsustainable development. More active and passionate engagement by professionals is needed to address current sustainability problems, often seeking new alternatives to the status quo. In this quest it is important for community development professionals to work actively with elected leaders, community organizations, businesses, and the general public to develop public understanding and political support for action.

So the concept of sustainability can be seen to have roots going back more than 35 years, a variety of different perspectives taken by different advocates, and some themes that can guide professionals in seeking real-world applications. With this background, some specific areas of sustainable community development planning are presented below. Since fully considering the topic would require a very large space, the intent here is simply to suggest some possible directions for community action.

Action Areas

Environment

Sustainability is often thought of as primarily an environmental concern, and certainly environmental initiatives are important within any sustainable community development agenda. These can be of many sorts, but one of the most timely and challenging types of initiatives aims to reduce greenhouse gas emissions. Global warming initiatives at the local level are increasingly common, thanks in part to the Cities for Climate Protection campaign coordinated by the international nongovernmental organization ICLEI: Local Governments for Sustainability, and require a very broad and interdisciplinary rethinking of many local government policies. In the US, some 27% of greenhouse gas (GHG) emissions stems from transportation uses, another 27% is related to building heating, cooling, and electrical use, and about 20% results from industry (World Resources Institute 2007). Local governments can affect all of these areas.

Communities can best reduce private motor vehicle use—and resulting GHG emissions—through three types of initiatives: better land use planning, better alternative travel mode choices for local residents, and revised economic incentives for travel. All of these types of initiatives are discussed later in this chapter. Local governments can also set an example by converting their own vehicle fleets, including buses, to cleaner technologies such as hybrid engines and use of compressed natural gas or biodiesel.

In terms of building heating, cooling, and electricity use, communities can modify building codes to require passive solar design of structures, higher degrees of energy efficiency, use of energy- and water-efficient appliances, and recycling of construction waste and debris. Subdivision ordinances can be modified to require solar orientation of lots in new subdivisions (with the long dimensions of lots and buildings facing south), and zoning codes can be amended to ensure solar access to each lot (by restricting the height of structures on adjoining lots near the southern property line). Other eco-friendly strategies such as handling rainwater runoff onsite, using graywater (lightly used wastewater) for irrigation or toilets, minimizing asphalt paving, promoting the use of alternative construction materials and green roofs, providing incentives for solar hot water or electricity, and encouraging shade trees to provide summer cooling can also be incorporated into these codes. In terms of electric power, communities can require that a certain percentage of electricity they purchase be generated from renewable sources. Some cities and towns have historically owned their own electric utilities which gives them an even greater ability to lower GHG emissions and promote green practices.

To reduce the 20% of GHG emissions stemming from industry, local governments can seek to identify such sources within their jurisdictions and work with them to reduce emissions, for example, by providing technical assistance, grants, or favorable tax treatment for eco-friendly practices. Giving priority to reducing emissions may also affect economic development policy choices, as discussed later in this chapter. In short, a local greenhouse gas reduction program must address many different aspects of policy, integrating these initiatives together. Each of these steps will have other sustainability advantages, however, whether in terms of reducing traffic congestion and driving, lowering home heating costs, or developing more efficient industry and businesses.

Adapting to climate change is an increasingly urgent priority at the local level as well. This may mean ensuring that low-lying communities are protected from flooding, taking steps to ensure long-term water supplies in the face of drought, programs to make sure that low-income households are protected from extreme heat, and urban greening programs to cool cities and improve comfort. A particularly important priority is to identify potential climate justice issues before they happen—so that neighborhoods aren't isolated, unprotected, or abandoned as was New Orleans' Ninth Ward following Hurricane Katrina.

Communities can meet environmental sustainability goals as well through programs relating to their use of materials. For example, of the three Rs—reduce, reuse, recycle—recycling has gotten the most attention in terms of municipal programs, but much greater energy and materials savings are likely in the long run from the first two. Reusing wooden shipping pallets or replacing them with more durable shipping materials offers many advantages over recycling them as chipped wood for mulch or throwing them away, as has been done in the US until recently. A system of washing and reusing glass bottles, as exists in many European countries and once existed in the US until the spread of plastic containers in the 1970s, offers far greater energy savings than collecting, crushing, and recycling them. Communities may want to eliminate some materials altogether. Cities such as San Francisco have banned use of non-biodegradable plastic bags. Portland, Oakland, and about 100 other cities have banned the use of styrofoam.

Ecosystem protection and restoration offers another main area for environmental initiatives. Whereas conservation was a main goal of previous generations of environmentalism, restoration has become a key objective in recent decades, especially in urban areas. Efforts to restore creeks, shorelines, and wetlands, replant native vegetation, recreate wildlife corridors, and preserve existing habitat can form centerpieces of local environmental initiatives. Traditional forms of local government regulation, such as zoning codes and subdivision ordinances, can be amended to ensure that such features are protected within new development. For example, a community can require a substantial buffer (30–100 feet or more) along waterways, thus preserving both ecologically valuable riparian corridors and opening up the possibility for a recreational trail system. Cities and towns can also require developers to preserve heritage trees and important areas of wildlife habitat on project sites.

Land Use

Local governments in the US have influence through regulation and investment over the development of land within their boundaries, and land use in turn can influence everything from how much people need to drive to how much farmland and open space remains near cities. Managing the outward expansion of communities is one main sustainability priority. "Smart growth" has been a rallying cry among U.S. local governments since the 1990s, especially since suburban sprawl often increases local costs for infrastructure and services (Transportation Research Board 1997; Ewing et al. 2002).

Smart growth is generally defined as development that is compact, contiguous to existing urban areas, well connected by a grid-like network of through streets, characterized by a diverse mix of land uses, and relatively dense. Internationally there is some debate over just what degree of density or compactness is desirable in order to create more sustainable communities (Jenks et al. 1996). Certainly cities and towns need not approve high-rise buildings, although Vancouver, British Columbia, provides a good example of how well-designed slender high-rises can work well within residential neighborhoods. However, there is little question that most U.S. communities can use land far more efficiently than at present. In many cases this will require local governments to guide much more precisely where development will go and in what form rather than maintaining a reactive mode to proposals. Two major ways that cities and towns can do this is through area plans that contain precise design requirements for new development and through subdivision regulations that require connecting street patterns, neighborhood centers, greenways, and other community design elements.

Infill development, which includes reuse of existing built land as well as construction on vacant or leftover parcels within urban areas, is one main smart growth strategy. The tens of thousands of old shopping malls, business parks, and industrial sites in American communities offer prime opportunities for infill and for creating new, walkable, mixed-use centers for existing neighborhoods. But infill is often more difficult for developers than greenfield projects and may require substantial municipal assistance. Community development staff can facilitate dialogue between developers and local constituencies, assemble land through redevelopment powers, develop design guidelines or a specific plan for the area in question, and provide infrastructure and amenities to complement new development. In the past, much urban redevelopment in U.S. communities was not done with sufficient respect for the historical context and existing residents, but more context-sensitive approaches in the future can help ensure that such intensification efforts work well. For example, ensuring historic preservation guidelines, affordable housing requirements, and public participation processes are in place can help ensure that redevelopment is done well.

A good mix of land uses is a further goal frequently cited within the sustainable communities literature as well as by advocates of the new urbanism and smart growth. Since the 1910s, Euclidean zoning has generally sought to separate land uses within North American communities, leading to the creation of vast housing tracts in one location, large commercial strips and malls in another, and office or industrial development in yet others. One result is that people need to drive long distances to get to basic destinations in life. Separation of land uses also makes it very difficult for anyone to walk anywhere, or for motorists to "trip-chain"—carry out a number of different tasks with one relatively short trip.

Improving land use mix requires fundamentally rethinking local zoning codes, and community and economic development. Better-defined neighborhood centers can be included within new development on the urban fringe, while downtowns and office parks can have new infill housing added. Zoning can be changed for existing neighborhoods to allow a greater variety of local uses, including home offices, second units within or behind existing homes, and apartments or mixed-use buildings along commercial streets. For example, the latter is especially important for allowing residential uses in the top floors of commercial buildings to use space more sustainably.

The scale of new development should be reconsidered as well. Bigness has been a defining feature of recent American land development whether residential, commercial, or industrial, but large-scale projects are not necessarily desirable from a sustainability viewpoint. Such development often provides little diversity, interest, or sense of place, and can generate community impacts such as large amounts of traffic. Local sustainability planning is likely to emphasize smaller local businesses, more incremental growth of new neighborhoods, and more detailed specifications for development. Such modest-scale land development can potentially create more diverse, interesting, and vibrant communities in the long run, with fewer long-distance commuting needs.

Park and greenspaces planning is a final area of land use that is essential for more sustainable communities. Although such planning has gone through a number of eras historically (Cranz 1982), many communities today emphasize networks of parks and greenways with a variety of environments for different user groups. Increasingly, native vegetation and restored wildlife habitats are part of the concept instead of the UK-style trees-and-lawn planting scheme. The idea is to reconnect residents to the landscape of their geographical community on a daily basis, both through small-scale parks and landscape design near homes and through larger networks of greenways and wildlife preserves throughout urban areas.

Transportation

A community's transportation systems determine much about its resource consumption, greenhouse gas emissions, civic environment, and quality of life. For the past 80 years, both infrastructure priorities and patterns of land development in North America, Australia, and many other places have emphasized mobility via the private motor vehicle and that the per capita amount driven annually has risen about 2% a year. In the 2010s this trend has finally begun to change in the US as many younger people want to live in more balanced, urban communities that require less driving. Retrofitting communities to make other modes of transportation more possible is now important. But everywhere some steps can be taken to encourage alternative modes of transportation.

Improving the pedestrian environment is one important step. This means not just adding or improving sidewalks in a given place, but coming up with a comprehensive package of street and urban design improvements to enhance the walking environment. Such a package may include street landscaping, pedestrian-scale lighting, narrower lanes and roadways, lower traffic speeds, improved medians, sharper curb radii at intersections, and better-connected street patterns within new development. Pedestrian-friendly boulevard designs can be employed in place of unsightly and dangerous arterials in some communities (Jacobs et al. 2003). Traffic-calming strategies can be employed in residential neighborhoods to slow traffic. These employ a range of design strategies including speed humps, traffic circles, chicanes (staggered parking), and extensive landscaping.

In general, the street design philosophy in many communities is shifting from one of increasing the capacity and speed of streets, common several decades ago, to one of promoting slow-and-steady motor vehicle movement. Street design nationally is also moving toward "context-sensitive design" that respects existing historical, cultural, and ecological environments and promotes walking, bicycling, public transit, and neighborhood use of streetscapes (Federal Highway Administration 2007).

In the past couple of decades, an increasing number of communities have developed bicycle and pedestrian plans to coordinate investment and policies for these two modes of transportation. Ever since the passage of the Intermodal Surface Transportation Efficiency Act of 1991 (ISTEA), federal transportation funding has been more flexible, allowing resources to be used for these purposes. An increasingly creative mix of public transportation modes is also appearing in cities and towns (Cervero 1998). Large-scale metro and light-rail systems have been built in cities ranging from Dallas to Denver, Portland to Phoenix. But "bus rapid transit" systems, in which high-tech buses provide light-rail-style service, provide a less expensive alternative to rail systems in places such as Los Angeles and Albuquerque. Some communities are experimenting with "ride-on-demand" service using small vehicles such as vans, while others have built old-fashioned streetcars with very frequent stops in urban areas. The ideal is to provide residents with a web of interwoven transit options. "Transit-oriented development" (TOD) land use strategies can then seek to cluster new development around transit routes, increasing ridership, and providing a range of destinations and residences close to transit.

Pricing is a final, and controversial, piece of the transportation planning puzzle. The aim is to make both transit and ride-sharing attractive and discourage long-distance drive-alone commuting. Car pools ("high-occupancy vehicles") are often given their own, toll-free lanes on urban freeways, while a few places have made transit use cheap or free. Portland, Oregon's, "fareless square," including most of the city's downtown, is one example. Cities and towns can provide economic incentives for residents to drive less. For example, some communities raise the cost of parking (Shoup 2005) and develop employer-based trip reduction programs. Internationally, a number of large cities, including London, have established toll zones requiring motorists to pay a substantial sum to enter city centers. Most European cities also have at least some parts of their downtowns that are pedestrian-only zones.

Housing

A community's housing stock affects its sustainability in several ways. For one thing, large amounts of energy and materials are required to construct and maintain housing. As previously mentioned, communities can revise local building codes to require more energy-efficient structures and appliances as well as water-efficient plumbing fixtures. But on a larger scale, imbalances of housing with jobs and shopping generate high levels of motor vehicle use, traffic, pollution, and greenhouse gas emissions. "Jobs-housing balance" has become a mantra for many communities. Ideally, communities will provide slightly more than one job per household (since many households have more than one worker). The price and size of the available housing must also balance with the needs of workers employed in the community. One typical problem is lack of affordable housing for service workers, teachers, firefighters, nurses, and other essential professions. These personnel must either pay a large percentage of their salary for housing or must commute from more affordable communities further away.

There is no easy solution to a community's housing affordability problems, but several strategies taken together can potentially make a difference. One basic step is to ensure that sufficient land is zoned for apartments, condominiums, townhouses, duplexes, and other forms of housing that tend to be less expensive. Another common strategy is "inclusionary zoning" in which developers are required to include a certain percentage (often 10–20%) of units affordable to households making a certain percentage (typically 80%) of the county median income. Other strategies include legalizing and encouraging creation of secondary units on existing single-family home lots, encouraging creation of land trusts that will lease housing units at below-market rates, and subsidizing nonprofit affordable housing providers to build affordable housing.

Economic Development

Economic development strategies are among the most challenging to revise from a sustainable communities perspective, in part because in the past they have been so often focused on what might now be seen as unsustainable development. Some cities and towns have traditionally sought any available form of economic growth regardless of impact—rapid land development, malls, big box development, and casinos. Although substantial municipal subsidies are often offered to such businesses, gaining them does not necessarily guarantee the community a stable and sustainable future. Multinational firms may move their jobs elsewhere. Big box retailers may negatively impact smaller locally owned businesses. Rapid suburban expansion can bring traffic, overburdened local services, and loss of local culture and identity.

Sustainable economic development is instead likely to emphasize the nurturing of green and socially responsible employers within a community. These businesses will use local resources, have clean production practices, pay decent wages, and contribute back to community through civic involvement. They will be of a range of sizes, including many relatively small, locally owned enterprises with deeper community roots than current employers (Shuman 1998, 2006). Rather than seeking rapid overnight expansion, such firms will add employment at a slower and more sustainable rate.

If this sounds like an unachievable ideal given the nature of the economy, it may well be. However, local community development efforts can help bring this vision about in a number of ways. One is by supporting the existing local businesses and encouraging them to undertake both innovation and greener production practices. Another strategy is to grow new businesses of desirable types, frequently through the creation of business incubators that provide affordable office space and shared services for start-ups, and the preferential issuance of public contracts to green businesses. Investing in public education and training is a further municipal commitment to its economic future. Finally, in recent years many jurisdictions have passed "living wage" laws requiring that workers be paid significantly more than the federal minimum wage. This policy improves both social equity and potentially increases workers' spending power within the community.

Social Equity

As a symbol of the integrating approach common within sustainable development, advocates have often spoken of the "three Es"—environment, economy, and equity. Of these, equity is by far the least well developed and perhaps the most difficult to bring about in practice. Such rising inequality brings about many sustainability problems—from the degradation of ecosystems by impoverished people struggling to survive, to the loss of social capital and mutual understanding essential for healthy democracies.

Ensuring social equity is in substantial part the responsibility of federal and state levels of government which can promote it through tax policy, funding of social services, establishment of decent minimum wages, and guarantees of fair treatment and civil rights. Local communities can promote equity goals as well. Providing adequate amounts of affordable housing, livable minimum wages, and a supportive environment for local businesses are among the ways to do this. Ensuring that underprivileged neighborhoods receive excellent services, schools, parks, and other forms of municipal investment is also important. Environmental justice is another key equity theme; too often lower-income neighborhoods and communities of color have borne the brunt of pollution, toxic contamination problems, and unwanted facilities. Cities and towns can address environmental justice through active steps to protect those most at risk, improve siting of hazardous land uses, bring about fairer and more transparent decision making, and include at-risk populations in local government processes.

Additional services important for social equity include adult education and literacy programs, preschool and after-school activities, drug and alcohol abuse treatment programs, and assistance for those with disabilities, mental health issues, or a history of homelessness. Good public education in general, of course, is also crucial. Such initiatives can help build the human capital important for healthy communities in the long run. The problem of funding always exists, but grant opportunities are available for certain types of programs, and creative, sustained attention by community development officials and political leaders can help build better support in the long run. Building a "healthy" community in all aspects supports long-term sustainability.

Process and Participation

A healthy democracy is an important element of sustainable communities in that it can enable informed decision making, meet the needs of diverse constituencies, and fulfill ideals of fairness and equity. For this reason, community sustainability groups have emphasized a variety of process indicators that reflect the health of our political system and society. Sustainable Seattle, for example, included "voter participation," "adult literacy," and "neighborliness" in its set of sustainability indicators (Sustainable Seattle 1998). The Jacksonville Community Council included not just voter registration, but "Percentage of people surveyed who are able to name two current City Council members" in its quality-of-life indicators, which have been updated for nearly 25 years now (Jacksonville Community Council, Inc. 2006).

For a healthy democracy, three things are needed: a clean, open system; real choices in elections; and an informed, active electorate. In particular, conflicts of interest, often around land development, are rife within local governments the world over. Historically, "growth coalitions" of developers, landowners, real estate interests, construction companies, and politicians have dominated local politics in many communities (Logan and Molotch 1987). These interests have often funded electoral candidates, and their members have frequently held elected or appointed office. Ending such conflicts of interest and improving the transparency and visibility of local government processes is important, as is making public office attractive to a wider variety of candidates including those without wealthy backers. Ensuring high participation rates in elections and citizen knowledge of development issues is a related challenge.

As anyone involved in community development knows, public participation in local government decisions is great in the abstract but difficult to ensure in practice. It can be hard to get people to turn out for meetings, ensure participation of diverse constituencies (especially lower-income groups and communities of color), and facilitate involvement that is constructive rather than oppositional. From a local residents' point of view, public involvement exercises often do not seem to include real opportunity to shape decisions.

From a sustainability perspective, it is vitally important to establish a creative and collaborative local government decision-making environment in which participants can agree on positive, proactive strategies, "think outside the box," and learn to respect each other's points of view. Too often in recent years community events have been oppositional in tone, involving mutual suspicion and animosity as well as NIMBY ("not in my backyard") groups simply trying to stop projects that are not in their own self-interest. In order to enable a constructive, collaborative planning environment instead, a number of procedural reforms can help. Transparent and well-publicized government processes can ensure that residents understand what is going on and do not feel excluded by back-room deals. Strong conflict-of-interest regulation can alleviate citizens' sense that officials are just out for special interests. Workshops and charrettes (design workshops) can be conducted with a collaborative and collegial tone rather than the top-down and patronizing styles sometimes found in reality. And local residents' ideas can be very consciously incorporated into planning alternatives and reflected back to them so that it is clear what their input has been.

That being said, it is very important for local residents to understand that they are not the only stakeholders involved in public decisions. "Community-based planning" is frequently seen as focusing just on the local neighborhood or town. But in line with the sustainability themes discussed earlier, any given decision affects multiple overlapping communities at different scales including regional, national, and global levels. From a sustainability perspective, the practitioners' role is to take into account the needs and concerns of *all* of these different communities, including the needs of the planet itself, and to help local residents understand this complex picture.

Operating as a professional with a concern for sustainable community development may require a great many skills. It may require active efforts to frame debates, develop background information, and outline alternative courses of action. It may require careful organizing both within government and within the community to pull different constituencies together and develop institutional and political backing. It may require specific intervention in debates to call attention to the long-term implications of decisions. It may require constant efforts to weave together of all aspects of community development, including physical planning, urban design, economic development, social welfare policy, and environmental planning, so that the public understands the interconnections. This is the challenge of working within local government and communities. Good communication, networking, facilitation, presentation, and political skills can help in this regard, as do passion and a sense of humor.

Conclusion

Sustainable community development is clearly a major challenge. We are in the early stages of a process that will take hundreds of years. But although the task can at times seem daunting, it is also a profoundly hopeful and inspiring one—helping to envision a future that truly can meet the needs of both human communities and the planet.

CASE STUDY: CITY OF SANTA MONICA, CALIFORNIA'S SUSTAINABLE CITY PROGRAM

One of the communities that understands the importance of integrating sustainability into overall community development is Santa Monica, California. Beginning in 1994, the City Council adopted a Sustainable City Program to address issues and concerns of sustainability. In 2003, the Sustainable City Plan was adopted, which is built on guiding principles of sustainability and focuses on eight goal areas:

1. Resource conservation
2. Environment and public health
3. Transportation
4. Economic development
5. Open space and land use
6. Housing
7. Community education and participation
8. Human dignity

2012 Sustainable City Report Card

The **Sustainable City Plan** was created to enhance our resources, prevent harm to the natural environment and human health, and benefit the social and economic well-being of the community for the sake of current and future generations.

Since that time, the community has increased its set of very aggressive goals as a sustainable city, and has been a factor in many positive changes. The Sustainable City Plan, which was developed by a diverse group of community stakeholders and lays out far-reaching sustainability goals for the community was updated in 2012 to ensure continued success in sustainability. The following lists the ten principles adopted in the latest version of the plan:

1. The concept of sustainability guides city policy.
2. Protection, preservation, and restoration of the natural environment is a high priority of the city.
3. Environmental quality, economic health and social equity are mutually dependent.
4. Decisions have implications to the long-term sustainability of Santa Monica.
5. Community awareness, responsibility, participation, and education are key elements of a sustainable community.
6. Santa Monica recognizes its linkage with the regional, national, and global community.
7. Those sustainability issues most important to the community will be addressed first, and the most cost-effective programs and policies will be selected.
8. The city is committed to procurement decisions which minimize negative environmental and social impacts.
9. Cross-sector partnerships are necessary to achieve sustainable goals.
10. The precautionary principle provides a complimentary framework to help guide city decision makers in the pursuit of sustainability.

The city promotes the idea that decisions have long-term implications for all, and that awareness, responsibility, participation and education are key themes to success.

An innovative approach to guiding and monitoring success toward these goals is a comprehensive community indicator system that provides measurements of goal attainment. Specific targets have been designed for many of the indicator areas. This allows for further progress tracking and provides information to consider adjusting policies or programs that may or may not be working as planned.

The city has devised a set of goals with indicators and targets for each area. These goals reflect their vision. The city explains the use of the indicators and targets as follows:

- For each goal, specific indicators have been developed to measure progress. Indicators are tools that help to determine the condition of a system, or the impact of a program, policy or action. When tracked over time, indicators tell us if we are moving toward sustainability and provide us with useful information to assist with decision making. Two types of indicators are tracked as part of the Sustainable City Plan. System level-indicators measure the state, condition or pressures on a community-wide basis for each respective goal area. Program-level indicators measure the performance or effectiveness of specific programs, policies, or actions taken by the city government or other stakeholders in the community.
- Many of the goals and indicators measure more than one area of sustainability. The amount of overlap shown by the matrix demonstrates the interconnectedness of our community and the far-ranging impact of our decisions across environmental, economic, and social boundaries.
- Specific targets have been created for many of the indicators. The targets represent aggressive yet achievable milestones for the community. Unless otherwise noted, the targets are for the year 2010 using 2000 as a baseline. For some indicators no specific numerical targets have been assigned. This was done where development of a numerical target was determined to be not feasible or where limits on data type and availability made it difficult to set a numerical target. In many of these cases, a trend direction was substituted for a numerical target.

(City of Santa Monica 2014)

Having the Sustainable City Plan as a guiding document, the Sustainable City Program strives to integrate sustainability into every aspect of city government and all sectors of the community (City of Santa Monica 2012). Collaboration among sectors as well as intensive evaluation and monitoring encourage desirable outcome achievement. This plan is also coordinated with overall comprehensive planning as well as capital budgeting and other local government tools to aid in decision making for investments and service provision.

Sources

City of Santa Monica. (2012). Sustainable City Report Card. See www.smgov.net/departments/ose/categories/sustainability.aspx for the full report and more updates on Santa Monica's efforts.

City of Santa Monica. (2014). Office of Sustainability. Downloaded February 25 from http: //smgov.net/Departments/OSE/progressReport/default.aspx

The Editors

Keywords

Sustainability, sustainable development, community development, planning.

Review Questions

1 Why is there "no single ideal" of a sustainable community?
2 How does a sense of place affect the context for local sustainability efforts?
3 Using the six actions areas, describe one recommendation from each and explain how it influences community development outcomes.

Bibliography

Barlow, M. and Clarke, T. (2001) *Global Showdown: How the New Activists are Fighting Corporate Rule*, Toronto: Stoddart.

Barnes, P. (2006) *Capitalism 3.0: A Guide to Reclaiming the Commons*, San Francisco: Berrett-Koehler.

Blowers, A. (ed.) (1993) *Planning for a Sustainable Environment: A Report by the Town and County Planning Association*, London: Earthscan.

Brown, L.R. (1981) *Building a Sustainable Society*, New York: Norton.

Capra, F. (1996) *The Web of Life: A New Scientific Understanding of Living Systems*, New York: Anchor Books.

Cervero, R. (1998) *The Transit Metropolis: A Global Inquiry*, Washington, DC: Island Press.

Congress for the New Urbanism. (1999) *Charter of the New Urbanism*, New York: McGraw-Hill.

Costanza, R. (ed.) (1991) *Ecological Economics: The Science and Management of Sustainability*, New York: Columbia University Press.

Cranz, G. (1982) *The Politics of Park Design: A History of Urban Parks in America*, Cambridge, MA: MIT Press.

Daly, H.E. (ed.) (1973) *Toward a Steady-State Economy*, San Francisco: W.H. Freeman.

Daly, H.E. (ed.) (1980) *Economics, Ecology, Ethics: Essays toward a Steady-State Society*, San Francisco: W. H. Freeman.

Daly, H.E. (1996) *Beyond Growth: The Economics of Sustainable Development*, Boston: Beacon Press.

Daly, H.E. and Cobb, Jr., J.B. (1989) *For the Common Good: Redirecting the Economy Toward Community, the Environment, and a Sustainable Future*, Boston: Beacon Press.

Devall, B. and Sessions, G. (1985) *Deep Ecology: Living as if Nature Mattered*, Salt Lake City: G.M. Smith.

Ewing, R., Pendall, R., and Chen, D. (2002) *Measuring Sprawl and Its Impact*, Washington, DC: Smart Growth America.

Federal Highway Administration. (FHWA). (2007) "FHWA and Context Sensitive Solutions" (CSS). Online. Available at: <www.fhwa.dot.gov/csd/index.cfm> (accessed April 1, 2007).

Goldsmith, E. (1993) *The Way: An Ecological World View*, Boston: Shambhala.

Goldsmith, E., Allen, R., Allaby, M., Davoll, J., and Lawrence, S. (1972) *Blueprint for Survival*, Boston: Houghton Mifflin.

Goldsmith, E., Khor, M., Norberg-Hodge, H. and Shiva, V. (1992) *The Future of Progress: Reflections on Environment and Development*, Bristol, UK: International Society for Ecology and Culture.

Hawken, P., Lovins, A., and Lovins, L.H. (1999) *Natural Capitalism: Creating the Next Industrial Revolution*, London: Earthscan.

Jacksonville Community Council, Inc. (JCCI). (2006) *2006 Quality of Life Progress Report*, Jacksonville, FL: JCCI.

Jacobs, A., Macdonald, E., and Rofe, Y. (2003). *The Boulevard Book: History, Evolution, Design of Multiway*, Cambridge: MIT Press.

Jenks, M., Burton, E., and Williams, K. (1996) *The Compact City: A Sustainable Urban Form?* London: E & FN Spon.

Leopold, A. (1949) *A Sand County Almanac*, New York: Oxford University Press.

Logan, J. and Molotch, H. (1987) *Urban Fortunes: The Political Economy of Place*, Berkeley: University of California Press.

Meadows, D., Meadows, D.L., Randers, J., and Behrens III, W.W. (1972) *The Limits to Growth*, New York: Universe Books.

Meadows, D., Randers, J., and Meadows, D. (2004) *Limits to Growth: The 30-Year Update*, White River Junction, VT: Chelsea Green.

Pearce, D., Barbier, E., and Markandya, A. (1989) *Blueprint for a Green Economy*, London: Earthscan.

Repetto, R. (ed.) (1985) *The Global Possible: Resources, Development, and the New Century*, New Haven: Yale University Press.

Shiva, V. (2005) *Earth Democracy: Justice, Sustainability, and Peace*, Cambridge, MA: South End Press.

Shoup, D. (2005) *The High Cost of Free Parking*, Chicago: Planners Press.

Shuman, M. (1998) *Going Local: Creating Self-Reliant Communities in a Global Age*, New York: Free Press.

Shuman, M. (2006) *The Small-Mart Revolution: How Local Businesses are Beating the Global Competition*, San Francisco: Berrett-Koehler.

Sustainable Seattle. (1998) *Indicators of Sustainable Community*, Seattle, WA: Sustainable Seattle.

Talberth, J., Cobb, C., and Slattery, N. (2006) *The Genuine Progress Indicator 2006: A Tool for Sustainable Development*, Oakland: Redefining Progress.

Transportation Research Board (1997) *Costs of Sprawl Revisited: The Evidence of Sprawl's Negative and Positive Impacts*, Washington, DC: National Academy Press.

Wheeler, S.M. (2004) *Planning for Sustainability: Creating Livable, Equitable, and Ecological Communities*, New York: Routledge.

World Commission on Environment and Development (1987) *Our Common Future*, New York: Oxford University Press.

World Resources Institute (WRI) (2007) "U.S. GHG Emissions Flow Chart." Online. Available at: <http: //cait.wri.org/figures.php?page=/US-FlowChart> (accessed March 28, 2007).

Connections

Explore "The Story of Stuff" and other related resources at Annie Lennard's site, http: //storyofstuff.org/movies. Short, animated movies drive home points about sustainability issues as well as those related to consumerism. There's also a toolkit on the site.

Calculate your footprint on the world at www.earthday.org/footprint-calculator. How many resources do you consume?

6 The "New" Local

Jeni Burnell and Rhonda Phillips

Overview

From policy making to grass-roots social movements, local thinking is enjoying a resurgence of interest and is influencing how cities are managed and planned on a variety of issues ranging from food sources to choices in shopping and spending. Examples include time banks gaining in popularity as people seek alternatives to financial exchange; mass localism and coproduction provide ways of bridging the gap between large-scale centralized control and small-scale local action; and socially responsible business and locally focused enterprise are changing how people work and the market operates. Locally focused social movements are also gaining momentum as people unite around a common cause or concern. Meanwhile, community-led development initiatives, such as the Small Change Approach, are drawing on local ingenuity to bring about positive change. With examples from the UK and the US, this chapter investigates the 'new' local landscape and how communities are revitalizing communities by focusing on local needs and opportunities.

Defining Local

In many cities and towns, people gather at local farmers' markets on a weekend to meet friends and buy locally sourced food for their busy week ahead. The growing popularity of farmers' markets is a testament to people's renewed love of local—but what is it? According to O'Riordan (2001: 37): "Definitions of the terms local, locality, localism and localization, all refer to place and the distinctiveness of that place." Social, cultural and institutional networks also help define this local identity as Meegan (see O'Riordan, 2001: 37) further explains:

> Social relations make places....This complex geography of social is dynamic, constantly developing as social relations ebb and flow and new relations are constructed. And it is this combination over time of local and wider social relations that gives places their distinctiveness.

Definitions of local are as much about community—be these of interest, culture, place/location, practice, or resilience (Hamdi, 2004)—as they are about place. Additionally, localization has been defined as an "adjustment of economic focus from the global to the local; rebuilding local economies around the meeting of local needs" (Hopkins, 2011: 51). It has also been seen as a positive way for communities to build resilience from global threats such as peak oil, climate change, economic crisis, and even terrorism (Hopkins, 2011). Its definition therefore extends to the political. Similarly, localism has become a term commonly used in UK politics to describe the process of decentralization and devolving greater control and decision making to local governments and communities. The following sections "Thinking Local" and "Acting Local" explore these ideas in more detail.

Thinking Local

Whether starting a local food cooperative or running a community renewable energy company, local actions require innovative thinking. Community assets and social capital form the foundations for these actions while the 'core' (nonmarket) economy, coproduction, time banking, and mass localism offer examples of how local thinking can transition to a larger, community scale. This section investigates some of the ideas that contribute to thinking local.

Assets and Social Capital

Generally, assets are the stock of resources people use to build livelihoods. They can be personal, household, or community-based and are often categorized as being tangible or intangible. The social anthropologist, Caroline Moser (2008), defines assets as: physical (stocks of plant equipment, infrastructure and other productive resources), financial (savings and supplies of credit), human (including education and health investment), social (reciprocity and trust embedded in social relations, social structures, rules, norms, etc.), and natural (soil, atmosphere, forests, water, minerals). Political assets have also been identified (Sanderson, 2009). Intangible assets are less grounded in empirical measurement including those which are "aspirational" and "psychological" (Moser, 2008). Hopes, dreams, and ambitions can also be defined as intangible assets (Burnell, 2013). Found locally, tangible and intangible assets are the foundations upon which people organize.

An asset-based community development approach (ABCD, or what some refer to as ABA) is the name given to community-led development that uncovers or unlocks assets within a community and in doing so generates opportunities for economic and social development. According to Mathie and Cunningham (2003: 474), an asset-based approach is built on the "premise that people in communities can organise to drive the development process themselves by identifying and mobilising existing (but often unrecognised) assets, thereby responding to and creating local economic opportunities." This approach aims to build capacity of people and households in order to reduce their vulnerability to external shocks (i.e. floods, earthquakes, etc.) and stresses (i.e. losing a job) (Sanderson, 2009). Local initiatives use a combination of asset types to spark action such as human, physical, and natural. Social assets or capital is arguably the most important in a local context as it determines how people organize to achieve a common goal. Gould (2001: 69) defines social capital as:

> the wealth of the community measured not in economic but in human terms. Its currency is relationships, networks and local partnerships. Each transaction is an investment which, over time, yields trust, reciprocity and sustainable improvements to quality of life. The assets of social capital are the skills and capacities, local knowledge and networks of each community.

Social capital is present when people work together to create, maintain, and sustain a locally focused initiative such as a community-operated bus service or simply helping a neighbor in need. The potential of social capital as a community and economic asset was identified by E.F. Schumacher (1973: 157) when he wrote in his seminal text *Small Is Beautiful: A Study of Economics as If People Mattered*: "Development does not start with goods; it starts with people and their education, organisation, and discipline." Schumacher's work helped define alternative economic thinking. His work has also inspired more recent actions with groups such as the UK's New

Economics Foundation (NEF) and the U.S.'s New Economics Institute, which has the mission of building "a New Economy that prioritizes the well-being of people and the planet" (New Economics Institute, 2013). With a focus on well-being and sustainability, both of these organizations provide alternative ways of thinking about how we measure and value productivity.

The Core Economy

The term "core economy" was coined by the U.S. economist Neva Goodwin to define and elevate the nonmarket economy. Social capital underpins the core economy as it relies on the relationships, mutuality, trust, and engagement that exists between people in their family, neighborhood and community units (Cahn: see Stephens et. al., 2008). Examples of contributions to the core economy include people volunteering to maintain a public park and grandparents sharing childcare responsibilities. These ideas are not necessarily new, but they are enjoying a comeback, as desires to establish more meaningful connection and local control over choices increases. According to Anna Coote (2010a: 3), the resources that make up the core economy are:

> embedded in the everyday lives of every individual (time, wisdom, experience, energy, knowledge, skills) and in the relationships among them (love, empathy, responsibility, care, reciprocity, teaching, and learning). They are "core" because they are central and essential to society. They underpin the market economy by raising children, caring for people who are ill, frail, and disabled, feeding families, maintaining households, and building and sustaining intimacies, friendships, social networks, and civil society.

The core economy is seen as a viable economic approach because it is based on the premise that "people helping people" has an economic value that can be exchanged. A part of the New Economic Foundation's research is exploring practical ways in which the core economy can have an impact on people's daily lives and on policy making in the UK. Coote (2010a: 4) argues for the "abundance of human resources and relationships" in the core economy to replace the "scarcity of economic resources" in the market economy as the basis for policy making. This approach would see dramatic shifts in how social, economic, and political structures operate. For example, how people prioritize and value the different types of work they achieve daily—both at home and at the "work"—would radically change. Achieving a balance between the market and nonmarket economies also requires a shift in how time is commodified. People have access to the same amount of time daily. Time, however, is not equal in terms of what people can contribute to the core economy. A single parent or person with caring responsibilities, for example, is often less rich in flexible time than those people without dependents. The core economy favors people who have greater control over how they can manage and spend their time. To overcome this inequality, Coote (2010a) suggests redistributing paid and unpaid time leading to shorter working weeks and prioritizing people's well-being in policy making. By placing a higher value on the core economy, there are greater opportunities for local actions to help define build stronger economies. If this is to be achieved, a balance needs to be struck between the market and nonmarket economies, leading to an understanding of growth, which, as E.F. Schumacher explains, is less focused on goods and more on people.

Coproduction

Coproduction was developed in the 1970s by the 2009 Nobel Prize winner for economics, Elinor Ostrom, and her team at Indiana University. The use of coproduction as a way of transforming public service delivery was further developed in the 1980s by the U.S. civil rights lawyer Edgar Cahn. The underlying principle of coproduction in service provision is that "people's needs are better met when they are involved in an equal and reciprocal relationship with professionals and others, working together to get things done" (Boyle et. al., 2010: 3). According to Cahn (see Stephens et al., 2008: 1), this requires a "fundamental partnership between the monetary economy (comprised of public, private and nonprofit sectors) and the core economy of home, family, neighbourhood, community and civil society." Within a coproduction process, service users are no longer passive recipients but active participants in the design and implementation of their care. To these ends, coproduction identifies people as assets and builds on their existing capabilities to create a positive change (Boyle et al., 2010: 9). With mutuality and reciprocity at its core, the practice of coproduction sees professionals facilitating the service process of which developing peer support networks is integral (Boyle et al., 2010: 9).

In countries such the US, Australia, and UK, coproduction has been used by the public and voluntary sectors to deliver services such as health, housing, and juvenile justice. In London, one charity is using coproduction to deliver support to people who are socially excluded because of physical disability or mental health issues. Working on behalf of the local authority, this charity provides people with vocational training along with a range of social programs such as dance movement therapy, art classes, and regular coffee mornings. Coproduction within this context is used both as a tool for monitoring and evaluation of the organization's programs, and as a way for service users to become more actively involved in program decision making. Mutual or cohousing schemes are also gaining in popularity in the UK. Coproduction is being used throughout the building process from concept design through to construction. Organizations, such as housing providers, are coproducing with older people and vulnerable groups supported yet independent housing schemes. Additionally, independent groups of like-minded people are coming together to design and build their homes. The Springhill Cohousing scheme, Stroud, was the first new-build cohousing project to be completed in the UK (see Figure 6.1). With an aim of fostering community spirit, one Springhill resident said:

> I like the fact that it is safe for my children to play outside the house in the pedestrian street ... I know everyone who walks past my door. I know the children's parents and even some of the grandchildren. It's also a big relief to eat at the common house and not have to cook every day of the week.
>
> (Architecture Centre, 2010: 2)

■ **Figure 6.1** *Springhill Cohousing Project, Stroud, UK, made up of 35 dwellings ranging from studio apartments to five-bedroom units. (Photo: C. Morton)*

The Springhill project is an example of how professionals can work collaboratively with community groups to coproduce housing solutions that meet both individual needs and the collective.

In the US, the Youth Court of the District of Columbia (YCDC) in Washington, DC is using coproduction as an alternative to mainstream juvenile justice system. According to the YCDC website, the program aims "to divert first-time youthful offenders, ages 13–17, away from the juvenile justice system and provide a meaningful alternative to the traditional adjudicatory format in juvenile cases" (YCDC, 2012). Along with support services and training opportunities, young people act as jurors in the hearings of their peers. This process is said to promote active citizenship among the jurors while building self-confidence and other essential life skills (YCDC, 2012). This coproduction process is having a positive outcome for many of the young people involved with the YCDC. The court has been operating for 16 years and has a 9% recidivism rate in comparison to 20% for the overall juvenile justice system (YCDC, 2012).

Time Banking

Edgar Cahn developed time banking as a way to "value the labour and contributions of those whom the market excluded or devalued and whose genuine work was not acknowledged or rewarded" (in Stephens et al., 2008: 2). Time banks work by making time the principal currency around which people and/or organizations exchange resources. Time transactions can be "person to person," "person to agency" (i.e. an organization or local authority using time banking to deliver aspects of its service), "agency to agency" (i.e. organizations trading resources and assets) or a combination of the three (Timebanking UK, 2012). Resources used in time banking can range from a professional sharing their skills for community benefit—such as a graphic designer creating a poster for a community fun day—to a group of parents helping run a childcare nursery. Organizations or agencies can also contribute to time banking by offering people access to their facilities—such as a local tennis court or meeting room—for a designated length of time. The time that people or organizations spend on community-focused activities is deposited into a time bank, which is often hosted by a local authority or a third-sector organization and managed by a *time broker* (Ryan-Collins et al., 2008). Members of a time bank can withdraw the time they have accrued and use it to support their needs. In Oxford, UK, a social enterprise company has set up a community time bank as part of its activities. The organization's time broker arranges swaps among members for tasks including clothing repair, ironing, DIY, and gardening. Regular coffee events are also arranged along with parents and toddlers mornings plus guitar and singing sessions (So Local, 2012). This time bank is an example of how social capital, local assets, and the core economy are benefiting people in Oxfordshire.

According to Cahn (2011), time banks are active in more than 34 countries around the world accounting for a combined membership of over 150,000 people. These schemes address a wide range of social issues from community development and regeneration through to health and criminal justice and neighborhood support. The success of time banks can be attributed to the scalability and versatility with which the time banking model can be adapted to meet local needs. A related idea is that of local currency, where shares are exchanged for services. Harkening to the days of bartering, these local currency systems provide a way to exchange value in a community. For example, the Slow Money movement is based on the following:

> In order to enhance food security, food safety and food access; improve nutrition and health; promote cultural, ecological and economic diversity; and accelerate the transition from an economy based on extraction and consumption to an economy based on preservation and restoration.
>
> (Tasch, 2008: 205)

Localism and Mass Localism

Localism in the UK is the term used by policy makers "to describe the transfer of political power toward local government and local communities, while also making central government smaller" (Hopkins, 2011: 51). With a focus on planning reform and local asset management, the Localism agenda includes new political rights and powers for communities. For example, the right for communities to express an interest in taking over a local authority service, or the right for communities to bid for ownership and management of local assets such as the village shops, library, or community center (Department for Communities and Local Government, 2011). The agenda also focuses on creating community-led neighborhood development plans that, according to the Department for Communities and Local Government (2011: 12), "allow communities, both residents, employees and business to come together ... and say where they think new houses, businesses and shops should go—and what they should look like." There is debate in the UK about whether localism has devolved enough power locally. Traditionally, the government has "found it difficult to support genuine local solutions which achieving national impact and scale" (Bunt and Harris, 2010: 3). Scaling up local initiatives without losing what made them effective in the first place can be a challenge for policy makers. When examining scalability of local projects one can ask the same question that Hamdi (2004: xviii) applies to development practice: "how much structure will be needed before the structure itself inhibits personal freedom, gets in the way of progress, destroys the very system which it is designed to service, and becomes self-serving?" When it comes to integrating local initiatives, mass localism proposes a federated approach to policy making. According to Bunt and Harris (2010: 31–32):

> Mass localism is an alternative approach to combining local action and national scale, by supporting lots of communities to develop and deliver their own solutions and to learn from each other....Instead of assuming that the best solutions need to be determined, prescribed, driven or "authorised" in some manner from the centre, policymakers should create more opportunities for communities to develop and deliver their own solutions.... Mass localism depends on a different kind of support from government and a different approach to scale.

There are five principles governing the mass localism approach (Bunt and Harris, 2010: 33–35):

1 Establish and promote a clear, measurable outcome at the start of a local initiative.
2 Presume a community capacity to innovate.
3 In the early stages, challenge and advice is more valuable than cash.
4 Identify existing barriers to participation and then remove them.
5 Do not reward activity, reward outcomes.

As a result of using the mass localism approach, project outcomes are often more organic and locally based in comparison to centrally determined and controlled initiatives. As part of this process, policy should provide a framework with inbuilt flexibility to support local needs and a network by which local innovation can be shared.

Acting Local

In recent years there has been a resurgence of interest in promoting locally focused businesses, community-owned businesses, local food systems, and related activities centered on community-led development approaches. An underlying notion is belief that a community's long-term success hinges on the idea of creating and owning its culture—economic, social, ecological, or any other dimension (Phillips et al., 2013). For example, the popularity of socially responsible business has increased remarkably over the last five to ten years, fostering a variety of organizations to serve those interested in participating. One of the most notable outcomes is the movement, in the sense that it is gaining popularity, "Local First." This reflects the notion that buying local will provide more support to the community economy more so than buying from establishments owned primarily by external interests.

Community-Owned Businesses

There are alternative ways that organizations can structure themselves to serve community-focused purposes. For example, local economic ownership can:

> improve local prosperity because these types of enterprises support the transition to a more sustainable economy (both local and global). When they're supported by government policy to focus on producing more needed goods and services that otherwise might be imported at a higher total cost (especially food, energy, and affordable housing), the economy becomes stronger and more resilient.
>
> (Phillips et al., 2013: 31)

Michael Shuman, author of *The Small-Mart Revolution* (2007), explains that local ownership fosters community prosperity because the businesses (or organizations) are vested in the community long term; fewer destructive exits occur; higher labor and environmental standards are evident with local ownership; businesses have higher success rates; and communities experience higher multiplier impacts when money is spent at locally owned establishments versus chains or big box retail.

The Main Street program of the National Trust for Historic Preservation provides information to communities interested in promoting community-owned businesses. They list four categories of these type of organizations as cooperatives (owned by its members), community-owned corporations (a for-profit corporation that integrates social enterprise principles), small ownership groups (ad hoc investor group operating as a partnership), and investment funds (community-based fund investing debt or equity in local ventures) (Bloom, 2010). Sometimes these types of businesses are established when other options are not available and community members step in to fill a need. Take for example the Clare City Bakery (a.k.a., "Cops & Doughnuts") in Michigan. After over 100 years in operation, the bakery was closing due to tough times, and the entire local police force of nine people decided to buy the business and operate it. It has been featured on CNN and the *Today Show*, and is a thriving business, encouraging other businesses to open in downtown Clare (Bloom, 2010).

Locally Focused Community-Led Development

The Small Change Approach to community-led development "explores how communities are strengthened by unlocking and using their existing assets and resources" (Burnell, 2013: 138). The approach is based on the common sense assumption: to achieve something big, start with something small and start where it counts (Hamdi, 2004). Pioneered by UK-based development practitioner and academic Nabeel Hamdi, Small Change explores how "small, practical and mostly low-budget interventions, if carefully targeted, act as catalysts for bigger, long-lasting change" (Burnell, 2013: 138). In his book, *Small Change: About the Art of Practice and Limits of Planning in Cities*, Hamdi (2004: xviii) explains that driving the Small Change process:

> is a simple, yet still challenging, premise: intelligent practice builds on the collective wisdom of people and organizations on the ground—those who think locally and act locally—which is then rationalized in ways that make a difference globally.

Small Change predominately focuses on place-based interventions and their potential to spark social and economic development. Starting with practice, Small Change draws on local innovation, creativity, and entrepreneurship to catalyze change (Burnell, 2013). The approach uses participatory methods to engage people about issues in their neighborhood. This process uncovers problems and opportunities and assists people in establishing goals and priorities for their project along with defining key resources and constraints (Hamdi, 2004). Burnell and Hamdi (2014: 160) describes this way of working as "reasoning backwards" by taking the "ideas, habits and lessons from practice in the everyday in order to inform behaviour and policy nationally, regionally and globally." The result is a placemaking approach based on action where local initiatives direct strategic policy making instead of the other way around. This way of working relies on strategic thinking about innovative partnering opportunities between the state, the market, and civil society (Hamdi, 2004). For practitioners using this approach, Hamdi (2004: xxvi) offers the following advice in the form of a code of conduct:

Ignorance is liberating
Start where you can: never say can't
Imagine first: reason later
Be reflective: waste time
Embrace serendipity: get muddled
Play games: serious games
Challenge consensus
Look for multipliers
Work backwards: move forwards
Feel good.

Small Change principles can be used in a wide range of development contexts such as community-led disaster risk reduction, housing, education, and health. Community arts organization Multistory has combined the Small Change approach with participatory arts to address development issues in the UK. This work is part of the Small Change Forum initiative, which has been developed by Multistory and the Centre for Development and Emergency Practice (CENDEP) based at Oxford Brookes University. Community artists working with a range of community groups have used digital storytelling, community journalism, and pop-up shops to explore issues such as revitalizing high streets, community-led regeneration and youth stereotyping after the 2011 riots in London. Using the arts to achieve development, education, and social impact is known as cultural action (Goldbard, 2006). Cultural action complements the Small Change approach as both processes rely on local action, imagination, and participation to catalyze change. Small Change also highlights the need for strategic thinking and creating new and innovative partnership opportunities between local stakeholders. By capturing local innovative and ingenuity, the Small Change approach offers a way in which local actions can catalyze change in communities; change designed to improve where people live and the opportunities available to them (Burnell, 2013).

Figure 6.2 *Committee of Lost Memories Pop-Up Shop in Charlemont Farm Estate, Birmingham, UK, created by community artists Katy Beinart and Torange Khonsari for Small Change creative programme. (Photo: K. Beinart)*

Local Food Systems and Community Development

Food is an essential component in community development practice. Whether fostering a local or regional food system, or trying to resolve issues of food insecurity, food is often a focal point within community development approaches. Farmers markets, community gardens, farm-to-school programs, and other food-centered initiatives have been used to foster community development processes across a spectrum of desired outcomes (Green and Phillips, 2013).

These activities usually occur in private local businesses, helping create micro-businesses as well as in social enterprises and cooperative organizations which further help to support the small farm movement. Some regions are even considering economic development strategies of food clusters to promote specialty food businesses and supporting programs (Green and Phillips 2013).

Food systems and food-based enterprises represent a significant contribution to an environmentally sustainable economy as well as inclusiveness and social components that in turn help foster a more resilient economy (Phillips et al., 2013). The interconnection between food and community development can be discerned clearly when placed in the context of a cooperative structure. Cooperatives are noted for serving the community of members, and beyond, with principles of practice to benefit those involved. Principles of practice underlie cooperative efforts, and include such areas as democratic member control, autonomy, and independence. The International Cooperative Alliance defines a cooperative as "an autonomous association of persons united voluntarily to meet their common economic, social, and cultural needs and aspirations through a jointly-owned and democratically-controlled enterprise" (ICA, 1995). While other forms of business enterprise such as private

> corporations, sole proprietorships, and/or partnerships where the owners might never be the patrons of the business (those who use its products and/or services), members of cooperatives come together both to control the cooperatives as member-owners as well as to produce collective benefits for themselves as patrons.
>
> Gonzales and Phillips, 2013: 6

BOX 6.1 THE HISTORY OF SLOW FOOD AND A DELICIOUS REVOLUTION: HOW GRANDMA'S PASTA CHANGED THE WORLD

It was 1986 and a McDonald's franchise was expanding its operations in the heart of Rome—adjacent to the Spanish Steps in Piazza di Spagna. Italian journalist Carlo Petrini was outraged. What would fast food do to the food culture of Rome? Would it threaten the local *trattorias* and *osterias*, the local dining establishments of the working class?

Petrini rallied his allies to take a stand against this intrusion of global industrialized food, and the social and culinary costs of homogenized eating. However, instead of picketing with signs, he armed the protestors with bowls of penne. Defiantly they declared, "We don't want fast food ... we want slow food!" The idea of Slow Food was born.

Soon after, Petrini realized that in order to keep our alternative food choices alive, it was crucial for an "eco-gastronomic" movement to exist—one that was concerned with environmental sustainability (eco), and the study of culture and food (gastronomy), to truly draw the connection between the plate and the planet.

With preservation of taste at the forefront, he wanted to support and protect small growers and artisanal producers, safeguard the environment, and promote biodiversity. Three years later, on December 10, 1989, the Slow Food movement was formalized with an event in Paris. The Slow Food Manifesto, drafted by co-founder Folco Portinari and endorsed by delegates from 15 countries, condemned the "fast life" and its implications for culture and society:

We are enslaved by speed and have all succumbed to the same insidious virus: Fast Life, which disrupts our habits, pervades the privacy of our homes and forces us to eat Fast Foods. ... A firm defense of quiet material pleasure is the only way to oppose the universal folly of Fast Life. ... May suitable doses of guaranteed sensual pleasure and slow, long-lasting enjoyment preserve us from the contagion of the multitude who mistake frenzy for efficiency. Our defense should begin at the table with Slow Food. Let us rediscover the flavors and savors of regional cooking and banish the degrading effects of Fast Food.

(Official Slow Food Manifesto, as published in *Slow Food: A Case for Taste*, 2001)

The concept of conviviality is at the heart of the Slow Food movement: taking pleasure in the processes of cooking, eating, and sharing meals with others. As a result, Slow Food's structure is decentralized: each chapter (or "convivium") has a leader who is responsible for promoting local artisans, local farmers, and local flavors through regional events, school gardens, advocacy, social gatherings, and farmers' markets.

Education is the first step in gaining the appreciation that can lead to preservation. In 2004, Slow Food co-founded the University of Gastronomic Sciences in Pollenzo, Italy, offering undergraduate and master's degrees in food studies. The University of New Hampshire, inspired by a visit from Petrini, launched the first U.S.-based "Eco-Gastronomy" major in 2008.

Today, Slow Food has over 150,000 members and is active in more than 150 countries, including national associations in Italy, the US, Germany, and Japan. There are more than 200 chapters and 2,000 food communities in the United States alone.

Every two years, Slow Food hosts the world's largest food and wine fair, *Salone del Gusto*, in conjunction with the Terra Madre world meeting of food communities, drawing over 250,000 visitors combined. Other international events include Cheese, a biennial cheese fair in Bra, and Slow Fish, a Genoan fish festival.

Richard McCarthy
Executive Director, Slow Food USA, www.SlowFoodusa.org

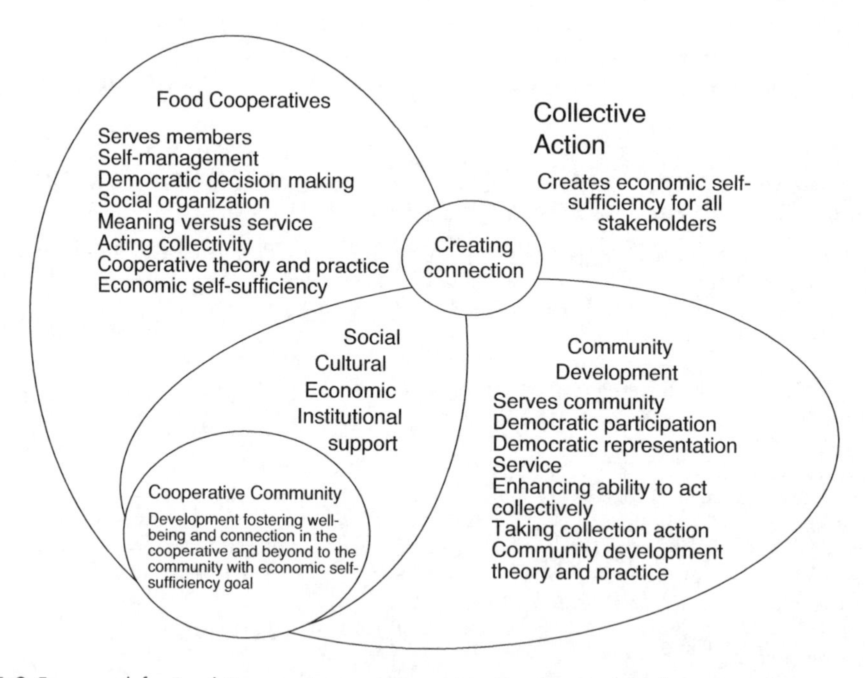

Figure 6.3 *Framework for Food Cooperatives and Community Development. (Phillips 2012: 195)*

Within community development, there exist subjective factors helping gauge involvement and relations. These can be thought of as capacity factors, and include such aspects as: positive attitude toward development and activities; participation in community activities; effective implementation of plans; unified vision and mission; and ability to work together and avoid factionalism (Pittman et al., 2009: 84). Perhaps more importantly is the impact on economic self-sufficiency and having local decision making; this is a central component of any successful community development approach, program, or policy (Phillips 2012). Figure 6.3 illustrates the importance of economic self-sufficiency, and the role of community development in collective action—all in the framework of a food cooperative. Note that the concept of cooperative community results from placing emphasis on economic self-sufficiency and the role of the cooperative in community (Nadeau and Wilson, 2001).The interaction is both supportive and effective in helping foster local economic self-sufficiency and the underlying principles of community development practice.

Conclusion

Assets, social capital, and the core economy offer alternative ways of thinking about growth and the human value associated with economics. Time banking, coproduction, and mass localism are examples of how local actions can be scaled up within frameworks that support local ingenuity and the unique character of a place and its community. Community-owned businesses, the Small Change Approach, and local food systems are all ways in which communities can foster more local input and decision making into community development outcomes. All these approaches can help communities' revitalization efforts by creating locally focused outcomes.

CASE STUDY: CREATING SMALL CHANGE IN STIRCHLEY PARK, BIRMINGHAM, UK

In 2011, the UK arts-led community interest company, Place Prospectors CIC, began a community arts program in Stirchley, South Birmingham, UK. Stirchley is predominately a residential area with a declining shopping street that also acts as one of the main artery routes into Birmingham city centre. Place Prospectors' work aimed to help define the cultural values and assets of the neighborhood and explore opportunities for community input into regeneration. One of their initiatives was the Making of Stirchley Park (MoSP). This brought together a group of local residents to investigate creative ways of improving the neighborhoods' underused and largely neglected park. Initial discussions with residents highlighted that poor access, seclusion, and a lack of adequate signage needed to be addressed if the park was to feel safer and therefore be better used. In response, Place Prospectors created a number of events and art installations in the space with the aim of raising public awareness about the park (see below).

The artwork "Cutting Through" explored the park's seclusion and lack of local visibility by creating a "puncture mark" through an adjoining vacant building. A large circular photograph of the park was hung on the building thereby revealing the park behind to passing motorists and pedestrians using the busy main road. During community events, red carpets were laid along the long, narrow park entrances making them more inviting—and intriguing—to passers-by. A local graffiti artist called Title was also commissioned to create a high-quality artwork on one of vast blank shop walls that bordered the park. This aimed to raise the standard of graffiti in the park while also inspiring young graffiti enthusiasts in the neighborhood. As one resident commented; "I think the graffiti is a fine idea as it gets people using the wall for some positive purpose…it is these small changes which encourage others to walk through the park."

Beyond their creative remit, Place Prospectors played an important role in helping establish the Friends of Stirchley Park (the Friends) group. Working in collaboration with the local authority, the Friends became instrumental to the ongoing improvements and management of the park. As such, new notice boards, seats, trees, and signage have been introduced along with a program of community events. From a Small Change perspective, Place Prospectors work in Stirchley Park acted as a catalyst for the park's continued physical improvements. Within this context, creative action gave local residents an idea about what could be achieved in their neighborhood and helped build a network of people committed to improving their surroundings.

Several principles can be taken from Place Prospector's work in Stirchley. There is a need for constant and consistent community engagement if local projects are going to inspire long-term social and economic impacts for communities. An inclusive participatory process and creating innovative partnering and governance opportunities is important if neighbourhoods are going to be managed more equitably. Action is also imperative to this process. Small yet strategic actions have the potential to spark change in communities including shifting people's perceptions about what they can achieve in their neighbourhood. As one Stirchley resident said: "Now I've seen it, I get what you're trying to do" (Larkinson and Murray, 2012: 6). The development process needs people with imagination and vision to share ideas of change. This process creates a platform for discussion between the various stakeholders in the community. If outside professionals are involved in this process, they often bring an independence and healthy ignorance that can challenge norms and deep-rooted assumptions about place. While community-led initiatives created by local people have the potential to sustain change through—as some argue—a more genuine participatory process (Eversole, 2010). However the process is carried out, community development should be accessible as well as fun, honest, challenging, and brave. Questions arise when assessing Place Prospectors' work in Stirchley. What extent has the MoSP helped define Stirchley's cultural values and create opportunities for community input into the wider regeneration process? How has the MoSP project created new partnership opportunities between civil society (such as the Friends group), the local authority and the business sector in Stirchley? Finally, how has Place Prospectors' work federated by influencing other professionals' work in the sector and the approach of similar community-led regeneration projects?

Source

J. Burnell and N. Hamdi (2014) "Small Change: The Making of Stirchley Park," *Community Development Journal*, 49 (1): 159–166.

Keywords

Local, localism, localization, assets, asset-based development, social capital, core economy, time banking, coproduction, mass localism, socially responsible business, locally focused enterprise, small change approach.

Review Questions

1. What are the differences between a traditional economy, and a "core" economy?
2. Why does the idea of "local" focus resonate with some, and not others?
3. What are current and potential future impacts of locally focused policies and activities in terms of community development outcome?

Bibliography

Architecture Centre (2010) "Springhill Cohousing, Stroud, Gloucestershire." Available at: http: //bit.ly/1fSyE8u (accessed October 22, 2013).

Bloom, J. (2010) "Main Street Story of the Week." Available at www.preservationnation.org (accessed May 25, 2014).

Burnell, J. and Hamdi, N. (2014) "Small Change: The Making of Stirchley Park," Community Development Journal, 49 (1): 159–166.

Burnell, J. (2013) "Small Change: Understanding cultural action as a resource for unlocking assets and building resilience in communities," Community Development Journal, 48(1), 134–150.

Bunt, L. and Harris, M. (2010) "Mass Localism: A way to help small communities solve big social challenges" Discussion paper. London: NESTA. Available at: http: //bit.ly/dle1Q0 (accessed July 4, 2013).

Boyle, D., Coote, A., Sherwood, C., and Slay, J., (2010) "Right here, Right Now: Taking co-production into the mainstream," London: NESTA and NEF. Available at: http: //bit.ly/bBO2eS (accessed August 9, 2013).

Cahn, E. (2011) "Time banking: An idea whose time has come?" November, 17. USA: *Yes! Magazine*. Available at: http: //bit.ly/tCKcrp (accessed July 4, 2013).

Coote, A. (2010a) "The Great Transition: Social justice and the core economy," London: New Economics Foundation. Available at: http: //bit.ly/1bldbDT (accessed July 25, 2013).

Coote, A. (2010b) "Ten Big Questions about the Big Society," London: New Economics Foundation. Available at: http: //bit.ly/166Wt4m (accessed July 31, 2013).

Department for Communities and Local Government (2011) "A plain English guide to the Localism Act," London: Department for Communities and Local Government. Available at: http: //bit.ly/YdWsea (accessed July 27, 2013).

Eversole, R. (2010) "Remaking participation: challenges for community development practice," *Community Development Journal*, 47 (1): 29–41.

Goldbard, A. (2006) *New Creative Community: The Art of Cultural Development*, New Village Press: Canada.

Gonzales, V. and Phillips, R. (2013) *Cooperatives and Community Development*, London: Routledge.

Gould, H. (2001) "Culture and Social Capital" in: F. Matarasso (ed.) "Recognising Culture: A series of briefing papers on culture and development," UK: Comedia in partnership with the Department of Canadian Heritage and UNESCO with support from the World Bank. Available at: http: //bit.ly/19zflhJ (accessed July 1, 2013).

Green, G.P. and Phillips, R. (2013) *Local Foods and Community Development*, London: Routledge.

Hamdi, N. (2004) *Small Change: About the Art of Practice and the Limits of Planning in Cities*, Earthscan: London.

Hopkins, R. (2011) *The Transition Companion: Making your community more resilient in uncertain times*, Chelsea Green Publishing: Vermont.

International Cooperative Association (ICA) (1995) Statement on the Co-op Identity. Available at: www.ica.coop/coop/principles.html (accessed September 5, 2010).

Larkinson, E. and Murray, J. (2012) Stirchley Prospects Evaluation Report (edited version).Report to funder. April 18, 2013.

Mathie, A. and Cunningham, G. (2003) "From clients to citizens: Asset-based community development as a strategy for community-driven development," *Development in Practice*, 13(3): 474–486.

Meegan, R. (1995) "Local Worlds" in J. Allen and D. Massey (eds.) *Geographical Worlds*, Oxford University Press: Oxford.

Moser, C.O.N. (2008) "Assets and Livelihoods: A Framework for Asset-Based Social Policy" in A.A. Dani & C.O.N. Moser, eds. (2008) *Assets, Livelihoods and Social Policy*. Available at: http: //bit.ly/18TJeK6 (accessed August 1, 2013).

Nadeau, E. G., & Wilson, C. (2001). New generation cooperatives and cooperative community development. In C. Merrett & N. Walzer (eds.), *A Cooperative Approach to Local Economic Development*. Westport, CT: Quorom Books.

New Economics Foundation (2013) Downloaded September 25, 2013, www.neweconomics.org

New Economics Institute (2013) Downloaded October 8, 2013, www.neweconomy.net

O'Riordan, T. (ed.) *Globalism, Localism and Identity*, London: Earthscan.

Phillips, R. (2012) Food Cooperatives as Community-level Self-help and Development. *International Journal of Self-Help and Self-Care*, 6(2): 189–203.

Phillips, R., Seifer, B.F., Antczak, E. (2013) *Sustainable Communities: Creating a Durable Local Economy*. Abingdon, England: Routledge.

Pittman, R., Pittman, E., Phillips, R., and Cangelosi, J. (2009) "The Community and Economic Development Chain: Validating the Links between CED Processes and Outcomes," *Community Development*, 40(1): 80–93.

PSMW (n.d.) "Time banking: Public service delivery with time currency," Wales: Public Service Management Wales. Available at: http: //bit.ly/1c3dG5T (accessed August 5, 2013).

Ryan-Collins, J., Stephens, L., and Coote, A. (2008) "The new wealth of time: How timebanking helps people build better public services," London: New Economics Foundation. Available at: http: //bit.ly/14bi4qz (accessed August 5, 2013).

Sanderson, D. (2009) "Integrating development and disaster management concepts to reduce vulnerability in low income urban settlements," Ph.D. thesis, Oxford Brookes University.

Schumacher, E.F. (1973) *Small is beautiful: A study of economics as if people mattered*, Blond & Briggs: London.

So Local (2012) "Publicising and growing the time bank," Oxford: So Local Ltd. Available at: www.solocal.info (accessed August 13, 2013).

Stephens, L., Ryan-Collins, J., and Boyle, D. (2008) "Co-production: A Manifesto for growing the core economy," London: New Economics Foundation. Available at: http: //bit.ly/13fekE9 (accessed August 5, 2013).

Tasch, W. (2010) *Inquiries into the Nature of Slow Money: Investing as if food, farms, and fertility mattered*. White River Junction, Vermont: Chelsea Green.

Timebanking UK (2012) "What is timebanking." Stroud: United Kingdom. Available at: http: //bit.ly/A95sW7 (accessed August 5, 2013).

YCDC (2012) "How does youth court work?" [online] US: Youth Court of the District of Columbia. Available at: http: //bit.ly/12wTmpJ (accessed August 12, 2013).

Wicks, J. (n.d.) "Local Living Economics: The New Movement for Responsible Business," Available at: http: //bit.ly/12NEiyj (accessed August 25, 2013).

Connections

Explore the ideas behind being a "localist" at the Business Alliance for Local Living Economies, http: //bealocalist.org

Learn about Transition Towns and initiatives globally at: www.transitionus.org

Multistory: www.multistory.org.uk

Centre for Development and Emergency Practice (CENDEP): www.architecture.brookes.ac.uk/research/cendep

Explore approaches and perspectives on alternative ways of thinking about economies at the New Economics Foundation: www.neweconomics.org

Learn about Timebanking UK: www.timebanking.org

Time Dollar Youth Court, Washington, DC: www.youthcourtofdc.org

7 Community Development Practice

John W. Vincent II

Overview

Community development is a wide-ranging discipline that encompasses economic development. Community development is a process whereby all residents are involved in the process of community change and improvement. Success in community development leads to more success in economic development. A set of values and beliefs and ethical standards has been developed that should always guide the community development process.

Introduction

As discussed in previous chapters, community development is a broad subject incorporating many different disciplines. As a process and an outcome, it is inextricably linked to economic development, which can also be defined as a process and an outcome. Because of its multidisciplinary nature, recent research on establishing a theoretical foundation for community development has drawn from different fields including sociology, economics, psychology, and many others. The focus of this chapter is on the practice of community development. It provides a broad overview of community development practice as a foundation for subsequent chapters dealing with specific aspects of community development.

First Step: Define The Community

As noted in Chapter 1, a community is not necessarily defined by geographical or legal boundaries. A community could involve interaction among people with common interests that live in a particular area. Or it could involve a collection of people with common social, economic, political, or other interests regardless of residency.

Professional community development (CD) practice could involve a community that comprises a single city, town, or village. On the other hand, it may involve working with a community comprising several cities, towns, and villages, or a community of counties that decide to form a regional development organization. When organizing a community development initiative, it is important to carefully define the community and who should be included in the process. The community should not be too narrowly defined.

Looking to the future, CD professionals may find themselves consulting in unique communities such as "virtual communities." With the great advances in computers and electronic communications, it is possible for individuals with common interests and concerns to have electronic communities without legal or physical boundaries. Their interactions, business dealings, and common concerns may result in an interdependency that requires support from a professional developer. This brings a whole set of unique problems and opportunities to the CD professional.

BOX 7.1 "PEACE CORPS FOR GEEKS" – TECHNOLOGY FOR COMMUNITY GOVERNANCE

Code for America, a nonprofit organization, has launched a program to help cities by connecting tech experts from Apple, Yahoo!, Google, and other companies in fellowships to address specific issues. Teams are constructed of fellows from the companies and assigned to spend a month in a city, listening to residents and officials discuss the challenge at hand. These range from transportation access and housing needs to food distributions to low-income neighborhoods. After the visit, the teams return to the Code for America's headquarters to develop a web-based program to address the challenge and help solve the problem.

Get the full scoop by watching an interview about the program at: www.pbs.org/newshour/rundown/2013/05/a-high-tech-solution-for-a-neighborhood-problem.html

The Editors

Practicing Community Development

As noted in Chapter 1, community development is both a process and an outcome. The practice of community development can be described as managing community change that involves citizens in a dialogue on issues to decide what must be done (their shared vision of the future) and then involves them in doing it.[1]

BOX 7.2 COMMUNITY AND ECONOMIC DEVELOPMENT TRAINING AND CERTIFICATION

As community and economic development have evolved into recognized academic and professional disciplines, training and certification programs in the field have grown as well. Thousands of local community and economic development training programs offer regional and local governments and organizations across the globe. Many universities offer undergraduate and graduate degrees in community and/or economic development. In the US, two major training programs are the Economic Development Institute offered through the International Economic Development Council at the University of Oklahoma (http: //edi.ou.edu) and the Community Development Institute offered through several universities and agencies around the country under the auspices of the Community Development Council (www.cdcouncil.com). The Certified Economic Developer (CEcD) program is administered by the International Economic Development Council (www.iedconline.org). The Community Development Council offers certification in community development to full-time professionals (PCED, the Professional Community and Economic Developer designation), and to community volunteers (CEDP, Certified Community Development Partner). There is also the American Institute of Certified Planners, with the AICP designation (www.planning.org/aicp). This designation is for the wide range of planning interests, although the continuing education component can be concentrated in housing and community development or economic development. See for example the Divisions of the American Planning Association, for economic development—https: //www.planning.org/divisions/economic; and for housing and community development—https: //www.planning.org/divisions/housing.

It should be noted there are several journals and other associations in community development, including the International Association for Community Development, a global community of community developers and organizations. See their site for a wealth of information about projects and approaches around the world: www.iacdglobal.org. *Community Development Journal*, published by Oxford Journals is allied with the organization and is an excellent source of research in the discipline, access the journal at: http: //cdj.oxfordjournals.org. The Community Development Society, based in the US, is an organization of scholars and practitioners focused on promoting community development. Their journal is *Community Development*, published by Taylor & Francis. See www.comm-dev.org for more info on the society, and to access the journal and other publications, including two books series on community development.

The Editors

Critical issues addressed in the CD process include: jobs and economic development (business attraction, expansion and retention, and new business development); education and workforce development; infrastructure development and improvement; quality of life, culture, and recreation; social issues such as housing, crime, teen pregnancy, and substance abuse; leadership development; the quality of governmental services; community image and marketing; and tourism development.

BOX 7.3 CROWDSOURCING FOR THE COMMON GOOD: A COMMUNITY DEVELOPMENT APPROACH

The advent of the internet allowed individuals separated by geography and by cultural differences to form robust communities online. Virtual societies now form around nearly any issue, and people with common interests have been able to exchange information and forge unlikely social ties in ways previously unimaginable. Innovations such as Wikipedia and Facebook further opened up the internet to users wishing to exchange knowledge and opinions with a broader community. And now, spurred by more participatory, user-oriented Web 2.0 applications, the phenomenon of "crowdsourcing," or engaging the web community in a particular project or problem, has become a popular means of online interaction.

Essentially, crowdsourcing describes the phenomenon of harnessing the collective knowledge of everyday internet users for a purpose that would otherwise be left to experts. Crowdsourcing can be used to mine established or original data and solutions, solicit feedback, improve organizational transparency and build consensus through a more interactive process, or even harness the labor of the crowd (for example, Amazon's Mechanical Turk). While crowdsourcing has been used by creative businesses, how can this strategy translate to the community development sector, including government and nonprofits?

There have been several successful attempts to employ crowdsourcing in a community development context. For example, local governments in San Francisco and Portland use a system called ParkScan to monitor and manage public space. The site allows users to post issues, which park department employees address in public responses. Interested citizens can also use the map-based application to locate and view detailed information about their neighborhood parks. Concerned citizens in the Broadmoor neighborhood in New Orleans created a tool on ning.com to crowdsource ideas for rebuilding after Hurricane Katrina. While activity on the site has waned over time, the 136 members offered useful input on neighborhood needs and preferences in the site's online forums. In New York City, the site Change by Us NYC collects resident ideas, such as infrastructure improvements and tree planting, and encourages users to create and promote project teams. The site also provides links to funding and other resources. A developer in Bristol, Connecticut, used the crowdsourcing concept to solicit ideas and even votes from citizens living or working within an hour radius on the preferred type of redevelopment for a 17-acre site of a former shopping mall. The Bristol Rising initiative used an online community platform with a voting tool to facilitate idea sharing with a strong visual component.

Crowdsourcing has also been used by academics and nonprofits for disaster response after the 2010 earthquake in Haiti. Sites like CrisisCommons, OpenStreetMap, Ushahidi, and GeoCommons collected on-the-ground information on survivors, shelters, and structural damage. In addition, nonprofits have used crowdsourcing during and after the 2011 Egyptian revolution through mapping geo-referenced, real-time Twitter feeds (HyperCities Egypt) and by creating an "interactive documentary" of user-uploaded content (18 Days in Egypt).

Crowdsourcing has many potential benefits. First, it derives creative solutions from a community—which may not have been conceived of by experts—virtually free of cost. It promotes communication within the community and with decision makers in an interactive and transparent manner. It also makes information and discourse constantly and conveniently available, unlike traditional, prescheduled public meetings. Crowdsourcing allows differing levels of participation based on individuals' levels of interest or engagement. Finally, crowdsourcing can increase buy-in through interactive commentary and voting. There are drawbacks with crowdsourcing, however. Most for-profit crowdsourcing uses payment to increase participation, which may not be feasible for government and nonprofit organizations. The fidelity of information may be poor if the crowd is not sufficiently large and representative. It also may be difficult to attract enough people to have an impact on a project with a small geographic scope. Participants may also self-select. In particular, because of the "digital divide" that limits access to the internet among poor, minority, and other disenfranchised populations, the crowd may not correspond to the demographic profile of the actual community and may trend more affluent and less diverse. Finally, technical expertise is needed to create an online interface for crowdsourcing and to interpret the feedback.

Clearly, there are strengths and weaknesses inherent in a crowdsourcing approach to community development. However, the innovations of the private sector and the communities noted should not be discounted. Crowdsourcing may prove to be a valuable tool for decision making among cash-strapped government and nonprofit entities. A crowdsourcing technique that promotes engagement and creative problem solving can empower citizens and maximize benefits to the community.

Ann Carpenter, Federal Reserve Bank of Atlanta
Partners' Update, Jan./Feb. 2012
www.frbatlanta.org/pubs/partnersupdate/12no1_crowdsourcing.cfm
(reprinted with permission of the author)

Sources

Ball, Matt. 2011. How do crowdsourcing, the Internet of Things and Big Data converge on geospatial technology? *Spatial Sustain.* www.sensysmag.com

Batty, Michael, Andrew Hudson-Smith, Richard Milton, and Andrew Crooks. 2010. Map mashups, Web 2.0 and the GIS revolution. *Annals of GIS* 16 (1): 1–13.

Brabham, Daren C. 2009. Crowdsourcing the Public Participation Process for Planning Projects. *Planning Theory* 8 (3): 242–262.

Bugs, Geisa, Carlos Granell, Oscar Fonts, Joaquin Huerta, and Marco Painho. 2010. An assessment of Public Participation GIS and Web 2.0 technologies in urban planning practice in Canela, Brazil. *Cities* 27 (3): 172–181.

Zook, Matthew, Mark Graham, Taylor Shelton, and Sean Gorman. 2010. Volunteered Geographic Information and Crowdsourcing Disaster Relief: A Case Study of the Haitian Earthquake. *World Medical & Health Policy* 2 (2): 7–33.

The Professional Community Developer's Role

The king (or queen) is dead! Plato's philosopher-king, described in Plato's *Republic*, may have been a good person, but his benevolent dictatorship is inappropriate in the community environment. Assuming control and telling people what they should do is a subtle form of dictatorship, regardless of the intent. Many consultants assume this role in such a sophisticated advisory manner that clients do not even realize that they are relegating their rights and personal responsibilities. Further, this type of consulting does little to foster local leadership development and self-sufficiency. To be successful, CD initiatives must originate from the citizens themselves. The people who live in a community must have "ownership" of the process and the will to work toward bringing about necessary change. Otherwise, the process will either fail or result in the creation of a plan of action that is never fully carried out.

A professional community developer's primary role is to facilitate the CD process. He or she is an agent of change, an individual in an advisory role focused on helping citizens assess their current situation, identify critical issues and options for the future, weigh those options, create a shared vision of the future, and make informed decisions about what they will do to make that vision a reality. If a professional community developer finds dependent citizens, he or she works to empower them so that they come to realize that they are not helpless and that their actions can make a positive difference in their communities. The professional developer moves citizens from a state of helplessness and dependency to one of self-directed interdependence. For a community developer, success is being "no longer needed." The professional works him or herself out of a job.

Professional community developers as defined in this chapter would include consultants trained in the field, but also local residents whose jobs involve community development activities such as economic development, planning, community health, and a host of other fields. The community development professional could also be a volunteer who is taking a lead role in the community development process. People who spend a large portion of their time in community development activities—whether paid or volunteer—can be considered professional community developers. As discussed below, professional community developers can obtain national certification as an indication of their experience and knowledge in the field. Whether considered a professional community developer or not, the more local citizens that understand the principles and practice of community development, the more successful the process and outcome will be.

Inclusiveness

CD is based on the idea that all people are important and should have a voice in community decisions, have potential to contribute, resources to share, and a responsibility for community action and outcomes. Citizens are entitled to make informed decisions about the factors influencing their community. The process is always open and transparent. Therefore, CD is an inclusive approach to working with people who participate in the process to the extent that they are capable and interested. The professional community developer uses many techniques that offer citizens an opportunity to be part of the development process. These techniques may include responding to a survey, being interviewed as part of a community assessment, and participating in a public meeting, task force, subcommittee or strategic planning committee. These actions greatly increase local citizen involvement and, thus, the potential to improve their quality of life.

CD is guided by a set of values and beliefs. These can be thought of as mental tools that professional community developers use to provide guidance and direction for their work (see Box 7.4).

Professional community developers believe in people being proactive within the framework outlined above. These values and beliefs help the process of decision making to be open, inclusive, interactive, and focused on the well-being of the entire community, not just one segment.

BOX 7.4 COMMUNITY DEVELOPMENT VALUES AND BELIEFS

1 People have the right to participate in decisions that affect them.
2 People have the right to strive to create the environment they desire.
3 People have the right to make informed decisions and reject or modify externally imposed conditions.
4 Participatory democracy is the best method of conducting community business.
5 Maximizing human interaction in a community will increase the potential for positive development.
6 Creating a community dialogue and interaction among citizens will motivate citizens to work on behalf of their community.
7 Ownership of the process and commitment for action is created when people interact to create a strategic community development plan.
8 The focus of CD is cultivating people's ability to independently and effectively deal with the critical issues in their community.

Community Development Principles of Practice

Over time, a set of principles has evolved that also act as guides to CD practice. These principles are:

1. *Self-help and self-responsibility are required for successful development.* No one knows better about what must be done or is more committed to change than those who live in a community. It is impossible for CD consultants to do all that is necessary for a community to realize its full potential. They do not possess all the knowledge and skills needed by their clients' community. The CD consultant's role is to organize citizens so that they realize their power, capabilities, and potential during the process of change.
2. *Participation in public decision making should be free and open.* While not everyone can attend or participate equally in every CD activity, citizens' opinions, ideas, and support can be sought throughout the process. Participation can come in many forms including the completion of surveys, volunteering for a particular project, attending a public meeting, or serving on a committee or task force.
3. *Broad representation and increased breadth of perspective and understanding are conditions that are conducive to effective CD.* The CD professional strives to organize a leadership group that is representative of the community and its stakeholders. Individuals representing all major groups within the community should be offered membership in this group, and all their constituents should be encouraged to participate to some extent in the many development activities created in support of their shared vision.
4. *Methods that produce accurate information should be used to assess the community, to identify critical issues, strengths, weaknesses, opportunities, and threats (SWOT analysis).* This assessment, or environmental scan, can result in the creation of a community profile that details all information collected about the community. It may involve a literature review, interviews with key informants, focus group meetings, citizen and business opinion surveys, and a review of multiple statistical and demographic data resources. Assessment results support creation of a strategic development plan and, eventually, marketing material.
5. *Understanding and general agreement (consensus) is the basis for community change.* Citizens need to make informed decisions. Community dialogue and assessment feeds that educational process. When making decisions, consensus should be sought. There is an important distinction between consensus and compromise. Compromise may not always result in the best alternative being selected. Usually one group gives up something important to get one thing it wants, and another group does the same. This usually results in neither side getting what they really want, nor the best alternative for everyone. In compromise, both sides have win-win and lose-lose outcomes.

   Consensus seeking, on the other hand, supports creation of an alternative that everyone can support, one that was developed through an open dialogue on options and is focused on producing the greatest good and positive outcomes. The process involves having all opinions heard, options discussed (including positives and negatives of each option), and developing a course of action that best addresses the identified critical issues. It is problem–solution focused versus group/individual focused. Consensus seeks win-win results that produce positive outcomes for all citizens.
6. *All individuals have the right to be heard in open discussion whether in agreement or disagreement with community norms.* Participants in community organizations owe it to

themselves to hear and consider all ideas and opinions. When everyone's views and ideas are heard, it provides more information for citizens to consider when making decisions. Often ideas from different individuals can be synergistically combined to produce even better solutions. However, individual rights should be exercised with respect and not be abused to the point of disrupting the activities and decisions of the majority of citizens. Individuals do not have a right to filibuster.

7 *All citizens may participate in creating and recreating their community.* The wonderful thing about community development is that there is a role for everyone. Young and old, rich and poor, highly educated and modestly-to-uneducated individuals, and citizens of all races and cultures can create or join a project in support of the shared community vision. Those with gardening as a hobby might get involved in community beautification. Others may have information technology skills that can be utilized in developing a website and creating community marketing material. Still others may have leadership and organizational skills that can be focused on organizing and leading groups. CD professionals and citizens should always be looking for new individuals to involve in the process and align their experience, knowledge, skills, and interests to the work that needs to be done.

8 *With the right of participation comes the responsibility to respect others and their views.* Every citizen should be treated with respect and kindness. The CD professional must work diligently to facilitate the process so that individuals feel positive about their own participation while respecting the rights of others. It is possible to be both fair and firm. Personal attacks or personalizing issues and ideas is counterproductive. It forces individuals into defensive posturing and conflict that disrupts progress.

9 *Disagreement needs to be focused on issues and solutions, not on personalities or personal or political power.* Disagreement is not a bad thing. It is a normal result of human interaction. As thinking individuals, humans naturally have opinions and develop alternatives based on their own life experiences, education, values, culture, and beliefs. CD professionals focus the group on what can be done instead of what cannot be done and on areas of agreement rather than on areas of disagreement. This focus avoids getting bogged down trying to resolve irreconcilable differences. The focus of CD activities needs to be on problem identification and solutions rather than on who did what to whom when or whose opinion or idea is the most important. The importance of ideas is not in their origins but in their utility for addressing community concerns.

10 *Trust is essential for effective working relationships and must be developed within the community before it can reach its full potential.* An old cliché states that trust must be earned before it can be given. There is generally a lack of trust among the very community groups that must work together in order for the community to be successful. This is particularly true in communities where there are many problems. Trust will only be developed if the preceding principles are practiced and supported within the community. Development of trust will take some time as the group forms and moves through a normal developmental cycle.

These principles are very practical and, when used, will protect a working group from selective participation that can create mistrust, rumors, misinformation, destruction of worthwhile efforts, and loss of key participants. Increasing the breadth of representation helps ensure that many different points of view are heard. This leads to the ability to implement decisions. General agreement also leads to a commitment to implement changes and a positive working relationship that supports long-term development initiatives. By hearing everyone, even those who disagree, it is possible to assemble a broad base of information. Sometimes those who disagree with the majority or general opinion may be right; their "weird" ideas or opinions may foster creative thinking in the group and innovative approaches to solving long-term or difficult problems.

The Community Development Process

The central theme of CD is that people, in an open and free environment, can think and work together to fashion their own future. However, when communities face serious problems, there is a tendency for citizens to feel frustrated and helpless. Anger develops and is often focused on each other. Certainly someone must have caused these problems. Differences among residents are exaggerated and passions run high. The community tends to splinter (e.g. rich vs. poor, white vs. black, long-term residents vs. newer residents, residents vs. non-resident property owners and business operators, and voters vs. elected officials). The CD professional often begins his or her work in a very contentious environment.

BOX 7.5 A PLACE-BASED COMMUNITY ECONOMIC DEVELOPMENT APPROACH VIA THE RESOURCE CENTER FOR RAZA PLANNING

The Resource Center for Raza Planning (RCRP) established itself in 1996 with the mission to "promote the sustainability and survivability" of traditional communities in New Mexico. Raza has many meanings around "people," and "community"; it is associated with ideas around advocacy for Hispanic communities. With its place-based commitment, the Center works alongside communities and builds long-term relationships. Applying planning processes and techniques, the Center facilitates "active community participation and just and effective decision making." Among its stated principles, Raza Planning advocates planning and development policies that "ensure economic equality political self representation, cultural preservation and ecological conservation for social and environmental harmony."

The Resource Center for Raza Planning, located in the School of Architecture and Planning at the University of New Mexico, worked over several years with a local community development corporation and was instrumental in the building of a small business incubator and commercial kitchen. This community economic development strategy was designed as a mechanism to promote self-employment as a strategy for "household wealth and community health" while circulating dollars through the community and providing access to local goods and services.

In the same community, southwest and adjacent to Albuquerque, Raza Planning recognized the significance of impending infrastructure development. RCRP became involved in shepherding the participation process to ensure the active engagement of community voices in the design and construction of water, sewer, and drainage systems as well as roadways. Through these RCRP-led processes, community concerns such as scale of development, cultural landscape preservation, small business enhancement, environmental quality and safety all became integrated into design and construction decisions.

The Resource Center for Raza Planning continues working with communities on issues of cultural and economic survival. More recently, RCRP is working with communities in Northern New Mexico on community economic profiles, conceptual visions, and economic development strategies. Again utilizing community participation processes, one such project involved developing a corridor plan with design and economic development ideas and recommendations for implementation.

The work of the Resource Center for Raza Planning demonstrates how a place-based values-driven center can contribute to community development efforts through the application of a wide range of planning processes and techniques.

Teresa Cordova, Ph.D.
Director, Great Cities Institute, University of Illinois at Chicago, USA

For more information on Raza Planning, see: https: //saap.unm.edu/centers-institutes/rcrp/index.html

One of the most difficult roles that a professional developer has is facilitating communication within the working group so that all views are heard and discussions do not degenerate into nonproductive complaint sessions and personal arguments. There is a fine line between facilitating meetings and manipulating meetings. This is a tightrope that all CD professionals must learn to walk. This is a particularly difficult task for those professionals who live and work full-time in the communities where they are consultants. The professional community developer has to continually reflect on his or her actions to determine whether they are facilitating or directing, providing information or solutions.

Information and professional experiences should be shared carefully so as to provide examples, options, and possibilities for consideration versus telling citizens what they must or should do. If actions and ideas come primarily from the professional developer, they are not likely be supported or implemented.

It is important to realize that people support what they help develop. The opposite is also true. Many perfectly written plans created under the direction of knowledgeable professionals with good intentions and at significant cost, lie unused in city hall bookshelves throughout the world. The consultants did part of their job. They were asked to create a plan and they did; but without citizen participation and ownership, the plan is useless. Participation promotes citizen ownership of and commitment to the actions that have been planned.

CD is a process through which people learn how they can help themselves. Self-help is the cornerstone of CD. Through self-help, people and communities become increasingly interdependent and independent rather than dependent on outsiders to make and implement decisions.

The CD process is a set of steps that guide the identification of a program of work and movement toward the ultimate CD goal. These steps require the involvement of community members and serve as a guide to problem solving, planning, and completion. The steps do not necessarily follow a sequential path. They may not follow the exact sequence below, and some can occur concurrently. The steps are as follows:

- *Establish an organizing group*—This might be a strategic planning committee or development task force. It could be a new, independent organization that has broad representation from many different community organizations and includes a broad cross section of community leaders. Or it might be an inclusive group sponsored by a successful community organization such as a Chamber of Commerce or economic development group. The most successful development organizations are public–private partnerships that involve a blend of prominent citizens, religious and neighborhood leaders, major community stakeholders (such as major property owners and managers of major businesses owned by external investors), elected officials, and local business leaders.
- *Create a mission statement*—This may be a mission statement for a strategic planning committee detailing why it was formed and what they aim to accomplish. This statement is important because it lessens the threat to and helps prevent role conflict with other community organizations by communicating its unique mission and role. Further, it keeps the group focused and helps prevent "mission creep," a loss of focus and drifting away from the primary purpose for which a group was formed. The following is a mission statement for a development organization in Phillips County, Arkansas:

> The Phillips County Strategic Planning Steering Committee will generate planned action that rebuilds the community on a shared vision of the future using a focused alliance of community groups, leaders, resource partners, and stakeholders.

- *Identify community stakeholders*—Who are the stakeholders that should be involved in the process? What roles should they play? When? For practical reasons, inviting all citizens to all meetings is not only an inefficient use of human resources, but prevents detailed analysis and discussion of critical issues and the development of strategies. Initially, a representative group of citizens has to be created. As its work progresses, more and more citizens are involved by serving on subcommittees, task forces, or project teams through which they provide information, opinions, ideas, and questions; challenges to the status quo, approval of the final plan, and help in implementing it.
- *Collect and analyze information*—Before beginning work, it is important to identify the current community environment. There are many methods of conducting this environmental scan, and many types of information that can be gathered in support of community development. It is often useful to assemble a community profile. This is a statistical overview of the current and past demographics of the community—income, population growth/decline, age of the population, community boundaries, population density, major employers, employment by sector, etc. Other useful tools include surveys of various types. A business opinion survey is useful in identifying economic issues and conditions that are impacting the community. This survey is also useful in retention and expansion program efforts. It surveys a sampling of local businesses and seeks opinions about the quality of local government, infrastructure, workforce, and other issues impacting business growth. Citizen attitude surveys are also useful in identifying a variety of issues that impact both economic and quality-of-life factors. Other assessments include comprehensive studies, surveys, and leadership workshops that examine all aspects of the community in order to identify critical issues and the strengths, weaknesses, opportunities, and threats (SWOT analysis) impacting development.
- *Develop an effective communications process*—It is extremely important to keep an open line of communication with the public. This will ensure that the process is inclusive and that activities are transparent. Many development activities are conducted by citizen representatives without a large public attendance. It will be important for them to keep their constituents informed, but this should not be the sole source of communications. If possible, involve the local press in the CD process. In addition to newspapers, it might be possible to use local television stations or public access cable channels as well as internet websites. Also, strive to establish two-way communications by providing a phone number and an email address which the public can use to send ideas, comments, and questions about what is being done.
- *Expand the community organization*—Once an organizing group is established, additional organizations and citizens can be involved in addressing specific problems that are of direct interest to them. For example, an economic development subcommittee might be created, and its membership expanded to include additional business leaders; representatives from development boards, airport, and port commissions; vocational educational institutions; and economic specialists from the state or other governmental agencies. Similarly, a tourism development subcommittee might be created where additional hotel, restaurant, gift shop, and tourist attraction operators would be invited to participate in discussions to determine how tourism might be expanded and supported within a community. This expansion continues the process of conducting a community dialogue that began with the environmental scan.
- *Create a vision statement*—As soon as the development group has identified the critical issues and conducted a SWOT analysis, they

can create a strategic vision. This forward-looking vision statement provides guidance and direction for the actions that will be taken to make improvements. It is usually one sentence that embodies the desired state of the community in the future. It must be a vision that can be realistically achieved within 15 to 20 years but should have enough "stretch" to challenge the community to achieve dramatic positive change. Below is a strategic vision for Kaniv, Ukraine, that was developed in 2005.

> By 2020, Kaniv will be: The spiritual capital of Ukrainians, where tourist attractions, communal infrastructure, education, culture, and scientific, ecologically-clean, high-tech industries are brought together for the well-being of people.

- *Create a comprehensive strategic plan*—After the development group has created the vision they would like to achieve, they begin to create a strategic plan to support the achievement of that vision. This can be done by subcommittees or task forces each working on one of the critical issues. Communication between the subcommittee chairs is important to avoid duplication of effort and to share ideas that may be useful for other groups. The typical structure of a strategic plan is a list of several goals, each supported by numerous objectives.
  An effective plan will be realistic and credible. Objectives are written so that they have specific completion dates and clear measurable outcomes. They are also supported by specific tasks and milestones that lead to achievement of the objective. Each objective should include the names of individuals responsible for its completion as well as the funding and resources to be used. If an objective does not have an individual responsible for managing it or funding or dedicated resources, it should not be included in the plan.
- *Identify the leadership and establish a plan management team*—From the very beginning of the process, the CD professional needs to identify those who will become leaders and champions of the process when he or she leaves. These individuals may come from the original organizing group; be the chairs of some of the committees, subcommittees, and task forces; or come from individuals included in the process as it expands to involve more citizens. Generally a group of seven to nine individuals should be identified and selected by the entire planning committee to become a plan management team.
  Since it is unrealistic to have a large group of individuals meet frequently, the plan management team is charged with acting on the planning committee's behalf. The team meets periodically to manage the ongoing process, and its membership should reflect the public–private nature of the group by including representatives from each of the critical issue/goal areas. The team's role is to manage the process, assisting those charged with implementing goals and objectives to move the process forward.
  The plan management team might initially meet twice a month to get an update on the overall progress of the plan. Later they might meet once every two months. They are also charged with making periodic written and verbal reports to the entire group, which might meet quarterly or semi-annually. Throughout the process, the work is continually being implemented by those responsible for goals and objectives.
- *Implement the plan*—Implementing the plan is when the "rubber meets the road." It is a crucial time that the plan management team needs to monitor very closely. Objective team leaders can be quickly discouraged if they do not see results. For that reason, it is important to build early successes into the planning process. When facilitating the process, the CD professional, plan management team, and committee chairs should include some objectives with the following characteristics:

- short time frame for implementation;
- highly visible;
- money and resources are available;
- popular with the vast majority of residents; and
- low risk of failure.

Ensuring early success is important because it builds momentum, helps attract additional volunteers, and instills the belief that things are changing for the better.

- *Review and evaluate the planning outcomes*—One key aspect of total quality management is the Plan Do, Check, and Adjust cycle. It is also important to realize that planning is a dynamic process and that the plan is a living document. Some objectives will be completed ahead of schedule. Others will be delayed, and their time schedule must be revised. Some will have to be eliminated because of changing environmental conditions or the loss or lack of anticipated funding. Most interestingly, however, is the fact that new objectives will be created and added to the plan as environmental conditions change and new challenges and opportunities emerge.

  The plan management team should perform periodic evaluations of the plan, including a review of each objective. What is going well? What problems need to be addressed? What needs to be changed? Their primary mission is to keep the planning activities and the community's progress moving forward so that the shared vision can be realized.
- *Celebrate the successes*—Winning events need to be built into the planning process. In addition to an annual report and celebration of the community's accomplishments, it is important to have smaller ongoing celebrations that provide reward, recognition, and continued motivation for volunteers and citizens. A comprehensive reinforcement process could be developed and implemented and could include having publicity in the newspaper; drinks and snacks, gifts and discount certificates, and/or awards for volunteers; and special T-shirts or hats for those who do the work. Since motivation is unique to each individual, leaders are charged with structuring celebrations into their activities that continue to recognize, reward, and reinforce volunteer efforts. Simply put, participation in the planning process should be fun, and citizens should want to participate.
- *Create new goals and objectives as needed*—A comprehensive plan usually has several goals, each supported by a number of objectives. It contains a multitude of projects that will be completed at different times. Some objectives actually lay the foundation for what must be done next for the community to realize its shared vision. As stated above, the community and the plan operate in a dynamic environment. In order to remain relevant, the plan must be continually updated and new goals and objectives added as others are achieved or completed.

How Does Community Development Practice Relate to Economic Development?

Traditional economic development (ED) focused on the attraction and location of businesses (industrial development) in a community. The focus was strictly on job creation and business investment. In recent years, economic development has been expanded to include the retention and expansion of existing businesses as well as the incubation of new businesses.

Attraction of new business development is a very competitive process. Each year thousands of communities in hundreds of countries compete on a continual basis for new business facilities. Location decisions are usually driven by economics, competition, and cost decisions. During the latter part of the twentieth century, much of the rural southern United States lost unskilled jobs to foreign workers. The same quality of labor for cut and sew operations, for example, could be found in Honduras, Vietnam, Sri Lanka, and China for a much lower cost.

An economic developer might promote a community to a business investor based on a very specific set of objective, but narrow criteria. However, at some point the investor will visit "short-listed" communities to select a location. It is at that point that the realities of the location are driven home.

Not every community can be successful in locating major industries. Often conditions other than available sites and buildings, labor costs, and location incentives impact location decisions. For the decision maker, these other factors might be very subjective and could be related to personal, cultural, sociological, and quality-of-life issues. Some of these factors are directly under the influence or control of local leaders and citizens.

A shortcoming of traditional ED activities is that they are focused on the "now" rather than on the future. What business can I locate here based on the current resources (e.g. labor, raw materials, location, infrastructure capacity, and available buildings or sites)? What is the best business development match for a specific community based upon its current strengths? Traditional ED activities were not focused on addressing problems such as social issues that, if solved, could result in making a community an attractive location for different industries. Initially, ED was asset-marketing focused on "picking the low-hanging fruit."

On the other hand, community development tends to be a long-term process. It is holistic in that it sees all aspects of the community as related and as effecting development. Citizens and business investors want more than just jobs or a good investment location. They know that their managers and technical workers will move there to work and live. Therefore other desirable traits, "soft factors," drive the final selection process. These include:

- effective community leadership;
- high-quality education for preschool, primary, secondary, higher education, and workforce training and retraining;
- economic development that creates a variety of quality job opportunities through the attraction of new businesses, the retention and expansion of existing businesses, and support of entrepreneurship within the community;
- attractive community known for its "curb appeal";
- high-quality, affordable housing and a variety of housing for citizens at all income levels;
- quality and affordable health care and emergency services;
- recreational and cultural activities (golf courses, resorts, parks, festivals, theaters, etc.);
- safe, crime-free community;
- honest and effective government that delivers efficient services;
- good infrastructure including roads, drainage, water and sewage services, solid waste disposal, etc.;
- reliable and competitively priced utilities (electric, gas, and telecommunications).

BOX 7.6 HOW COMMUNITY DEVELOPMENT CREATED ECONOMIC DEVELOPMENT IN SLOVAKIA

As an example, representatives of a major Japanese electronics manufacturer were visiting a location in Trnava, western Slovakia. They were in Trnava during the Christmas season (St. Nicholas Day). At night, the historic town center was brightly decorated and lighted. A fresh snow had just fallen, and children and parents were on the street celebrating. Saint Nicholas was making his rounds visiting homes and delivering gifts to children. It was a totally unplanned coincidence, but one that demonstrated many aspects about the city's quality of life. After witnessing this event, the selection team told the economic developer, "We will visit the other cities, but we have already made our decision." It was obvious that the site selection team was greatly impressed with the spirit of the town and its people, which was very visible during the holiday event.

What Do Community Developers Do?

Positive change in communities is driven by many factors. Therefore, positive change will be difficult to bring about by working on only one or two projects. Community development has to take a holistic approach to organizing, planning, and implementing change. The CD process identifies and organizes local leadership, involves the public, identifies critical issues, creates a plan, and implements actions to solve problems across a broad spectrum of areas so that desired change occurs.

As part of their practice and by the very nature of the community environment, professional community developers perform "grass-roots" economic development. A community may have a very desirable building or green field site available. Initiating a marketing campaign to attract businesses to these sites may fail, however, if some underlying development issues are not addressed before or concurrently with the marketing initiative. The local workforces' skill level may need to be addressed. Social issues, such as a high percentage of the working age population abusing illegal substances and alcohol, may prevent successful business attraction. The community's "curb appeal" may also detract from its competitiveness. The local infrastructure may be at or near full capacity or be in need of major renovations.

CD professionals draw on many knowledge and skill bases in order to be successful in their work. Some professionals are generalists, while others specialize in one or more areas of the CD practice. In their practice, professionals may be found to perform one or more of the following functions (while not an exhaustive list, the following provides an overview of the broad range of knowledge, skills, and abilities exercised by CD professionals):

- *Community assessment*
  - Perform statistical and demographic research.
  - Organize citizens to conduct assessments (environmental scans).
  - Design and conduct surveys.
  - Prepare and present assessment reports.
- *Strategic planning*
  - Organize a planning committee.
  - Facilitate the planning process.
  - Assemble and edit the plan.
  - Assist in plan implementation and management.
- *Organizational development*
  - Identify stakeholders.
  - Recruit volunteers.
  - Organize community groups.
  - Recruit and work with volunteers.
- *Leadership development*
  - Recruit leaders.
  - Conduct learning needs assessment; create and evaluate training and development activities.
  - Deliver training and development activities.
- *Economic development*
  - Create and manage business attraction programs.
  - Support expansion and retention of existing businesses.
  - Create and deliver entrepreneur training and business incubation activities.
  - Foster technology transfer to help businesses regain/retain their competitiveness.
- *Public and private development financing*
  - Identify governmental grant opportunities and write applications.
  - Identify foundation grant opportunities and write applications.
  - Seek venture capital sources and write applications.
  - Assist businesses in writing bank and financial institution loan applications.
- *Community land use planning and research*
  - Plan industrial parks, commercial and residential developments.
  - Plan and design utility infrastructure.
  - Plan and design roads.
  - Plan and design seaports and airports.
  - Plan and design parks and recreational centers.

Professional Standards of Ethical Practice

Every profession has a set of ethical standards that it expects its members to follow. Certified Professional Community and Economic Developers (PCED) subscribe to the following ethical standards adopted by the Community Development Council (CDC). These are based largely on the Standards of Ethical Practices of the International Community Development Society from which the CDC professional certification process originated. Failure to adhere to these principles and ethical practices may result in the International Community Development Council rescinding a professional's certification.

Professional Values

The following values guide the professional practice of community development:

- honesty
- loyalty
- fairness
- courage
- caring
- respect
- tolerance
- duty
- lifelong learning.

Professional Principles

- The purpose of community development is to raise living standards and improve the quality of life for all citizens.
- Community development seeks to build initiatives around shared values and critical issues after identifying existing strengths, weaknesses, opportunities, and threats.
- Positive change begins with creating a shared vision that can be transformed into reality by the actions of citizens using goals, objectives, and action plans.
- Community development is inclusive and involves developing leaders and building teams across class, gender, racial, cultural, and religious lines.
- Community development is more than social service programs and "bricks and mortar" construction. It is a comprehensive initiative to improve all aspects of a community's interdependence: human infrastructure, social infrastructure, economic infrastructure, and physical infrastructure.
- Community development involves consensus building that looks for the best solutions to community problems rather than only those that are politically expedient or popular or that benefit a few citizens.
- Community development is directed toward increasing a community's leadership capacity for solving problems and moving citizens from dependence to interdependence.
- Leadership in community initiatives is shared so that responsibility and commitment are encouraged across a broad base of the population.
- Development leaders work to transform their communities for the better and inspire others to do the same.
- Community development is an educational process that helps citizens understand the economic, social, political, environmental, and psychological aspects of various solutions.
- Community development is focused on action that improves communities by transforming learning into performance.
- Community development includes economic development initiatives that help bring high-quality employment, career, and business opportunities to citizens.
- Successful initiatives at the community level lay the groundwork for regional alliances and cooperation directed toward solving common problems.

Ethical Standards

Ethical standards adopted by the Community Development Council are as follows:

- Establish and maintain a professional and objective relationship with the client community and its representatives, one that advances the ethical standards of practice.
- Always perform in a legal and ethical manner.
- Immediately disengage from activities when it becomes apparent that they may be illegal or unethical, reporting illegal activities to appropriate authorities as required.
- Adhere to the professional principles outlined above.
- Clearly and accurately detail personal knowledge, experience, capabilities, and the outcomes of past consulting when requested.
- Clearly and accurately detail the scope of work to be performed (and its anticipated outcomes) and the fee for and terms of that work prior to engaging in consulting.
- Avoid conflicts of interest and dual relationships, especially those that could result in or appear to result in personal benefit (outside the scope of work) at the expense of a client community or its representatives.
- Disengage from activities that might result in one group or individual unethically or illegally benefiting at the expense of another.
- Adhere to all professional principles and practices regarding selecting, administering, interpreting, and reporting community assessment measures.
- Keep confidences and only reveal confidential information at appropriate times and with proper authority.
- Maintain confidential records in a secure location and under controlled access.
- Discuss ethical dilemmas that arise with other Certified Professional Community and Economic Developers or the Community Development Council to solicit guidance and opinions regarding possible actions.
- When feasible, confidentially consult with professional colleagues whose behavior is in question or when the colleague requests assistance in resolving ethical or legal dilemmas at a personal level.
- Notify the Community Development Council when ethical and legal dilemmas are unable to be resolved at a personal level.

Getting Started

1. *Ground rules*: It is often helpful to make copies of the values, principles, and process of CD and distribute them to participants in the process. They can be used as a checklist or reference for guiding the process. You may also help the group create a set of ground rules that will guide their group activities. This might involve such topics as confidentiality, conflict of interest, and resolving disagreements. It may also be useful to guide the group through the creation of a credo that lists their beliefs and values. These actions will help guide the CD process and serve as a tool for the facilitator to refocus the group when problems occur.
2. *Ice breaker*: Ask people to make a list of all the topics they know well enough to teach someone else. Either post the results on a flip chart page or white board, or project the results on a wall or screen using a digital projector and a laptop so that all participants can see the results. This exercise can help citizens come to realize the wide range of resources within the group. You may want to record this information for future reference as a guide for involving individuals in later activities.

Conclusion

This chapter has provided a broad overview of the practice of community development. It has discussed community development values and beliefs, practice principles, the community development process, community development tasks, and professional standards of ethical practice.

Professional community developers include consultants and those whose jobs include community and economic development activities. Volunteers who devote a significant portion of their time to community development and community activities might also be classified as professional community developers. People who practice community development should be aware of its underlying beliefs, values, and ethical standards. The more familiar all local citizens – not just professionals – are with the principles of community development practice, the more success a community is likely to enjoy.

CASE STUDY: MAYVILLE AND LASSITER COUNTY

Mayville is a town with a population of 50,000 in the south central portion of the United States. It is the county seat of Lassiter County. In its early years, Mayville grew as an agricultural market, a transportation hub, and a central location for regional services and retail. German immigrants came into Mayville in the early twentieth century and established many of the retail businesses. To this day, the town still has several excellent restaurants founded by the German immigrants. On one side of the menu is traditional German fare and on the other side is traditional southern fare.

In the latter part of the twentieth century, Mayville attracted several manufacturing operations paying good wages and the town prospered. Most of these industries located within the city limits because of good municipal services. Despite the fact that people commuted from all over the county (and surrounding counties) to Mayville and these good jobs, county elected officials often seemed envious of the jobs and industries in Mayville. Some of them even began to believe that the city, which had a strong economic development program and staff, was actively steering prospects and jobs away from the county and into the city.

Despite this mistrust, the economies of Mayville and Lassiter County continued to prosper because of the good local industries. Amid this prosperity, many county elected officials remained suspicious that the city did not have the county's best interests at heart. Some county officials and residents even came to believe that city residents looked upon them as uneducated "rubes" who did not deserve high-wage jobs.

One day both the city and the county received notice that the local U.S. military base was on the Base Realignment and Closing (BRAC) list to be considered for downsizing or closing. For the first time, the city and county faced a common imminent danger and were required to work together. They formed a Local Re-Use Authority (LRA) to begin contingency planning in case the base did close, although the initial job of the LRA was to make the case to the Department of Defense that the local base should not be closed.

The LRA hired an outside consultant to help develop the contingency plan, which included a strategy for economic development in the city and county to replace the jobs that might be lost through base closure. Following the principles of community and economic development practice, the consultant conducted an assessment of the community identifying the strengths and weaknesses and performing a SWOT analysis. The consultant conducted interviews with business leaders, elected officials, and other community stakeholders; and public meetings were held to solicit feedback and ideas.

City officials forewarned the consultant about the negative attitude and lack of cooperation on the part of the county, saying they were at a loss to explain this behavior. Despite this warning, the consultant was shocked at the first meeting with city and county officials together in the same room. Not only would certain county officials not cooperate and help develop the strategic plan, they were outright hostile to city officials, essentially claiming that the city did not respect them. It was obvious that there was absolutely no mutual trust, and, as a result, county officials always assumed that the city had an ulterior motive for anything they suggested and the county's response was always "No!"

The consultant realized that to accomplish anything, his team had to restore some semblance of mutual trust and let the county officials know how much their lack of cooperation was jeopardizing the continued strength of the local economy. With the city and county openly disagreeing and even feuding, they would not be able to develop an effective case for keeping the military base open or an effective strategic plan for economic development in the future. And, as the consultant pointed out, it was unlikely that any businesses would choose to move to the area in the future with such open intergovernmental hostility.

Through one-on-one confidential interviews with city and county officials, the consultant attempted to dispel the rumors and misguided notions that the city was trying to undermine the county. In the written report, the consultant pointed out (as diplomatically as possible and without including names) the harm resulting from the city/county distrust and hostility, including the fact that it would jeopardize the military base outcome even more and significantly interfere with economic development in the future.

Following the confidential interviews and report, many of the county officials began to realize the consequences of their actions and pledged to work together with the city for the good of all local residents. They realized that they had been putting personal issues and disputes before the good of the city, the county, and the residents that had elected them. However, one county commissioner continued to point fingers at city officials and refused to change his hostile ways. In the next election, he was the only county commissioner not reelected. The county residents and the other county commissioners had certainly gotten the community developer's message, even if the lone commissioner had not.

A year after the consultant's work, it was announced that the local military base would not be downsized or closed after all. The city and county had worked together and combined their resources to make the case that the military base was vital to national security and demonstrated that it was one of the most productive and efficient in the country. Not only did they not lose the military base, but the city and county gained a joint strategic plan to enhance future economic development and a strong working relationship with which to implement it.

Keywords

Community, community development, CD values and beliefs, economic development, ethical standards.

Review Questions

1 Define the practice of community development.
2 Describe the professional community developer's role.
3 What are the community development practice values and beliefs?
4 What are the community development practice principles?
5 List and describe the steps in community development practice. Do you agree with the order in which they are listed in the chapter?
6 List some of the tasks professional community developers perform.
7 What are some of the ethical standards to which professional community developers should adhere?

Note

1 Circa 1989, a working definition developed by Bill Miller, retired Director of the Community Development Institute Central in Arkansas; George McFarland, retired Community Development Manager, State of Mississippi; and the author.

Bibliography

Block, P. (1981) *Flawless Consulting: A Guide to Getting Your Expertise Used*, San Francisco: Jossey-Bass/Pfieffer.

Community Development Council. (2014). Principles of Good Practice. Accessed February 1, 2014 at: www.comm-dev.org

Dyer, W. (1994) *Team Building: Current Issues and New Alternatives*, 3rd ed., Reading, MA: Addison-Wesley.

Schein, E. (1987) *Process Consultation: Lessons for Managers and Consultants*, vol. 2 (paperback), Reading, MA: Addison-Wesley.

Connections

Crowdsourcing for the Common Good: A Community Development Approach www.frbatlanta.org/pubs/partnersupdate/12no1_crowdsourcing.cfm

Learn more details about practice with the University of Missouri's guide, "Practicing Community Development," at: http: //extension.missouri.edu/p/dm7616

Spend some time exploring two professional community development association websites:

International Community Development Association, at www.iacdglobal.org (global network for community development action, based in Scotland).

Community Development Society, at www.comm-dev.org (provides a journal, newsletter, and conferences. Based in North America).

PART II
Preparation and Planning

8 Community Visioning and Strategic Planning[1]

Michael McGrath

Overview

There are many examples of communities that have faced highly complex issues and reached their goals through sheer determination and a collaborative spirit. These communities succeeded in large part because they were willing to convene different players and undergo an extensive planning process. Not allowing the plan to sit on the shelf, these communities continued on and persevered throughout the plan's implementation. All sectors—government, business, nonprofit, and the citizens themselves—participated in the development of a common agenda. In addition, the community at large received ample opportunity to provide input. Because all sectors of these communities were involved in the creation and ongoing development of programs for the future, such programs received widespread support and encountered minimal resistance.

Introduction

Some communities allow the future to happen to them. Thriving communities recognize the future is something they can create. These communities take the time to produce a shared vision of the future they desire and employ a process that helps them achieve their goals. Achieving the desired future is hard work. Yet successful communities understand that the things they dream about will only come true through great effort, determination, and teamwork.

One way of achieving these community goals is through a community-visioning project. Such a process brings together all sectors of a community to identify problems, evaluate changing conditions, and build collective approaches to improve the quality of life in the community. The participants must define the definition of a community. Some projects define their community as a neighborhood; others a whole city or town; many projects have focused on regions that include multiple cities, towns, and counties.

Process Principles

Collaboration and consensus are essential—successful community efforts focus on ways in which business, government, nonprofits, and citizens work together. In reviewing successful collaborative efforts around the country, it has been found that all possess the following ingredients:

- People with varied interests and perspectives participated throughout the entire process and contributed to the final outcomes, lending credibility to the results.
- Traditional "power brokers" viewed other participants as peers.

- Individual agendas and baggage were set aside so the focus remained on common issues and goals.
- Strong leadership came from all sectors and interests.
- All participants took personal responsibility for the process and its outcomes.
- The group produced detailed recommendations that specified responsible parties, timelines, and costs.
- Individuals broke down racial, economic, and sector barriers and developed effective working relationships based on trust, understanding, and respect.
- Participants expected difficulty at certain points and realized it was a natural part of the process. When these frustrating times arose, they stepped up their commitment and worked harder to overcome those barriers.
- Projects were well timed—they were launched when other options to achieve the objective did not exist or were not working.
- Participants took the time to learn from past efforts (both successful and unsuccessful) and applied that learning to subsequent efforts.
- The group used consensus to reach desired outcomes.

These ingredients make up the essence of collaboration itself. True collaboration brings together many organizations, agencies, and individuals to define problems, create options, develop strategies, and implement solutions. Because they typically involve larger groups, collaborative efforts help organizations rethink how they work, how they relate to the rest of the community, and what role they can play in implementing a common strategy. Many times it becomes clear that no single organization has the resources or mandate to effectively address a particular issue alone. A group effort can help mobilize the necessary resources and community will.

Though a consensus-based decision-making process takes more time on the front end, it can save time during the back end of the implementation phase of a visioning project where blocking ordinarily occurs. If citizens are provided a forum in which their ideas and opinions are heard, seriously considered, and incorporated into the action plan, they will be less inclined to resist or ignore new initiatives. Community ownership of a plan and willingness to assist in its implementation often corresponds directly with the public's level of participation in the plan's development. As a result, projects can be completed in timely fashion through the consensus-building process. In collaborative processes, the sharing of information and pooling of resources build understanding and lead to better decisions. Special interests are not as inclined to block implementation of the plan, since it reflects their own interests and efforts. While collaborative problem solving is not appropriate for every issue and situation, it is an absolute necessity for a community-visioning project. Collaborative problem solving should be used when:

- The issues are complex or can be negotiated.
- The resources to address the issues are limited.
- There are a number of interests involved.
- Individual and community actions are required to address the issue effectively.
- People are interested and willing to participate because of the importance of the issue.
- No single entity has jurisdiction over the problem or implementation of the solutions.

Community visioning is both a process and an outcome. Its success is most clearly visible in an improved quality of life, but it can also give individual citizens and the community as a whole a new approach to meeting challenges and solving problems. Citizens of all types who care about the future of their communities conduct community-visioning projects. These people are collectively called "stakeholders." The stakeholders in successful visioning efforts represent the community's diversity—politically, racially, geographically, ethnically, and economically—lending different "stakes," or personal and group interests,

BOX 8.1 HEART & SOUL COMMUNITY PLANNING

Vision without action is a dream. Action without vision is simply passing the time.
Action with Vision is making a positive difference.

Joel Barker

The Orton Family Foundation is a nonprofit with the mission that "empowering people to shape the future of their communities will improve local decision-making, create a shared sense of belonging, and ultimately strengthen the social, cultural and economic vibrancy of each place." A variety of resources are available, including the *Heart & Soul Community Planning* handbook. A series of guides are provided as well, each focusing on an essential component of effective community planning.

Source: www.orton.org

to the process. They form the core planning group for the visioning effort, perform community self-evaluation, set goals, and develop the action plan and implementation strategy. To ensure the success of the stakeholders' work, effective process design and structure are essential.

Phase One: Initiation

Providing the Groundwork for the Process

The process of building a solid foundation for an effective community-visioning project includes a number of key tasks. The first is the selection of an Initiating Committee—a small group of 12–15 individuals that represent a slice of the community. Their job includes:

- selecting a stakeholder group that reflects the community's interests and perspectives;
- designing a process that will reach the desired outcomes of the community effort;
- forming subcommittees that will play key roles throughout the project;
- addressing key logistical issues such as staffing, siting, scheduling, fundraising, and the project name.

The Initiating Committee focuses on process, allowing the broader stakeholder group to work on content (identifying problem areas, formulating action plans, etc.). Preparation and completion of logistical tasks can send the visioning effort on its way toward success.

These individuals must be willing to invest a substantial amount of time over roughly three months in the development phase of the project. They may or may not wish to continue on as members of the stakeholder group for the planning effort itself. The Initiating Committee needs to reflect the community's diversity in terms of race, gender, economic sector, and place of residence and employment. Each member of the Initiating Committee should wear "multiple hats" or represent multiple interests. The Initiating Committee will make the first statements about the visioning initiative to the community, so it must be credible and well balanced. The two crucial attributes of the Initiating Committee are diversity and credibility. A good question to ask while forming the group is: "Will any community member be able to look at the Initiating Committee membership and say, 'Yes, my perspective was there from the beginning'?" If this isn't the case, then the missing perspective must be identified and a credible individual recruited to participate. The purpose of the Initiating Committee is to focus on the process and logistics required to move the project forward. The content of the community vision will be developed during the broader stakeholder-planning phase. The diverse voices on the Initiating Committee must create and agree to methods by which stakeholders can equitably address complex and controversial issues.

In order to create a safe environment for discussion of difficult issues, the Initiating Committee must complete a number of tasks. These tasks are made up of the following 15 actions:

1 Identifying who must be at the table

Using a "stakeholder analysis," the initiating committee must identify a group of 100 to 150 individuals to serve as the core planning group. The stakeholder group must be as diverse as possible and represent every major interest and perspective in the community. Even more than the initiating committee, the stakeholder group must represent the community's demographic diversity in terms of age, race, gender, preferences, and places of residence and employment. In selecting stakeholders for the community-visioning process, the initiating committee must consider the diverse sectors and various interests and perspectives of the community. The committee must avoid "rounding up the usual suspects" or forming a "blue ribbon panel" of the same community leaders and organizations that have always been involved in past community efforts. These active people are valuable contributors, but this type of project must tap into populations and people that are traditionally excluded from community processes. A balance of the "old guard" and "new blood" is useful. Further, it is important that participants act as citizens with a stake in the quality of life in the *whole* community, not simply as representatives of a particular organization, part of town, or issue. In this process, stakeholders should be effective spokespersons for their interests and perspectives, but they should not simply serve as advocates for their agencies and organizations.

One of the most critical groups of stakeholders will be those who have a stake in the future of the community but have little political or financial power. It will also be important to include both "yes" people and "no" people in the stakeholder group. It is easy to pick positive people who have the power to get things done. It is harder, but no less important, to pick people who have the power to stop or delay a project. As with the initiating committee, it is useful to look for people who wear multiple hats or fall into a number of categories: for example, a single parent with kids who is a banker and lives in a northwest-quadrant neighborhood, or a small business owner who is on the planning commission and serves as a soccer coach for his child's team.

A sample of categories for identifying the stakeholders in the community may include:

- pro-growth/no growth
- business (small, corporate, industrial)
- old/new resident
- conservative/liberal/moderate
- geographic location
- age
- ethnicity/race
- service provider (seniors, different abilities, youth)
- income level
- education reform/back to basics
- elected/appointed leadership
- single-parent/dual-parent house
- institution type (schools, police, etc.)
- inside/outside city boundaries.

2 Designing the process

It is important to note two fundamental premises about the community-visioning process. First, key leaders and the community as a whole must empower the stakeholder group to make decisions. Citizens are too knowledgeable to accept the role of only advising officials and community leaders, who may or may not choose to accept their advice. Although elected officials clearly have legal authority over issues such as taxes and the provision of services and corporate leaders are free to determine their own business development strategies, they must participate in this process as peers and agree to honor, while not necessarily rubber-stamping, the stakeholder group's conclusions. If the process works correctly, honoring the conclusions should not be a problem since the "power" people were a part of building the same conclusions.

Second, the orientation of the entire process, from the very beginning, has to be proactive. Too many community task forces have been convened over the years with marginal results. The goal of this effort is not to conduct interesting discussions or forge new relationships, though these will certainly result. The goal must be to develop a broad, able to be implemented, community-owned action plan that will truly serve the whole community and then to put that plan into action. The process must be customized to fit the community's needs and desired outcomes. It must take into consideration local realities (budget, time constraints, etc.) and complement other useful community efforts. Also, the outreach process of the project must take into account the community's political, social, cultural, and geographic characteristics and fit the specific language, literacy, and accessibility needs of the local population.

3 Setting the project timetable

The experience of successful efforts has shown that a comfortable schedule for the visioning project is to have the stakeholders meet once every three weeks over 10 to 12 months. Some extra time may be taken to work around major holidays or significant community activities. The initiating committee may choose to meet more frequently, such as once a week, in its preparatory work to speed up the process. The time frame will depend on the nature and needs of the community, local scheduling realities, and the urgency surrounding issues in the community. The timing of stakeholder meetings is an important factor. Successful visioning projects have made accessibility and participation in the project a priority. Therefore, stakeholder meetings often took place in the evenings to allow working people to participate on a regular basis.

4 Designing structure to coordinate the project

The project should have a project chairperson, at least three small subcommittees, and adequate staffing. Stakeholders, not those individuals staffing the project, must lead committees. Though initiating committee members may take leadership positions on subcommittees in the early phases of the project, new leaders may be available after the project kickoff once stakeholders are more involved and further recruitment can take place. The coordinating committee and the outreach committee are the best places to involve stakeholders who want to contribute.

5 Selecting a chairperson

All successful community projects have strong and fair leadership. Therefore, the selection of the project chairperson is critical. She/he must be (and must be perceived as) open, fair, neutral, and likeable. The chairperson's duties include:

- formally opening and closing every stakeholder meeting;
- chairing the meetings of the coordinating committee;
- appointing the chairs of the other committees, representing the project in the press;
- leading the fundraising effort;
- being the spokesperson for the project to the broader community;
- resolving any disputes within the group, and putting out any fires that may flare up during the course of the project;
- working with the facilitation team to assure the meetings run effectively and a safe environment for discussion is maintained.

The chairperson also submits recommendations for the composition of the coordinating committee to the initiating committee. Every process goes through challenging periods, and heated discussions may take place during meetings. The chairperson has a crucial role to play during these periods. She/he must work closely with the project facilitator to remind stakeholders of the project purpose and goals and to keep the environment safe for discussion from all perspectives. Above all, the chairperson is a role model for the whole group and must have a strong commitment to the project and participants. If she/he is accountable, the entire group is more likely to be accountable. She/he must be willing to devote a substantial amount of time to the community-visioning project.

6 Forming a coordinating committee

The first subcommittee is the coordinating committee. This group of 10–15 stakeholders manages the process, but not the content, of the project. Its members guide the plan and schedule; serve as liaisons with the stakeholders; fundraise; supervise the other committees and the project staff; and generally keep the effort on track. They will also "own" the project on behalf of the entire community to ensure that the visioning process does not become merely an advisory effort. The coordinating committee will need to hold a planning/debriefing meeting for each meeting of the larger stakeholder group. Work will often have to be done between sessions, and the coordinating committee, with the support of staff, will need to ensure its completion. Some members of this committee, which begins its service at the kickoff and continues into the implementation phase, will likely have served on the initiating committee and, in some cases, may continue on into the implementation committee.

7 Forming an outreach committee

The second subcommittee is the outreach committee. This group of 10–12 stakeholders will take ownership of the community outreach process, ensuring an active exchange of information between the stakeholders and the community at large. If the outreach strategies are successful, the community as a whole will have played a large role in developing the vision and action plans. All individuals will have had opportunities to provide input, and their interests, perspectives, and concerns will have been represented within the stakeholder group.

8 Forming a research committee

The third subcommittee is the research committee. Its purpose is to provide the stakeholders with information to help them determine current assets and challenges the community faces. This group of three to five individuals joins project research staff to develop at least two sets of documents:

- preliminary materials for the external environmental scan on global, national, and regional trends that influence community quality of life;
- local indicators and a profile of where the community is today (e.g. growth, population, crime rates, employment rates, etc.).

This information can also be used to educate the general public. Outreach committees in some projects have used the information to provide the public with a rationale for certain strategies.

It is important to make the distinction between primary and secondary research. Primary research involves the collection of raw data in the field. Such research should only be conducted if the desired information is not already available from other sources (i.e. through secondary research). Most information can be gathered from local health departments, census data, government agencies, nonprofit organizations, chambers of commerce, local colleges and universities, and so forth. The research committee's work must begin with the initiating committee to assure availability of appropriate materials for the presentations during the stakeholder planning phase.

9 Staffing the project

Administrative staff plays a crucial role in the visioning process. The staff's ability to coordinate and complete the many logistical tasks involved often makes or breaks the overall effort. Administrative staff handles the following types of tasks:

- general communications (phone and written correspondence with stakeholders, committee members, and the community);
- coordination of mailings and meeting-reminder postcards;
- coordination of speaker and information requests;

- preparation of meeting room and other meeting logistics (refreshments, supplies, etc);
- taking of attendance at stakeholder sessions;
- preparation of meeting materials;
- taking of meeting notes;
- copying and other general administrative tasks.

Should staff members come from the chamber of commerce, city government, or other influential body, it is critical that citizens, not staff, direct the stakeholder planning and outreach effort to avoid accusations that individuals with a hidden agenda developed the recommendations.

10 Selecting a neutral, outside facilitator

In visioning projects, it is helpful to have an experienced outside facilitator run the community-visioning meetings. Such a facilitator or facilitation team can assist in several ways including:

- helping to design the process;
- keeping the effort true to its purposes and values;
- ensuring that the process stays on track and on schedule;
- helping to identify experts from around the state and nation on various issues of priority importance to the community;
- facilitating the large group stakeholder meetings—including encouraging wide participation and discouraging any personal attacks or group domination.

It is essential that the facilitator(s) be a neutral third party not connected with any organization in the process and possessing no specific stake in the outcome. As the project progresses, stakeholders can facilitate the small groups and task forces.

11 Identifying funding sources

Visioning projects require financial resources and in-kind contribution of other resources where possible to cover administrative, logistical, research, outreach, and facilitation costs. Successful visioning efforts have made a point of gathering financial and other resources in cooperative fashion from throughout the community to ensure broad ownership of the project. Developing these resources early can help ensure success in the planning phase and guarantee the availability of adequate funding for those portions of the action plan requiring financial investment and other resources. Community-wide visioning projects usually range from $75,000–$200,000 when all costs are taken into account. In developing a project budget, a community must consider the following questions:

- What types of resources are required (and in what amounts) for the successful completion of the planning phase of this project? Costs may include:
  - staffing ($20,000–$30,000);
  - facilitation costs ($30,000–$75,000);
  - food ($7,500–$15,000);
  - printing, copying, and office/administrative costs ($4,000–$7,500);
  - travel ($1,500–$7,500);
  - community-meeting-related costs ($4,000–$7,500);
  - outreach-related costs ($3,000–$10,000);
  - research-related costs ($1,000–$10,000);
  - equipment and meeting materials ($2,500–$7,500);
  - the final report ($3,500–$15,000); and
  - the community celebration ($3,500–$10,000).
- What money and in-kind resources can be raised from within and outside of the community for implementation of the various action plans determined by the stakeholder group?
- Who will take the lead on resource development?

12 Creating a project name

Giving the visioning process a name is an early way to develop project identity and a following for the project. Some names of visioning projects from around the US include:

- Our Future by Design: A Greater Winter Haven Community—Winter Haven, Florida;
- Invent Tomorrow—Fort Wayne, Indiana;
- Envision 2010—Dubuque, Iowa;
- Vision North—Clay and Platte County, Missouri;
- Clear Vision—Eau Claire, Wisconsin;
- Envision Palm Desert—Palm Desert, California.

Project names should give the stakeholders a sense of ownership and enable the general public to identify with the effort.

13 Selecting a neutral meeting site

An accessible and neutral meeting site with a large and open layout, available parking, and supporting facilities is a must. If possible, avoid governmental and organizational facilities to prevent the perception that the effort is being driven by that entity. The site should have quality lighting, good acoustics, and no pillars to block the sight of participants. The room should have adequate wall space for the hanging of flip charts. The building should have adequate parking, restrooms, air conditioning, tables, chairs, a kitchen, and separate rooms for child care needs. Community centers, schools, or churches typically serve as good neutral sites for meetings. In considering a site, room layout considerations must be taken into account. The 100–150 stakeholders typically sit around moveable round tables arranged comfortably around the room. One end of the room should be reserved for the facilitators, flip charts, screen, and an overhead projector. With large groups, two or three wireless microphones are crucial to aid people whose voices do not carry well.

14 Recruiting the stakeholders

A broad-based community-visioning effort should start with an initial list of 300–400 prospective stakeholders. This list will be whittled down to a committed stakeholder group of 100–150 individuals who will attend all regular planning sessions. Past visioning projects have regularly shown that 50–70% of prospective stakeholders initially agree to participate in the effort. Of these, 5–10% never attend stakeholder meetings. An average of 15% of those invited turn down the request because they are unable to attend a regular session at any given time.

15 Planning for the project kickoff

The final tasks of the initiating committee are to ensure that all logistical details are covered and that significant public awareness of the community planning effort exists leading up to the kickoff. All staff and committees—especially the research and outreach committees—should be in place and carrying out their tasks by that time. Composition of the stakeholder group and committee composition may require fine-tuning through the first one or two stakeholder meetings. Also, the stakeholders will be strongly encouraged to assist in the outreach effort by spreading the word to other community members and through other strategies developed by the outreach committee. The initiating committee must devise a plan to bring early attention to the project and focus media and public attention on the kickoff. Press conferences, public events, and other communication means have proven to be effective in building community awareness.

Creating a Parallel Outreach Process

An essential key to the success of the community-visioning process is an active community outreach effort. Despite all efforts to recruit a stakeholder group that is representative of the community's diversity, there will be some gaps. For a variety of reasons, certain groups cannot or will not participate in stakeholder meetings. If certain groups cannot come to the stakeholder meetings, then the means must be developed to go out to them. Different strategies must be employed simultaneously to ensure that all sectors and segments of the community's population are kept informed throughout the life of the project. An effective two-way dialogue between the stakeholders and the community is a critical component in creating a relevant, widely supported, and effectively implemented action plan. An outreach effort running parallel to the stakeholder planning process, with activities at several key steps along the way, is necessary to test current thinking within the community and allow citizens to have input on an ongoing basis. An outreach committee of 10–20 stakeholders coordinates the effort. To attain its goals and objectives, the outreach committee will need the active support of project staff and the stakeholder group as a whole. The community outreach continues the principles ingrained in this community-based planning model by emphasizing an all-inclusive approach. An indication of a thorough outreach program is the absence of surprise and backlash when the action plan is released to the public. This is because people are already knowledgeable of the plan's content due to the ongoing information loops established by the outreach committee.

A primary contact on the outreach committee should be designated for each major task area. One press-oriented person should serve as the media contact. Another should be recruited to liaison with community residents who may have questions regarding any of the activities or strategies. Others should be made responsible for the town meetings and speakers bureau. In addition, the outreach committee should have strong contact with the coordinating committee, whose assistance it will need from time to time. It should be well organized and develop and use a plan of action to cover both regular and special needs. Outreach is no different from any other communitybased effort to the extent that one strategy alone is not enough to ensure success. A multiple-strategy approach is the only one that works consistently. A creative, highly prepared, hardworking outreach committee can attract positive attention and useful input to the community-visioning process. The success of the community's initiative depends on it. Approaches to community outreach include the following actions.

Project Kickoff

The project kickoff has two primary audiences. The first consists of the stakeholders, who hold their first regular session and become familiarized with the project purpose, the planning process, and their colleagues. The second audience is the community as a whole. The kickoff can be the most effective way of introducing the visioning initiative to the media and the citizens whose support will be required throughout the project. Visioning teams often hold a public event/press conference prior to the kickoff to generate publicity. They may invite media representatives, key community leaders, and the public to a 30- to 45-minute presentation on the project given by three or four spokespersons (perhaps including the chair and/or a member of the initiating committee or coordinating committee). The guests would be able to learn about the work completed to date and receive detailed information about the participants and the planning and implementation effort. The presentation might be followed by a 10- to 15-minute press conference wherein reporters would be able to ask additional questions. It is essential to prepare project fact sheets and media kits in advance.

Surveys

At certain points throughout the community-visioning process, the stakeholders will need specific feedback from the community in order to direct or refine their planning actions. Surveys and focus groups are common instruments for gathering such information. An entire industry centers on the effective use of these very powerful research tools. In this limited space, therefore, the subject can only be introduced and participants encouraged to seek professional assistance or to read further about these tools before using them to enrich the community project. There are many types of surveys, and any number of them may be used depending on the information needed. Standard surveys characterize a given problem after it has been identified but before a solution has been selected and implemented. Surveys should contain specific questions about individual topics, although multiple topics can be addressed in a single survey. Survey data may provide guidance on the most appropriate methods to use in addressing a given issue. In addition, surveys may be applied during any phase of the process to monitor the effectiveness of approaches being used.

Survey questions must be specific and designed to minimize the chances of misinterpretation by respondents (something that can skew the results). Moreover, questions must be relevant to the target population or, again, the results will be inconsistent. Finally, the analysis of the survey results will be invalid if it does not take historical patterns into account. Surveys can be administered in person, over the phone, or through forms filled out anonymously by large numbers of people. It is often effective to code the forms by the respondent's area of residence, income group, organization, and/or other characteristics.

Focus Groups

Focus groups are a form of survey designed to identify and solve problems. Surveys help communities determine a course of action once a problem/issue has been identified; focus groups help communities find what problems/issues actually exist and how they should be defined. Focus groups are in-depth, specific interviews with people representing a cross section of the community based on ethnicity, race, age, socioeconomic status, perspective, and so forth. Focus groups are time-consuming, usually requiring a minimum of one month to assemble and conduct. It is critical to ask the right questions of the right people and then base the conclusions on historical trends and community background. The focus group leader must make sure the respondent pool reflects the demographics of the community to ensure a valid sampling of perspectives. Because focus groups (and surveys) must be designed carefully if they are to achieve sound results, it is advisable to look carefully at your group's capacity before undertaking such projects without guidance. If no one on the planning team has extensive experience with surveys, college research departments or outside professionals should be consulted or even hired to do the job. A few tips for surveys/focus groups include keeping the language on the survey simple to allow participation by people of all levels of literacy/language proficiency. Also, surveys should be translated into the language of non-English-speaking residents; focus groups for non-English-speaking residents will need a translator. Finally, allow sufficient lead time for each method to give the designers a quality sampling of the community.

Town Meetings

Town meetings are large gatherings at which the stakeholders and planners can inform the public about the project and receive valuable feedback from community residents. Anyone may attend to listen, learn, and voice his or her opinions, interests, and concerns. An effective town meeting includes presentations by the planners, but most important it allows for public input. Individuals from all sectors of the community are encouraged to attend through carefully planned, highly proactive recruitment strategies. People tend not to come to meetings without a strong sense of their importance—especially the types of folks whose input is most critically needed. It is precisely the most marginalized community members who typically do not participate in such activities. It is recommended that at least three major town meetings be held during the planning phase of the visioning project. The first meeting should take place after the "current realities and trends" stakeholder session, just prior to the first visioning session. This meeting is intended to get the word out about the purpose and nature of the project and to solicit ideas from citizens on their visions for the future. The second town meeting should come after the visioning sessions and prior to the

Key Performance Areas (KPAs) sessions. At this gathering, stakeholders present their consensus on the vision and receive community input on KPAs and ideas for "trend-bending" action strategies. The third meeting takes place after the stakeholders have reached a rough consensus on the action plan and implementation strategy but before they have finalized that work. The community has the opportunity to give suggestions and help fine-tune the strategies prior to final consensus.

A strong turnout by community members and interested parties is crucial for town meetings. The outreach committee can employ various strategies to ensure adequate participation representative of the many sectors of the population. To begin with, the stakeholders themselves can spread the word. In addition, the outreach committee can send press releases to print, radio, and television media; mail flyers to key contacts or place them in conspicuous places; translate written materials for non-English-speaking populations; offer assistance with transportation and day care; and so forth. Neighborhood meetings are a variation on the town meeting theme. Such gatherings can target specific parts of a community whose residents might not attend larger meetings in other parts of a city.

Press Releases

Communities must enlist the aid of local experts in working with the media. Their knowledge of how to approach and follow up with news organizations can be crucial in effectively getting the word out about the community effort. A first step in publicizing a town meeting, the kickoff, or any major part of the visioning process is to maintain regular contact with the media. The most common tool in this effort is the press release, a very specific document announcing an event or major benchmark. Press releases will frequently need to be drafted and sent to pre-developed contacts at each print, radio, and television news organization in the region. This mailing should always be followed up by a phone call to answer any questions and to lobby for coverage of the news item in question. The press release should be accurate and succinct. Media will cover events that are well supported (i.e. those with large attendance numbers, community leader participation, etc.). Press releases should be delivered to local, regional, and even statewide news organizations if appropriate. The papers that print the announcement will sometimes translate the text for specific non-English-speaking populations.

Flyers

Flyers advertise upcoming events on single, brightly colored sheets of paper that give the group's name, date and time of the event, location, nature of the event, a contact phone number, and specifics regarding refreshments, transportation, and child care. Flyers may be posted in public places and/or handed out to individuals on busy street corners. Flyers that catch the eye, are positive, and evoke an atmosphere of importance and fun are most effective.

Speakers' Bureau

Stakeholders can utilize their public speaking talents to spread specific messages to the community about the progress of the community-visioning project. This is an effective way to receive input, share information, and promote visioning efforts in the community. Within certain pockets of the population, such as communities of color, it may be best to have a face-to-face meeting with elders or other community leaders to explain the program. Once they buy in, they may be able to inform their community and recruit new participants more effectively than "outsiders" could on their own. If these individuals do not have time to assist the outreach committee, ask for names of other people within the community who may be available. It is important to train members of the speakers' bureau together and provide them with good fact sheets and an overview of frequently asked questions so they will deliver a consistent message to the public. In addition, preparation will assist in effectively reaching the targeted population. Consider the following:

- Accountability and follow-up plans should be addressed. The group needs to ask itself the following questions: "How can we ensure that people will show up for the meetings?" "How can we keep their attention once they are present?"
- A sign-in sheet for attendees should be used. The individual's name, address, and phone number may be valuable as the group attempts to recruit new members and to keep the community updated. Such information also supplements the record of the meeting itself.
- Finally, a contact person or persons should be designated so those who didn't give feedback at the town meeting may do so at a later date, if desired.

Op-ed Articles

Opposite-editorial articles ("op-ed" stands for "opposite editorial," as in "opposite the editorial page") are written by non-journalists, usually community leaders and citizens, and are printed periodically by newspapers. They offer insight into local happenings, express grass-roots perspectives and interests, and update ongoing community programs. A newspaper's ultimate goal is to sell papers; for this reason, publishers want articles that are of high quality, are timely and in the public interest, and are positive in nature. They want to produce something the public will want to read. To get an op-ed article published, begin with a query letter to the editor. This letter should be short and to the point, including such facts as what inspired the effort, who is involved, how the project arrived at its current stage, where it is headed, and how specific plans will be implemented. This correspondence needs to be well written—the editor will look upon the letter as a sample of the author's writing ability. The language of the article itself should be positive, focusing on the action the group is taking. Write about specifics—the obstacles the project has overcome, how breakthroughs were achieved, changes in team members' thinking. Focus on what the project is about, what it has accomplished, and what it will accomplish in the future. The writing should be inspiring for readers and leave them wanting to be a part of the effort or, at the very least, highly supportive of and informed about its progress. Finally, the article must be concise and should not be more than 1,000 words long (the newspaper will probably shorten the text of the article anyway). High-quality writing is critical to acceptance of the op-ed article. If the writing is good, the editor may ask for more articles to print in the future.

Public Service Announcements

Radio and television stations were once required by the Federal Communications Commission (FCC) to provide public service announcements (PSAs). Many stations still do so as a community service. A PSA is a 30- or 60-second spot, provided free of charge, that informs the public about a cause, issue, program, service, or opinion. When contacting a broadcast outlet, ask for the individual in charge of PSAs, ask what the station's preferred PSA format is, and follow it carefully. Many PSAs are not broadcast because they do not follow station format. When working with the media, always strive to minimize the amount of work they must do.

Internet and New Media

In our new age of media and technology, citizens have an almost inexhaustible capacity for communication and information sharing. On the other hand, the world of technology and media is growing and changing so rapidly that it is often difficult to know where things are going. It is not just a question of how information is gathered and disseminated, but also the new relationships and networks that are being formed through social media and new media. Think of the way social media sites were used by crowds during the "Arab Spring" or the way political campaigns are using the internet and mobile devices to reach and mobilize new audiences.

New media and social media can be invaluable tools for recruiting stakeholders, providing information about meetings, surveying residents, and publicizing the findings and action plans. Communities now have an amazing repository of information that can be used to facilitate understanding and more knowledgeable discussions and dialogues on local affairs. An individual can carry more information in one iPhone, marketed by Apple Inc., than was contained in the entire library in Alexandria of the classical era. Last year, Olathe, Kansas, recently was named one of the top "digital cities" by the *Center for Digital Government and Government Technology* magazine. Like most communities, the city has public meetings to discuss budget issues and, to get citizens to attend, and holds them in different venues. "In our experience, budget hearings at city hall were dwindling," says Erin Vader, the city's manager of communications and public engagement. "So you take it on the road and do road shows."

But even going out to the neighborhoods and bringing meetings to the people didn't seem to get the crowds, so the city's communications and public engagement department went in search of new ideas. So they decided to hold an E-Town meeting in the studio of the local government access cable station and to drive interest and participation with social media. Chris Hernandez, a Kansas City TV news personality, hosted the meeting, which was cablecast and live-streamed, and members of the public asked questions to city council members via email, the city's budget web page, Twitter, and Facebook.

The city launched an online forum six days before the scheduled e-meeting, asking citizens to submit questions. Questions could also be submitted live during the meeting. Local officials consider the experiment a success. The city's Facebook page saw an increase of about 60% in post views during the live-cast of the event and traffic on the city's budget web page increased ninefold. "We're trying to meet the citizens where they are, which is online," explained Chris Kelly, the city's IT director.

The rise of new media has raised new concerns however about the need for equity. Some subgroups within communities have much more access than others to fast broadband connections and digital media. It's not just a question of access. There is also a question of knowledge and familiarity. Some communities are finding new ways to get media training to young people and others who find themselves on the wrong side of the "digital divide."

Free and accurate communication is indispensible to democracy and effective governance. For as John Dewey noted in his book, *The Public and Its Problems* (1988: 167), there "can be no public without full publicity." Whatever obstructs and restricts publicity, "limits and distorts public opinion and checks and distorts thinking on social affairs." The tools of democracy and self government could only be "evolved and perfected" though application and "this application cannot occur save through free and systematic communication."

Today, citizens are capable of accessing information through a dizzying variety of means—the traditional news media, the internet, open public records, networks of community-based organizations, or even through smartphone applications. There is a clear link between information access and citizen participation. When community information sharing and communication are working optimally, citizens understand the vital issues of their communities and make informed decisions. Just as it is the responsibility of media, government, and other community sources to make information readily available to the public, citizens have a responsibility to search out a true cross section of opinions and viewpoints.

The Final Report

The report on the work of the community-visioning process serves many of the same objectives as the celebration (i.e. acknowledging contributions to date, building momentum, and enrolling new implementers). At the same time, it is a flexible tool that can be used to inspire organizations and companies to embrace the community vision and frame parts of their own strategic planning around it. The report also serves to remind implementers and the community of their commitments and provides future efforts with something on which to build. The important thing is that it is used, not simply published, bound, and left to gather dust on the shelf.

Phase Two: The Stakeholder Process

Community Visioning

Many communities begin their visioning project by determining the vision or desired future. Others look at where the community currently finds itself before identifying the desired future. Both approaches have produced quality results in visioning projects around the country. However, starting with the vision statement is preferred because it sets a positive tone for the process from the very start. This process convenes residents holding very diverse perspectives who come into the process with personal agendas. By starting the process with the development of vision themes, participants recognize early that despite the different views, there are many areas on which they all agree. Experiencing such a "win" early in the process sets the tone for participants to work toward agreement throughout the process.

A Vision Is a "Stretch"

In spring 1961, President John F. Kennedy, seeking increased funding for space exploration, described a most ambitious vision: to land a man on the moon before the end of the decade and return him safely to Earth. At the time, the United States had only launched an astronaut into "sub-orbital" space, let alone gone to the moon. The vision, in the midst of the space race, was inspiring and motivating. The country vowed to move ahead on the vision and the ambitious timeline. Achieving the vision had its costs. In 1967, three Apollo 1 astronauts perished during a launch practice session because, some say, the timeline was too demanding. Staff within the space program learned from the tragedy, changed their approach, and continued working toward Kennedy's goal. On July 20, 1969, Neil Armstrong and Edwin Aldrin walked on the moon and returned safely to Earth with fellow astronaut Michael Collins. Kennedy's clear vision with specific outcomes, the timing of the space race, the program's ability to bounce back from loss, the enthusiastic commitment of the masses, and a number of other variables produced a technological achievement for the ages. On a summer day in 1963, Dr. Martin Luther King, Jr. addressed the masses at the Lincoln Memorial in Washington, DC. His "I Have A Dream" speech stirs as many souls today as it did on that memorable afternoon. Communities continue to struggle toward the future he described for all of the country's children and people.

Picture the Desired Future

A community vision is an expression of possibility, an ideal future state that the community hopes to attain. The entire community must share such a vision so that it is truly owned in the inclusive sense. The vision provides the basis from which the community determines priorities and establishes targets for performance. It sets the stage for what is desired in the broadest sense, where the community wants to go as a whole. It serves as a foundation underlying goals, plans, and policies that can direct future action by the various sectors. Only after a clear vision is established is it feasible to effectively begin the difficult work of outlining and developing a clear plan of action. A vision can be communicated through a statement, a series of descriptions, or even a graphic depiction of how the community would look in the target year. Communities have used a number of methods and media to create and express their visions, their desired futures. The following ingredients are crucial to generating an exciting community vision:

(A) Include a Hefty Dose of Positive Thinking

In developing a community vision, it is important not to be constrained by either political or economic realities. For many people who think negatively, it is challenging to focus their energy on how things *can*, rather than on why things cannot, happen. People who have been through successful visioning projects have challenged themselves to move beyond the constraints and to dream about what their ideal community would be like. In developing the action plans, they focus their thinking on what must happen to ensure that the vision becomes a reality. It is always better to aim too high than too low. The positive thinking will be reflected in the vision statement itself. The statement should be entirely in positive terms and in the present tense—as if it were a current statement of fact. The vision and its components should be stated in clear, easily understood language that anyone in the community could understand. The vision statement must be reached by consensus and encourage the commitment of diverse community members. It is the vision that will drive the entire planning process—every action plan will be designed such that, when implemented, it will help bring about the desired future.

(B) Strong Visual Descriptions

In the visioning process itself, stakeholders can literally ask and answer such questions as:

- What words do you want your grandchildren to use to describe the health of the community?
- If the very best quality of life existed in the community, what would be happening?
- What common values exist across all perspectives and interests within the community, and how do they manifest themselves?
- How are people interacting with one another in this desired future? How are decisions being made?
- What is unique to our community that no other community has, and what does it look like 20 years from now?

(C) A Long Time Frame

The stakeholders select the time frame of the vision project. It is probably more useful to set a vision for a point at least ten years in the future. Though communities would like to be able to achieve a desired future in the short term, the reality is that many changes will take a great deal of time to bring about. An effective vision typically addresses a period stretching 15 to 25 years into the future. A quality vision statement has these important ingredients:

- positive, present tense language;
- qualities that provide the reader with a feeling for the region's uniqueness;
- inclusiveness of the region's diverse population;
- a depiction of the highest standards of excellence and achievement;
- a focus on people and quality of life;
- addresses a time period 15 to 20 years in the future;
- language that is easily understood by all.

Step One: Developing the Vision Statement

The process of refining the vision statement and its component points can be lengthy and arduous. There is no shortcut to working through the process as a group. Though groups often get caught up in "wordsmithing" the statement, it is more important to get agreement on the themes of the vision. The stakeholders may have to be reminded that the vision is the "end state," the final result. They will determine the specifics of how the vision will be reached later in the process, during the action-planning phase. The time required to generate a clear vision statement that expresses explicit themes can vary widely from one community to the next. It is unlikely that a broad group of citizens would complete the process in fewer than 8 hours of working time, but they should not require more than 15. One effective format is a weekend visioning retreat. Typically, however, stakeholders work on vision statements over two nonconsecutive evenings. Through the visioning process, people draw heavily on the values that are important to them. The process translates these individual and collective values into a set of important issues that the community wants to address. With a clear vision statement articulated and the component points serving as a beacon for the future, the stakeholders can shift to determining their priorities.

Step Two: Understanding Trends, Forces, and Pressures

A community scan is a brief but important step in the community-visioning process. It enables stakeholders to develop a shared understanding of the major events, trends, technologies, issues, and forces that affect their community and/or will do so in the future. National and global realities often have a significant impact on a community's ability to meet its challenges. It is not necessary for the group to reach true consensus on these observations, but all participants should recognize how their community relates to the world around it and how broader issues affect local choices. The research committee presents its first piece of work during this phase by providing to the stakeholder group a preliminary list of key present and future trends. The factors might include:

- the influence of population growth, age, and funding trends on the educational system;
- the affect of in- or out-migration on housing quality and affordability;
- new technologies, their costs, and the impact on jobs and the community's quality of life;
- changes in funding and/or policies of national, state, and local government programs;
- global trends regarding trade, the environment, and labor.

At an early initiating committee (IC) meeting, the research committee members with the assistance of the IC, should generate a list of issues for a preliminary scan. However, this should only be considered a first step; the preliminary scan should spark further discussion of the influence of these factors on the community's current and future quality of life. The final environmental scan must reflect more than merely the "experts' view." Community knowledge and perceptions of these larger issues must be considered during the stakeholder process. Following the presentation by the research committee, the stakeholders can discuss the issues in a large group format and then work in small groups to encourage greater participation. The small groups then report back to the larger group, discussing priority areas in greater detail. This step combines the findings of the research committee and community perceptions in general. While some of the issues raised during the environmental scan are beyond local control, their influence must be addressed if the community is truly to move to a new level. The discussion of these regional, national, and global forces sets the stage for identification of *local* realities and trends.

(a) Scanning the Community

The community scan consists of local indicators of how well the community is doing at a variety of levels. Developed from secondary data, the profile depicts assets (those areas/programs in which the community is doing well) and challenges (those areas in which the community is struggling). In many visioning projects, the research committee conducts a survey asking for residents' perceptions of their community's assets and challenges. These survey findings are combined with stakeholder perceptions to assist in identifying areas of focus during the action-planning phase. The research committee collects community indicators from secondary sources. For instance, crime statistics can be obtained from local sheriff and police departments. Real estate, business, and other economic indicators can be collected from the local chamber of commerce. Health-related figures (teen pregnancy, sexually transmitted diseases, immunization rates, etc.) can be collected through social service and health departments. Effective research presentations have compared the latest data with baseline data from a number of consecutive years to display annual changes and illustrate local trends. The community scan combines the survey results, the research data, and stakeholder perceptions into a single, powerful tool that the stakeholders can use in their own discussions and decision making. The combination allows stakeholders to base their deliberations on information and perceptions, a scenario that is both common and healthy in visioning processes. For instance, the community may perceive that violent crime has increased, but statistics may show that the opposite is true. Such occurrences build understanding of both the issues and the perspectives that exist within the community. In addition, the data and discussions provide the stakeholders and the community with a "likely future"—that is, the probable outcome of current trends and pressures if the community does not intervene. Stakeholders can identify areas of strength and those needing improvement by breaking into small groups and asking the following questions:

- What is the "likely future" of the community?
- Which elements of that direction are good or bad?
- Which aspects of it do we wish to maintain, and which should be altered?
- What are our most important opportunities and dangerous threats?

Once the likely future has been evaluated, new scenarios may be considered under different starting assumptions.

(b) Sample Indicators of a Community

In 1992, Jacksonville, Florida, conducted a quality-of-life project to monitor annual progress within the community. These indicators included the following:

- public high school graduation rate;
- affordability of single-family homes;
- cost of 1,000 KWH of electricity;
- index crimes per 100,000 population;
- percent reporting feeling safe walking alone in their neighborhood at night;
- compliance in tributary streams with water standards for dissolved oxygen;
- resident infant deaths per 1,000 live births;
- percent reporting racism to be a local problem;
- public library book circulation per capita;
- students in free/reduced lunch program;
- tourism/bed-tax revenues;
- taxable real estate value;
- tons per capita of solid waste;
- people reporting having no health insurance;
- employment discrimination complaints filed;
- people accurately naming two city council members;
- percent registered who vote;
- bookings of major city facilities;
- average public transit ridership per 1,000 population.

Jacksonville set goals for each indicator and had mechanisms in place to measure progress for each year. Included with the indicators were community action steps for achieving the target goals. For more information on Jacksonville's Quality of Life Project or specific indicators, contact Jacksonville Community Council Incorporated at (904) 396–3052. A key factor in strategic planning is good information based on data and research available throughout the community.

Information will be interpreted in different ways, therefore it is important to have sessions where facts and perceptions can be studied, analyzed, and discussed. Having a common perception about how the community is currently doing will assist in developing the desired future and identifying key areas in which to focus the action planning. There are several ways in which the Community Scan can be developed with each way requiring a different investment of time and resources. Many of these decisions can be resolved by the Initiating Committee during the early phases of the project. Whatever approach is decided on, the desired outcome of this phase of the project is to leave the stakeholders and the participants with a shared understanding of:

- the community strengths/assets;
- the difficult challenges the community faces;
- the realities the community faces that are both within and outside its control;
- the community's likely future should no interventions take place.

Step Three: Gauging the Community's Civic Infrastructure

Scholars and practitioners of urban and community affairs argue that associations and traditions play an integral role in the health of the communities, whatever their size. The National Civic League refers to the formal and informal processes and networks through which communities make decisions and solve problems as "civic infrastructure." Successful communities honor and nurture their civic infrastructures. They do not look primarily to Washington for money or program guidance. Rather, leaders in America's most vital communities recognize the interdependence of business, government, nonprofit organizations, and individual citizens. In particular, these communities recognize that solving problems and seizing opportunities is not the exclusive province of government. They carry on an ongoing struggle through formal and informal processes to identify common goals and meet individual and community needs and aspirations.

What accounts for the different experiences of different communities in addressing problems? In each case the strength or weakness of the civic infrastructure, the invisible structures and processes through which the social contract is written and rewritten in communities, determined success or failure. The Civic Index (see Box 8.2) is one way to analyze civic infrastructure. A comprehensive look at civic infrastructure is a fine point of embarkation—both for that conversation itself and for a reframing of the social contract.

BOX 8.2 THE CIVIC INDEX: A TOOL TO QUALITATIVELY ASSESS CIVIC INFRASTRUCTURE

The National Civic League developed the Civic Index to help communities unleash their power and abilities by evaluating and improving their civic infrastructures and measuring and developing the skills and processes that a community must possess to deal with its unique concerns. Whether the specific issue is a struggling school system, an air pollution problem, or a lack of adequate low-income housing, the need for effective problem-solving and leadership skills is the same. Since the index was first introduced, hundreds of towns, cities, countries and regions have used it to apply for the for the All-America City Award or to begin a long-range visioning, strategic planning or problem-solving effort facilitated by NCL staff. It was a way communities could begin discussions on how to improve themselves and how to develop action plans to make their visions of a better future come true. The Civic Index provides a framework that allows communities to do this efficiently and effectively. It is a method and a process for first identifying weak and strong points and structuring collaborative solutions, accordingly. It creates a setting or an environment within which communities can undertake a self-evaluation of their civic infrastructures. In developing the original Civic Index, the National Civic League board members and staff developed the concept of civic infrastructure to describe the societal and political fabric of communities—how decisions are made, how citizens interact with government and how challenges in the community are confronted. Creating civic infrastructure is not an end in itself. It can be a community's first step in building its capacity to deal with critical issues.

The components of the Civic Index are used to identify and measure those capacities and qualities that explain why some communities seem more able than others to address tough challenges. Whether the issue is a struggling school system, an environmental hazard, a natural disaster, or a lack of affordable housing—the necessary leadership and problem-solving skills are the same. A community must have strong leaders from all sectors who can work together with citizens to reach consensus on strategic issues that face the community and its region. Committed individuals help develop a community's capacity to solve problems, and communities must resolve to increase their problem-solving capacity. Outside consultants can make recommendations, but sustained action is unlikely without local ownership of a strategy and an implementation plan. The most recent iteration of the Civic Index identifies seven primary components or capacities for communities to consider: (1) community leadership; (2) participation and civic engagement; (3) diversity and inclusiveness; (4) networking, information, and communication; (5) decision making and consensus building; (6) partnerships and collaboration; and (7) community vision and pride.

Incorporating the Civic Index Into Your Visioning Project

A community must have strong leaders from all sectors who can work together with citizens to reach consensus on strategic issues that face the community and its region. Committed individuals help develop a community's capacity to solve problems, and communities must resolve to increase their problem-solving capacity. Outside consultants can make recommendations, but sustained action is unlikely without local ownership of a strategy and an implementation plan. The Civic Index provides a framework within which communities can increase their problem-solving capacity. It provides a method and a process for first identifying strengths and weaknesses and then structuring collaborative solutions to problems.

The index offers a framework within which communities can undertake a self evaluation of their civic infrastructures. Creating civic infrastructure is not an end in itself. Rather, it is a community's first step toward building its capacity to deal with critical issues. Since a strong civic infrastructure is the foundation for any kind of collaborative community work, the Civic Index can be usefully applied in many ways. The Civic Index is most effective when used with a group of citizens of diverse interests and perspectives.

Some communities have used the Civic Index as a "stand-alone" project to enhance the community's civic infrastructure. Others have incorporated it into their visioning process. The steps taken to build the community's problem-solving capacity can easily be integrated into the action plans developed during the visioning process, particularly in areas where networks and communication mechanisms are identified as need areas from the Civic Index. These must be in place before longer-term action steps can be implemented. The Civic Index's results are greatly enhanced when a large, diverse group such as the stakeholder group uses the Civic Index. Perceptions of the different components will vary among group members, and the discussion provides a great opportunity to build understanding and trust—the key ingredients of civic infrastructure. Discussion of the components takes place both in small groups (to ensure participation) and in large groups (to enhance the small group findings). Once the seven component areas have been assessed collectively, it is important to focus on each component individually. Stakeholders then develop benchmarks that indicate progress toward the desired level. The benchmarks, with steps to reach them, are incorporated into the action plan during a later phase.

Most communities craft their own questions for the Civic Index, matching the conditions within their jurisdictions with the seven components of the index to ensure relevance. Table 8.1 provides an example of a sample worksheet with the following instructions: Please consider the following statements and indicate the level of your agreement by assigning a letter grade from A to F to each statement listed below. Assigning "A" would mean a very high level of agreement with the statement and "F" a very low level of agreement. Give an average letter grade to each component of the Civic Index and average the grades of the various components to get an overall community Civic Index grade.

Table 8.1 *Civic Index Worksheet*

Community leadership

Our community has active programs to encourage the development of emerging leaders.	
Our community has specific programs to encourage the leadership development and community engagement of young people.	
Our local leadership programs provide multiple avenues for new leaders to apply their skills.	
Our leadership programs seek to develop and encourage a diverse and inclusive group of leaders.	
Community leaders listen to the views of others and encourage the public to be engaged in local problem-solving and planning efforts.	
Community leaders encourage and seek collaboration in local planning and decision-making efforts.	

Average	

Public participation and civic engagement

Residents of our community feel their participation matters in solving community challenges.	
Most residents are more committed to problem-solving to address community issues rather than assigning blame.	
It is easy to get people's engagement in community issues.	
It is easy to get qualified people to run for public offices.	
The community has adequate public opportunities for residents to engage in public planning, decision making or problem solving.	
Residents have ample opportunities to learn and practice their rights and responsibilities as members of the community.	
Public spaces in schools, libraries, and local government buildings are readily made available to community residents as welcoming spaces to meet.	
Public participation efforts in our community typically lead to positive outcomes.	

Average	

Diversity and inclusiveness

Our community recognizes and celebrates its diversity and inclusiveness (ages, ethnicities, genders, cultures, religions, and sexual orientation and expression).	
Our community takes the extra steps to ensure the broad diversity of residents is included in local and regional planning and actions that directly impact them.	
Our community is welcoming and inclusive of immigrants and refugees, seeing them as assets and not liabilities.	
Our community does a good job of addressing equity issues.	
Our community does a good job of creating democratic spaces for young people where they can actively participate with other community residents in addressing community issues.	

Average	

Networking and communication

The community is well informed of the plans of the local governing bodies.	
Most community residents know how to access information on public issues.	

The local news media report credible information about public issues.	
Community residents use social networking tools to organize and/or communicate on important issues.	
There are opportunities for community residents from all walks of life to access communications and information technology.	
Local governments and school districts have a strong commitment to openness and information sharing.	
There are both online and face-to-face public forums where community residents can engage in civil conversations and dialogue about their interests and concerns	

Average	

Decision making and consensus building

The community addresses challenges directly and quickly, instead of deferring or postponing difficult decisions.	
Community members can disagree about ideas and issues without differences typically leading to a breakdown in progress.	
Community leaders usually resolve controversial issues fairly with practical compromises and solutions.	
The community does have neutral conveners and forums to resolve pressing conflicts and challenges.	
The community is willing to try new ideas to solve problems.	
The community holds local governments accountable for the decisions that are made.	

Average	

Partnerships and collaboration

Local governments in the region work well with each other to address local and community-wide challenges.	
Most community leaders are able to set aside their own interests for the broader community good.	
Community agencies and organizations do a good job of coordinating their activities.	
Our local governments seek partnerships with nonprofit groups and private sector leaders.	
Nonprofit groups manage "turf" issues well and collaborate with each other in seeking resources.	
Nonprofits work together to jointly address community problems.	
Local businesses partner with nonprofits and schools to achieve better community outcomes.	
Local businesses encourage community voluntarism and charitable giving on the part of their employees.	

Community vision and pride

Our community regularly engages in strategic planning and other actions to help achieve a common vision.	
Our residents have strong positive identification with the community and a clear sense of what makes the community unique.	
Residents feel a strong sense of attachment and pride in their neighborhoods and community.	
Our community has a shared vision of what it wants to look like in the future.	

Average	

Total Civic Index average	

Step Four: Selecting and Evaluating Key Performance Areas

By this point in the process, the stakeholders will have discussed and reached consensus on where their community is today, where it is likely heading, and where they would like it to go. The next step in this results-oriented process is to decide how the community can get from where it is today to where stakeholders want it to be in the future. This step involves the selection and development of Key Performance Areas (KPAs). KPAs are highly leveraged priority areas for which specific actions will be developed to redirect the future of the community. Implementation of the strategies developed for the KPAs will bend the trend from the likely future (as determined by the community profile) toward the desired future (as articulated by the stakeholder group). Successful community-visioning projects have prioritized their visions into four or five KPAs. They reasoned that only some issues are of high-level priority; moreover, not everything can be done at once. Choices must be made. Secondary priorities can be tackled later. The KPAs can be broken down in a variety of different ways—by sector (e.g. business), by issue area (e.g. homelessness), or by project (e.g. community center).

The Key Performance Area Process

Forming Task Forces

Successful visioning projects have formed task forces either by assigning interested stakeholders or by choosing members at random from the stakeholder group. Either way, additional expertise and perspectives are usually added to help balance the group and develop comprehensive plans. Task forces vary in size from as few as 15 people to as many as 50. Each KPA task force should:

- Assign a convener who is responsible for convening the sessions, keeping the group on task and focused, and reporting the updates back to the large stakeholder group.
- Assign a facilitator to run the meetings (s/he may or may not be the convener) and a recorder to keep minutes and write up the work in a presentable format.
- Plan a number of meeting sessions (how many depends on the timeline) around the large stakeholder meetings.

For each Key Performance Area, the task force and the stakeholder group as a whole will complete the following tasks:

Recruiting outside expertise. One of the task force's first assignments is to look at the group's composition and ask, "What interests and expertise are missing from our group?" The task force members should generate a list of people who can fill in the gaps and recruit those individuals to participate. Just as balance was important in filling out the large stakeholder group, the same consideration must be given to the smaller task force groups. Although presentations to the larger stakeholder group will often "safeguard" any domination within the task forces by individuals with special interests, developing the plan with diverse perspectives always enhances the plan's credibility and likelihood of implementation.

Evaluating the community's current performance within the KPA. Task forces will assess the community's current performance in each priority area using the work of the research committee, surveys, and past discussions in the stakeholder group. This is also the time to integrate the findings of the Civic Index, if utilized in the visioning process. Much of the work from this stage will provide the rationale for proposals to address this key area. It will also help members identify what benefits they want to result from implementation of the action plans. These benefits should be developed into a "mini-vision" that will drive the action planning in this specific KPA.

Developing goals. Task forces will develop specific goals to reach the desired future for each

KPA. There may be numerous goals and objectives within a specific KPA. For instance, for a KPA of economic development the goals may be:

- starting an incubation program for small business development;
- attracting new corporations to headquarter in the community;
- retaining and enhancing current businesses based in the community;
- building and retaining the skills of the labor force in the community through mentorships and scholarships.

It will be up to the task force to prioritize the goals and make recommendations on which ones should receive the greatest emphasis.

Specifying "who will do what by when and how." Task forces must delineate specific action steps, identifying what resources will be required and the options for acquiring them, where they will come from, what the time frame for action is, and who will be responsible for ensuring that implementation occurs. It is during this step that the specific benchmarks and actions of the Civic Index will be integrated into the appropriate KPAs to build community capacity. By now the vision has been translated into practical and attainable outcomes to be achieved through specific tasks and actions. This step crystallizes the vision into a tangible program.

Reporting back to the stakeholder group and receiving feedback. As the KPA task forces proceed through the plan development, they will periodically meet with the larger stakeholder group to share their findings and coordinate overlapping efforts as appropriate. When reporting back to the large group, each task force should hand out written summaries of the work done to date, with highlights transferred to overhead transparencies for viewing by the group. The task forces should incorporate feedback from the stakeholder group into their planning to ensure agreement on the direction being taken. Although many of the action plans developed will require financial and other resources, sometimes in significant amounts, communities can take certain actions to increase cooperation or shift to approaches that require little or no financial resource outlay. As the whole stakeholder group reaches consensus on the work of each KPA task force, the high-priority projects must be identified and a rough consensus reached on their inclusion into the final action plan. If successful to this point, the stakeholders will have reached a general agreement on the individual goals, objectives, action plans, implementers, resource needs, and time frames identified by the task forces. Once the KPA evaluations have been completed, it is necessary to integrate all of the goals and recommendations into a final action agenda with a formalized implementation strategy. Certain goals and action steps will be complementary and will need to be combined in some way to create a coherent overall strategy. The stakeholder group should publish a report on its community-visioning process and final action plan, but it is essential that the work not stop here. Too many visioning projects end with a report that eventually gathers dust on the shelf. The community-visioning process is designed to produce action and results. Reports do not, in and of themselves, assure any action.

Building a final consensus. Consensus on the final action plan is the final—and occasionally most difficult—phase of the community-visioning process, the phase in which previous agreements are tested and a final community consensus is reached. The stakeholders meet in large and small groups to confirm the soundness of their goals and plans and the projected results of their implementation strategies. Some action plans may require initiation of new projects. Others may involve support for existing efforts. Some may entail the termination of an existing activity. Because the actions will be varied in nature, it is essential that the entire community and its diverse sectors be behind them. Some action plans may embrace policy initiatives or changes; some may involve significant financial investment; and others might simply pose new approaches to current practices.

All may involve the development of new cross-sectoral partnerships. As the visioning action plans and implementation strategy are finalized, the stakeholders must specify who will take responsibility for what. Some issues will clearly fall within the purview of a specific government agency or nonprofit service provider. Other action steps might not immediately suggest a "champion," and the group will have to engage some entity to take the lead. Though an accountable organization or group of organizations may not initially be found for every action step, this is an essential part of the process that cannot be left incomplete. It may be necessary to assign a group of entities to locate a champion for a specific action area. The general rule is: there will be no action without an implementer. As the formal planning steps of the process draw to a close, stewardship of implementation becomes the responsibility of the stakeholder group and the community as a whole. If the process has been effective at developing a sense of ownership and true consensus, it will be possible to hold the whole community and all its citizens and organizations accountable to their commitments. This point highlights the importance of the community outreach process. The investment of time and resources made earlier to ensure full community representation and participation comes to fruition here.

True consensus and rough consensus. At the end of this phase, the stakeholders should have reached consensus on the content of each KPA. Sometimes a full consensus cannot be reached. If a large number of stakeholders cannot live with the plan, then the group must take the time to discuss the reasoning of the disagreeing viewpoint and look at ways to fine-tune the approach so all participants can live with the final plan. If one or two people continue to dissent after all discussion and alternatives have been addressed, it is important to move ahead, while making sure the differing viewpoint is noted and placed in the final report.

The community celebration. Celebration is an essential part of a community-based visioning project. There should be a celebration to acknowledge the commitment of individuals involved in the planning phase of the initiative and the results they achieved. Such an event brings citizens together around shared values and aspirations and nurtures the seeds of change in building a better community. The city of Lindsay, California, held a celebration that residents are sure to remember for years to come. At the conclusion of its long-term visioning process, the community held a grand festival at the City Park and community center, attracting approximately 1,500 people. Featured events included a games arcade for kids; live entertainment on the main stage led by Mariachi Infantil Alma de Mexico; a winding parade through the park; a canine fashion show; a decorated bicycle contest; a "kiss-the-pig" contest in which nearly two dozen of the city's leading citizens gave Blossom, a pot-bellied pig, a big smack on the lips; drawings for a television and a blimp ride for two; and a food booth serving burritos, corn on the cob, strawberry shortcake, and watermelon. The celebration was a great success, allowing the citizens of Lindsay to take pride in their accomplishments and enjoy the fruits of their labor. The celebration should acknowledge the planning work of the stakeholders and various contributors, announce the action plan, and—most importantly—be seen as the commencement of the implementation phase of the project.

Step Five: Implementation of the Action Plans

In the community-visioning effort, a minimum of two years following completion of the planning process is recommended for intensive focus on project implementation. For many communities, this will be a multi-decade effort. Successful implementation processes have the following ingredients:

- the establishment of implementation structure such as a committee with staff that oversees and ensures that a variety of areas (that follow this bullet) are addressed;

- clarity of goals/desired result for both the implementation committee and implementers;
- criteria (established by the stakeholders or implementation committee) that will be used to prioritize projects;
- prioritized projects based on the applied criteria;
- implementers/champions for each project;
- identification of barriers to implementation and steps to overcome them;
- an overall timeline based on the prioritized goals, barriers, and resources;
- coordination of all efforts being implemented from the action plan;
- ongoing community outreach of successes and ideas.

Community and outside resources will be needed to implement the action plan. The Initiating Committee should have laid the groundwork for this resource development process, but more work will likely remain. The implementers named in the action plan will need to champion these efforts. Resource development will be most effective if it begins immediately, capitalizing on the momentum from the publishing of the report and the community celebration.

Choosing or establishing an implementation entity. Implementation efforts should follow the plans created during the planning process. Lead implementers must confirm the commitments already agreed upon and begin their work, drawing on the momentum created by the celebration and publication of the final report to facilitate rapid progress. From the kickoff until this point, the coordinating committee has provided process management for the community effort. Some of its members will be ready to leave the committee, and others will be ready to serve in a more active manner. This process should leave current participants with a strong sense of accomplishment and invite the participation of others. The coordinating committee may retain its original form and become an implementation committee, or it may choose to change its structure as well as its membership. Typically, retaining the cross-sector, broad-based citizen form is the most successful approach as it avoids controversy and keeps the focus on community-wide participation. Some communities choose to create a separate nonprofit organization to serve the ongoing effort. The coordinating committee might also be embraced by an existing entity deemed neutral and inclusive, although this can be risky if the organization attempts to hoard the effort or takes actions that dampen community-wide ownership in implementation.

Monitoring and tracking. There are three primary areas where active, ongoing monitoring and tracking are required in order to:

- Ensure follow through on the implementation of action plans and policy recommendations.
- Provide ongoing support for implementers.
- Measure changes in the community quality-of-life indicators developed earlier in the community scan effort.

During the first two years, the implementation committee or other implementation entity should consider providing updates at least quarterly to the community on project and policy actions. In subsequent years such updates can be made annually.

Conclusion

It should be stated that the community visioning and implementation process described in this chapter is an overview of a model. This model has been successfully used and tested in different forms in many communities around the nation in recent years. Each community should work closely with experienced facilitators to adapt the model presented here. Use it as a guide to the design of a local process; customize it to match specific needs, priority areas, and available resources.

CASE STUDIES

Eau Claire, Wisconsin

In March 2007, city and county officials held a meeting to discuss the tough fiscal challenges facing the community in its efforts to provide much-needed services to the public. The city brought in the National Civic League to assist in the planning and implementation of Clear Vision, a visioning and strategic planning process for the greater Eau Clare region. Funding for the process came from the city, the county, the school district, the University of Wisconsin-Eau Claire, Chippewa Valley Technical College, United Way, and the local chamber of Commerce.

The process was designed, as former City Manager Mike Huggins later wrote in the *National Civic Review*, to:

> bring all sectors of the community together to create a broad community vision and strategic plan, aimed at building a reinvigorated sense of community purpose with clear community priorities for the future. The Clear Vision process did not replace the formal planning, decision-making and budgeting processes of the city, county, and school governments but was intended to strengthen the community's civic capacity for effective collaboration by providing an integrated and coherent community perspective essential for action on community problems and improved coordination among government organizations and institutions. The initiative's operating premise was that active and meaningful participation of community residents in building their individual public lives will strengthen and sustain a connected and collaborative community where individuals, families, and businesses thrive and flourish.

The conveners recruited a 15-member initiating committee to design the planning process, identify and recruit a diverse group of participants, and do the planning for the stakeholder meetings. The initiating committee created work committees to support the process and identified about 500 community members to invite to the process. The list of 500 spanned a broad cross section of age, gender, location, race employment and income differences. The committee reached out to underrepresented groups such as the area's large minority of Hmong refugees and their families, African-Americans, and members of trade unions.

The stakeholder meetings began October 2007 in the community room of a local congregation. The meetings were open to the public, but a group of about 150 were regular attendees. Participants, according to the Huggins article, included:

> members of local not-for-profit community organizations, including faith-based groups; environmental and housing activists, health-care providers; business groups; neighborhood associations; students; retirees; and a limited number of government professional staff and elected officials. Many stakeholders were not affiliated with any formal organization, but participated as individual citizens interested in community issues.

The process resulted in the issuing of a report written by the participants laying out a community vision and a set of specific action plans to address six specific areas with an emphasis on action that would have dramatic impacts on the community. Three themes emerged: Preserving the quality of life, transforming the local economy, and empowering individuals. Six key performance areas were identified:

- civic engagement
- economic development
- education
- health
- quality of life
- transportation.

During the first four years, the action plans were implemented through a variety of projects, including the creation of a countywide "adventure pass" to provide access to historical and cultural centers, declaring the

City of Eau Claire an official "eco-community," organizing a free book program for all second and third graders, creating a "jobs road map" to assist unemployed residents, and developing a civic engagement toolkit to support Clear Vision efforts.

Despite challenges in maintaining the momentum in the implementation process, the community continued to innovate and create new opportunities for public participation. In 2009, Clear Vision began convening small civic work groups to be trained in problem solving and relational organizing skills to help in addressing specific community issues. In 2010, a professional facilitator from the University of Wisconsin-Eau Claire worked with the implementation committee to develop a five-year strategic plan to support KPA activities. Among other things, the plan called for the creation of a 501(c)3 nonprofit corporation, which was accomplished in 2012. Later that year, the organization held an empowerment summit to bring together community members to come up with a new set of priorities. Within the next year, 15 community work groups were organized to work with the nonprofit board of directors on the goals identified at the summit.

Dubuque Iowa's Vision

In 2009, during the trough of a severe national recession, IBM announced it was opening a technology center in Dubuque, Iowa, a move that would bring 1,300 jobs to the region. After word got out, an editorial by a TV commentator in Madison, Wisconsin, said: "IBM could have located here, and chose Dubuque. That's just not right." What seemed remarkable to this understandably proud Madisonian was that a small, farm-belt city would be selected instead of a more recognized technology "triangle" or "corridor." The editorialist obviously was unfamiliar with the "new Dubuque."

What the city lacked in glitz or cachet, it more than made up in pluck, public participation, organization, vision, and civic spirit. "It all started in the 1980s when people decided we had reached the bottom and collectively wanted to make it a better community," said Mayor Roy Buol. "The new Dubuque, that's what I call it. People really bought into the idea. There was a common desire to better the community and make it a place where everybody has opportunities, a place people want to come, and when they do come, to stay."

In 1985, Dubuque had one of the highest levels of unemployment in the country, upward of 23%. The city's largest employer, John Deere, recently had shut its doors, and residents were leaving in droves. Old-timers remember when a joker put up a billboard outside town that read: "Will the last person to leave Dubuque please turn out the lights?" A few years later, the city undertook an ambitious public planning process called Vision 2000, in which citizens from across the region met to lay out a road map for economic recovery: The vision that emerged was a "diverse and balanced economic base that provides job security for all segments of the community ... secured through the support, retention, recruitment of retail, manufacturing, hi-tech, services, year-round tourism, recycling businesses and industries."

Focusing on bringing in new industry—insurance, technology, publishing, health care, education, and tourism—Dubuque rose to number one among Iowa's metro centers for job growth. A revitalized waterfront with hiking trails, restaurants, a museum and an aquarium reconnected the city with one of its great resources, the Mississippi River. Vision 2000 was the first of four strategic planning processes that took place in Dubuque during about a dozen years, the latest being Envision 2010 in 2005.

A grass-roots effort by citizens of Dubuque's tri-state (Iowa, Illinois, and Wisconsin) region, the Envision 2010 was kicked off at a community breakfast. Visioning toolkits were distributed and instructions on how to form a work group were provided. Community groups worked over the summer to discuss the questions: "Where do we go from here?" and "What's next for Dubuque?" Each group was asked to submit their ten best ideas for making Dubuque a better place to live, work, and play. Within the next six months, residents met in small and large groups to discuss ideas. More than 2000 ideas were submitted. A diverse, 20-member selection committee was empanelled to review the various ideas. Each idea was submitted to a selection committee, which reviewed the suggestions and edited the list down to ten ideas.

In 2006, community groups volunteered to become owners of the various "ideas" and push forward projects to implement them. Although not all the goals were fully realized, significant progress was made on each of the ten projects. For example, the goal of having fully integrated bilingual programs in all area learning institutions proved elusive, but the local Catholic schools were leading the way in establishing a new curriculum. In the example of the performing arts center, Envision discovered that the University of Dubuque was planning to build a new performing arts center. The group has worked with the university to complete the project by 2013. As Cynthia Gibson wrote of these implementation groups in a *National Civic Review* article, "Their commitment and energy to the process led to a striking achievement: In just two years, every one of the ten ideas had come to fruition in one way or another—often, very close to what was envisioned."

The city's success has not gone unheralded. A partial list of its recent accolades would include three All-America City Awards (2007, 2012, and 2013). The Equal Opportunity Project named it first among 58 cities of similar population when it comes to upward mobility. *Forbes* magazine ranked the city fourteenth in the nation in its list of "Best Small Places for Business and Careers" in August, 2013. Harvard's Kennedy school gave the city one of its 25 Innovations in Government Awards in 2013.

Keywords

Strategic planning, community visioning, civic infrastructure, implementation, evaluation.

Review Questions

1 What is civic infrastructure, and what is its role in community visioning?
2 What are the five major steps of the community-visioning process?
3 How does community visioning relate to community development both in general terms and specific applications?
4 Can you identify an example of a visioning process in your community or another that focuses on community development?
5 What are the major considerations in starting a community-visioning process, and under what conditions could you recommend a community to do so?

Note

1 Parts of this chapter were adapted from *The Community Visioning and Strategic Planning Handbook* authored by Derek Okubo and developed based on the work of staff and board members past and present of the National Civic League.

Bibliography

Chrislip, D. (2002) *The Collaborative Leadership Fieldbook*, New York: John Wiley.
Dewey, J. (1988) *The Public and its Problems* (reprint), Athens, OH: Swallow Press and Ohio University Press.
Doyle, M. and Straus, D. (1982) *How to Make Meetings Work*, New York: Jove Books.
Gibson, C. (2013) "Case Study: Dubuque Community Foundation," *National Civic Review*, 102 (3): 25–30.
Huggins, M., (2012) "Community Visioning and Engagement: Refreshing and Sustaining Implementation," *National Civic Review*, 101 (3): 3–11.
Larsen, C. and Chrislip, D. (1994) *Collaborative Leadership: How Citizens and Civic Leaders Can Make a Difference*, San Francisco, CA: Jossey-Bass.
Leighninger, M. (2006) *The Next Form of Democracy: How Expert Rule Is Giving Way to Shared Governance—and Why Politics Will Never Be the Same*, Nashville, TN: Vanderbilt University Press.
National Civic League (1999) *The Civic Index*, Denver, CO: National Civic League Press.
Okubo, D. (2000) *The Community Visioning and Strategic Planning Handbook*, Denver, CO: National Civic League Press.
Peck, M.S. (1993) *A World Waiting To Be Born: Civility Rediscovered*, New York: Bantam.
Straus, D. and Layton, T. (2002) *How to Make Collaboration Work: Powerful Ways to Build Consensus, Solve Problems, and Make Decisions*, San Francisco, CA: Berrett-Koehler.

Connections

Learn more about Appreciative Inquiry, a way to gauge and connect with community desires, assets and vision: http: //appreciativeinquiry.case.edu

Explore how to integrate sustainability into community visioning at: www.sustainable.org/creating-community/community-visioning

Delve deeper into the community-visioning process with this excellent guide from the University of Wisconsin-Madison's Extension service guide at: http: //oconto.uwex.edu/files/2010/08/G3708-BuildingOurFuture-AGuidetoCommunityVisioning.pdf

9 Establishing Community-Based Organizations

Monieca West, Patsy Kraeger, and Timothy R. Dahlstrom

Overview

By definition, community development is about organizing people and resources to accomplish common goals. Therefore, one of the most fundamental questions facing community developers is, "How should we be organized?" This chapter identifies some important issues that should be addressed before a community-based organization (CBO) is formed and provides an overview of different types of CBOs. Entrepreneurial organizations are one type of CBO while others take more traditional formats. We look at traditional community development organizations as well as entrepreneurial organizations. This chapter will introduce the need for a wider organizational lens to be considered. We also look at new actors and organizations in the social economy. These new actors are called social entrepreneurs and new hybrid organizations are driving change in community development. For those new to the field, students and even experienced community developers, this chapter can serve as a "how-to" guide for establishing CBOs. Examples of different types of CBOs are provided.

Introduction

So, you want to start a community-based organization (CBO) or a community development corporation (CDC). You've looked around the community and found something you think should be or could be made better, be it economic development, health care, housing or transportation. You may start a venture on your own if you are an entrepreneur to meet a community need. You might decide to open a hybrid organization which is a new organization in the community developer's toolbox. You are all committed to doing something positive alone or in concert with other community members. That usually leads to forming an organizational committee that will lead to a formal community-based organization that will lead to... Stop! Before the first bylaw is written, before the first officer is elected, there are some basic organizational structuring steps that, if taken, will save time and effort and create a more productive and sustainable organization. A little thought and deliberation at this juncture will pay great dividends over time.

Starting a community organization should be approached with as much care as starting a small business...because that's exactly what you are contemplating. Both must plan, market, manage staff, create revenue streams and maintain cash flow. The important point for CBOs is to give as much attention to the process of running the organization as it has passion for the project.

Fundamentals of Forming a Community-Based Organization

Whether the CBO is organized by a social entrepreneur, an informal steering committee, a traditional board/committee, or as a complex network of organizations, there are several fundamental questions that should be answered by all start-up organizations. Get these fundamentals right and most everything else will fall nicely into place.

What Do We Want to Do?

To determine this, make sure you're asking the correct question. The initial question should not be, "How do we become a CBO?" Community leaders should ask questions, such as, "How can we do something better?" or "How can we do it differently?" Community developers can also ask, "How can we use our unique resources?" or "What can we do that no one else can do?" By asking these questions, community actors stimulate entrepreneurial ideas which will allow for real change to occur.

Keep focused on the destination, not the vehicle in which you will be traveling. This is not always as easy as it sounds, especially in the formative stages of group process when strangers come together to pursue a passion. Some may get bogged down in the mechanics. Others are driven by passion and may want to skip the mechanics of building the infrastructure. While others, may get lost in the debate and go elsewhere with their time and talent. Community-based organizations are different than for-profit firms because there are a myriad of stakeholders and viewpoints bringing complexity to structuring the organizations. What is the community change that actors want to see happen? Articulate the change by developing an organizational mission which can then be actualized through vision and plans for execution. Once you understand the mission, the vision, and who the stakeholders are, this chapter can be used as the compass to guide all future decisions.

Is There Anyone Already Doing This?

The next logical question would then be, "Is there anyone else already doing this?" Determine if there are organizations already addressing your concerns because the best interests of the community are not served when too many organizations compete for the same resources. You will want to look inside of your community to see not only who the actors are driving change but what types of organizations are carrying out the work? Are they grass-roots member organizations without a structure that seek to formalize infrastructure? Are they traditional nonprofit entities who are working on the community issues? Are new entrepreneurial ventures established through social enterprises called "hybrid corporations" by social entrepreneurs? For community developers, considering these various organization types may hold promise for local development. If there are others, you may elect to simply join their group, before proceeding with establishing your organization. Community actors may decide to form a collaborative arrangement with the existing organization or to focus on another issue in community development. Just be sure you are not fragmenting scarce community resources with your admirable zeal to do good.

Community actors also need to understand that they are working in the context of the larger social economy. The social economy captures economic growth as inclusive of both commercial, including new hybrid companies, and noncommercial activities such as accounting for unpaid work along with paid work.

Mission and Purpose

A clearly defined mission statement is critical, but even more critical is its constant use to guide strategic decision making. It establishes what will be done, who will be involved, and how the community will be affected. The mission statement is what keeps your ship in the correct ocean.

While crafting the mission statement, it is important to recognize that there are six different types of community development functions and people—entrepreneurs, organizers, developers, planners, resource providers, and social entrepreneurs—and that each has different purposes and expected outcomes.

Entrepreneurs are not usually the creative geniuses with revolutionary products of entrepreneurial lore. Rather they are mostly regular people who see a way to do something differently or to meet a need and want to make a living with their idea. The typical entrepreneur looks like your next door neighbor. While this may not be an exciting portrait of an entrepreneur, it does show that entrepreneurs are not limited to a certain type of person or a certain geographic area. Rather, entrepreneurship potential exists in every community.

Organizers are about advocacy and empowerment, about influence and being heard, about applying political pressure, or staging protests. Organizers are likely to congregate in neighborhood coalitions organized around social issues. These groups can focus attention on a need previously ignored and provoke resource commitments previously unavailable.

Developers are project-centered and about doing, creating, and building. Development projects can require substantial resources to deliver the products or services making the organization dependent upon outside support and relationships. Developers require a wide range of technical and administrative skills to accomplish complex and time-consuming, long-term projects. Development organizations vary widely in scope and can be independent or networked, simple or complex.

BOX 9.1 CHANGE AGENTS: SOCIAL ENTREPRENEURS: MEET DOUG RAUCH, FOUNDER OF THE DAILY TABLE

Doug Rauch, the former President of Trader Joe's Company, a supermarket chain, decided to look at food slightly past the sale date as a source of abundance for poorer communities. He is opening the Daily Table grocery store and restaurant in Dorchester, Massachusetts. Rauch was with Trader Joe's for 31 years, 14 as president where he grew a small local chain of nine stores into a national chain. Rauch has now moved onto to community change ideas with his new project the Daily Table where he will serve food slightly past the sell date at junk food prices.

In 2012, Rauch became aware that 40% of food was thrown away by consumers. "Rauch told the Boston Globe that many inexpensive meals available are often unhealthy. For that reason, the Daily Table will be aimed at lower-income consumers in the Boston area" retrieved from huffingtonpost.com. (Doug Rauch, Trader Joe's Ex-President, to Launch Store Selling Expired Food, *Huffington Post*, September 23, 2013).

Junk food is cheap. Rauch likens what $3.00 can buy in junk food and the relating 3,700 calorie count in a recent profile in the *New York Times*. Rauch recognizes that the sell by date is often artificially manufactured and he can offer a better alternative in a poorer community. He was able to get community buy-in by holding meetings to see if there was not only a need but a desire by the community for such a product. He disagrees that he is selling poor people the scraps from the rich. "I've been down to Dorchester and held a number of neighborhood meetings. When people hear that I'm interested in only recovering wholesome, healthy food and using that to bring affordable nutrition, it has actually received a very positive response" (retrieved from www.nytimes.com/2013/11/10/magazine/doug-rauch-wants-to-sell-outdated-food-at-junk-food-prices.html).

Rauch is taking a calculated risk to create a new equilibrium in the food market—selling slightly outdated food that is still good at junk food prices. Rauch read reports documenting that sell by dates were not scientifically valid and led to too much food waste. He used this new information to bring a new product to the community, a food desert, to create abundance at junk food prices. His skills in the retail market will serve him well with his new project the Community Table.

Planners are about visioning and charting a course of action. They compare the past to the present and develop a road map for the future. Planners examine possibilities and pull the pieces together into a larger picture. While elements of planning are found in all CBOs, some organizations form purely for the purpose of strategic planning and must be able to bring together diverse groups to create a shared vision. It is not uncommon for planning organizations to complete the planning process and then regroup into development organizations.

Resource providers are just what the name implies. They are about giving and assisting. Providers include private and public charities, nonprofit organizations, government agencies, private individuals and businesses, faith-based organizations, and human service agencies.

Social entrepreneurs bring private sector business practices to community development because they embrace and seek change. They take calculated risks to shift resources from low to high productivity in order to create something new. Social entrepreneurs are often seen as individuals but they can also be organizations

Is the Organization Feasible?

A worthy cause must be supported by a workable organization plan. The plan should include competent leadership and sufficient funding in order to actualize the organization. Capacity includes:

1 Short- and long-term strategic plans.
2 Professional and/or volunteer staffing with effective leadership and management skills.
3 Sufficient available facilities and equipment.
4 Sustainable financial resources and sufficient cash flow to provide for the organization's operations.

Can We Pay the Bills?

"It takes money to make money," is as true for nonprofits as it is for private business. Capital in the nonprofit world takes many forms; it is often more than money. It may be cash, in-kind donations, volunteer time contributions which can be monetized through social counting methods or other institutional support. The bottom line is that the organization must have sufficient start-up support to get organized and secure initial and sustained funding. Most small businesses fail because of inadequate cash flow and CBOs are no different. Another way to look at the internal/external capacity of the organization or whether the organization can pay its bills is to conduct a SWOT analysis as part of the initial strategic planning. A SWOT analysis looks at the strengths, weaknesses, opportunities, and threats. This type of analysis allows the organization to look at what it does better or worse than others.

Leadership of the organization can then look for new and perhaps entrepreneurial opportunities for solving community problems. It allows a community organization to see if it is duplicating what is already being done. This is important when the organization thinks about opportunities relating to funding its mission and programs. Threats to organizational viability include an environmental scan of funding opportunities from grants and donations to social enterprise fees for services, etc.

This type of organizational capacity analysis leads to the credibility and influence of those leading the organization. A strong track record of due diligence coupled with feasible plans and a constituency base will serve the organization well. If the organization and/or its leaders also show a demonstrated ability to achieve results, they will achieve much buy-in from the community when rolling out operations.

These questions have driven community leaders to think about the business and entrepreneurial skills that they can bring to community development. Entrepreneurial skills are the range of skills necessary to bring an idea to market and manage the business. They include product/service, management, marketing, finance and accounting, organizational development, self-discipline, and self-awareness. These community actors will now also recognize that there are a variety of new organizational structures such as the benefit corporation (BCorps) and the low profit limited liability (L3C) company to support community change through venture capital.

Is There a Business Plan?

An important development process for any new venture is business planning, and producing a written business plan is a common way to formalize that planning. Putting a business plan together forces an entrepreneur to take an objective, critical, unemotional look at the project in its entirety, and it becomes the roadmap to the future (Robinson, 2009). The most important use of a business plan is to ground the organization in reality and provide operational guidance for daily decisions. A business plan should include:

1 A description of the organization (what it will do and how it will help).
2 A marketing plan (who will be served and how they will be reached).
3 A financial plan (start-up financing, expenses, revenues, and cash flow).
4 A management plan (how the idea will be implemented and by whom).
5 Key metrics (important indicators of success).
6 Future projections (the financial and operational future for the next few years).

While experts disagree about the length of a business plan, they generally agree that the content should include the items above. These items can be covered in detail, in multiple pages, or in a shorter bullet-point-type format. The choice depends on a few important considerations. Who will use the plan? A plan to obtain financing, for example, may be more detailed about financial projections than a plan that will be used for operating the business. Did the entrepreneurs who started the venture develop the plan? It is more important for entrepreneurs to work through the planning process than simply to have a plan written for them. Does the content fit the venture need? A business plan must show knowledge of the proposed business and its environment and be able to communicate that to others. But the business needs should drive what goes into the plan.

A business plan has many benefits. It increases the chances of securing funding, helps identify the strengths and weaknesses of the venture, and provides a way to measure actual results against what was planned. There are some who will resist writing a formal business plan, seeing it as a waste of time. However, a business plan is a management tool that orients all parties in the same direction. It should be updated as changes occur and used as a benchmark to gauge success. There are a large number of good resources to assist entrepreneurs in writing a business plan. These include websites for the U.S. Small Business Administration (www.sba.gov) and SCORE (www.score.org/resources) and the book by Bill Robinson (2009).

Local Community Development: Sectors, Organizations, Networks and Partners

Community-based organizations do not exist in a vacuum. They must establish local relationships as well as partnerships with state and federal agencies and organizations throughout the public, private, and nonprofit sectors.

Public Sector

Public Sector Federal Programs

A review of federal agencies indicates that resources from federal agencies most commonly flow to their own regional branches, to state and local government entities, to regional economic development authorities, to educational institutions, or to nonprofits. These entities may then distribute the resources further until they are received by a local development organization, an intermediary linking government funding with a private sector entity, a nonprofit, or an individual consumer.

The Catalog of Federal Domestic Assistance (cfda.gov) is an exhaustive listing of federal programs searchable by function, agency, program title, eligibility, and beneficiary available to state and local government which includes the District of Columbia as well as federally recognized Indian tribes and territories and other entities including nonprofit organizations. Federal agencies with a significant community development mission and some of their most recognizable programs are delivered through the federal Departments of Health and Human Services, the Interior, Agriculture, Housing and Urban Development, and Justice. It is a basic reference source to help identify users of the site to find and align specific program offerings with work in the community.

Public Sector State Programs

Every state has an agency that delivers economic and community development programs using a

Table 9.1 *Some Federal Programs for Community Development*

Federal department	*Programs and focus areas*
Department of Agriculture:	Office of Rural and Community Development
www.usda.gov	National and State Rural Development Councils
	National Rural Development Partnership
	Rural EZ/EC/CC*
Department of Commerce	Economic Development Administration
www.doc.gov	Minority Business Development Agency
	Economic Development Districts
	University Centers
	Small Business Innovation Research Program
Department of Housing & Urban Development	Community Renewal Initiatives
www.hud.gov	Rural and urban EZ/EC/RC*
	Rural Housing and Economic Development
	Community Development Block Grants
Department of Health & Human Services	Food and Drug Administration
www.hhs.gov	National Institutes of Health
	Administration for Children and Families
Department of Labor	Office of 21st Century Workforce
www.dol.gov	Center for Faith-Based Community Initiatives
National Science Foundation	Division of Grants and Agreements
www.nsf.gov	R&D in science and engineering fields
Small Business Administration	Small Business Development Centers and sub-centers
www.sba.gov	Women's Business Centers
	Office of Business and Community Initiatives

Sources: Individual agency websites.

*Enterprise Zones, Enterprise Communities, Champion Communities, Renewal Communities.

variety of organizational models. The superagency structure combines all functions of primary economic and community development activities —such as economic development, workforce development, and tourism—out of a single centralized administrative unit. The regional structure is also a single department but services are delivered through a clearly defined network of regional offices and regional advisory councils. The umbrella-structured agency centralizes policy and administration with daily operations executed by a network of private nonprofit corporations and related state agencies (tourism, workforce development, etc.). The private-sector-structured agency basically outsources its programs through a state-level public–private partnership.

State development agencies also provide community betterment programs with awards and recognition when a community meets certain levels of achievement in community development. These programs may be run entirely by the state agency or in partnership with other organizations. It is also common for an independent organization to take the lead in these programs with support coming from the state agency.

Public Sector University-Based Programs

University Centers are partnerships between federal government and academia that mobilize the vast resources of universities for development purposes. Partially funded by the Economic Development Administration, these centers perform research to support development of public policy and economic programs. They may also initiate special development projects and consult with government agencies, businesses, media and the general public. Local universities play important roles in business development through business incubators for technology or biotechnology and are the funnel through which extremely important federal research and development dollars floor. Fully accessing the resources of colleges and universities should be a priority when organizing the community for development.

BOX 9.2 UNIVERSITY—COMMUNITY DEVELOPMENT CONNECTION: THE GREAT CITIES INSTITUTE

Great Cities Institute at the University of Illinois at Chicago (UIC) is the articulation of the university's ongoing commitment and responsibility to serving Chicago and the surrounding metropolitan region. As a hallmark of UIC's Great Cities Commitment, the Institute provides research, policy analysis, and community development support to a range of constituencies. Great Cities Institute (GCI) is an example of how universities can be of service to organizations and agencies dedicated to improving the quality of life in their communities.

In its early years, Great Cities Institute received funding from HUD's Office of University Community Partnerships' Community Outreach Partnership Center (COPC) grants. These funds have since been discontinued but represented the federal government's commitment to incentivize the linking of the intellectual capital of universities with the efforts of community-based organizations.

GCI's Neighborhood Initiative was born out of its COPC grant. Technical support, leadership training, direct service, and collaborative community development projects were some of the many efforts connecting university resources and community needs. More recently, as part of its work with neighborhoods, Great Cities Institute is fully engaged with the Participatory Budgeting Project throughout the city of Chicago. GCI works directly with aldermen and community members to facilitate participatory processes that engage residents with local government decision making through a community-based proposal and voting process to identify the best use of capital improvement funding at the neighborhood level. In seeking citizen empowerment and transparent and responsible government, the Participatory Budgeting Project benefits from Great Cities Institute program support including that related to nonprofit management, civic engagement, and the technical side of municipal budgeting.

Great Cities Institute convenes researchers, policy makers, and advocates on issues related to employment and economic development, local and regional governance, and the dynamics of global mobility. The Institute addresses these issues at multiple scales from the neighborhood to the mega-region. By engaging a range of partners, Great Cities Institute provides a space to analyze issues and formulate directly applicable solutions to tackle the multifaceted challenges facing communities. GCI represents UIC's commitment to "engaged research," thus highlighting the value of a university research center for contributing to community development practice and addressing today's urban problems.

Teresa Córdova, Ph.D.
Director, Great Cities Institute
University of Illinois at Chicago, USA
https: //greatcities.uic.edu

The Cooperative Extension Service was founded in 1914 and is connected to land-grant universities. CES programs have grown far beyond the original county extension agents and home economists, with programs including Master Gardeners, 4-H, leadership development, business development, and building healthy communities for the twenty-first century.

Public Sector: Small Business Development

The Small Business Development Centers are partnerships between the U.S. Small Business Administration (SBA), state or local governments, and universities. The SBDC network has a central statewide office and also includes regional sub-centers which provide assistance to business owners. SBDCs offer entrepreneurial training and counseling. They also provide access to financing and other referral services. They are usually housed in community colleges, so they also have connections to formal entrepreneurial education programs and many state-level programs.

A nonprofit association called SCORE offers free business counseling to aspiring or active entrepreneurs. SCORE can help with starting a business or with solving a need in an existing business. They also conduct a number of seminars at low cost on business topics relevant to new or small businesses. SCORE offices are often located within offices of the SBA. In addition to in-person counseling, counselors can work with entrepreneurs by phone or email. SCORE counselors have experience in entrepreneurship or business management and have received training in business counseling and business plan development. Finally, the SBA has a large number of training and referral opportunities. Each district office has a great deal of information about the state business climate, and how to start and manage a small business. Information is available both in person and online. The SBA also has a large amount of assistance information and tools on their website: www.sba.gov.

Chapter 17 focuses on entrepreneurship in the community development context. It is important to consider for this chapter how entrepreneurship directly informs community action. The mindset of an entrepreneur is typically the mindset of a community developer—one who seeks to solve complex problems with new solutions either rapidly or over time. In the private sector, we have seen a new actor—the social entrepreneur—come on to the scene. As part of the discussion about entrepreneurial talent in the community context, let us look at social entrepreneurs. Social entrepreneurs create programs and resources that benefit our communities and our lives. Often, this work is done through a nonprofit organization or informal community or neighborhood association. Social entrepreneurs need skills in planning their enterprise, marketing their product or service, earning revenues or obtaining funding to keep the organization financially solvent, and creating value. As with business entrepreneurs, they perceive and act upon opportunities.

Private Sector

Private sector companies and organizations participate in community development through grants, technical assistance, and staff involvement in community leadership roles. Utilities, banks, and private developers are the most common private enterprises to undertake community development activities. Private and community foundations are also involved in community development in order to co-create community change. These organizations are generally involved in activities that relate to their specific business mission and/or that will create revenue-generating opportunities for them. Electric utilities in particular offer extensive support in prospect leads and responses, community preparation, and leadership development. Corporate grants and contributions, while often small in scale, can be valuable supplemental resources for community organizations. Community actors should seek to establish relationships with the local businesses, chambers of commerce, and other economic development players. While it is important to seek relationships with larger establishments, community actors will find that often entrepreneurs share the same community goals that they do for sustainable communities. Changing economic times have seen more community organizations fail due to increased demand and lack of traditional funding. Entrepreneurs and social businesses are now a part of this landscape.

Types of Community-Based Organizations

Community-based organizations are too numerous to fully describe, but they can generally be categorized as independent organizations including social benefit firms and foundations, networks, partnerships, or regional initiatives.

Independent Civic Organizations

The local chamber of commerce is perhaps the most pervasive community-based organization and represents the fundamental organizational structure for CBOs—a volunteer board of directors with committees. Chambers pursue a variety of interests including economic development, leadership programs, community promotions, and governmental affairs, and are as diverse as the communities they serve. Because chambers are such familiar organizations, there is little need to describe them in detail, but they should be recognized for the direct and indirect impact they have on a community. A community with a strong, progressive chamber is likely to be a very successful and competitive one.

Other local civic organizations include Business Improvement Districts, Downtown Partnerships, and Main Street programs. The Rotary Club, PTA, and the Boys and Girls Club are other examples of local CBOs that need to be brought into the mainstream of community development at the local level.

Community Development Corporations

A community development corporation (CDC) is a nonprofit organization that serves a particular geographic area and is normally controlled by its residents. A board of directors, usually elected from the membership, governs CDCs and may also have board positions reserved for representatives of key local institutions such as banks, city government, or the hospital. A paid staff and volunteers execute the programs of work.

Some CDCs focus on one issue, such as housing, while others pursue a wide range of activities from grass-roots advocacy to job creation. CDCs may vary in focus but they have community in common. In many ways, CDCs are the purest example of a grass-roots development organization.

CDCs depend upon collaborative efforts and must be very astute in developing relationships

BOX 9.3 COMMUNITY DEVELOPMENT CORPORATIONS (CDCS) IN MASSACHUSETTS

Community Economic Development Corporation of Southeastern Massachusetts
Structured with a board of directors, paid staff, consultants, and volunteers, CEDCSM is representative of a CDC formed primarily for organizing and advocacy. Originally founded by a community action agency, CEDCSM has created partnerships between private sector employers, immigrant workers, organized labor, grass-roots activists, entrepreneurs, banks, local and state government and many others. Programs include a microenterprise network, computer centers, workers support network, training and support for all-volunteer grass-roots groups, and affordable housing (CEDCSM).

Massachusetts Association of Community Development Corporations
MACDC is the statewide policy- and capacity-building arm of the CDC movement in Massachusetts. Its board of directors includes representatives from member CDCs and its extensive staff provides support in the areas of economic development, affordable housing, community organizing, advocacy, and building CDC capacity. One relatively new tool that MACDC has deployed is the Individual Development Account (IDA). IDAs are income-eligible savings accounts that are matched anywhere from a 1:1 to 4:1 ratio from a combination of private and public sources. The funds are then used for one of three purposes—education for themselves or their children, business capital, or the purchase or repair of a home. MACDC and the services it provides are greatly responsible for the successful CDC movement in Massachusetts (MACDC).

and building partnerships. The importance of a strong support system for local CDCs cannot be overstated. Typical partners include faith-based institutions, nonprofits, government departments and agencies, private developers and businesses, banks, national and state intermediaries, social service agencies, schools and colleges, and others related to the specific mission of the CDC. These organizations also provide the volunteer base that is critical to CDC success. CDCs defy a single definition because they are custom-made to fit each community.

Networks

A network is formed when two or more organizations collaborate to achieve common goals; to solve problems or issues too large to face independently; to leverage the power of numbers in exercising influence or flexing political muscle; to maximize limited financial and human resources of a community by reducing duplication of organizations; or to operate more efficiently in concert with others.

Information Exchange Networks

Networks may be as simple as an arrangement to exchange information among organizations or businesses that share common interests such as trade associations or chamber-based Leads Groups. Leads Groups are structured so that there is only one company per business type represented in the group—one bank, one utility, one lawyer. Group members share business opportunities, and doing business with one another is encouraged.

Service Delivery Networks

Other networks, such as the International Council of AIDS Services Organizations, are very sophisticated and can be very high maintenance. ICASO is a massive network of numerous CBOs providing prevention and/or treatment services for AIDS patients and is organized from the local to international levels (ICASCO 1997). Networks can be very effective in addressing social issues, especially those that are multidimensional.

Flexible Manufacturing Networks

Flexible manufacturing networks (FMN) gained popularity in the US in the 1980s and are still an important economic development tool. FMNs may limit their partnerships to simple exchanges of information or may go so far as to jointly own production facilities or share IT infrastructure. They can be structured on the strength of a handshake or through legal, contractual arrangements. The catalyst for the FMN can be an existing organization such as a trade association, a government agency that organizes new private or nonprofit organizations, or a group of firms that organize themselves in response to market conditions. In most cases, an FMN is built with the services of a broker, someone who helps companies form strategic partnerships, organize network activities, and identify business opportunities. The broker may be an employee, a consultant, a government agency, or a nonprofit.

The wood products industry was among the first to embrace the FMN concept. In Arkansas, it was a nonprofit—Winrock International—that was the catalyst for formation of the Arkansas Wood Manufacturers Association (AWMA), which now offers both youth and adult apprenticeship programs and provides collaborative marketing and support for member companies. In Kentucky, it was the state legislature that enacted legislation to develop the Kentucky Wood Products Competitiveness Corporation to develop secondary wood industry business networks.

Business Cooperatives

A business cooperative is special-purpose for-profit business. It exists for the betterment of the community rather than for a limited number of owners or shareholders. It allows members of a community to pool their talents and resources in order to give a community a sense of belonging and an economically viable business. The unique feature of a business cooperative is that it is jointly owned and democratically controlled. As such, cooperatives are member-owned businesses that operate for the benefit of members. Cooperatives often pursue social as well as economic goals and are active in many industries, including telecommunications, agriculture, insurance, art, and credit unions.

BOX 9.4 THE ACENET COOPERATIVE

In 1985, a small group of community members in southeastern Ohio established a number of worker cooperatives to help low-income people start worker-owned businesses. This evolved into ACEnet, the Appalachian Center for Economic Networks, and was based upon models found in Spain and Italy. In 1991, ACEnet started a small business incubator to serve the food sector market niche. The network clustered food processing companies, trucking firms, and restaurants and provided support services, access to capital, and cooperatively owned equipment.

Over time, ACEnet spun off freestanding organizations to further enhance network services. The Food Ventures Center provides technical and start-up assistance to local organizations and entrepreneurs, and access to shared warehouse and equipment and an automated distribution hub. TechVentures helps business owners integrate computers into their business operations and also opens up the computer training center to the general public. The Computer Opportunities Program trains teachers in local schools to offer entrepreneurship classes and sets up student-owned businesses. A separate nonprofit subsidiary, ACEnet Ventures, provides venture capital to companies in the network. For more information on cooperatives, see Box 9.5.

The Editors

BOX 9.5 COOPERATIVES AND COMMUNITY DEVELOPMENT

Essentially, cooperatives are people organizing a business activity for mutual benefit (Bloom 2010). It is not a new concept (think thirteenth-century craftsmen guilds), and organizing cooperatively has a long history across cultures. Cooperative organizational structures have enjoyed a resurgence of interest, with over 800 million members of cooperatives around the world, in many different industrial sectors. While there are a number of ways to characterize the diversity of cooperative forms, traditionally cooperatives have been divided into five broad types: (1) consumer; (2) producer; (3) marketing and purchasing; (4) agricultural; and (5) social cooperatives (Gonzales and Phillips 2013). Cooperatives and community development intersect at several junctures, including foundational tenets such as cooperative management principles of collective action, democratic decision making, and the desire for improving or progressing to a better situation. This improvement impetus is "a primary motivator for organizing and taking action (and) while cooperatives focus on improvement/meeting the needs of members, community development desires to improve the well-being of the overall community" (Gonzales and Phillips 2013: 5). The International Co-operative Alliance, housed in Switzerland, provides the following core principles of cooperatives, many of which overlap with foundational principles of community development:

1 Voluntary and open membership.
2 Democratic member control.
3 Member economic participation.
4 Autonomy and independence.
5 Education, training and information.
6 Cooperation among cooperatives.
7 Concern for community.

(ICA 2013)

An example of a cooperative that is community owned and has generated community development benefits is Casa Nueva Restaurant and Cantina in the small town of Athens, Ohio. Organized as a worker-owned cooperative business, it is owned and run by 25 members each providing equity investments. Casa Nueva is committed to promoting healthy, locally sourced food at affordable prices. After 17 years in business, the cooperative provides positive local impacts with over $1 million in annual sales.

The Editors

Reference

International Cooperative Alliance (2013) "Core Principles." Available at: http://ica.coop/en/whats-co-op/co-operative-identity-values-principles (accessed February 24, 2014).

Foundations

Private, nonprofit foundations are a modern phenomenon based on the ancient notion of social capital and community good. The formal, professional, and bureaucratic organization of today represents institutionalized nongovernmental philanthropy. This section discusses both private and public foundations including the community foundation. Modern institutions came into being when the likes of Andrew Carnegie and John D. Rockefeller, Sr. created organizations with broadly defined community and social purposes. "Their governance was to be private; relying on self-perpetuating boards of trustees or directors, but their missions would be to serve the public good" (Smith 1999). These new philanthropic trusts or foundations sought to build social capital for the betterment of the community. Private foundations are primarily funded by the private wealth of a single individual or a family. Both private and community foundations are nonprofit, tax-exempt public charities. The Council on Foundations (COF) is a national organization that supports foundations development.

BOX 9.6 COUNCIL OF FOUNDATIONS

Independent foundations are distinct from other kinds of private foundations like family or corporate foundations in that they are not governed by the benefactor, the benefactor's family, or a corporation. They are usually funded by endowments from a single source such as an individual or group of individuals.

Family foundations are usually funded by an endowment from a family. With family foundations, the family members of the donor(s) have a substantial role in the foundation's governance.

Corporate foundations (or company-sponsored foundations) are philanthropic organizations that are created and financially supported by a corporation. The foundation is created as a separate legal entity from the corporation, but with close ties to the corporation. Companies establish corporate foundations and give to programs to have a positive impact on society. Corporate foundations tend to make grants in fields related to their corporate activities or in communities where the corporation operates, or where their employees reside. Corporate foundations are usually set up as private foundations, but can be created as public foundations, particularly if they will be largely publicly supported. Rather than establishing a separate foundation, a company can also make gifts and grants directly to charitable organizations through a program within the company itself. This is called a corporate giving program.

International foundations typically are foundations based outside the United States that make grants in their own countries and overseas. The term "international foundations" also can refer to foundations in any country that primarily engage in cross-border giving. Not all foundations that engage in cross-border giving are private foundations; many are established as public charities. Under U.S. law, contributions from U.S. donors and corporations are not eligible for a charitable deduction if the organization is not formed in the United States or recognized by the United States as charitable. For more information on international foundations, visit the Council's international grant-making website at www.usig.org; for information on Council of Foundations, see www.cof.org/templates

As private philanthropy was developing in the United States at the turn of the nineteenth century, Frederick Goff, a banker started the first Community Foundation in Cleveland in 1913. Community foundations, on the other hand are primarily funded by contributions from individuals, corporations, government units, and private foundations. Community foundations are public charities which include a multitude of organizations, hospitals, and schools that make grants to other organizations.

Both private and community foundations focus more on grant making than on providing direct charitable services. Most importantly, through an organized and deliberate effort, they increase awareness of philanthropy and provide the community with a systematic approach for charitable giving. Community foundations are endowed by individuals or groups and may be for general or specific purposes. It is common for there to be a statewide community foundation that makes grants in addition to providing support infrastructure for the network of local foundations.

BOX 9.7 COMMUNITY FOUNDATIONS

The Arkansas Community Foundation is a good example. ACF coordinates the work of 26 affiliated foundations across the state. In addition, ACF provides staff support and manages the investment portfolio for many of these organizations (see http: //arcf.org).

Regional foundations are also established such as the Foundation for the Mid-South (FMS). These regional foundations create partnerships between statewide associations of nonprofits, community development corporations, faith-based organizations, and other CBOs (see www.fndmidsouth.org/home.php).

Table 9.2 *Charitable Foundations*

Private foundations	*Description*
Family	Donor's relatives plan a significant governing role.
Independent	Makes grants based on charitable endowments.
Operational	Uses the bulk of its resources to provide a service or run a charitable program of its own.
Company	Corporate giving programs.
Public foundations	
Community foundations	Funded by public giving and focuses more on grant making than on providing direct charitable services.

Source: Council on Foundations.

Table 9.3 *Faith-Based Organizations*

Type	*Description*
Congregational	Individual congregations and their denominational organizations, such as the Roman Catholic archdiocese, and their service organizations, such as the Jewish Federation or Notre Dame University.
National networks	National networks of special-purpose providers formed to mobilize energies of individuals and congregations around specific projects such as Habitat for Humanity or the Christian Coalition.
Freestanding	Independent organizations that are not part of a congregation but have a religious connection organized to pursue special development objectives.

Source: U.S. Department of Housing and Urban Development.

Social Enterprise Firms or Hybrid Corporations

Social entrepreneurship or hybrid firms are currently in vogue. While we do not know what community impacts these firms produce, they are new organizations in the community development matrix of organizations that are a mix of for-profit and community benefit typically run by entrepreneurs. The contributions of social benefit firms suggest that they are free of the constraints that community benefit firms face because they self-sustaining. They can receive venture capital and they have the ability to make a profit to support the social mission rather than rely on institutionalized philanthropy and individual donors.

Benefit Corporations

Benefit corporations, or B corps, share three characteristics: (1) a requirement of a corporate purpose to create a material positive impact on society and the environment; (2) an expansion of the duties of directors to require consideration of non-financial stakeholders as well as the financial interests of shareholders; and (3) an obligation to report on its overall social and environmental performance using a comprehensive, credible, independent and transparent third-party standard. "There are currently 521 B Corporations with approximately $2.9 Billion in Revenues across 60 Industries. BLab, a nonprofit organization has set out standards based on community impact. Corporations are assessed and then accept the

BOX 9.8 NEW IDEAS: SOCIAL BENEFIT AND HYBRID SOCIAL CORPORATIONS

Hybrid social ventures are emerging organizational phenomena. In the past if you wanted to make money, you opened a private firm. If you wanted to do good, you started a nonprofit or a community-based organization. The new dominant view is the shareholder return on investment (ROI) theory that is moving into the community sector. We must consider multiple bottom lines to shareholders and community stakeholders.

An example of multiple bottom lines: It could be their community, it could be their suppliers. For example, Ben and Jerry's could hit cream crises (an oversupply of cream making prices collapse). When one of those cycles hit, and the price of cream fell below the cost of maintaining the farms, then Ben & Jerry's continued to pay the previous year's prices to support the family famers.

(www.businesspundit.com/hybrid-companies-and-the-future-of-the-economy-an-interview-with-criterions-andrew-greenblatt)

suggested legal framework. Approximately nine states have B corporations and they are: California, Hawaii, Maryland, New Jersey, New York, Virginia and Vermont, Delaware and Arizona" (retrieved from www.bcorporation.net).

The L3C: Low Profit Limited Liability Company

The L3C is a hybrid between nonprofit and for-profit firms. The firm has a double bottom line. It is a profit-seeking (not maximizing) firm that has a social orientation. In other words, it is an organization that combines or supports a social mission with market-oriented methods (i.e. low profit/return on social measure). Vermont has enacted legislation allowing for the L3C. Components of the statute are: (1) the company significantly furthers the accomplishment of charitable or educational purposes and would not have been formed but for the company's relationship to the accomplishment of charitable or educational purposes. The company must comply with the established by the Internal Revenue Service. (2) The company does not have any significant purpose to produce income or the appreciation of property; if this does occur, it is not conclusive evidence of a single purpose on income production. (3) No purpose of the company is to accomplish one or more political or legislative purposes. Legislation for L3Cs has been enacted in the following states: Illinois, Louisiana, Maine, Maryland, Michigan, North Carolina, Rhode Island, Utah, Vermont, and Wyoming.

Faith-Based Community Organizations

Faith-based organizations (FBOs) are a strong presence and resource in a community. Some are nonprofit organizations while others are grassroots organizations without formal structures. There are some hundreds of thousands of congregations in the United States with estimated yearly expenditures in the billions of dollars. Places of worship have long been noted for their prominence in providing food, clothing, and shelter to people in need. More recently, however, FBOs have expanded their areas of involvement and are now quite active in workforce training and housing initiatives and, with federal funding of FBOs gaining wider acceptance, the trend is likely to continue. Community development corporations are common examples of congregational sponsorship of a freestanding organization.

BOX 9.9 FAITH-BASED CDCS

The Abyssinian Development Corporation was established by the Abyssinian Baptist Church in Harlem and received major in-kind support from the church including office space and infrastructure, donated management services from skilled congregational members, and extensive volunteerism from among church members.

Allen Methodist Episcopal Church in Jamaica, New York was originally structured as a coalition of urban and suburban churches following the 1968 Newark riots and has grown to offer its own affiliate organizations that provide economic development and social services initiatives.

Other examples of faith-based projects also include economic development projects such as commercial real estate developments, full service credit unions, microloan funds, and workforce training programs. The Jobs Partnership began in Raleigh, North Carolina when the owner of a construction company had to turn down business because he didn't have enough qualified employees and the pastor of his church had congregants desperate for work. Beginning very informally, the Partnership has now been replicated in at least 27 cities and has placed over 1,500 individuals in well-paying positions.

Faith-based programs have several advantages including access to volunteers, access to financial and other types of resources resident within the members of the congregation, and a reputation as trustworthy and working for the public good. FBOs also have the disadvantage of being narrowly viewed as "church" and have the potential for conflict between religious views and the secular marketplace.

The best-case scenario for probable success is when faith-based organizations secure the services of seasoned community developers or non-faith-based partners. In 1989, the Lilly Endowment launched a national program that encouraged congregations to form partnerships with experienced community development organizations. Lilly funded 28 programs and evaluated progress in 1991. Within this time period, 1,300 housing units had been built rehabilitated or were under construction; 11 revolving loan funds with almost $6 million in assets had been established; eight new businesses had been created; and seven faith-based credit unions had increased their assets by $500,000. Perhaps as important, and certainly consistent with the principles of good community development practice, evidence was found that the bringing together of religious institutions and community organizations opened up increased possibilities for bridging barriers across racial and class lines (U.S. Department of Housing and Urban Development 2001).

Public–Private Partnerships

Public–private partnerships (PPPs) are collaborative arrangements between government and the private sector that involve the public partner paying, reimbursing, or transferring a public asset to a private partner in return for goods or services. Government today struggles to deliver public services, and it is often forced to choose between harmful reductions or significant tax increases. PPPs can provide a welcome alternative. Table 9.4 depicts some different levels of PPPs.

Outsourcing and privatization are fairly straightforward, but the blending of the two is much more complex. PPPs are not always well received, with strong opinions on both sides. Supporters cite cost savings of up to 40% while realizing more innovation and improved quality of service. Detractors fear violations of constitutional or statutory law, private sector greed, lack of accountability, and an increase in unemployment. The debate can be very polarizing, so it is important to focus on actual case studies when exploring the potential benefits of implementing a local public–private partnership.

Table 9.4 *Public–Private Partnerships*

Level of partnership	*Description*
Outsourcing	Contracting by a public agency for the completion of government functions by a private sector organization. For example, contracting janitorial services for a city hall or contracting the design and maintenance of a city's web presence.
Public–private partnership	Means of utilizing private sector resources in a way that is a blend of outsourcing and privatization—an interactive, working partnership.
Privatization	The sale of government-owned assets to the private sector. For example, when government turns over prison functions to private providers.

Source: National Council for Public–Private Partnerships.

PPPs often pair up competitors to create a win-win partnership. For example, the U.S. Postal Service recently awarded a contract to Federal Express for the transportation and delivery of its international Global Express Guaranteed mail service, with about 7,400 USPS locations offering the co-branded service. It may not be as extreme as a Coke machine in the Pepsi plant, but it does seem a bit unusual at first thought.

Union Station in Washington, DC was condemned property when Congress passed the Union Station Redevelopment Act to restore the building and create a functional transportation center. The $160 million price tag did not include one taxpayer dollar. How was this done? Through a public–private partnership, the facility is owned by the U.S. Department of Transportation and is managed by a private development firm that leases Union Station space to 100-plus retail shops and restaurants with annual sales exceeding $70 million. Retail rents pay for the operation of the public facility and for its debt service.

Given the dire conditions of most K-12 public education facilities, PPPs may offer the best opportunity for major relief. Through a PPP between the District of Columbia public school system and a national real estate development company, a new state-of-the-art school and a new apartment building were constructed on an existing school property. The District approved a tax-exempt bond package to be repaid entirely with revenues generated by the apartment building. The children attend a brand new school built at no cost to the taxpayer.

Similar success stories can be found in other public service areas. The first transcontinental railroad was the product of a PPP when Congress chartered private companies which issued stock to finance construction of the railways. Federal lands along the route were granted to the railroads for private development, helping the private companies further recoup their investments. More recently in New Mexico, a private firm supported by issuance of state bonds is expanding a major highway. The private firm also holds a 20-year contract to maintain the road instead of using the state highway department (National Council for Public–Private Partnerships).

One of the most visible community-based organizations is usually the economic development agency. Perhaps the most successful model for economic development is the public–private partnership, which comes in various forms. Some communities have separate nonprofit economic development organizations funded by local government monies and private-sector contributions. In other communities, the local government sector may provide funding to the economic development agency housed in the chamber of commerce. Nationally, the average funding share for local economic development is approximately 50% public and 50% private.

The litmus test for assessing effectiveness of PPPs is this: Has the partnership enabled government to act more efficiently and better utilize its limited resources to meet critical societal needs? Has the public been well served by the public–private partnership? While there is clear evidence supporting the use of PPPs, there are important cautions to take when delivering essential public services:

- Provide for public involvement in the process through community meetings or public hearings to educate and gain approval from affected constituencies.
- Allow considerable time for the planning process.
- Provide for public disclosure of the PPP agreement including financial arrangements and guarantees.
- Clearly define performance guarantees and associated penalties and/or incentives in contracts to minimize the risk of disreputable contractors in the marketplace.
- Finally, don't eliminate the potential of PPPs because of their complexity or newness. Consider them as one possibility among many that can be utilized for community development.

Regional Initiatives

Regional initiatives are important because they provide a framework for addressing society's most complex problems. Communities are finding that issues of air quality, transportation, infrastructure investments, and economic development are well beyond the ability of one municipality or one organization to handle alone.

Regionalism involves formal institutional arrangements, shared decision making and participation of governing institutions throughout the region, and it varies structurally according to objective, project scope, who is involved, and time requirements. Regional collaboration among government units can take many forms such as consolidated government functions (combined police forces with equal authority across multiple jurisdictions), metropolitan planning councils, special service taxing districts, and joint service agreements. Regional collaboration is also possible among citizens groups, area coalitions, and alternative planning organizations (National Association of Regional Councils).

Regionalism can be especially effective in economic development. More and more cities and counties are banding together to create regional marketing organizations in order to "get on the radar screen" for investment projects. Regionalism can be one of the best tools for rural communities that otherwise would not have the resources to market themselves, or even to develop infrastructure by themselves.

Regional Leadership Programs

In order for regional initiatives to succeed, leaders must adopt a regional attitude. Similar to local leadership development programs, regional ones focus on increasing understanding about issues of the region, developing collaborative skills, and building personal and trusted relationships throughout the region. Programs come about due to a variety of reasons.

The Kansas City Metropolitan Leadership Program was created by a number of civic leaders who were upset that progressive public issues, such as bond issues and school assessments, never seemed to pass. In Atlanta, Georgia, the Council of Governments was about to launch an extensive regional visioning process and it created the Regional Leadership Institute to prepare a cadre of citizens to provide leadership for the project. The IDEAL Program was developed in California's Central Valley to provide skills to a growing immigrant population that was assuming leadership positions. In Denver, a group was butting heads over a controversial issue and discovered that the process was much easier once they got to know and trust one another. From that emerged the Denver Community Leadership Program.

Regional programs are just as varied in their organizational structure. Some are run by regional government groups, by regional nonprofits or citizens groups, some by a local university, some through a consortium of other civic groups. The underlying foundation, however, is an understanding of the importance of regionalism and building the capacity to achieve positive results at that level (Alliance for Regional Stewardship).

Metropolitan Planning Organizations

Metropolitan planning organizations (MPOs) act as development intermediaries for federally funded programs and are governed by city, county, and state governments as well as representatives from various community stakeholder groups. Federal highway and transit statutes require, as a condition for spending federal highway or transit funds in urbanized areas, the designation of MPOs, which have responsibility for planning, programming, and coordination of federal highway and transit investments. These MPOs are composed of local elected officials and state agency representatives who review and approve transportation investments in metropolitan areas.

Regional Government Initiatives

Portland Metro is a directly elected regional government, serving 1.3 million residents in 24 cities in the Portland metropolitan area. The Metro Council has a president elected region-wide and six councilors elected by district. The Council is responsible for growth management, environmental conservation, transportation planning, public spaces, solid waste and recycling, and owns and operates the Oregon Zoo.

Planning and Development Districts

With the passage of the Economic Development Act of 1965, a statewide system of planning and development districts was put into place in many states to provide a single system of planning, development, and programming from a regional approach. Because of its multifunctional capability, the PDD is used by many federal and state agencies as a delivery organization for those programs. Cooperative relationships are maintained through formal and informal contacts, partnerships, and memoranda of understanding which focus on areas such as technical assistance, education, industrial development, recreation, social services, environment, tourism, zoning, housing, agriculture, communications, consulting, and workforce development.

Special Service Taxing Districts

Special service taxing districts are regional efforts such as regional water and sewer districts, fire districts, postsecondary vocational and technical education, library systems, and transportation systems. Districts are legally constituted and can levy taxes for a specific community improvement. These taxes are collected and then redistributed across the district rather than to the jurisdiction where the tax originated. Districts are administered by a board that can be elected or appointed.

Conclusion

So, do you still want to start a community-based organization? If you choose to proceed, there is one final point to keep in mind. Community development organizations should always remember that they are only part of a broader community.

The strength and diversity of these groups, their relationship to one another, and their ability to form alliances with organizations outside their community greatly influence the level of success experienced locally. Cities with comparable development organizations will experience different results based upon how development is organized and the working relationships among the CBOs.

The concept of comprehensive community development recognizes the varied sectors of a community—social, economic, physical, governmental, cultural, educational, and environmental —and the need to address them in holistic rather than piecemeal fashion. This chapter also recognizes that new entrepreneurial forces are changemakers in community development with the likes of the social entrepreneurs and new hybrid organizations that are interested in economic and social development simultaneously.

Now elect a chair, build a budget and get organized for a most rewarding experience—effecting positive community change.

CASE STUDY: THE FUTURE MELTING POT: FROM COMMUNITY ACTION TO THE FORMATION OF A COOPERATIVE COMMUNITY INTEREST COMPANY BRINGING CHANGE TO NEET YOUTH IN BIRMINGHAM, ENGLAND

> Right from the start, the philosophy of the Future Melting Pot has always been about creating the services and opportunities that young people want as determined by them themselves, not just with them in mind. This will continue as we move forward on what has proved, so far, to be a very exciting journey.
>
> Estella Edwards, Founder[1]

Estella Edwards, with 20 years of experience as a mentor, trainer, and leadership coach and a background in social work and community development saw a gap in the Birmingham employment market. It was a youth employment gap for women, minorities and other marginalized groups. The gap is best understood by its acronym, NEET (not in education, employment or training). "I had decided to set up an event called Change for Success which brought together a lot of different networks and disciplines with a largely youth audience," she said. She explains:

> Following on from that I was approached by around 30 young people requesting personal mentorship and training. There was obviously a gap in the market—a hole needing to be filled by an organization skilled and expert enough to be able to offer disadvantaged young people the support they needed, building their confidence and developing the skills we know they already have.[2]

Edwards began to hold community meetings with the disenfranchised youth and others to see what could be done to make bold change. Community meetings were followed by trainings. Trainings were followed by events with keynote speakers. The momentum kept rolling until today where the movement is now a one-stop shop for young people to learn about the skills they need to set up their own business and thrive.

A one-year pilot program was developed to leverage the original community work. In order to participate in the program, young people had to be over 18 and receiving welfare benefits for at least three months. After an interview and selection process, participants were offered the chance to: improve personal development; nurture their entrepreneurial 'mind'; start the business they had always wanted to start; create their own work and become their own boss; and make a difference for themselves, their family and their community. An action plan was drawn up with a mentor and participants received support in developing business ideas from initial design to completion. It provided the opportunity to explore the option of self-employment in an environment which was led by the needs of individuals and where feedback was incorporated into the project.[3]

It was the young people who were involved in the original community-building movement who wanted to look at the past, capture key learnings, and build on it for the future. These youth were inspired to not only spread the word of empowerment but to create change with a structure to get rid of the NEET problem in Birmingham. The youth disbanded to volunteer on the movement's board and build upon its strengths and successes, "providing a "one-stop" support network for women and young people seeking advice and looking to develop their business skills, career prospects and awareness of the business world. Over time, closer links were forged with other youth and business organizations and a board was created, where professionals shared skills with young people. More volunteers were recruited and, in 2010, the Future Melting Pot became a Community Interest Company.[4]

The Co-Operative Community Interest Company is a hybrid company which allows private companies to engage in social enterprise. Hybrid companies in the United Kingdom and the United States and elsewhere are a new phenomenon where private enterprise uses part of its earnings to promote social good. The Co-Operative Community Interest Company,[5] although formed, is still a work in progress according to the founder. She says that "the new co-operative model will be more democratic[6] and our young people will have a bigger share of direction as members."[7]

The key features of the success for the Future Melting Pot have been:

1 A strong client-focused approach which addressed the needs of specific individuals—the NEET generation.
2 Meeting with key local stakeholders and potential service recipients to effectively target resources to the needs of a particular group or individual.
3 Developing a wide range of outreach and contact strategies to effectively engage with groups or individuals. Here it went from skill-based trainings to conferences to the development of a formal community organization.
4 Continuity to ensure an authentic collaborative approach to allow for opportunities to build on past learnings—successes and failures to drive the future.

A key component of the rapid success since its implementation in 2009 is that the founder allows youths to have an authentic voice to determine what the best solutions are for them. Edwards adds, "The philosophy of the Future Melting Pot has always been about creating the services and opportunities that young people want. This will continue as we move forward on what has proved, so far, to be a very exciting journey."[8]

Patsy Kraeger, Ph.D.
School of Public Affairs, Arizona State University

Keywords

Community-based organization, community development corporation, business plan, entrepreneurship, social entrepreneurship, nonprofit, public–private partnerships, hybrid companies, community foundations, faith-based programs.

Review Questions

1 What are some fundamental questions that should be addressed before a community-based organization (CBO) is established?
2 What are some of the public and private sector programs that can support the establishment and operation of CBOs? Where do these programs reside?
3 What are the different kinds of public–private partnerships and how can they be used in community development?
4 How has entrepreneurship affected community development? Who are social entrepreneurs and what types of organizations do they run?
5 What are the different types of foundations and how can they help with community development?

Notes

1 www.thefuturemeltingpot.co.uk/wp-content/uploads/TFMP_Coop-News-article.pdf
2 www.thefuturemeltingpot.co.uk/wp-content/uploads/TFMP_Coop-News-article.pdf
3 www.wilcoproject.eu/wp-content/uploads/2013/10/Birmingham_report-on-innovations.pdf
4 www.thefuturemeltingpot.co.uk/wp-content/uploads/TFMP_Coop-News-article.pdf
5 www.uk.coop/sites/storage/public/downloads/co-operative_community_interest_company_clg.pdf
6 Co-operative Principles are the principles defined in the International Co-operative Alliance Statement of Co-operative Identity. The principles are those of voluntary and open membership, democratic member control, member economic participation, autonomy and independence, education, training and information, co-operation among co-operatives and concern for the community (UK Companies Act 2006, Private Company Limited by Guarantee, Co-operatives, UK Co-operative Community Interest Company Model).
7 www.thefuturemeltingpot.co.uk/wp-content/uploads/TFMP_Coop-News-article.pdf
8 www.voice-online.co.uk/article/future%E2%80%99s-melting-pot

Bibliography

Alliance for Regional Stewardship (2002) *Best Practices Scan: Regional Leadership Development Initiatives.* Online. Available at: <www.regionalstewardship.org/Documents/LeadershipDev_BestPract.pdf > (accessed June 2, 2004).

Allston Brighton Community Development Corporation. Online. Available at: <www.allstonbrightoncdc.org> (accessed April 14, 2004).

Appalachian Center for Economic Networks. Online. Available at: <www.acenetworks.org> (accessed June 1, 2004).

Arkansas Community Foundation. Online. Available at: <www.arcf.org> (accessed June 1, 2004).

Arkansas Wood Manufacturers Association. Online. Available at: <www.arkwood.org> (accessed June 1, 2004).

Association of Metropolitan Planning Organizations. Online. Available at: <www.ampo.org>

Bloom, J. (2010) "Community-owned Businesses." *Main Street Story of the Week*, March/April 2010, www.preservationnation.org/main-street/main-street-now/2010/marchapril-/community-owned-businesses.html#.Ue2JgY04vt4 (accessed June 5, 2013).

Bromberg, A. (2011) A new type of hybrid. *Stanford Social Innovation Review,* 9(2), 49–53.

California Center for Regional Leadership. Online. Available at: <www.calregions.org>

Center for Community Change (1985) *Organizing for Neighborhood Development.* Online. Available at: <http://tenant.net/organize/orgdev/html> (accessed May 11, 2004).

CEOs for Cities. Online. Available at: <www.ceos-forcities.org/index.htm>

Community Building Resource Exchange. Online. Available at: <www.commbuild.org>

Community Development Society. Online. Available at: <www.comm-dev.org>

Community Economic Development Corporation of Southeastern Massachusetts. Online. Available at: <http://members.bellatlantic.net/vze3h2jm> (accessed April 14, 2004).

Community Foundations of America. Online. Available at: <www.cfamerica.org>

Council on Foundations. Online. Available at: <www.cof.org> (accessed November 15, 2013).

Dahlstrom, T. (2013) *The Role of Business Counselors in the Human Capital Resource Acquisition of Entrepreneurs.* Ph.D. thesis, Arizona State University. Retrieved from ProQuest Digital Dissertations. (3559606).

Foundation Center. Online. Available at: <www.fdncenter.org>

Foundation for the Mid South. Online. Available at: <www.fndmidsouth.org> (accessed June 11, 2004).

Franklin County Community Development Corporation. Online. Available at: <http://fcdc.org> (accessed April 14, 2004).

Fructerman, J. (2011) For love or lucre, *Stanford Social Innovation Review* 9(2), 42–47.

Galaskiewicz, J. (1989). Corporate contributions to charity: Nothing more than a marketing strategy? In R. Magat (Ed.), *Philanthropic Giving: Studies in*

Varieties and Goals. New York, NY: Oxford University Press, pp. 246–260.

General Services Administration. *The Catalog of Federal Domestic Assistance*. Online. Available at: <www.cfda.gov> (accessed April 26, 2004).

Givens, R.J. (2004). Social Capital: The currency of grant professionals. *Journal of the American Association of Grant Professionals*, 2 (Spring/Summer): 29–34.

Gonzales, V. and Phillips, R. (2013) *Cooperatives and Community Development*. London: Routledge.

Holley, J. *Creating Flexible Manufacturing Networks in North America: The Co-Evolution of Technology and Industrial Organizations*. Online. Available at: <www.acenetworks.org/juneholley/docs.html/concept.htm> (accessed June 1, 2004).

International Community Development Council. Online. Available at: <www.cdcouncil.com>

International Council of AIDS Service Organizations. (1997) *HIV/AIDS Networking Guide*. Online. Available at: <www.icaso.org> (accessed May 5, 2004).

James Irvine Foundation. (1999) *Getting Results and Facing New Challenges: California's Civic Entrepreneur Movement*. Online. Available at: <http: //www.calregions.org/publications.html> (accessed May 29, 2004).

Jay, J. (2013) Navigating paradox as a mechanism of change and Innovation in hybrid organizations, *Academy of Management Journal* 56(1), 137–159.

Lane, B. and Dorfman, D. (1997) Northwest Regional Educational Laboratory. *Strengthening Community Networks: the Basis for Sustainable Community Renewal*. Online. Available at: <www.nwrel.org/ruraled/stengthening.html> (accessed 1 June 2004).

Massachusetts Association of Community Development Corporations. Online. Available at: <www.macdc.org> (accessed April 14, 2004).

Missouri Department of Economic Development. *Starting a Nonprofit Organization*. Online. Available at: <www.ded.mo.gov/business/startabusiness> (accessed June 5, 2004).

Moore, J. (2000, October 5). Restoring Americans' civic spirit: 'Bowling Alone' author sees key role for charities in renewing social ties, *Chronicle of Philanthropy* 12(1), 37–41.

National Association of Development Organizations. Online. Available at: <www.nado.org>

National Association of Manufacturers. *Manufacturing Networks: Partnerships for Success*. Online. Available at: <www.nam.org> (accessed April 14, 2004).

National Association of Regional Councils. Online. Available at: <www.narc.org> (accessed June 4, 2004).

National Council for Public–Private Partnerships. *For the Good of the People: Using Public–Private Partnerships to Meet America's Essential Needs*. Online. Available at: <http://ncppp.org/presskit/ncpppwhitepaper.pdf> (accessed June 4, 2004).

New Jersey Turnpike Authority. *History of Metropolitan Planning Organizations*. Online. Available at: <www.njtpa.org/Pub/Report/hist_mpo/default.aspx> (accessed June 10, 2004).

Parzen, J. (1997) Center for Neighborhood Technology. *Innovations in Metropolitan Cooperation*. Online. Available at: <http: //info.cnt.org/mi/inovate.htm> (accessed May 13, 2004).

Portland Metro. Online. Available at: <www.metro-region.org> (accessed June 10, 2004).

Robinson, R. (2009) *Basic Business Planning for Entrepreneurs: Your Roadmap to Success*. Phoenix, AZ: Acacia Publishing.

Shuman, Michael (2007) *The Small-Mart Revolution: How Local Businesses Are Beating the Global Competition*. San Francisco: Berrett-Koehler Publishers, Inc.

Smith, J. A. (1999) The evolving American foundation. In C. T. Clotfelter and T. Ehrlich (Eds.), *Philanthropy and the Nonprofit Sector in a Changing America*. Bloomington, IN: Indiana University Press.

U.S. Chamber of Commerce. Online. Available at: <www.uschamber.com>

U.S. Department of Housing and Urban Development (2001) *Faith-Based Organizations in Community Development*. Online. Available at: <www.huduser.org/publications/commdevl/faithbased.html> (accessed May 15, 2004).

U.S. Government's Official Web Portal. Online. Available at: <www.usa.gov>

Connections

Explore a website—International Co-operative Alliance, http: //ica.coop

Learn more about Benefit Corporations at: http: //benefitcorp.net/about-b-lab

Learn about the Daily Table at: www.digitaljournal.com/article/359261 and www.npr.org/blogs/thesalt/2013/09/21/222082247/trader-joes-ex-president-to-turn-expired-food-into-cheap-meals

Learn more about L3C corporations at: www.fastcompany.com/1526568/l3cs-hybrid-way-do-well-doing-good

Learn about the Main Street Program at: www.preservationnation.org/main-street

10 Leadership and Community Development

Alan Price and Robert H. Pittman

Overview

Perhaps there is no topic more discussed, more analyzed, yet more confounding than how to be a good leader, and how to create and train good leaders. Yet it is through effective leadership that most work is accomplished, great visions are implemented, and our finest moments occur. Still in all, rarely is the essence of highly effective leadership practically defined. This chapter provides an overview of leadership approaches as well as an extended case from the private sector that holds relevance for learning how to lead under pressure.

Introduction

Internet searches using phrases such as "leadership development" and "effective leadership" yield tens of millions of results. Even searches using "leadership and community development" yield millions of results. Within this vast body of information, there is a myriad of definitions of leadership and recommendations on how to develop leadership skills. Much of the information is oriented toward private sector applications, although research on effective leadership in the public sector, including communities, is growing, such as that provided by the Community Action Partnership. It is certainly not feasible in one book chapter to provide a meaningful review of this vast literature. Instead, in this chapter we will mine certain gems from the research and combine them with the authors' experience in organizational leadership in both the public and private sectors.

BOX 10.1 YOUTH LEADERSHIP DEVELOPMENT: THE UNESCO CHAIR PROGRAM IN RURAL COMMUNITY, LEADERSHIP AND YOUTH DEVELOPMENT

Strong communities equal stable, civil, and just societies. Through the United Nations Educational, Scientific and Cultural Organization (UNESCO) Chair Program in Rural Community, Leadership, and Youth Development we're rethinking the way we look at community and youth engagement to achieve this vision. The program is uniquely positioned to offer high-quality, research-based information, programming, and best practices to the world. We contribute to a better future, where local people are engaged and actively involved in shaping the future of their communities. We work with those who feel the same and are willing to step up to the challenge. Our work is focused on four programmatic themes: governance, democracy, and civil societies; health and nutrition; economic and social innovation; and natural resource management, and sustainability.

Why Focus on Rural Youth and Communities?
The number of young people in the world is at an all-time high. According to the UN Department of Economic and Social Affairs (2011), 27% of the 6 billion people on Earth are age 15 or younger, with 85% of all youth living in less-developed countries that are heavily dependent on agriculture. Today, millions of adolescents and communities in developing countries face challenges brought about by limited access to resources, health care, education, training, employment, and economic opportunities. They face early marriage and childbearing, inadequate education, and the threat of HIV and AIDS. This reality creates an environment in which the active engagement of young people in rural communities is essential to capacity building, gender equality, and equal access to education for women and girls and a host of socioeconomic development outcomes.

What We Do
There is a well-documented need for rural community leadership and youth development research, teaching, and programming, particularly in developing countries of the southern hemisphere (Africa, Asia, and Central and Latin America), where an increasing share of the total national population is youth. We work to create opportunities for youth worldwide to improve their lives and communities. Through proven research, education, and programming, we focus on personal youth development community capacity building in which youth play a central role, and foster activism oriented toward social justice, change, and equality.

An Example of Our Work: Youth Leadership Development in Africa
In 2013, we worked with our partners at National University of Ireland-Galway, Foróige (the National Youth Organization of Ireland), and youth in Kenya and Zambia to develop leadership and community-building capacities. This proven model and programming is transforming communities and providing opportunities for youth that were unimaginable just a short time ago.

Mark Brennan, Ph.D.,
Professor and UNESCO Chair in Rural Community, Leadership, and Youth Development
Pennsylvania State University, USA
http: //agsci.psu.edu/unesco

Reference

UN Department of Economic and Social Affairs (2011) "The Report on the World Social Situation 2011: The Global Social Crisis," New York: UN Department of Economic and Social Affairs (UN-DESA).

What Is Leadership?

There is much to be learned concerning public sector or community leadership (which is our concern in this book) from the large literature on leadership in organizations—public or private. However, there is a fundamental difference in the environment for leadership in an organization with a hierarchical structure vs. a democratic community or other political entity where all citizens theoretically have an equal voice and actions are largely based on consensus or majority vote. In the authors' experience, being an effective leader in a community environment can be even more challenging than doing so in a company or organization. Most of us at some point in our lives have encountered ineffective leaders. Many of them rely on command and control instead of empowerment and engagement. The difference seems obvious on paper, but in practice many "leaders" do not seem to understand it.[1] Ask yourself this simple question: Would you rather be in a situation where your leader(s) told you what to do instead of asking your opinion and including you in a group decision process where your voice is heard? One author and consultant (Kruse) nicely characterizes effective leaders by identifying what they are *not*. Let's consider these (in italic) and the implications for communities (comments after italics):[2]

Leadership has nothing to do with seniority or one's position in the hierarchy of an organization or titles. Leaders in communities are not necessarily elected officials or high-level administrators. The authors have encountered communities where generally accepted and respected leaders have included such diverse people as retirees, football coaches, newspaper editors, company executives, and simply community volunteers.

Leadership is not just about personal attributes. A take-charge person barking orders is not a leader. Charisma can be a positive leadership asset, but charismatic persons are not necessarily leaders.

Leadership isn't management. Managers plan, measure, monitor, coordinate, and solve problems, among other things. Typically, managers manage *things*. Leaders lead *people*.

Kruse provides some interesting and meaningful definitions of leadership from various well-known people:

> The only definition of a leader is someone who has followers.
>
> (Peter Drucker)

> Leadership is the capacity to translate vision into reality.
>
> (Warren Bennis)

> As we look ahead into the next century, leaders will be those who empower others.
>
> (Bill Gates)

> Leadership is influence—nothing more, nothing less.
>
> (John Maxwell)

BOX 10.2 LEADERSHIP IN COMMUNITY DEVELOPMENT

> Community building requires a concept of the leader as one who creates experiences for others—experiences that in themselves are examples of our desired future. The experiences that are created need to be designed in such a way that relatedness, accountability, and commitment are every moment available, experienced, and demonstrated.
>
> (Peter Block, *Community, the Structure of Belonging*, 2008, San Francisco, CA: Berrett-Koehler, p. 86)

BOX 10.3 WHAT DOES IT MEAN TO BE A LEADER?

It is difficult to discuss the development of a stronger leadership base without common agreement of what "leadership" means. Unfortunately, the definition of "leader" or "leadership" is frequently misused or misunderstood. For that reason, it is important to begin by attempting to get a better grasp of what these terms mean.

Leadership Defined

An extensive review of the literature reveals that no one has satisfactorily defined what leadership is, especially in the context of a community. The term often refers to anyone in the community who has relatively high visibility. However, a leader should be identified as someone who is more than a widely recognized individual or a local official. Recognition alone does not constitute leadership.

Most articles and books on the topic of leadership conclude that it involves influencing the actions of others. According to Vance Packard, "leadership appears to be the art of getting others to want to do something you are convinced should be done" (in Kolzow 2009: 120). Leadership implies "followership." A community leader emerges when he or she is able to get a number of community residents and/or business people to strive together willingly for leaders' goals.

How does one become a leader in a community? This doesn't occur by simply declaring oneself a leader; others need to acknowledge that leadership. The followers actually determine whatever "real" power the leader may have. No simple formulas or models exist to guarantee that one can achieve leadership. So, how does one attract followers? It would appear that the level of credibility of an individual is the single most significant determinant of whether he or she will be followed over time. Leaders create followers because they are able to bring about positive change in others' understanding. This is in contrast to leaders who emerge from a group of citizens who react to an adverse situation, often by doing nothing more than getting in front of the parade.

(David Kolzow (2009) Developing Community Leadership Skills. In R. Phillips and R. Pittman (eds.) *Introduction to Community Development*, London: Routledge, pp. 119–132).

These definitions describe effective community leaders very well. As a matter of fact, if a person does not embody these traits, especially translating vision to reality and empowering others, they are unlikely to become true leaders in a democratic community setting. An elected official or administrator may or may not be an effective leader according to these criteria.

Case Study

On the morning of September 11, 2001, a small group of Delta Airlines senior executives faced an unprecedented decision—should they ground Delta airlines? American Flight #11 had struck the North Tower of the World Trade Center. Moments later, United Flight #175 crashed into the South Tower, thereby removing all doubt that this was indeed an act of terrorism. Captains Dave Bushy and Charlie Tutt, Delta's Senior VP and Director of Flight Operations, immediately recommended to Delta's CEO, Mr. Leo Mullin, that he ground the airline.

Delta had no plan to ground the fleet, nor was there any guidance on how to go about such a thing. Airlines only make money when airborne, and Delta had never conceived of the need for a plan to put its fleet on the ground. After a few moments to consider the ramifications, Mr. Mullin agreed. Word went out to the crews of almost 500 aircraft, airborne at locations all around the world, to "land immediately at the nearest suitable airport."[3]

While the decision itself is remarkable, several factors surrounding it are even more so. Mr. Mullin gave his order some moments before the FAA reached the same conclusion. Delta did not follow, it led. Later, during on-the-ground inspections,

Delta personnel found box-cutters discarded in the seat backs on several of its aircraft. We are left to wonder what might have occurred had Delta not put its fleet on the ground ASAP?

Even more remarkable, there had been no discussion of the "cost" to ground the fleet until days after the event. The senior staff at Delta had made their decision based solely on the Flight Operations priority of safety first...Safety of the passengers and crews was the first and only consideration used when making this momentous and courageous decision.[4]

Leadership Lessons from 9/11

Following the horror of 9/11, one of the authors, the Atlanta Chief Pilot for Delta, was asked what his secret was for working with a great diversity of groups toward a common goal. After a moment's hesitation Captain Price replied "nothing bad comes from great communication." What are some key principles of leadership and the values and culture at Delta Airlines that guided them during the 9/11 attacks and in the immediate aftermath? There are three in particular that were important to Delta and in the authors' opinion can help individuals become more effective leaders and help communities create the right environment to nurture them. The three principles are:

1 Principle-centered decision making.
2 Expecting the unexpected—envisioning the future.
3 Continuous improvement—the secret of great leadership.

Below is a discussion of these core leadership principles in an organization in the words and from the notes, speeches, and other communications of Captain Alan Price and other Delta executives who guided the airline through its most turbulent event. The chapter concludes with an application of these principles to community and economic development.

1 Principle-Centered Decision Making

Leadership is a team function and both leaders and teams need to know, understand, and abide by principles consistent with their organization's purpose and beliefs. These principles should be thoroughly discussed, easily understood, broadly accepted, and consistently and widely publicized for they will guide organizational decision making both in normal times and times of crisis.

If an organization fails to fully understand, enunciate, and rely upon its core principles, or if these principles are not widely embraced, they will prove to be poor or inconsistent decisional tools. When an organization understands the power of core principles, however, they will invest time and effort to insure these beliefs comport with their fundamental values and are consistently accepted throughout the organization.

U.S. airlines train as they wish to fly. Great effort is expended by these organizations to insure their training practices are consistent with their core principles. Such core beliefs embody valued organizational concepts, capture historical wisdom, and give strong and consistent guidance to those faced with decisions in good times and in times of peril.

Organizations and teams will all face times of crisis. The core principles relied upon in everyday operations, must be those we can rely upon in times of turmoil. Relying upon clear and consistent corporate values provides the best opportunity to survive and thrive in any circumstance. Lacking such guidance, organizations will often fail or flounder, especially in time of great stress.

At Delta Airlines, senior leadership could rely upon clear and consistent core values as they made the momentous decision to ground the fleet. Safety first is a core value, which served as a decisional guide and led to a highly successful outcome. Delta could not control the situation on September 11, but it could and did provide the means for its leaders and teams to successfully navigate uncharted waters. They knew, accepted, and understood principle-centered decision making.

2 Expecting the Unexpected—Envisioning the Future

Assumptions are only as good as the information upon which they are based. When asked, leaders will say timely and accurate information about future events is their greatest need. While we would all like to believe that the past is prologue to the future, it can be a point of deception. Using the past as a guide to the future is a common and often costly mistake.

The future is all about change, and necessitates that we adapt past practices to changing circumstances. It is our human nature to stick with what works, but in truth, change is constant. Dr. Albert Einstein is credited with observing that the definition of lunacy is "doing the same old things and expecting different results."

This, then, is the leader's dilemma. How do we understand the future and comprehend what adjustments are appropriate to a changing set of circumstances? In combat, we call this lack of accurate information the "fog of war," and Gen. Colin Powell is known to have said a field commander never knows more than 75% of what he needs to know to make effective decisions. When Mr. Leo Mullin was CEO of Delta Airlines, he once shared that leaders often know less than 50% of what they would like to know about future contingencies.[5]

This information vacuum, the inability to know what tomorrow brings, is a huge issue for leaders and teams. How are we to deal with uncertainty while positioning our organization for the greatest success? In truth, there are several highly effective techniques to gauge future events, to expect the unexpected, and the leader who understands their power will be well ahead of their peers. Let's focus on the three of these.

(A) HOW THE FUTURE WORKS—UNINTENDED CONSEQUENCES

Dr. Earl Wiener's 29th Law, the Law of Unintended Consequences, states that "Any time you solve a problem, you create a problem. The best you can hope for is that the problem you solved is of greater magnitude than the one you created."[6]

Far too often, we are surprised when a planned change creates problems. Any disturbance in a system in equilibrium will produce unintended consequences. While we many not know the exact nature of these changes, we can know with certainty they will occur.

The men who hijacked U.S. airliners on September 11 knew enough about commercial cockpits to turn off the IFF (Identification Friend or Foe) Beacon. The IFF is an electronic black box, which broadcasts a discrete 4-digit code hundreds of times/second to Air Traffic Control (ATC). Thereby, each aircraft is uniquely identified by ATC radar. When these beacons were turned off, the hijacked aircraft became almost invisible to radar.

In April of 2002, six months after September 11, an MD88 aircraft was towed from the Delta Jet Base to a passenger gate in preparation for a scheduled flight from Atlanta to Reagan National Airport in Washington, DC. The aircraft had been specially modified with a system which, when activated by the pilot, would in fact make it impossible to turn off the IFF beacon while in flight. The designers of this system thought that in this way a hijacker could never prevent ATC radar from "seeing" the aircraft electronically.

Capt. Mark Pass, an excellent young line check airman, was in command of the flight. Mark knew there was a prototype system on his aircraft, and that it was not yet ready to be tested. Neither Mark nor his First Officer (FO) had been trained on the system—neither knew there was a guarded toggle switch on the FO's side console which activated the system, nor did they know this switch had been inadvertently left in the "ON" position by maintenance.

All was normal until the flight lifted off during takeoff. At that point, electronic sensors in the nose gear indicated the aircraft was now airborne and the modified IFF system now began to broadcast an electronic *hijack* signal to Atlanta tower and departure control. Atlanta tower began to ask generic questions over the radio, fearing the crew was under duress with hijackers in the cockpit. Since Mark and his FO had no clue this was happening, their answers were non-specific and less than helpful.

Quickly, a plan was devised to get the aircraft on the ground and then sort out the details. Atlanta Tower told Capt. Pass that Delta needed him to come back and land so they could load some high-value last-minute cargo bound for Washington. This sounded reasonable to Mark, so he readily complied with instructions to return to Atlanta.

A 1,500′ overcast layer reduced pattern visibility until Mark's aircraft descended below the overcast during the landing approach. At that time, two F16 fighters became visible on either wing! Had Mark not complied exactly with ATC instructions, his aircraft could have become the first commercial aircraft shot down by friendly fire.[7]

Wiener's 29th Law teaches that change will bring unintended consequences. While we may not know the exact nature of these consequences, the knowledge that they will occur allows us to better prepare for them. *Anticipating the unknown allows us to begin to exercise control over the unknowable.*

(B) Red Flags

Many post-accident studies in the aviation, healthcare, and other team-centered environments show several consistent factors preceding each and every accident. There were, in fact, warning signs (Red Flags), which can be readily grouped into seven major categories:

- confusing inputs;
- preoccupation;
- not communicating;
- confusion;
- violating policy or procedures;
- failure to meet targets; and
- not addressing discrepancies.

In studies in both aviation and health care, researchers find a striking similarity—each accident was preceded by no fewer than four with an average of almost seven Red Flags. At any point along the way, team members could have prevented the accident had they seen, communicated, and acted upon the warning presented by these Red Flags.[8]

In practical terms, leaders and teams can be taught to understand and recognize Red Flags, and act when they are present. When a team member sees a Red Flag, they cannot know if it is the first or last in a sequence so it is essential that this information be shared with team members in order to "connect the dots." The key skill practiced by high-performing teams is embodied in the simple admonition to *see it/say it/fix it.* When recognizing a Red Flag, communication followed by action is essential to change the trajectory of the team before an accident occurs.[9]

(C) Briefing Behavior

A National Aeronautics and Space Administration (NASA) simulator study in the early 1990s used volunteer airline crews flying a B727 simulator (three persons in cockpit) on a multi-leg simulated airline flight. The crews were given information about their aircraft, weather, destination airports, etc., so as to replace a normal airline environment. Unknown to these volunteer crews, nothing was to go as expected. Induced conditions were radically different from those originally planned. Observers recorded crew actions with written notes and video cameras. Later, as researchers analyzed these scenarios, a very interesting trend emerged: Crews that briefed well, even though they encountered conditions radically different from anticipated, performed at a much higher level than crews that briefed poorly or not at all.

In point of fact, a timely and relevant briefing has the power to prepare the team to operate at a very high level even when encountering circumstances very different from those expected. That's a powerful thought. Preparing for the expected prepared teams to better deal with the unexpected. An understanding and comfort level results within the team, which then empowers it to deal with unanticipated circumstances.

Intentional and focused communication (briefing) is a leadership function which is a huge factor in insuring appropriate team formation. If a team is formed well, it coalesces around a common understanding of team objectives—a Shared Mental Model (SMM). Briefing is an intentional and focused formative act early in a team's life that helps produce a SMM, and is an under-appreciated, underutilized secret weapon in helping teams prepare for the unexpected.

3 Continuous Improvement—The Secret of Great Leadership?

When groups are asked what one single characteristic distinguishes ordinary from exceptional performance, they give a variety of answers. Rarely, someone will note that leaders, teams, people who are exceptional all have this in common—they have learned how to profit from their experiences to become "*Life Long Learners* (L3)." L3s understand that excellence is not a moment in time, but a continuous and intentional process, a process in which a person looks at their performance self-critically so as to be measurably better next time.

The single greatest skill utilized by L3s is debriefing. Learning from others about our own performance is the key element that L3s harness to put themselves on a path to continuous self-improvement. Organizations, individuals and teams who leverage debriefing as a foundational skill consistently outperform those who do not.[10]

Organizations where debriefing is a cultural expectation create and encourage high-performing teams. Following completion of any major event, process, or procedure, teams that debrief are more able to understand their own performance and the necessary steps to improve. Debriefing also becomes a great means to recognize and analyze positive performance, not just a time to discuss things that could have gone better.

In the US, many are familiar with the Thunderbirds and the Blue Angels, the USAF and U.S. Navy aerial demonstration teams. What many of us do not know, however, is that they spend twice as much time debriefing as they do flying. The world's best pilots know and understand debriefing is a path to excellence.

The debriefing process is not complicated, but executing it consistently is not easy. The process begins with a commitment to "just do it"! After the commitment, time must be found for the team to gather and spend precious moments asking themselves these simple questions:

- What went well?
- What can be improved?
- How can this happen? (Who is responsible/ what is the timeline?)

A basic rule involving normal debriefs is to focus on performance, not on personality. Keep the discussion non-threatening and non-personal. Individuals have to feel it is all about maintaining or improving performance, not about individual criticism.

When a team commits to solve or improve items identified during a debrief, verbal and written recognition is critically important. Public acknowledgment of these efforts holds the team accountable before its peers. This then contributes to a cultural expectation of excellence = continuous improvement.

Some years ago, co-author Price worked with an older gentleman who was one of the best leaders he had ever encountered. "Bob" seemed to move effortlessly through his daily tasks and others loved working with and for him. As I observed Bob's behavior, I began to realize that far from being a natural-born leader, he worked very hard at making teamwork a first and essential priority in all he did. The appearance of "naturalness" in his work occurred as he faithfully adhered to the essentials discussed above:

- Bob knew and embraced the core values of our organization and made a point of sharing examples that illustrated both his commitment to and support for these values. He held himself to a standard he expected the team to embrace.
- Bob seemed to never be surprised by unexpected events. He did not have perfect foresight, but had learned to understand the environment and how change begot change, how Red Flags were predictors of future events, and how to always insure his team knew timely and relevant information as he conducted interactive briefings. Everyone felt involved and valued, as Bob shared his vision for the team's performance.
- Lastly, Bob never failed to take advantage of a chance to involve others, after task completion, through a debrief where everyone felt comfortable offering their observations. Never, ever did Bob allow a debrief to become personal. It was always about the team's performance, never about personalities. Bob also realized that debriefings are an excellent means to recognize and reinforce positive performance, not just things that "need improvement."

> Leadership is a group or team function. The leader's job is to create the conditions for the team to be effective.[11]
>
> (Dr. Robert Ginnett)

The key concepts of highly effective leadership discussed above and illustrated by the quote by Dr. Robert Ginnett provide the foundation for great decision making, offer a means to understand and manage the future, and allow us to be on the path to continuous improvement.

Conclusions and Implications for Community Development Leadership

The experience of Delta Airlines during the 9/11 crisis and the conclusions derived by its management as described above help illustrate some core principles of community leadership.

Principle-Centered Decision Making

Delta's core values of passenger and crew safety helped Delta make the right decision to ground all its flights in the immediate minutes after the 9/11 attacks even before the FAA issued the order to all airlines to do so. Communities need core principles as well to help them build consensus around key issues, prioritize, and implement their decisions. Chapter 8 of this book discusses the importance of community strategic planning and visioning. A vision and plan can serve as a much-needed rallying point or decision filter for communities. Core values can help a community act in a more unified way to pursue the vision and achieve its goals and objectives. Where to start with community core values? As a first step, a community can disseminate and embrace the core values of community development given in Chapter 7 of this book and reproduced in Box 10.4. During the planning and visioning process, other beliefs and values can be agreed upon and added.

Good community leaders adhere to the principles and encourage others to do so as well.

BOX 10.4 COMMUNITY DEVELOPMENT VALUES AND BELIEFS

1 People have the right to participate in decisions that affect them.
2 People have the right to strive to create the environment they desire.
3 People have the right to make informed decisions and reject or modify externally imposed conditions.
4 Participatory democracy is the best method of conducting community business.
5 Maximizing human interaction in a community will increase the potential for positive development.
6 Creating a community dialogue and interaction among citizens will motivate citizens to work on behalf of their community.
7 Ownership of the process and commitment for action is created when people interact to create a strategic community development plan.
8 The focus of CD is cultivating people's ability to independently and effectively deal with the critical issues in their community.

Envisioning and Anticipating the Future

This is a key to successful strategic planning for community and economic development as discussed in Chapter 8. The "big three" techniques of unintended consequences, red flags, and briefing behavior to envision and anticipate the future as applied to Delta's response to the 9/11 attacks also have direct relevance to community development, and they can be translated into several of the key principles of successful community development and community leadership as discussed by Paul Mattessich in Chapter 4 including:

- flexibility and adaptability;
- good systems of communications;
- systematic gathering of information;
- training to gain community-building skills;
- use of technical assistance when needed; and
- leadership training.

Good leaders understand their community's strengths and weaknesses, and often engage in community assessments, sometimes referred to as SWOT (strengths, weaknesses, opportunities, and threats) analysis. An example of the importance of envisioning and anticipating the future by community leaders which the authors have seen repeatedly is the "If It Ain't Broke, Don't Fix It" mentality. For example (and one that the authors have seen often), community leaders might think they have no need for economic development planning because their unemployment rate is very low. The reality may be that key industries generating most of the community's employment could be "sunset" industries such as textiles and apparel that have migrated in recent years from developed to less developed countries. A simple analysis of key industries and employers in a community to gauge vulnerability (SWOT analysis) could turn "ain't broke" into "let's fix it before it breaks."

Continuous Improvement

As Delta Airlines learned and practices today, debriefing and learning from failures makes for much more productive leaders and organizations and better outcomes. This is equally true in community development. In the experience of the authors, open review and debate on a community's successes and failures (e.g. not receiving a grant that was applied for, or not winning the "site selection" contest with other communities to attract a given company) keeps community leaders and citizens aware of changes that might need to be made and helps build a sense of teamwork and morale. Just as a commitment to continuous improvement helped make a good reputation for companies like Motorola and Toyota which in turn contributed to growth and profitability, it can help communities create their own successful futures.

While the analogy between the 9/11 crisis and Delta's reaction and community development and leadership is not exact (although global terrorists can strike anywhere), there are many parallels that illustrate important principles that can help foster more effective leaders and successful communities. Just as successful private sector organizations invest in leadership development, communities can follow suit and move ahead and meet their community and economic development goals. Good leadership, leadership development, and visioning and planning separate the "what happened to us" communities from the "we made it happen" communities.

Keywords

Collaboration, civic capacity, leadership, skill development, community leadership.

Review Questions

1 What are the characteristics of good leadership?
2 How can principle-centered leadership be useful?
3 How can leaders get involved in community development?
4 What can leaders do to facilitate community development?
5 Can you think of an example in your community where leadership impacted an outcome?

Notes

1 "Are You a Leader or Just Bossy?" *Forbes Magazine*, March 28, 2013. www.forbes.com/sites/aileron/2013/03/28/are-you-a-leader-or-just-bossy
2 "What Is Leadership?" Kevin Kruse, *Forbes Magazine*, April 9, 2013. www.forbes.com/sites/kevinkruse/2013/04/09/what-is-leadership
3 From Captain Alan Price's personal interviews with Mr. Leo Mullin and Captain Charlie Tutt following September 11, 2001.
4 Order of decision-making priorities for pilots from Delta's Flight Operations Manual (FOM): Safety/Passenger Comfort/Public Relations/Schedule and Economy.
5 From Captain Alan Price's personal interview with Mr. Leo Mullin following September 11, 2001.
6 Wiener's Laws, from a talk to Delta Airlines Flight Operations on "Aviation Automation," March 1996. Dr. Earl Wiener, Emeritus Professor of Management Science, University of Miami.
7 From Atlanta Chief Pilot interviews, April 2002, Captain Alan Price.
8 National Transportation Safety Board (NTSB). 1994. "A Review of Flight Crew-Involved, Major Accidents of U.S. Air Carriers, 1978 through 1990." NTSB/SS-94/01, Washington, DC.
9 From Atlanta Chief Pilot interviews, April 2002, Captain Alan Price.
10 Lifewings Patient Safety Course—How to Use Red Flags. Lifewings, Inc., Memphis, TN (Safer Patients.com).
11 "Leading a Great Team: Building Them From the Ground Up, Fixing Them on the Fly," a presentation by Dr. Robert Ginnett, April 20, 2001.

Bibliography

Block, P. (2008) *Community, the Structure of Belonging*, San Francisco, CA: Berrett-Koehler.

Community Action Partnership. (2014). Accessed July 1 at: www.communityactionpartnership.com

Kolzow, D. (2009) "Developing Community Leadership Skills." In R. Phillips and R. Pittman (eds.) *Introduction to Community Development*, London: Routledge, pp. 119–132.

Nwokorie, Ndidi, Deborah Svoboda, Debra K. Rovito, and Scott D. Krugman (2012) "Effect of Focused Debriefing on Team Communication

Skills," *Hospital Pediatrics* 2(4): 221–227 (at: http://hosppeds.aappublications.org/content/2/4/221.abstract?related-urls=yes&legid=hosppeds;2/4/221).

Reason J. (1997) *Managing the Risk of Organizational Accidents*. Burlington, VT: Ashgate.

Connections

See resources for developing community leadership at The Heartland Center for Leadership Development, www.heartlandcenter.info

Explore *The Community Leadership Handbook, Framing Ideas, Building Relationships, and Mobilizing Resources* by James Krile with Gordy Curphy and Duanne R. Lund (Fieldstone Alliance Publications, Saint Paul, MN). See www.FieldstoneAlliance.org for more details.

11 Community Development Assessments

John W. Vincent II

Overview

Before a community develops a strategic plan, before it develops a marketing plan to attract new jobs, before it develops action steps to address community problems—in short, before it does anything—it should complete a community development assessment. A good assessment forms the foundation for a successful community and economic development effort. A good assessment helps communities determine the feasibility of their strategic planning and marketing programs.

Introduction

Community leaders often approach the community development process with a "let's get started" mentality. "We know what's wrong. Why delay getting started? We have talked about this for years; that's all we do is talk." They believe that an assessment is not needed to decide when to begin implementing changes in their community. They are anxious to get started and usually want to begin by creating a strategic plan or implementing some very specific initiatives. These often address the most visible of problems, to improve the community. While they may be correct, sometimes they may not be. Whether or not their actions will result in positive change is another matter.

Why Conduct an Assessment?

Before beginning a community development effort or creating a strategic plan, an assessment should be performed to determine what assets are present for development and what liabilities exist that need to be addressed in order for desired improvements to occur. This assessment will identify the strengths on which planned development can be built and identify the weaknesses that need to be eliminated or mitigated as much as possible to give the community the best probabilities for success. The community development assessment process is directed toward supporting the creation of a strategic plan that will guide a comprehensive development effort and involve other citizens in the process. This assessment also provides specific information that helps leaders identify opportunities to be exploited and threats that need to be considered when creating and implementing a strategic plan.

Beginning a community development process without conducting an assessment is like a doctor prescribing prescription medicines without first giving a patient a thorough examination in order to get a correct diagnosis. He or she could be treating symptoms rather than root causes of problems, and the treatments may interact with other conditions to cause additional problems. In the same way, community leaders who act without conducting an assessment can spend valuable time and scarce resources treating only the symptoms. They may not see the long-term improvements they desire if they do not identify and work on root causes of problems. Their efforts could simply become another failed community effort, added to those that preceded it.

A common goal in many communities, for example, is the creation of new, high-quality jobs that offer advancement and career opportunities. Citizens want job opportunities for themselves and their children. In support of that goal, a community development group may identify and market available buildings and greenfield sites and tout their transportation and services infrastructure. They may also offer development incentives. Yet the desired development may still not come because the leaders may not have addressed the need for workforce development, the poor appearance of their town, mismanagement of public revenues and government operations or problems in the local housing stock. There could be problems in the local school system that make the community a less desirable location for workers and management who would transfer in to operate a new facility. There may also be a poor quality of life due to a lack of recreational and cultural activities. Considering and working on only a few of the factors related to successful development will probably not result in bringing in the new jobs that are sought after.

Other Benefits of Assessment

In addition to the data produced for decision making, a formal and comprehensive community development assessment can provide major secondary benefits in support of the community development process. It can help initiate a community dialogue in which citizens discuss problems and issues and agree on the future direction of their community. Many citizens are often surprised to find that they share similar values with others who they had perceived to be quite different from them. This realization helps citizens focus on problem solving versus focusing blame within the community. This initial dialogue feeds directly into the production of a vision statement toward which all citizens can identify and work.

Many citizens may not even be aware of their community's problems or how poorly their community compares to others. Feelings of frustration and dissatisfaction are often focused inward, and this disunity can dissuade outside investors and new residents from investing in or moving into the community. Community forums held in support of the assessment help identify problems and help citizens understand what is wrong with the status quo. This awareness creates additional momentum for planned change when it is properly focused on problem solving rather than on personal attacks or hidden agendas. Further, the final assessment report also serves as a data source for creating a marketing plan or a community profile for responding to inquiries for information.

Project Planning vs. Strategic Planning

Assessments are performed in communities for many reasons. Some are performed by internal groups in support of a particular community project, program, or initiative such as a Main Street program. Others are performed by individuals from outside the community such as business representatives seeking sites for relocation or expansion. These external assessments, usually performed by site selection specialists, are likely to be very specific, examining the economics of a specific building or greenfield location and the area labor force. It could involve a general overview of all community factors related to a company's operations such as workforce availability, prevailing wage rates, transportation infrastructure, utility capacities and costs, business development incentives, and tax rates. The assessor attempts to answer specific questions related to that one project or initiative.

Community development assessments, however, are comprehensive reviews of the community aimed at supporting a host of initiatives, programs, and projects. They are normally performed for (and possibly by) community leaders to help guide the creation of a comprehensive community development plan. That plan usually includes several goals and a multitude of objectives (the programs, projects, and initiatives that support the plan). This chapter will provide an overview of a community development assessment and provide some guidance as to how one might be performed.

Quantitative and Qualitative Data

Traditional economic development assessments primarily focus on quantitative data. They include population demographics (e.g. education and income levels), tax rates, wage rates, and other objective data on which business decisions can be made. These focus on such business-related topics as cost–benefit ratios, return on investment, cost of operations, and profitability of operations.

However, many business location decisions toward a community's successful development are also influenced by more subjective or qualitative factors. Because they are subjective, however, does not necessarily mean that they cannot be measured. Citizen opinions, for example, can be measured and tracked over time. Community spirit and a progressive "can do" attitude can also be observed and measured. Qualitative factors include the underlying attitudes in a community, the way citizens feel about themselves and their community, and how those feelings are interpreted into visual expressions of pride, cleanliness, friendliness, pro-business attitudes, and can-do spirit.

Many times qualitative factors are what drive a business decision about locations with similar quantitative benefits. The location selected may be chosen because it is perceived to be a quaint, historic, and safe community. It could be that the community has wonderful curb appeal due to its cleanliness, well-landscaped public places, and private property. It may offer a variety of recreational and cultural activities that make it a great place for leisure activities. Often, a community's qualitative factors can tell an assessor far more than the objective data alone.

Defining the Community

So what is a community? It is important to define the community before beginning the assessment. At first, many communities performing assessments or conducting strategic planning define themselves by legal boundaries. However, that is usually not an accurate description of their community. It defines the community too narrowly and results in artificially excluding resources and allies that have a vital stake in the community's success. A community is as large as the area it impacts or from which it draws its existence.

Many sister cities, such as St. Paul and Minneapolis, Minnesota, are really one large community. Similarly, West Memphis, Arkansas and Memphis, Tennessee, are part of the same community even though they are in different states. Legal boundaries only play a small role in determining what needs to be considered when conducting a community development assessment. It should be remembered that the word "unity" is literally a part of the word "community." Community development is not about "him and me" or "us and them." As John F. Kennedy said in his inauguration speech, "A rising tide raises all boats." What affects me also affects my neighbors.

An individual who lives outside a town's boundaries yet owns and operates a business in town is a stakeholder in that community. A business located just outside the town's boundary, but using the town's water, wastewater, and utility system, is also a stakeholder in the community. Rural residents who rely on the town for shopping and services are also stakeholders. Being a resident of a town is different from being a citizen of a community. As a community goes, so go the fortunes of all its citizens.

The assessment and planning process needs to be broadened to include all stakeholders—those within the legal boundaries as well as those within the community's "impact" boundaries. Limiting the community to the legal boundaries of a town, city, or county also limits its ability to claim nearby resources as part of its assets. Is the large plant just outside the city limits in the community? What about the interstate highway that is five miles away?

A more appropriate definition for a community would be the geographic area with which people identify themselves as well as the area served by a town's retail and service sectors. The actual community might be a portion of one or more counties. It could include farmland and unincorporated villages and may span a state line.

Comprehensive Assessment

Assessment is basically a process of asking and answering questions about key factors that influence the community's potential for planned growth and development. A community development assessment is a broad assessment since many factors within a community are interrelated and influence each other.

If a community seeks new residents, for example, it must consider such development factors as housing availability, construction time for new housing, price, and quality. If it is seeking families with school-age children, it must consider the quality of local schools since housing location decisions are often driven by school districts and their perceived performance. When retirees are sought as new residents, other factors such as health-care facilities and recreational alternatives become more important. The existence of such resources as retirement homes, assisted living facilities, organized leisure activities, physical therapists, waterways and lakes, and golf courses can also influence retirees' location decisions.

If a community is seeking new business or expansion of existing businesses, community leaders must consider the availability of labor and their skills as these skills relate to the type of work required by the industries being sought. In addition to the current skills possessed by the workforce, communities planning to attract business must consider workforce training capabilities. Are there educational institutions that can provide specialized training to support the industry types being sought? Will local educational institutions work with business to develop and deliver new training courses?

Data Collection Methods

Community development assessments use a variety of methods for collecting and analyzing data. Any data collection technique can result in misinformation and errors. It is not the author's intent to provide details of all scientific data collection methods but to provide an overview of some of the more common ones. For most community development activities, qualitative data collection does not need to meet strict scientific research standards. What is usually being sought is information that provides general direction and that identifies broad categories of community advantages or problem areas. To minimize the chances of collecting incorrect information, it is suggested that multiple methods of qualitative data collection be used and the information collected by one method be verified through another method.

Whenever data is collected from individuals, respondents should be told that their identities will be protected. They can be told that while some of their comments or suggestions may appear in the final assessment report, their identity will not be associated with the comments. In that way, the assessor can provide a safe environment for the respondent to share candid opinions. Specific quotes and comments that represent the general feelings and opinions found in the community can also be included in the report on an anonymous basis. Some data collection methods are listed and described below.

Research and Read

The internet provides a huge amount of information that supports the community assessment process. It puts the power of professional research in the hands of volunteer community leaders. A professional community developer can use the internet to get a great deal of information about a community before he or she ever visits it. The federal census site, www.census.gov, provides a vast array of demographic data on housing, income, race, education and a variety of other areas. When compared with past census data, the information can provide trends that also help identify strengths and weaknesses. Similar data can be found on the sites of other federal agencies such as the Department of Energy and the Department of Labor.

A search engine can also provide links to other sources of data including a state's department of education and department of economic development. For example, under the "no child left behind" initiatives, many state education departments publish online data about school systems and individual school performance. Another good source of data is the state's labor department or job service office which also provides current information on unemployment statistics and workforce demographics. If there is a topic of interest, usually an internet search can result in the identification of a reliable and respected source of data, and much of it is free.

Other sources of data include a community's public utilities, enforcement agencies, and local employers and merchants. If a community assessor calls and introduces him- or herself over the phone, many utility representatives will provide data in support of the community development initiative. For example, utilities can provide data on number of customers as well as growth and decline in various indices such as power, water, residential and commercial telephone subscribers, and wastewater use.

Observe and Listen

This author often drives into and all around a community on a "windshield tour" the afternoon before the day he is expected. He visits local coffee shops and stores. By politely eavesdropping on conversations and asking a few questions during the visit, a great deal of information is collected. These methods often produce many qualitative insights about a community's strengths and weaknesses.

If "a picture is worth a thousand words," then a drive around all areas of the town can be an eye-opening experience. Be certain to travel not only the main streets but back streets as well. Later, you will likely find that even some locals do not know all that can be found on their own town's back streets.

Use a Camera

Smartphones and digital and 35 mm cameras are valuable tools to record support for visual observations. Later, when making public presentations or when compiling reports, it is possible to use pictures to reinforce major findings. Photographic images give residents the ability to visually visit all areas of the community including areas that most citizens do not visit or see. The images provide them with a more comprehensive understanding of their community's strengths and weaknesses.

One-on-One Interviews

The benefit of one-on-one interviews is that interviewees often feel free to be open and honest, especially if they do not have to be concerned about their identity being compromised. Usually there are a number of key informants in a community who are extremely knowledgeable. This usually includes the school superintendent, school board members, the mayor, council members, the police chief, the fire chief, the city engineer, major property owners, chamber of commerce executives, economic development executives, ministers and priests, public housing officials, neighborhood and civic organization leaders, major business owners/managers, long-term residents, bankers, and leaders of minority groups.

Before beginning the interviews, it is best to develop a structured interview form. This provides the interviewer with a consistent set of questions for all respondents. Asking these same questions often produces different answers and perceptions that can be explored further. Usually, one or two responses to questions will lead the interviewer to ask follow-on questions. On the structured interview form, list some open-ended questions that allow respondents to tell you what they want you to know. Leave some blank spaces to record comments and notes. Comparing responses when tabulating the results also helps corroborate qualitative data.

Some interviewees don't like to just respond to questions from your interview form; they have several key issues they want to "vent" about. You can usually tell from body language and clipped responses when they are chomping at the bit to talk about their issues. In these situations, it is usually better to let the interviewee lead the discussion. Let him address his major issues (he will appreciate the fact that someone is listening) and try to get answers from him on some other areas of interest from your survey form if you have time. Often these one- or two-issue interviews will give you a wealth of in-depth knowledge about the community.

Community Meetings

Community meetings open to all interested citizens can be held to gather information, opinions, and ideas. These meetings need to be planned and managed very carefully so as to avoid their becoming divisive and disruptive. At the very beginning, an agenda and topics of discussion need to be laid out as well as ground rules that govern the meeting. An announcement can be made that a major community revitalization effort is being launched and that the citizens' ideas and opinions are being solicited so that the most important issues can be addressed. First, participants might be asked to identify the things that they believe are community strengths. Then they may be asked to identify things they believe are weaknesses. These topics can be categorized, and the group can be asked for ways to build on or take advantage of strengths and what needs to be done to eliminate or lessen the impact of weaknesses. Participants might also be asked to identify factors outside the community that present opportunities that should be pursued or threats that should be considered.

The focus of these meetings should be positive, not on, "How did we get where we are?" but on "What do we want our community to become?" and "What do we need to do to make it that way?" If a professional facilitator is not used to manage the meeting, it is recommended that a well-respected community leader lead the meeting so as not to become a target for any dissatisfaction or frustration that may relate to the topic. This individual should be very tactful, focused, outcome oriented, and experienced at managing meetings. Above all, he or she should not be perceived as strongly on one side or another of a critical community issue.

Focus Group Meetings

Focus group meetings are similar to community meetings but are directed toward one topic or a few related topics such as job creation and economic development, tourism and recreation, or education and workforce development. Citizens who attend these meetings usually have sincere and deep feelings about the importance of the topics being discussed. Many may be experts in these fields. They also have ideas and suggestions for what needs to be done to solve the problems related to the topic.

The strategy for these meetings should be similar to that of community meetings. The leader needs to have a specific agenda and objectives for the meetings. This should be shared with the group along with any ground rules for managing the meeting, such as "no personalizing any comments or attacking any individuals." Focus group meetings need to be problem identification and solution oriented. The agenda should involve a set of questions or topic areas in which participants can give their opinions about problems, and offer observations about and solutions to problems.

These meetings also help identify individuals who may later work on strategic planning committees, provide information or services to strategic planning committees, or volunteer to help work on projects, programs, and initiatives developed during a strategic planning process that follows the assessment.

Questionnaires and Opinion Surveys

Questionnaires and opinion surveys are another data collection method that can be used to collect qualitative information from citizens. While there are data collection problems associated with these types of methods, they still provide another way to collect data and involve citizens. These methods might also appeal to those who would not normally attend a meeting or participate in an interview.

Distribution and collection of questionnaires and opinion surveys should be planned carefully so as not to exclude any particular group of citizens and to ensure a good cross section of community representation. It is sometimes possible to distribute them through prominent citizens who bring them to civic organization and club meetings where they are completed, collected and returned. The local newspaper may even publish the survey so that it can be mailed in by respondents for tabulation. Pastors may distribute them at churches.

Thanks to modern technology, these can also be posted on the internet at a very low cost. As of this writing, for example, a short questionnaire can be set up at www.branenet.com for under $5 a month. As each vote is submitted, the results are tabulated and reported in a graphic format. Caution should be used with online surveys due to the fact that those without computer access cannot participate and anonymous "ballot box stuffing" can occur.

Traditional questionnaires and surveys can also produce a large volume of paper that must be processed. Care must be taken in tabulating the results to be sure that all responses are accurately captured and reported. However surveys are useful since they offer confidentiality to respondents and may reach a whole group of individuals who, for many reasons, will not or cannot participate in the assessment process through other means.

Citizens groups should not be overly concerned about collecting data through written surveys, meetings, and interviews. Care and thought should be given to designing the questions to be asked or in planning and managing the meetings to be held. There are many books available at libraries and bookstores that deal with tests and measurements (in education) and surveys and research methods (in business and science). Many of these provide chapters on question design and ample guidance on response options. All things considered, it is always best to use a number of data collection sources and methods so that information collected during the assessment can be corroborated through multiple sources.

Community Assessment Topics

As previously mentioned, assessment is a comprehensive process that involves a review of all major sectors of a community. The process attempts to involve a broad cross section of stakeholders in identifying the factors to be considered in planned growth and development. It examines four broad areas: physical infrastructure, social infrastructure, economic development infrastructure, and human infrastructure. Within each of those major categories, many factors are considered. Each of these factors can be reported as chapters in the assessment report and include both the quantitative and qualitative data collected from the many methods described above.

Physical Infrastructure

When considering the physical infrastructure, community leaders need to examine the factors that will influence a business's operations in their area. Shown below are the areas that are evaluated and a sample of the type of questions that might be asked about each factor.

Transportation System for Moving Goods

Highways—Are the highways to and from town in good condition and well maintained on a regular basis? Are they two lanes or four lanes? Are they primary or secondary roads? Do they have shoulders on which disabled trucks may pull over? Are any future improvements planned? How far is the community from the nearest interstate highway?

Rail—Is there rail? Is it a main or secondary line? Is there a siding on which cars may be stored? Are the rail rates competitive?

Airfreight—Is there a local general aviation facility or international airport? Is local airfreight available at that facility (other than from FedEx, DHL and UPS)? Are their rates competitive? What are the heaviest items that service providers can take?

Transportation Network for Business Travel

Is there an international or hub airport nearby? Is there a general aviation facility (GAF) nearby? What is the driving distance and time? What is the length of the runway at the GAF? What flight and landing services are provided at the GAF? Does the GAF have instrument flight rules (IFR) support equipment? Is it lit at night? Is the GAF tower manned 24 hours? Does the GAF have aircraft maintenance services? Does the GAF have hangar space available? What leasing arrangements are possible at the GAF?

Weather and Geography

Geography and weather often drive location decisions. Many northern companies have chosen southern locations over the past 30 years. The Sunbelt, particularly the southeast, is the leading growth region in the United States. When moving to new locations, residents and business alike often look for areas with sunny weather and weather that lacks extreme snow and rain events. Businesses in particular often seek sites with land that is well suited for development and that does not require large sums of site preparation money. They also usually seek locations in or near major metropolitan centers so that they can access the amenities and workforce skill found in these areas. Therefore, geographic and weather conditions can play a major role in relocations and expansions. Community leaders would do well to ask the following type of questions about their areas:

> Is the community isolated or near other towns? Is the area near a Standard Metropolitan Statistical Area (SMSA)? Is the area prone to flooding? Are there other hazards to structures in the area such as earthquakes, forest and grass fires? Does the area have relatively flat land that can be easily developed? What is the average rainfall and snowfall for the area? What are the average high and low temperatures? How many days a year, on average, does the temperature fall below freezing? How many days, on average, does the temperature exceed 100 degrees Fahrenheit? Will the weather lead to transportation and construction delays or work stoppages? What are the soil and subterranean conditions in the area (e.g. will there be higher construction costs related to site preparation due to the soil conditions)? Are there any EPA emission restrictions that govern the area (e.g. is the community in an ozone

attainment zone)? Are there sources of fresh water? Are they available for distribution or accessible by private well? Do water sources need to be heavily treated or processed with simple filtration and chlorination?

Social Infrastructure

Listed below are some of the key social infrastructure factors influencing the development of communities.

Availability of Quality Health Care

With drops in rural population and migration to major metropolitan areas, fewer doctors are opening practices in rural areas. This can make it difficult for residents in rural areas to get quality health care. As the world's baby boomers age, gerontological health care is becoming increasingly more important. Many now have all their children leaving home, but find themselves taking care of elderly parents, many of whom have health problems that need regular management. Community leaders need to explore their community's health-care services by asking the following type of questions:

> Are there local clinics and doctors? How many? Does the community have ambulance service? Are emergency medical technicians present on the ambulances? Does the community have dentists? Are there medical specialists practicing in the community? Are there physical therapist and rehabilitation therapists in the community? Is there a hospice? How far is it to the nearest hospital with an emergency room/trauma center?

Safety of Investment

Business location as well as home-buying decisions are driven by economics. While it may be possible to get an excellent low-cost building or piece of land, the operating costs of that location need to be examined. High insurance rates and losses due to vandalism or theft can quickly erode the initial cost advantage.

> How safe is our investment at this location? Are there hidden costs of doing business related to threats in the community? Community leaders need to know the answers to these questions and others below when planning a development initiative or marketing the community to investors.
>
> What is the public's perception of safety in their community? Is it still an environment where many people do not lock things up or is there an obvious abundance of burglar alarm company signs, home window bars and other indicators of criminal activity or fear? What is the fire insurance rating for the area? What is the crime rate? Are there signs of vandalism or gang activity? Do individuals gather on street corners for illicit activities? Is there an obvious drug problem in the community? Is the community well lit? How many police officers are on duty? How many firemen serve on the local fire department? Is the fire department a volunteer unit or does it have some full-time personnel?

Quality of School System

The quality of local schools not only supports workforce capabilities but often drives residential growth. Given a choice, parents will usually seek a residential location in a school district with a good reputation for academics and administration. Many state education departments have data available on school performance by district, and sometimes on individual schools. The Census Bureau, at www.census.gov, can also provide data on citizens' educational levels. Listed below are some of the questions that community leaders should ask about their local schools:

> How well do students in the local system perform on standardized tests? What is the

dropout rate compared to state and regional rates? What is the condition of the schools' physical plant and facilities? What is the pupil–teacher ratio? Do the high schools offer a full range of academic subjects? In addition to academic courses, do the high schools offer vocational training? Where do most major employers' high school graduate employees come from? Is the community a residential demand area because of its school system?

Parks, Recreational, and Cultural Opportunities

Community leaders need to determine what their community offers compared to other communities of its size. This can be done by asking the following type of questions:

> Where do most residents go for recreation? Are there periodic cultural events in the community for locals and visitors alike? Is there a community center? Are planned recreational and sporting activities held for youth and adults? Are the playgrounds well maintained, neat and clean? Are facilities well lit? Is the play equipment safe, attractive and functioning? Are boundary fences in good repair? Are neighboring buildings and structures in good repair and attractive? Are nearby and adjacent properties residential or compatible with commercial use? When making a visual inspection of the location, are people present? What are they doing? Are recreational areas defaced and vandalized? Are there recreational activities for people of all ages?

Availability, Affordability, and Quality of Housing

A drive though most communities will tell you a lot about their housing stock. Since many residents travel the same routes to and from their home and work locations, there may be conditions on other streets or in other areas of the community of which they are not aware. Realtors can provide very good insight to the condition of and demand for area housing. Another good source of somewhat dated housing information is the Census Bureau (at www.census.gov). Listed below are some questions that community leaders should ask about their area's housing:

> What percentage of the county's housing consists of mobile homes? What is the value of housing compared to the adjacent counties? Is there zoning in the community? Are local ordinances consistently and strictly enforced? What percentage of residents rent compared to adjacent counties? What is the occupancy rate of local housing? What is the cost of a three bedroom home with two baths? What is the condition of vacant buildings and housing? What is the time needed for construction of a new home? What is the cost of construction per square foot for residences and commercial structures? What percentage of local housing is in need of renovation or demolition? Are there abandoned commercial buildings in the community? Are they potential brownfield sites? Are there abandoned commercial buildings that could be quickly rehabilitated and put back into use? Are there housing subdivisions with infrastructure in place in which new homes could be constructed? Are there quality apartments or rental properties in the community that could serve as temporary housing until individuals buy or build homes? Are there existing commercial buildings that could be converted into multi-family housing? Is there public housing? In what condition and how well managed are the public housing units?

Quality College/University Nearby

Institutions of higher learning provide the skilled workforce needed for most of today's higher-paying jobs. Being located in close proximity to a college or university can be an advantage to a community because it provides a ready source of candidates for skilled jobs and provides locals with an opportunity to upgrade their skills for advancement. Listed below are some of the questions that community leaders should ask about nearby institutions of higher learning:

> Is there a community college, four-year college or university nearby that serves the community? What type of technical and vocational degrees or training do they offer? Are there technical and scientific courses offered that can support high-tech manufacturing and businesses? How many graduates are produced in each of the programs annually? Do the institutions have an active placement office serving new and past graduates? Where are these graduates employed? Are the institutions willing to and capable of adding more programs in support of new businesses coming to the area? How far away are the institutions? From where do most of the students come? Where do most of the students live? Are there knowledgeable and expert professors who can provide research support for business operations?

Economic Development Infrastructure

Low Cost of Living

The cost of living in the United States varies greatly from region to region. Housing, taxes, food, insurance, taxes, and other expenses are significantly higher in places like San Francisco, Los Angeles, and New York. Employees relocating to these areas often receive salary bonuses or higher salaries to help compensate for the cost associated with moving there. Communities with a lower cost of living have an advantage when recruiting new residents and businesses from areas with higher living costs.

Community leaders working for planned development need to ask and answer questions similar to those that follow:

> What is the cost of living in the community? Will those moving into the region experience much higher living expenses than the area from which they are moving? How does the cost of living here compare to similarly situated communities in the regions from which we hope to draw businesses and residents?

Quality and Competitiveness of Public Utilities

Energy is becoming a major concern for most businesses. It is often less expensive to operate in foreign countries where energy costs can be lower, particularly when oil, coal, or natural gas are needed for production processes. In spite of recent major outages, the United States still has an advantage over the rest of the world with regard to reliability of electric and gas distribution systems. In time, however, that reliability advantage might also decline. When assessing their community's utilities, leaders should ask questions similar to the following:

> Who are the current utility providers serving the community? How well maintained are the current distribution and transmission systems? Is the community on a transmission and distribution network? What are the reliability percentages for the electrical and gas service? Has there been a history of prolonged interruptions? Are the costs competitive? Are there any factors that will result in significant increases in the near future? Are long-term contracts available? What is the potential for long-term rate stability? What are the sources of generation (nuclear, oil, gas, or coal)?

Availability of Water and Waste Systems

Many fast-growing communities have outpaced their community's public water and wastewater system's ability to expand. As in many older communities, their systems may not be in compliance with recent EPA standards and may soon require major upgrades that will result in rate increases.

Further, new restrictions by the Environmental Protection Agency may have limited the capacity of existing systems so that the community has no growth capacity. Additional residents and businesses coming to a community can result in the inability of these systems to meet demand. Some areas of the US are experiencing water shortages and rationing (particularly arid areas of the west southwest and, recently, even the southeast US).

Similarly, solid waste processors and facilities are raising their rates, and some landfills are closing because they have either reached the end of their useful life or can no longer operate due to environmental regulations. It is not uncommon for many communities to export their solid waste many miles for disposal.

Listed below are some of the questions leaders should ask about their community's water and waste systems:

> What are the current capacities of the water and wastewater system and, if they are not operating at capacity, how much surplus capacity do they have? Are the current water, sewer, and solid waste disposal rates competitive? If the community owns its systems, are they subsidized by taxes or operating on their own at a "profit"? If they are operating at a "profit," are the funds being wisely managed? Are the rates being charged reflective of fair market value or are they artificially low? What are the long-term operating outlooks for the systems (replacement, major renovations, or closure)?

Telecommunications

From hard line (fiber-optic and copper) telephone and cable systems to internet service providers and wireless broadband, telecommunications are becoming increasingly important for both business operations and personal convenience. Businesses use these systems to transfer electronic data and to maintain contact with other offices and customers. Communities with a variety of telecommunications services and providers are more competitive as sites for business relocation and expansion than those with limited services. Listed below are the type questions that business leaders should ask about their community's telecommunications capabilities:

> Is cell phone coverage and service good in the community? Does the community have a cable provider? Is broadband (high-speed) access to the internet available in most areas of the community through cable, DSL, or perhaps satellite? Is the community served by fiber-optic cable service? Are rates for these services competitive? Is there a loop network supporting these systems?

Available Commercial Buildings

It has long been recognized that communities with available commercial buildings are much more likely to locate a new or expanding business than those with greenfield sites. Many businesses seeking relocation or expansion want to make the move quickly and find greater value in existing buildings that can be quickly refitted to their needs. Community leaders would do well to work with their area's commercial realtors to identify and maintain an inventory on available commercial buildings. Here are some of the questions that leaders should ask about the available buildings in their community:

> How many commercial buildings are available in the community? What size are they? What are the features of each (office space, warehouse space, truck bays, ceiling height, floor load, utility capacities, etc.)? Who is the current owner? What is the asking price? Is the price negotiable? Will the owner enter into a right of first refusal? Will the owner sell or lease? Is owner financing available? [Note: the International Economic Development Council provides a template for collecting building data at www.iedconline.org]

Lack of Governmental "Red Tape"

Complicated building permits, compliance with local zoning and other ordinances can make attracting new businesses and residents difficult. Local leaders should carefully examine the process for setting up new businesses, building new commercial buildings, and building new residences. The following are some of the useful questions that will provide insight about the process:

> Who needs to be contacted to apply for permits, approvals, and licenses? Is there a one-stop shop that can guide an applicant through the process? Is there one document that clearly walks an applicant through all requirements and provides contact names and numbers? What is the cost of each? Are the forms easy to understand? What support needs to be provided in addition to the forms? How long do the forms take to process? Once approved, can they be revoked or amended? In the event of a mistake on the part of a governmental agency, who will pay the cost of bringing the project into compliance? Are hidden kickbacks and payoffs a hidden part of the system? Are government clerks and officials "customer focused" or "process focused?" (Do they seek to help the applicant comply with the law or seek themselves to comply with the law?)

Tax Rates on Business

When considering a location for an expansion or relocation site, business managers look at the total cost of doing business at the prospective location. While the location might have all that is needed with regard to location, utilities, building, and workforce, the tax structure can be such that it makes a particular business unfeasible. Some taxes originate at the state level. However, others, such as local sales and franchise taxes, are under local control. Local leaders need to examine the impact of their tax structure on the types of businesses that are the best fit for their location by asking questions such as the following:

> Do businesses pay a grossly inequitable share of taxes compared to other taxpayers? Do the public services and infrastructure supported by business taxes run efficiently, using the tax income specifically for services delivered? Does the local tax structure have a chilling effect on the location of companies for whom the community would otherwise be an excellent location? Is the current business tax structure competitive compared to other locations?

First-Rate Scientific Community in Area

As the United States transitions from an industrial to an information service and high-technology environment, more emphasis will fall on emerging technology and scientific research that supports business competitiveness and opportunities. An assessment should consider what resources are available to a community to support its businesses research and scientific needs. Questions similar to those that follow should be asked during the assessment:

> Are high-tech businesses or scientific research facilities nearby? Are there any high tech or scientific companies that are seeking partners for testing or marketing the work in which they are involved? Is there a technology transfer center that can assist businesses in adopting new technologies that will lower operating costs and increase competitiveness? What fields of excellence are available at local universities and major hospitals? Are these facilities willing to partner with businesses to bring new technology to the marketplace?

Availability of Fully Developed Sites

Communities with a fully developed industrial or business park have a distinct advantage over those without. That advantage can increase when the park is owned or managed by a government or quasi-governmental organization responsible for economic development. Where necessary, rather than relying on private property owners to market and develop their commercial property, local leaders can take the lead in developing a site ready for development that can be marketed to existing and new businesses. If a community has an industrial or business park, all utility, lot, and building information should be included in the assessment. Shown below are the type of additional questions that should be answered about the park:

> Is the property available for lease or sale? Can speculative buildings be constructed? Is there a revolving loan fund resulting from park revenues that is available for new businesses? What incentives can the park offer for businesses locating there? What is the cost of lots and buildings? Are those prices competitive with other buildings and sites? Is the property governed by a restricted covenant? What types of businesses are being sought for the park? What types of businesses are not of interest to the community?

Strong Existing Businesses

A healthy local economy and tax base are heavily dependent on existing local businesses. While existing business owners and operators may be concerned about new businesses coming and competing for their employees, they can also provide support for and benefit from the building of new or expansion of existing businesses in the community. Local retailers, for example, often fear the competition from major retailers such as Wal-Mart. However, the savvy retailers realize that they can benefit from the increased traffic to the community if they can realign their operations to take advantage of the change. Having knowledge of a community's businesses can also help identify the type of businesses that should be recruited.

A community assessor should examine local businesses, make judgments about their strength, and include his or her findings in the assessment report. Are they competitive and successful? Has there been a lot of turnover in local businesses (i.e. a lot of openings and closings) or have many of the local businesses been in business for a long time? Is there a large inventory of commercial buildings due to the many closed businesses? What is affecting the health of local businesses (e.g. population decline)? What types of businesses can use the services and products delivered or made by local businesses?

Human Infrastructure

Broad-Based "Can-Do Spirit"

Many times the prevailing local attitudes can poison the development well. Communities that have a history of failed development activities or ill-conceived projects often see themselves as helpless. One of the most important things that a professional community developer can do is help instill the belief that citizens are not helpless and that by conducting an assessment and carefully planning and working together, they can see positive change in their community.

Positive attitudes are a powerful force for planned change. Community leaders and citizens who believe that they can make things happen and see opportunities and possibilities rather than problems and threats, often help separate their community from those that fail.

A community assessor should ask questions about the community's past projects, programs, and initiatives to determine what citizens and leaders believe. When citizens offer ideas about what can be done or discuss the history of past events, what are their responses? Is there a history of failed events? Has there been a lack of action to address community problems? Are local leaders hesitant to try new things or do they lack the political will to make hard and sometimes controversial decisions? If comments such as the following are heard, in all likelihood the community has a poor "can-do spirit":

> "That will never happen here."
> "We discontinued the town festival because of lack of support."
> "We have tried to get organized several times in the past."
> "We had an assessment and developed a plan once but I don't really know what happened to it."
> "We don't have the local leadership to get that done."
> "We tried that once and ..."
> "We just don't have the resources like other small towns."

Conversely, if comments such as the following are heard, in all likelihood citizens and their leaders have the confidence needed to bring about positive change:

> "We have a very successful local festival every year."
> "We were able to leverage public and private money to improve the local community center and playgrounds."
> "Our bankers and government officials can work together to put a financial and incentive package together for a major new business."
> "We know that grants and funding are available for developing our business park."
> "Our local government is well run and efficient. Not only do we have a surplus, but our utility rates and fees are among the lowest in the area."
> "Our elected officials, business and civic leaders have a history of working together on projects. We have..."

Desire for Development

It is important for a professional community developer to remember that his or her mission is not to direct the community development process or tell citizens what their community should become. The professional developer's role is an advisory one. Professional developers listen and reflect back the values and desires of the community; they provide options, ideas, alternatives. It is also appropriate to describe strategies and projects that have worked well for other communities. However, in the end, citizens have to decide for themselves what they want their community to become and the actions that they will take.

There is often an interesting paradox found in quaint and attractive rural communities. Many citizens want the amenities and conveniences associated with larger communities, but they do not want to do what is needed in order to have them or deal with the side effects that come as a result. They want more retail variety, for example, but local business leaders do not want the competition from chains and franchise operations and many citizens do not want new residents that contribute to traffic congestion and overcrowding.

Desire for development sounds like a very "soft" subjective measure. However, a community's collective desire for development can and must be determined as well as the nature of the development that citizens would like. This determination can be made quite quickly and easily by asking some very simple questions such as, "What would you like to see take place here?" or "What do you think needs to be done to make the community a better place for citizens?" If the majority of comments are similar to those below, then low-impact activities and strategies such as residential and retirement community development, recreational and tourism development, Main Street specialty and theme retail, or a "bedroom" community strategy in support of a larger nearby community are the likely development goals.

> "I moved here because it is clean and green and that's the way I want it to stay."
> "We don't want this to become like a city."
> "We don't want more people moving in."

> "We like it just the way it is, there is nothing that we should change."
> "I moved from the city because this was a quiet and safe community. I don't want the growth and problems associated with the city."
> "We don't want any smoky smelly businesses here and all the truck traffic related to their operations."

On the other hand, if the majority of the comments are similar to those that follow, then a community's citizens are very supportive of change and desire positive change and growth.

> "We need more highly paid technical jobs in our community, jobs that help keep our children here and that bring in new residents."
> "We need to increase the tax base with more business development that will help support the public infrastructure."
> "We need to develop a business park and actively market our community to business investors as a location."
> "It would be great if we could get a Wal-Mart and some chain restaurants in town."
> "We really need to improve our streets and public utilities to support new growth. Our systems are old and do not have the capacity for new businesses and homes."
> "We really need to improve our town's appearance and be stricter in enforcing abatement regulations."

Competitive Wage Rate/Salary Rates

Wages and personnel costs usually represent a significant portion of most manufactured goods and services. Regardless of all the political rhetoric, it is a simple economic fact that in order for businesses to stay competitive in a world marketplace, they must seek locations with the lowest possible operating costs for a given labor force availability. This economic principle not only applies to businesses but to consumers. Given a choice between buying the same quality goods in their community or another location, consumers will make their purchases where they get the most value for their dollar.

Not only do communities compete against those in other states or regions of a country for business investments, but they compete against the labor force in other nations. The wage rates paid in a local market must not be so great that they discourage new businesses from relocating or opening a new venture there.

As an example, the petrochemical industry tends to be clustered in locations such as Louisiana, New Jersey, and Texas, where there is oil and/or gas production or importation or where the refineries are located. The wages paid to workers in most of these unionized facilities are quite high. These operations are able to attract the most qualified and highly trainable employees from the regions in which they are located. Those attempting to locate other businesses within these high wage rate areas will face tough competition for the best employees and may be forced to pay a premium above what they would in other locations. This can have a chilling effect on the attraction of new businesses for which labor costs are a significant percentage of their operating costs.

Usually a state's department of labor and its job services office can provide demographics on wages and salaries in an area and provide insights into the competitiveness of local wage rates.

Labor Availability

Not only must labor be affordable, but it must be available in order to attract new businesses. Not only must the unemployment rates be identified, but an estimate of those in the workforce who are underemployed (having skills greater than those required for their current positions) should be made. When researching the labor force available, a community assessor should consider the community's own citizens, and those in nearby communities and counties from which workers might realistically be drawn in support of business operations. As discussed above, the state's labor department and local job service offices can provide a good deal of data. Similarly, the U.S. Census Bureau can provide additional workforce data.

Quality of Workforce

"Cheap labor" is not always the answer for successful business expansion and attraction. Workforces in developed countries cannot compete with the wage rates offered in developing countries such as those in South and Central America, and Asia. Over the past several decades, the developed countries have lost many low- and semi-skilled jobs (such as cut and sew operations and unskilled manufacturing assembly) to developing nations whose wage rates are less than $1 per hour.

Much information about the quality of labor can be provided or found in the sources previously listed. However, local employers are often an excellent source of information about the capabilities of a community's workers. Asking local employers about their workers can provide anecdotal evidence regarding the ability of employees to learn new skills; their dependability; honesty; drug abuse; and overall performance. About five years ago, for example, the author was told by the branch manager of a national police equipment manufacturer that when the company went through a cost-cutting effort, another plant was closed and the work transferred to his location, which then expanded. The reasons workers were more productive (as measured by the cost of production including raw materials consumed) were that their absentee rate and turnover were low, and the quality of their work (as measured by their rejection rate) was very high.

Labor Climate

Most enlightened business managers and labor leaders realize the need for a team environment rather than a combative one that adds to the cost of operations. For labor, management, and shareholders to benefit, they must work together. This not only makes them competitive with other domestic and international operations, but makes them likely candidates for reinvestment, expansion, and continued operations.

When conducting the assessment, local leaders should be asked questions about work stoppages, sabotage, strikes, and other labor unrest. Is this a location where production and services can be reliably provided? Or is this a location that will consume valuable management time and corporate resources, continually distracting attention from the primary function of business operations? Communities with a reputation for high employer/worker teamwork will have an advantage to retain and expand business operations over those locations that are more disruptive and combative.

SWOT Analysis

For many years, business leaders have used the "SWOT analysis" as a method of identifying the *strengths*, *weaknesses*, *opportunities*, and *threats* impacting their commercial ventures. The SWOT process provides them with a systematic approach for analyzing options and making decisions. By researching and laying out the SWOT factors, they can prioritize their actions and focus their efforts for the greatest impact.

The same process can be used in support of the community development assessment. Strengths and weaknesses are the direct factors impacting a community, those that can be directly controlled or influenced by the local leadership. The condition of local streets, water and wastewater systems, fire protection, and emergency services are all within the control of local leaders. They are examples of areas in which strengths and weaknesses can be found.

Strengths describe the assets that are already present and upon which development can be

Table 11.1 *Subjective rating of location factors for anytown: possible ratings—Good (2) Average (1) Poor (0)*

Location factors	*Ratings*	*Point value*
Physical infrastructure		
Transportation system for moving goods	Good	2
Transportation network for business travel	Average	1
Weather and geography	Good	2
Social infrastructure		
Availability of quality health care	Average	1
Safety of investment (police, fire, emergency services)	Average	1
Quality of school system	Average	1
Parks, recreational, and cultural opportunities	Average	1
Availability, affordability, and quality of housing	Poor	0
Quality college/university nearby	Good	2
Economic development infrastructure		
Low cost of living	Poor	0
Quality of public utilities (electricity and gas)	Good	2
Availability of water and waste systems	Good	2
Telecommunications	Good	2
Available commercial buildings	Poor	0
Lack of governmental "red tape"	Average	1
Tax rates on business	Good	2
First-rate scientific community in area	Good	2
Availability of fully developed, publicly owned sites	Poor	0
Strong existing businesses	Good	2
Human infrastructure		
Broad-based "can-do" spirit	Poor	0
Desire for development	Good	2
Relatively low wage/salary rates	Good	2
Labor availability	Average	1
Quality of workforce	Average	1
Labor climate	Good	2
Overall rating	Average	1.28

Note: Overall rating scale: 0–0.69 = Poor 0.7–1.39 = Average 1.4 or higher = Good.

built. What do we have here in abundance? Who needs it? For example, a community may have a bountiful supply of fresh water that requires very little processing before distribution. It may also have excess capacity in its municipal wastewater system. With regard to solid waste disposal, it may have a large, well-managed sanitary land fill with many years of capacity. Such communities are an attractive location for businesses in need of fresh water supplies and whose operations result in considerable solid waste, such as those of food and beverage processors.

Weaknesses are also factors that influence business location decisions. Will the assets mentioned above be countered by major community weaknesses such as poor housing; a declining and aging population; high tax rates; unreliable utilities; badly deteriorated streets; a lack of land use planning and zoning; and poor local elected leadership?

Strengths and weaknesses are primarily under the control of local leaders. Once identified, strategies can be implemented to build on existing strengths, to turn weaknesses into strengths, and to minimize if not eliminate weaknesses.

Opportunities and threats tend to be factors outside the control of local leaders but that can impact the community's development efforts. Geopolitical events and economic trends, national political events, environmental factors, and corporate realignments tend to present opportunities and threats outside the control and possibly the influence of communities. However, once such factors are identified, communities can create new strategies, or adapt their existing strategies, to manage threats to the extent that the negative impacts either do not occur or are minimized when they do.

If, for example, a town is a "company town" with one major employer providing most of the local jobs and supporting most of the local service businesses, this presents a threat. The community does not have control over the corporate management's decisions. If the plant becomes less cost-effective to operate, management may decide to close the facility. Prior to that threat becoming a reality, local leadership might have regular meetings with plant and corporate managers to identify potential problems so that they could be mitigated. For example, they might seek state incentives and tax abatements to help the facility expand or modernize. Further, they might initiate an economic diversity strategy to attract new, unrelated businesses to the local economy so that the community is no longer a "company town."

Other threats are environmental. In the early 2000s, West Nile Virus entered and began spreading throughout the United States. This event created a threat to summer tourist areas that had large populations of mosquitoes. While action should be taken at the local level to reduce the threat, the problem was still largely out of local leaders' control. It is a threat that would have to be managed by increased mosquito control efforts, reminiscent of those against malaria and yellow fever. Some communities also increased their advertising budgets and changed their advertising messages.

Sometimes threats can also produce opportunities. On September 11, 2001, terrorists attacked and destroyed major offices in the World Trade Center in New York. Some of the firms there, such as Kantor-Fitzgerald, had their entire operations disrupted and lost a significant percentage of their total employees through the attack because all their corporate functions were at one location. As a result, other companies began to consider dispersing their vital business functions to different and, in some instances, less threatening locations.

The 2001 terrorist attacks, as tragic as they were, created opportunities for communities that might be destinations for dispersed corporate operations. The realignment of the federal government and creation of the Homeland Security Department (HSD) will result in the creation of regional HSD offices, another opportunity for some communities.

The Assessment Report

The product of a community assessment is usually a written report. This report generally consists of four parts. First, it provides an analysis of each of the development factors listed and described above and possibly others that are specific to the assessment location. This section is a blend of objective data, observations, and subjective findings and may include pictures supporting the findings. Second, the report should include a SWOT analysis that describes the strengths and weaknesses identified during the assessment and the opportunities and threats identified by local leaders. Third, the report provides a section on "possibilities." This is a compilation of wants and desires as expressed by citizens (what they would like to see happen in the future) and possible courses of action (programs, projects, and initiatives) to help the community reach its full potential. In addition to describing possibilities, it may also include pictures of projects from other communities, so that citizens and leaders can see what similarly situated communities have accomplished. Finally, the report would include appendices of statistical data that supports the findings including opinion surveys, demographic data, and other evidence on which the findings are based.

It has been this author's experience that community leaders and citizens also want a summary of the findings. They seek a quick reference and a numerical rating that they can use to make comparisons between themselves and other communities. Often, statistical data is used to compare a community with other similar or sometimes "competing" communities in the SWOT analysis. Communities also want a benchmark from which to measure improvements. Table 11.1 is an overview of the development categories for a small community as it would appear in the report. This table would typically appear in the first part of the assessment report, possibly at the end of the findings. The ratings would be supported by the descriptive statistics and observations of the assessment team. There are potentially hundreds of community factors that could be included in an assessment. Each situation is unique. The SWOT analysis categories in this chapter and Table 11.2 are illustrative of many of the potential categories but are not exhaustive. The categories chosen in an actual SWOT analysis (see the Case Study below) will depend in part on the exact nature and purpose of the assessment.

Conclusion

Starting economic development and community development initiatives without an assessment can often lead to addressing problems' symptoms rather than root problems, misidentifying strengths and weaknesses, or failing to identify opportunities and threats. Assessments are an important precursor to strategic planning and can help start the community dialogue that fosters planned change. They can be performed by community leaders themselves, allies such as state and federal agencies, and professional developers.

The objectives of community development assessments are:

- to support a community dialogue that involves citizens in determining how they would like to see their community develop;
- to identify the factors that influence the potential development of the community;
- to identify the strengths, weaknesses, opportunities, and threats that are or can influence the community's development;
- to provide information for those leading a strategic planning process so that a workable and focused set of goals and objectives can be created to help the community achieve its potential as well as help citizens realize their shared vision of the community.

CASE STUDY: BAYSHORE COUNTY COMPETITIVE ASSESSMENT

Bayshore County, located in a coastal county, has a strong tourist industry as well as a variety of manufacturing and service industries. The county realizes that tourism is highly competitive and jobs in this industry are not high-paying. The Bayshore County Economic Development Agency hired a consultant to assess the county strengths and weaknesses for recruiting manufacturing and service industries. The consultant gathered and analyzed demographic, economic, and social data for the county and compared it to similar counties that might compete for the same types of industries. In addition, the consultant conducted over 30 interviews with local company executives, elected officials, and other community stakeholders. In these confidential interviews, the consultant asked questions and probed to determine what the interviewees considered to be the county's strengths and weaknesses from the economic development standpoint.

Based on the data analysis, interviews, and a tour of the county, the consultant rated the county in terms of its strengths, weaknesses, or neutral on several key factors relevant to attracting new manufacturing and service industries. Table 11.2 shows the summary factor rating matrix. The consultant's report was comprehensive and included a narrative on each factor and explanation of how the rating was determined. Excerpts from the consultant's report are also included below.

K-12 Education: Strength

- K-12 education in Bayshore County is a strength in comparison to the rest of the state.
- The biggest issue facing K-12 education in Bayshore County is the perception of the local community. Individuals who come from school systems outside of the state seem to have the perception that the school system is not very good.
- The dual enrollment program offers high school students the ability to earn a two-year degree prior to graduating from high school. This program provides a mechanism for postsecondary training for students entering the workforce upon graduation from high school.
- The International Baccalaureate program at Bayshore High School is a nationally recognized program that provides a competitive college preparatory curriculum.
- Offering choice of school in the county is a positive attribute and has lead to a strongly competitive environment within the high school system. Each school must be able to recruit students in order to maintain levels of funding. This provides for a strong education system since the best and brightest are being recruited to choose a high school rather than simply being districted to one school or another.
- There is a countywide school system. This is a strength since there are no competing school districts within the county.

Highways and Roads—Weakness

- No access to an interstate or interstate-quality roadway is a weakness for Bayshore County. This limits the type of industries that might be recruited to the area.
- There are only three main highways into and out of the county: U.S. 79 and U.S. 262. The east–west route of U.S. 24 is bottlenecked at the Harrison Bridge crossing in West Bayshore. The bridge is currently under construction for expansion.
- Traffic congestion could pose a problem in recruiting industry to the area. It is not currently a critical factor in the county since a majority of trips take, at the most, 20 minutes. Adequate planning for traffic growth could cause additional problems in the future.
- Along with the new airport it is understood that a four-lane, potentially limited access highway is planned to connect the new airport to the interstate highway. This would greatly improve Bayshore County's position in regard to highways and roads.

Table 11.2 *Factor Rating: Bayshore County*

Summary Strengths and Weaknesses Matrix

Category	*Strength*	*Neutral*	*Weakness*
Available land and buildings			
Available land		X	
Available buildings		X	
Labor			
Labor cost		X	
Labor availablility	X		
Labor–Management relations		X	
Labor productivity and work ethic		X	
Utilities			
Electricity	X		
Water		X	
Sewer		X	
Natural gas		X	
Telecommunications	X		
Transportation and market access			
Air		X	
Roads and highways			X
Rail		X	
Water	X		
Market access			X
Business, political and economic climate			
Local taxes and incentives		X	
State taxes and incentives		X	
Local Government support		X	
Economic base			X
Permitting			X
Zoning and land use planning			X
Image and appearance			X
Economic development vs. tourism			X
Regional cooperation	X		
Education			
K-12	X		
Vocational and technical training		X	
Higher education	X		
Quality of life			
Cost of living		X	
Housing availability	X		
Health care		X	
Cultural activities		X	
Recreational activities	X		
Other			
Economic Development Program	X		

Keywords

Community development, asset identification, economic development, data collection, assessment report, appreciative inquiry, SWOT analysis, human infrastructure, community profile.

Review Questions

1 What is the purpose of a community assessment? Why conduct one?
2 How do you collect information for an assessment?
3 What are some of the characteristics of a community an assessment should cover?
4 What is a SWOT analysis and why is it useful?
5 What should be included in the assessment report?

Bibliography

Clark, M.J. (2007) *Community Assessment Reference Guide*, Upper Saddle River, NJ: Prentice Hall.

Cornell-Ohl, S., McMahon, P.M. and Peck, J.E. (1991) "Local Assessment of the Industrial Development Process: A Case Study," *Economic Development Review*, 9(3): 53–57.

Luther, V. (1999) *A Practical Guide to Community Assessment*, Lincoln, NE: Heartland Center for Leadership Development.

Stans, M.H., Siciliano, R.C. and Podesta, R.A. (1991) "How to Make an Industrial Site Survey," *Economic Development Review*, 4: 65–69.

U.S. Census Bureau. Online. Available at: <www.census.gov>

U.S. Department of Energy. Online. Available at: <www.energy.gov>

U.S. Department of Labor. Online. Available at: <www.dol.gov>

Williams, R.L. and Yanoshik, K. (2001) "Can You Do a Community Assessment Without Talking to the Community?," *Journal of Community Health*, 26(4): 233–248.

Connections

Explore a useful resource for community-based service project assessments with the Rotary International's guide at: www.rotary.org/en/document/578

Learn more about assessments with the University of Kansas' guide at: www.communityaction.org/files/HigherGround/Community_Needs_Assessment_Tool_Kit.pdf

12 Community Asset Mapping and Surveys

Gary Paul Green

Overview

Although a variety of approaches can be used to begin an asset-based community development plan, those used most often are either a needs assessment or an asset mapping study. This chapter focuses on several techniques for mapping community assets such as asset inventories, identifying potential partners/collaborators, and various survey instruments and data collection methods. Much of the discussion is also appropriate for conducting needs assessment studies and covers some issues organizers may want to consider in conducting surveys of individuals, organizations, and institutions.

Introduction

The community development process is often initiated with some form of needs assessment or asset mapping. Needs assessment is a method for identifying local problems or issues. These "needs" in turn become the basis for a strategic action plan (Johnson et al. 1987). Community organizers mobilize residents around the specific issue in order to seek new resources, obtain information and expertise, or to pressure local officials to solve the problem. It is assumed that residents will act to address the perceived deficiencies in their neighborhood or community. Organizers hope that residents will gain a sense of efficacy by ameliorating problems, which in turn will help them to build the confidence to address other issues as well.

Kretzmann and McKnight's (1993) asset-based development model, however, turns this approach on its head and starts with a map of local resources which provide the basis for a community vision and action plan. This model focuses on the strengths rather than weaknesses of the community. Asset-based development attempts to leverage local resources to obtain additional support for projects. Another distinguishing characteristic is that asset-based development relies heavily on developing partnerships and organizing across the community in order to identify ways of building on local assets to improve the quality of life. The goal is to develop trust among various partners and to operate on the basis of consensus.

Both models of community development rely heavily on survey data. Needs assessment uses surveys to obtain individuals' evaluations of local services, social problems, and other local issues. Asset-based development relies on surveys of local residents to identify skills and talents that may be underutilized or not often recognized. Surveys can also be used to conduct assessments of organizational and institutional resources within the community. With both models, the survey is more than just a means of obtaining information—it is also considered a way of securing public participation.

This chapter focuses on several techniques for mapping community assets, but much of the discussion is also appropriate for conducting needs assessment and covers some of the issues organizers may want to consider in conducting surveys of individuals, organizations, and institutions.

Asset Mobilization

Kretzmann and McKnight (1993) describe several steps in mobilizing community assets. First, it is important to map the capacities of individuals, organizations, and institutions within the community. This process helps identify the resources that are available for development. Not only are these assets frequently overlooked, but residents have a tendency to focus more on how external resources can address local deficiencies. Second, organizers build relationships across the community in order to generate support. Most often community organizers work through existing organizations and associations (Chambers 2003). The goal is to identify common values and concerns that can form the basis of strategic action. Third, the community develops a vision and an action plan for achieving its goals. The vision should be based on the values of local residents and the resources that are available to them. Finally, communities can leverage their resources to gain outside support. Although the asset building approach relies on mobilizing local resources, it does not ignore the importance of tapping into external resources and sources of information. However, the focus is on how community action can build on the expertise, experiences, and resources that are already available.

A core premise of asset-based development is that local resources are often overlooked in community development. One of the best examples of asset-based development is the Dudley Street Initiative in Boston (Medoff and Sklar 1994). Like many inner city areas, this neighborhood had numerous vacant lots that had become illegal dumping sites. Rather than viewing these vacant lots as a problem, residents worked with city officials to give the neighborhood association the right of eminent domain to claim this property for affordable housing projects. They formed a land trust for the property that essentially gave the community control over the development of the land. Rather than letting developers decide the best use for the land, residents found ways of developing affordable housing that was sorely needed.

Organizations and associations play a critical role in the community development process. In addition to their resources and membership, they can provide community organizers with legitimization in the community. Rather than working to start new organizations, it is possible to build on existing ones. This strategy is typically used by the Industrial Areas Foundation (IAF), which relies heavily on mobilizing churches, unions, and community-based organizations. Mark Warren's (2001) book, *Dry Bones Rattling*, does an excellent job of demonstrating how schools and churches worked together to build a powerful network of organizations in Texas. The network not only crossed various denominations and organizations, but also formed a multiracial constituency.

Finally, local institutions are important actors in communities. They facilitate regular interaction among residents and link the locality to the broader society. Local institutions can also affect the neighborhood or community through its practices, and many institutions have resources that remain untapped by local residents. For example, schools may purchase goods and services outside the area, and banks may be investing their resources in non-local investments as well. With greater understanding of these resources, community residents can potentially shape local institutions so that they serve local residents better.

Methods for Mapping Assets

Communities can use several different methods for mapping assets. The purpose of the project should guide the decisions about which method to use and what specific information should be collected. The goals may range from promoting local economic development or community health to supporting youth programs. The community and/or organizations need to clearly state the purpose of the project.

Next, it is important to define the territory of the neighborhood or community. This decision will affect almost everything else that is done and should specify which individuals, organizations, or institutions should be included in the project as well as what issues face the community. In most cases there will not be a consensus regarding the boundaries of the community or neighborhood. Natural barriers, such as rivers or lakes, often serve as a boundary. Major streets or highways can help define a neighborhood. Many people will define neighborhoods by key institutions such as schools and churches. School districts provide a useful way to define a community because the population often is fairly homogeneous with regard to socioeconomic status and home ownership. Schools generate interaction among residents which can facilitate the community development process. They also have the advantage of having clearly demarcated boundaries. However boundaries are chosen, they should reflect residents' perception of the community and promote interaction on issues of common interest.

After identifying the purpose and the geographic boundaries of the asset-building community development project, it is important to consider the appropriate method(s) for conducting the project. There are several issues to consider including the available resources, timing, geographic area, etc. Weighing different factors can be difficult, and there is no easy way to balance various considerations. One of the most difficult trade-offs is between cost and quality. For example, how much will the increased cost of a second wave of surveys improve the quality of the data? Each community needs to decide how they want to balance this trade-off.

Most communities rely on surveys to document local assets. However, it may be more appropriate to use other methods such as focus groups. Focus groups have the advantage of being relatively inexpensive and can be conducted more quickly than surveys. They can also provide more in-depth information on why people feel the way they do on various issues. On the other hand, focus groups do not give most residents an opportunity to participate in the process, and the findings may not be very representative of the larger population.

Regardless, it is not necessary to choose between surveys and focus groups. In fact, it might be useful to use both focus groups and surveys. Some communities may initially use focus groups to identify what types of experiences and skills may be available locally. Alternatively, conducting focus groups after a survey may permit organizers to ask follow-up questions about issues and questions raised through the findings of the survey.

Surveys can be administered face to face, in group settings, over the phone, or through the mail. Conducting a survey requires time and a financial commitment from community members. They need to ask themselves several questions before embarking on such a project: Do we want to conduct a survey or use some other technique for obtaining public participation? What is the best way to obtain the information that is needed? What do we want to know? How will this information be used? Is there sufficient time and financial commitment on the part of residents to conduct a survey? Does the information already exist through bureaucratic records, census data, or some other survey that has recently been collected?

When Is a Survey Appropriate?

Most communities use surveys to map community assets. If the goal is strictly to obtain public participation on a policy issue, there may be a variety of other techniques that may be more appropriate or cost-efficient. For example, it may be quicker and easier to hold public meetings or to conduct focus groups. Focus groups may be more appropriate in a situation where it is necessary to understand why people feel they way they do about particular issues. Public meetings provide an opportunity for residents to voice their opinion about issues and listen to the perspectives of their neighbors. A survey instrument may not provide the type of information obtained from these two other techniques.

Communities also need to consider whether they have sufficient resources for conducting a survey. There is always a tradeoff between the cost and quality of conducting surveys. By conducting a survey as cheaply as possible, communities may end up with a low response rate, results that are nonrepresentative, or poor data. As discussed below, there are several low-cost strategies that can significantly improve the quality of data that are collected.

What Is the Best Technique for Conducting a Survey?

There is no single "best" technique for conducting surveys. The appropriate technique depends on the resources available, the type of information that is desired, and the sampling strategies. Probably the best resource for conducting surveys and constructing questionnaires is Don Dillman's (1978) *Mail and Telephone Surveys: The Total Design Method*. In this book, Dillman provides a step-by-step procedure for conducting mail and telephone surveys. His procedure has been used extensively by survey researchers, and he provides several excellent ideas for obtaining the highest response rate possible. However, most researchers do not completely follow his procedure for mail surveys, which calls for several waves of mailings with a certified letter after several unsuccessful attempts to obtain a response, because most communities cannot afford the complete procedure he recommends. Another good resource for conducting surveys is Priscilla Salant and Don Dillman's (1994) *How to Conduct Your Own Survey*. This book is more accessible to residents who have little experience with survey research.

The advantages and disadvantages of four commonly used survey techniques—face-to-face interviews, mail surveys, telephone surveys, and group-administered surveys—are discussed below. While most communities tend to rely on a single method of conducting surveys, it may be possible to combine several methods.

Face-to-face interviews generally provide the best response rate (usually more than 70%) among the four survey techniques considered and permit the interviewer to use visual aids and/or complex questions. This technique is often used with very long or complex questionnaires. Face-to-face interviews are also used to obtain information from groups that would not likely respond to other methods. For example, it may be easier to contact low-income residents through this method than a mail survey or some other technique. Also when interviewing employers, it may be preferable to use face-to-face interviews rather than mail surveys. In all cases, interviewers can follow-up on responses to get a better understanding of why a given response is provided.

However, face-to-face interviews are the most expensive of the four techniques, and there may be problems with "interviewer bias." Interviewers will need training, and there is often more coordination of those involved in face-to-face interviews than with other techniques.

Web-based surveys are now one of the most common methods of administering surveys. Web surveys are used increasingly because of their low cost and ease of administration. There are several software packages, such as Survey Monkey, that are quite easy to use and help tabulate results. There are a few limitations, however. The return rate on web surveys tends to be very low. Obviously this method requires access to email addresses. In many neighborhoods this could be administered through a listserv. It is still the case, however, that many people (such as the elderly or low-income residents) may not have access to the internet.

Mail surveys are frequently used for conducting community surveys. Mail surveys are usually shorter in length than face-to-face surveys and can include maps and other visuals aids, but the instructions need to be concise and understandable. The response rate for mail surveys will vary depending on several factors such as how many follow-up letters are sent, the extent to which the material is personalized, the length of the survey, and whether or not incentives are provided. Many communities mail only one wave of questionnaires which generally produces a response rate of between 30 to 50%. A follow-up postcard can yield another 10%, and a replacement questionnaire will generate another 10 to 20%. However, there are several disadvantages to using mail surveys. They are more limited in regards to the length of the survey than in other techniques. It is also very difficult to ask complex questions, and there are no opportunities for follow-up questions or clarifications.

One effective approach in limited situations is to drop off the questionnaires and have respondents either mail them back or have volunteers pick them up at a later date. This approach requires coordination of volunteers as well as local knowledge of the neighborhoods involved. With this technique it is be possible to explain to respondents the purpose of the survey and how it will be used and to clarify any issues or questions. Also, when respondents know that someone will be dropping by to pick up the survey, it improves the likelihood that they will complete it. The response rate is typically higher than that of the standard mail survey.

Telephone surveys can be completed quickly and generally have a higher response rate than mail surveys. The cost may vary depending upon whether or not respondents are randomly sampled. The response rate for telephone surveys is not as good as face-to-face interviews, but they have the advantage of possible follow-up by interviewers. One of the chief disadvantages is that the interviewer cannot use any visual materials or complex questions. Phone surveys can be difficult to organize when using volunteers to conduct the surveys, and it is more difficult to manage interviewer bias with this survey technique.

Group-administered surveys can be used in situations where the targeted population is likely to attend a meeting where the survey could be administered. For example, a survey could be administered at a neighborhood meeting of residents. The chief advantage is that a large number of respondents can be reached quickly with very little cost. This approach to administering the survey can introduce several problems in terms of the representativeness of the results. For example, the people attending the meeting may not be representative of all members of the association or organization. Similarly, if the goal is to provide information on residents, members of a group or association in the area may not be representative of all residents. However, group-administered surveys can be a cost-effective way to conduct a survey under certain conditions. It is also possible to provide complex instructions and use visual aids, such as maps, with this method.

These methods of conducting surveys should not be considered mutually exclusive. In many cases it may make sense to mix methods. For example, it may be possible to conduct a mail survey of neighborhood residents and then supplement it with either a phone survey or face-to-face interview with people who have not responded to the mail survey. This strategy of combining survey techniques usually improves the response rate and enables communities to collect information from various groups that may not respond to one particular survey approach.

What Is the Best Way to Draw a Sample for a Survey?

Often communities struggle with developing a random sample of residents for their survey. The problem is that there is no easy way of identifying the population in a neighborhood or community. Using telephone books or even random digit dialing is becoming a major problem for survey research. Many low-income residents do have telephones. A growing number of households use cell phones and/or have unlisted numbers. Caller ID makes it more difficult to complete telephone interviews because many residents can screen their calls. These issues make it increasingly problematic to obtain random samples from telephone surveys. And for the purposes of neighborhood organizations, it is somewhat difficult to use this method to interview residents.

Property tax records are inadequate in settings where there are a large number of renters. It often takes a lot of work to get the lists in a useable form because business and absentee owners will be included in the list. Also, there may be multiple entries with the same names (i.e. individuals who own several properties).

Because many people have utility hookups, these records are probably one of the best sources for drawing a sample of households. However, these records can be difficult to obtain, and, in some multi-family units, there will only be one name for the entire housing unit. But almost everyone has electricity, and this source can be supplemented with others to provide a good list of the population.

So, what is the best strategy for developing a sample of households? One approach is to combine lists or methods to draw the sample. For example, many communities rely on property tax records to identify property owners, supplement these lists by locating rental units in the neighborhood/community, and then conduct face-to-face or drop-off surveys among these households. It is possible to purchase a random list of residents from firms that compile these lists, but this can be too expensive for many community groups.

Another possibility is to generate a list by identifying all housing units in a neighborhood or community. From this list of housing units/structures, a sample can be drawn. This is a very time-consuming process, and, obviously, this strategy cannot be used on a large-scale area if there are limited resources available.

If at all possible, the community may want to simply conduct a survey of all residents. One of the questions I always get when presenting the results of a survey is: How could this sample be representative if I did not receive one? Some people do not believe it is possible to develop a scientific sample of a population that does not have a bias. If one of the goals is to obtain public participation in the process, it is probably better to conduct the survey among all households. Surveying the complete population will put to rest the concerns about the representativeness of the sample and promote interest in the community development project.

Finally, in some circumstances it is appropriate to draw what is referred to as a purposive sample. For example, it might be possible to conduct interviews with residents at a neighborhood event. It must be recognized that this is not a random sample and may not be representative of the community at large, but it may be sufficient in many cases.

How large should a sample be? The main goal should be to develop a sample that is sufficiently large enough to provide an adequate number of responses for each group in the community that need to be considered in the analysis. For example, in order to compare the responses of youth and adults, it is necessary to have a sufficient number of responses from both categories of respondents. Depending on the types of comparisons that will be made, the sample should be larger. The larger the sample size, the smaller the margin of error in the results.

Sometimes it may be advantageous not to use a random sample at all. If, for example, one of the goals is to looks at the assets of the working poor, a random sample may not pick up enough residents in this category. If this is the case, it may be useful to develop a stratified sample that has a disproportionate number of residents in the groups under consideration. This approach can work if it is known where the working class residents are most likely to live in the neighborhood and then that area is targeted. It may be more difficult to use a stratified sample using other characteristics.

One important issue that is often neglected by communities is the *unit of analysis* of the survey. Is the focus of the study on individuals, families, or households? If the goal is to obtain an accurate random sample of individuals, it may be necessary to conduct a random sample of adults in the household/family. One method of obtaining a random sample is to conduct the interview with the person in the household/family who has had the most recent birthday. Although this strategy reduces the problem of gender bias in responses, when conducting a phone survey, it will increase survey cost because it may be necessary to make several calls before reaching the person who is to be interviewed.

Decisions about the best way to develop a sample of households in a neighborhood or a community are intimately tied to the resources available for the project. Clearly, the best method for sampling households and conducting a survey would be to identify each household and conduct face-to-face interviews with a random sample of individuals. This strategy would be very expensive and impractical for most community organizations. In most cases, there is a need to balance competing demands of cost, data quality, and resident participation.

The quality of community surveys can be significantly improved with some preparatory work. One of the most important things to do in advance is to set up an advisory committee to help construct the survey and build support for it in the community. An advisory board can help raise funds as well as possibly recruit volunteers for the survey. Another role for the advisory committee is working with the media to publicize the survey. The advisory committee can also help plan the feedback sessions to residents.

Pretests are essential to a successful survey. Typically, volunteers can administer this survey face to face. It is important to do this face to face in order to assess whether the respondents are confused about the meaning of any of the questions. Pretests can help communities avoid the embarrassing situation of collecting data that has limited usefulness.

Marketing the survey improves the response rate and helps residents understand how the survey information will be used. One strategy is to place an advertisement in local newspapers that explains the purpose of the survey. A cover letter should accompany the survey and explain the objectives of the survey and identify supporting organizations and/or institutions. This letter should also identify a contact person if residents have any questions about the survey. Contacting local organizations—such as churches, schools, and civic organizations—may be another way of explaining the purpose of the survey and gaining support for the effort in the community.

Providing feedback to the neighborhood or community can be a useful way to gain some additional insights into the results of the survey and to reward residents for participating in the survey. It is preferable to provide residents with a written report of the results. Some discussion of feedback in the cover letter may improve the response to the survey.

Mapping Individual Capacities

A central premise of the asset-based community development approach is that all individuals have a capacity to contribute to community well-being. However, assets of youth, seniors, and people with disabilities are frequently ignored. The most obvious assets are formal labor market skills such as work experience, leadership, and organizational skills. Other assets include experiences that individuals may have had outside the formal labor market such as care-giving skills, construction skills, or repair skills. Abilities also include "art, story-telling, crafts, gardening, teaching, sports, political interest, organizing, volunteering and more" (Kretzmann et al. 1997: 4). Another component of individual capacity is interest in participating in various community organizations and working on local issues. All of these capacities need to be documented and analyzed for their potential contribution to the community.

An excellent resource for such an analysis is the workbook on mobilizing community skills of local residents authored by John Kretzmann et al. (1997). This workbook provides several sample surveys that have been used in the past to document the skills and experience of local residents. Although these surveys can be a useful beginning point, it is important to consider the characteristics and dimensions of any unique population in the community. In other words, do not assume that these sample surveys will necessarily work in all communities or that they adequately tap all the individual assets among various groups in the population.

The economic capacities of individuals in the community may also be considered. Residents often purchase goods and services outside of their community. These purchases could contribute to the local economy. Part of the mapping process is to evaluate where consumers purchase goods and services and how much they spend in- and outside of the community. When aggregated, these figures should provide community organizations with a sense of what goods or services could be provided locally. For example, if most neighborhood residents purchase groceries outside the area, this may be a signal that there is a potential for establishing a grocery store in the neighborhood. It should be understood that there are many factors that affect the potential for retail establishments and this type of information is just one of those considerations. Aside from proximity, consumers may choose to purchase goods and services based on quality, service, and/or convenience (they may work close to the establishment). Some sample surveys for documenting these assets can be found in Kretzmann et al. (1996b).

How is this information used? First, the data can be used to help existing retail establishments identify goods and services that local residents consume but do not purchase locally. Second, this information is useful to potential entrepreneurs interested in starting businesses in the neighborhood. Finally, survey data collected on expenditures can be used as an educational tool to help residents understand the power their local expenditures have on the neighborhood's economy.

BOX 12.1 CAPTURING COMMUNITY WITH STORYTELLING

Storytelling is a powerful tool to capture the essence of communities, both reflected in individual stories as well as those told collectively. Capture Wales was a project that was run from 2001 to 2008 by the British Broadcasting Corporation to capture people's own digital stories. Nearly 600 were recorded after providing workshops on digital media. These two-minute vignettes cover a wide range of topics reflecting community members' histories, viewpoints, and perspectives. According to the BBC, the goal of the project "was to give production tools over to the public who traditionally viewed themselves as consumers, rather than producers, of media. As a result, individuals were able to contribute unique, personal stories to a multimedia collection of histories." See the website for more details: www.bbc.co.uk/wales/arts/yourvideo/queries/capturewales.shtml

Storytelling is powerful in the context of the collective as well. In the rural community of Colquitt, Georgia (US), residents take on the role of folk life performers in renditions of stories handed down through the generations. Residents from different ethnic, race, and socioeconomic backgrounds as well as across the ages join together to tell their stories.

For more details on the power of storytelling in community development approaches, see the white paper by Barbara Ganley in conjunction with the Orton Family Foundation's Heart and Soul Project at: http://storytellingwhitepaper.digress.it

The Editors

Mapping Associations

Associations and organizations can facilitate community mobilization. Many efforts to mobilize communities begin with existing organizations because they have established relationships, trust, and resources that can be used in the asset-based community development effort. Although formal organizations are often visible and well established, there are many more organizations without paid staff that are not as easily identified. Some examples of these informal organizations are block clubs, neighborhood watches, garden clubs, baby-sitting cooperatives, youth peer groups, recreation clubs, and building tenant associations (Ferguson and Stoutland 1999).

How are these informal associations mapped? Most local organizations and associations do not show up on any official lists of nonprofit organizations because they are not incorporated or have no paid staff. Probably the best way to identify these associations is to conduct of a survey among residents to identify any associations/groups they have heard of or belong to in the community. This method not only enables communities to identify nonprofit organizations but also the interorganizational networks that exist. For example, it may be useful to understand how organizations are linked through overlapping membership as a means of creating potential partnerships and collaborations.

Another reason it is useful to map organizational resources is to help identify who should/could be mobilized in the asset-based community development project. One of the keys to the success of community development projects is inclusion of a wide range of residents. It may be possible to use a list of organizations and associations as a means of checking which community groups and interests are represented in the project. Obviously, it would be impossible to invite representatives from all organizations and associations to participate, but it may be useful to at least ensure that individuals from various areas, such as environmental, health care, economic development, or other areas, are selected.

What type of information should be collected about associations and organizations? The most important issue is identifying individual participation. Individuals should be asked what associations and organizations they belong to inside and outside the community. This exercise should produce a relatively comprehensive list of organizations and associations. It is also important to identify an individual's leadership role (e.g. served as an officer) and level of participation.

Based on the list that is generated from the surveys of individuals, it also is possible collect information from these organizations and associations. A list of board members and officers is useful to identify potential leadership in the community. It is helpful to collect information on the issues and concerns the organization has and what types of programs they have developed or implemented. Finally, some basic information on the resources of the organization will be helpful in order to identify assets that can be mobilized.

Mapping Community Institutions

Community institutions hold important resources that could potentially contribute to asset-based development projects. Institutions that are typically most important include parks, libraries, schools, community colleges, police, and hospitals. For example, each of these institutions purchases goods and services that could be directed at local businesses to improve the economy. They may have facilities and equipment that could be used by residents for community events. These institutions could adjust their employment practices so as to benefit local residents. Or, they could offer programs that could be redirected toward local residents. The main goal of mapping local institutions is to identify their resources and mobilize them in a way to benefit the community.

The first step in mapping community institutions is to develop an inventory of the institutions in the community. In most cases this is fairly straightforward, but it needs to be done systematically so that nothing is overlooked. In a small neighborhood, this can be done quickly; but it may take more work and time in a larger region.

The next step is to identify institutional assets. Depending on the goals, this may involve identifying the spending patterns of the institution, i.e. where goods and services are purchased. This typically involves conducting surveys of these institutions and identifying key underutilized resources. One example of this would be the school-to-farm programs that have developed across the country. Community groups are assessing where schools are purchasing food and whether this food could be purchased locally through farms in the region. School-to-farm programs improve the markets for local farms and the quality of food in the schools. However, there can be institutional obstacles such as cost, scale, and price to use locally produced goods in the schools.

Another example of institutional mapping is the community reinvestment activities that many neighborhoods have been involved in over the past 20 years. These neighborhoods often begin with a thorough analysis of the lending patterns of local financial institutions to understand the capital flows in their area. Analyses of capital markets permit local residents to understand how well financial institutions are meeting the needs of local residents and whether their savings are being invested elsewhere. With these analyses, residents can work with local institutions to invest more in the neighborhood, use the data to challenge bank mergers, and identify the potential for other community-based lending institutions in the neighborhood.

Mapping institutions could require documentation of the talents and skills of the personnel who may or may not be community residents. It is also useful to assess other resources, such as meeting space, services, or equipment, that could be used for community purposes. One area that is often overlooked is the hiring practices of institutions. In many cases local institutions have little control over hiring because it is done through some central location such as a city or state office. But there may even be opportunities at these institutions to hire local residents for jobs that are available.

Finally, organizations can build strategies based on the identified resources in the community. These strategies need to be consistent with the broad set of goals and vision established at the beginning of the project. Kretzmann and McKnight's (1993) workbook, *Building Communities from the Inside Out,* is an excellent resource for identifying examples and methods of capturing local institutions for community building.

Conclusion

Asset-based development differs from traditional community development strategies in several key ways. Local assets drive the community's plan for development and mobilization rather than problems or needs. It relies heavily on building relationships and local leadership. The beginning point for the asset-based approach is a map of the community's assets. This mapping effort requires residents to go beyond their preconceived notions of what they believe exists in the community and asks them to identify resources that could be used to achieve their vision for the future of their community. Developing an accurate assessment of resources is critical to the success of asset building. This also shifts the focus from the problems to the opportunities that face the community.

It should be stressed that mobilizing communities around assets can be a difficult process. Most residents want to move quickly to identifying solutions without adequately assessing the issues, understanding the resources that are available to them, or developing a vision of what the community should be in the future. While organizing communities around issues and problems often works for the short run, it is difficult to maintain in the long run. Mobilizing communities around partnerships and developing new leadership should provide a basis for long-term community action.

A final note regarding the quality of community surveys. One of the surest ways to halt a community development project is to start with questionable data. Because everything that follows is based on an accurate assessment of community resources, it is essential to conduct a survey that residents can trust. Many communities outsource the survey to professional or support staff. This has the advantage of improving the perception that the survey will be neutral. Yet the survey is never actually value-free, and residents may still feel that the process is biased. By contracting outside the community for the survey, residents are missing out on an opportunity to build interest and support for their neighborhood. Residents may gain more insights into how they might promote asset-based development by creating their own questionnaire and conducting some of the interviews and/or pretests.

CASE STUDY: HAZELWOOD COMMUNITY ASSET MAP: ASSESSING THE SERVICES, NEEDS, AND STRENGTHS OF HAZELWOOD'S COMMUNITY SERVICE PROVIDERS

Communities frequently struggle to identify the available services and service providers in their area. Several factors contribute to this problem. Residents are often bewildered by the complex bureaucracy involved in social services. There is a frequently great deal of overlap in responsibilities across organizations and institutions. At the same time, many residents are not aware of the array of local resources available to them. In 2005, the Hazelwood neighborhood (in Pittsburgh, Pennsylvania) decided to map their assets to better understand the existing organizational resources in their area as well as to document the concerns among residents. In collaboration with the University of Pittsburg Community Outreach Partnership Center (COPC), the neighborhood conducted face to face and phone interviews with the contact person from every organization, church, and agency responsible for providing social services. The neighborhood was able to successfully complete 28 out of 40 interviews.

One of the most important findings was that there was very little overlap in social services provided to the neighborhood. While on the survey it appeared that several programs distributed emergency food to residents, they offered them at different times and places. Similarly, after-school programs were fairly well distributed across the neighborhood. Most residents were not aware of where and when these services were provided.

Thus, there was a need to establish a more formalized method of coordinating services. One of the conclusions from the asset mapping exercise was that most of the service providers were located outside the community and some effort should be made to provide a common space for these providers in the neighborhood. Another "need" that emerged from this study was that residents discovered that public transportation was a major hindrance in accessing the services that were identified. The asset mapping exercise also identified some gaps in services, especially homebound services, transportation inside and outside the community, and health needs of the residents. For more information about the project, see: www.pitt.edu/~copc/Hazelwood_Asset_Map.doc

Keywords

Needs assessment, data collection, surveys, data analysis, asset-based development, asset mapping, institutional mapping, asset mobilization.

Review Questions

1 Why is it important to map assets?
2 What is the process for mapping individual capabilities?
3 What are some techniques for surveying communities?

Bibliography

Beaulieu, L.J. (2002) *Mapping the Assets of Your Community: A Key Component for Building Local Capacity*, Mississippi State, MS: Southern Rural Development Center. Online. Available: HTTP <http: //srdc.msstate.edu/publications/227/227_asset_mapping.pdf> (accessed 8 June 2004).

Chambers, E.T. (2003) *Roots for Radicals: Organizing for Power, Action, and Justice*, New York: Continuum.

Dillman, D.A. (1978) *Mail and Telephone Surveys: The Total Design Method*, New York: Wiley.

Ferguson, R.F. and Stoutland, S.E. (1999) "Reconceiving the community development field," in R.F. Ferguson and W. Dickens (eds.) *Urban Problems and Community Development*, Washington, DC: Brooking Institution Press, pp. 33–76.

Green, G.P. and Haines, A. (2007) *Asset Building and Community Development*, 2nd ed., Thousand Oaks, CA: Sage.

Johnson, D.E., Meiller, L.R., Miller, L.C., and Summers, G.F. (eds.) (1987) *Needs Assessment: Theory and Methods*, Ames: Iowa State University Press.

Kretzmann, J.P. and McKnight, J.L. (1993) *Building Communities from the Inside Out: A Path Toward Finding and Mobilizing a Community's Assets*, Chicago: ACTA.

Kretzmann, J.P., McKnight, J.L. and Puntenney, D. (1996a) *A Guide to Mapping and Mobilizing the Economic Capacities of Local Residents*, Chicago: ACTA Publications.

Kretzmann, J.P., McKnight, J.L. and Puntenney, D. (1996b) *A Guide to Mapping Consumer Expenditures and Mobilizing Consumer Expenditure Capacities*, Chicago: ACTA.

Kretzmann, J.P., McKnight, J.L. and Puntenney, D. (1996c) *A Guide to Mapping Local Business Assets and Mobilizing Local Business Capacities*, Chicago: ACTA Publications.

Kretzmann, J.P., McKnight, J.L. and Sheehan, G. (1997) *A Guide to Capacity Inventories: Mobilizing the Community Skills of Local Residents*, Chicago: ACTA Publications.

Medoff, P. and Sklar, H. (1994) *Streets of Hope: The Fall and Rise of an Urban Neighborhood*, Boston, MA: South End Press.

Salant, P. and Dillman, D.A. (1994) *How to Conduct Your Own Survey*, New York: Wiley.

Warren, M.R. (2001) *Dry Bones Rattling: Community Building to Revitalize American Democracy*, Princeton, NJ: Princeton University Press.

Connections

Watch informative presentations on asset mapping and development at:

www.youtube.com/watch?v=wYw14uCGbkw (Asset Mapping vs. Needs Assessment)

www.youtube.com/watch?v=pSwpQWAUQAc (John McKnight—Asset Based Development, pt. 1)

www.youtube.com/watch?v=SAmpUDayWpk (John McKnight—Asset Based Development, pt. 2)

13 Understanding Community Economies

Hamilton Galloway

Overview

Other chapters in the book have discussed the inextricable link between community and economic development. A strong economy helps make a community a better place to live and often helps mitigate problems such as substandard housing and poor nutrition. For economic development purposes, a critical first step toward developing a strategy or plan is the fundamental understanding of how the local economy functions. In other words, a local area should know what industries, occupations and other assets contribute to the sustainable functionality and employment of the economy. Once this level of understanding is achieved, decisions and strategies for enhancing and diversifying the local economy are better informed and can align with the area's capacity and capabilities. This in turn can increase the prosperity of community residents. In this chapter, the focus is on economic analysis such as learning how industry and occupation data is classified; different metrics for evaluating industry and occupation characteristics; sources for obtaining economic data; and the types of industries that bring new money into an economy versus industries that circulate existing money. Furthermore, understanding the potential positive impacts (or pitfalls) due to changes in the local economy is also critical. An introduction to input–output modeling is provided in this chapter to provide further understanding of decision-making outcomes. Finally, the chapter concludes with a case providing a different approach to looking for opportunities in a community's economy with leakage analysis.

Introduction

The purpose of this chapter is to introduce four critical areas of assessing an economy, whether local, regional, state, or even multi-state. The first two areas focus on industry and occupation classification and standard measures of performance and competitiveness. Knowing your economy's industry and workforce assets allows for greater perspective in decision making, business recruitment, business expansion and/or business retention. Furthermore, having industry and workforce insight enables the economic development community to better engage other community and regional leaders, especially those in the private sector, education and workforce development. The third topic discussed is the use of economic base theory—a common assessment of an area's export-oriented industry sectors—and understanding what forces drive new money into an area's economy.

Industry Classification

In order to better organize and track changes throughout the national economy, the U.S. Census Bureau developed what is known as the North American Industry Classification System, or NAICS. This system categorizes industries based on similar market attributes, starting with broad industry classifications such as manufacturing and more narrowly defines specific industries within that category, for example, semiconductor and related device manufacturing. The NAICS system has a hierarchy of five levels, starting with what is commonly referred to as 2-digit NAICS, the highest level, and progressing down to 6-digit NAICS, the most detailed level. Currently, about 1,195 detailed 6-digit NAICS are reported across 20 different industry categories. Table 13.1 illustrates the hierarchy involved for credit and lending industries.

Table 13.1 *NAICS Hierarchy*

48–49	Transportation and warehousing
51	Information
52	Finance and insurance
522	Credit intermediation and related activities
5222	Nondepository credit intermediation
52229	Other nondepository credit intermediation
522291	Consumer lending
522292	Real estate credit
522293	International trade financing
522294	Secondary market financing
522298	All other nondepository credit intermediation
53	Real estate and rental and leasing
54	Professional, scientific, and technical services

Occupation Classification

Similar to industry classification, the U.S. Bureau of Labor Statistics developed a system of categorizing occupations called standard occupation classification, or SOC. Typical users of the SOC system include: government program managers, industrial and labor relation practitioners, students considering career training, job seekers, vocational training schools, and employers.[1]

The SOC system classifies workers at four levels: (1) major group; (2) minor group; (3) broad occupation; and (4) detailed occupation.

Occupations are categorized into 23 major groups, containing approximately 840 detailed occupations. Table 13.2 shows a sample of the levels for life scientists.

Table 13.2 *Example of Levels in NAICS Coding*

17–0000	Architecture and engineering occupations
19–0000	Life, physical, and social science occupations
19–1000	Life scientists
19–1020	Biological scientists
19–1021	Biochemists and biophysicists
19–1022	Microbiologists
19–1023	Zoologists and wildlife biologists
19–1029	Biological scientists, all other
21–0000	Community and social services occupations
23–0000	Legal occupations

Types of Data and Analysis

Community- and regional-level data can provide insight into what is happening in a given area and what an area has to offer for residents and businesses alike. To add value to simple statistics, specific analysis and modeling can be undertaken, which provides varying perspectives of an economy and answers questions such as: Is the economy growing, shrinking? What types of economic specializations exist? What are the workforce capabilities? What types of workforce specializations exist? Is the workforce young, middle-aged, or older?

To provide context, this chapter evaluates three areas of data and analysis—demographic, industry, and occupation. Within each area, several different types of analysis can be undertaken. Typical analysis includes historical change, projections, location quotient (a measurement of industry/occupation concentration) and shift share (a measurement of competitiveness).

BOX 13.1 THE CREATIVE CLASS AT THE LOCAL LEVEL

This case study of the Washington, DC area provides an excellent overview of the creative industries, and builds on regional growth theories focused on knowledge and technology-based growth clusters. The methodology is easily replicable and can be used in sub-regional or local levels.

Source

Terry Holzheimer, AICP, CEcD, Benchmarking the Creative Class at the Local Level, *Economic Development Journal*, 5(3), 2006 pp. 34–39.

Demographic Data

Data on an area's demographic make-up comes in all shapes and sizes. Typically gathered from the U.S. Census Bureau, demographic data is separated into five-year groups (e.g. 20–24 years old), also known as cohorts. These groups are further broken down by sex and race. Evaluating an area's demographic structure allows stronger insight into the employment diversity of the workforce as well as potential assets or hurdles to overcome (Center for Community and Economic Development 2014). For example, an aging workforce could pose problems of replacing retiring workers. Likewise, an aging community may also require planning activities to support accommodation and needs of the elderly (see: www.economicmodeling.com/2010/07/28/rural-counties-are-getting-older-with-some-notable-exceptions). As such, understanding your demography will allow for focused strategies to address economic challenges and opportunities.

Another type of demographic analysis considers the education attainment of the area's population. Data collected on an education attainment usually only includes the population that is 25 years and older, since many postsecondary education programs can take significant time to achieve. Education attainment data can be evaluated based gender and race, providing valuable insight on education talent as well as shortcomings. Schools and state policy makers can use this information to target specific demographic groups for program engagement, such as scholarships for Hispanic women or community outreach programs for underrepresented ethnic groups. Figure 13.1 shows a sample of less than high school diploma attainment between 2002 and 2010. From this sample, we can conclude that overall the white, non-Hispanic and black or African-American populations are increasing their education attainment levels while the bulk of the remaining groups—predominately white Hispanic and non-white Hispanic—have increased high school dropout rates.

Industry Data

Having covered how industry data is categorized, understanding the types of industry data available is critical to understanding an area's economy. Specifically, knowing how the area functions and what drives job creation or job destruction (i.e. layoffs) allows for better planning of the area's future. For example, in the face of significant economic downturn in certain areas of traditional manufacturing, some communities have undertaken a rebranding approach to drive investment and job creation in high-tech manufacturing, clean energy and research and development.

Conventional industry data and analysis includes evaluating historical, current and project growth/decline of an industry, at various levels of the NAICS hierarchy. However, mistakes are often

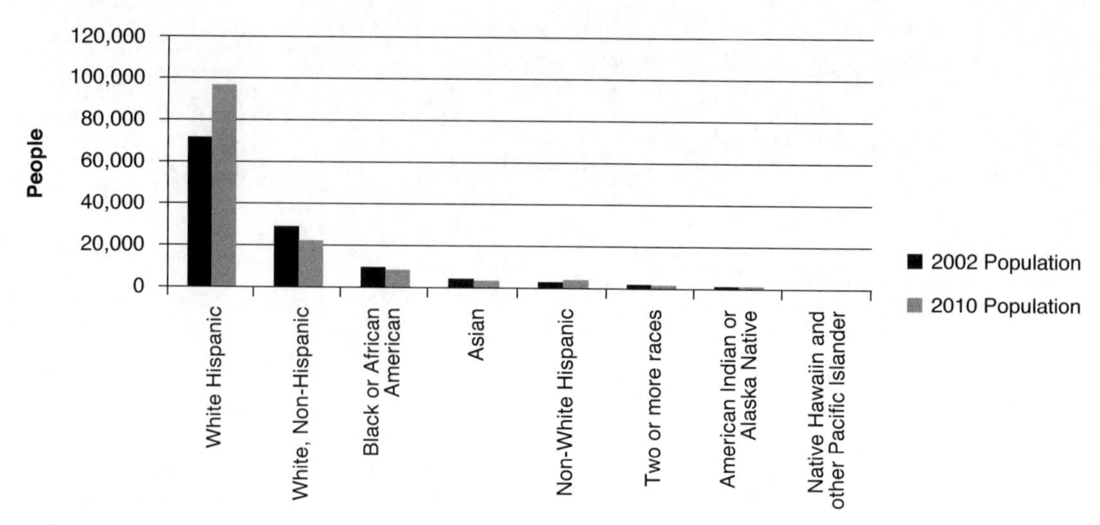

Figure 13.1 *Less than High School Education Attainment*

made when specific detailed industry performance and information is not taken into account. For example, the perceived decline in the manufacturing sector may have merit on an aggregated level. Nevertheless, specific pockets of manufacturing growth can be identified in most areas. These manufacturers may have leveraged specific regional characteristics and/or amenities in order to outperform industry expectations—a quality that should not be overlooked. The following are core data features used to evaluate the economic composition of a given area.

- Employment: historical, current and industry projections provide an understanding of the past and present performance of an industry, as well as future expectations of industry growth/decline.
- Percent change: enables further insight of the rate of change within an industry, which can include employment, earnings, or output.
- Earnings: evaluates industry productivity and value. Two basic measurements include total industry earnings—a measure of income generated by the industry—and earnings per worker (EPW) a measurement of productivity associated with the employment within the industry.
- Location quotient: a typical measurement of industry concentration and potential competitiveness. Commonly referred to as LQ, a location quotient is a comparative statistic used to calculate relative employment concentration of a given industry against the average employment of the industry in a larger geography (e.g. nation). Industries with a higher location quotient (usually greater than 1.2) indicate that a region/state has a comparative advantage or specialization in the production of that good or service. LQs are calculated by dividing specific industry employment in the region (*r*) by total employment in the region. The result is then divided by industry employment in the nation (*U.S.*)—if looking at national LQ—divided by national employment

$$\frac{Industry\ i_{employment_r} / \sum All\ Industry_{employment_r}}{Industry\ i_{employment_{U.S.}} / \sum All\ Industry_{employment_{U.S.}}}$$

The Σ sign stands for summation, in this case over all industries. This equation simply calculates the percentage of all total employment in an area accounted for by one particular industry (denominator is the sum of employment over all industries and numerator is employment in industry I).

Shift share: provides another perspective that regional economists and economic developers use to gauge competitiveness. A shift share analysis separates regional job growth into its component causes. The three main causes identified are "national growth effect," which is regional growth that can be attributed to the overall growth of the entire U.S. economy; "industrial mix effect," which is regional growth that can be attributed to positive trends in the specific industry at a national level; and the "regional competitiveness effect," which is growth that cannot be explained by either overall national or industry-specific trends. A positive value indicates that an industry has a competitive advantage compared to the nation. Note: Positive shift share values do not explain why an industry has a competitive advantage, only that there are potential factors that contribute to the industry's ability to outperform the national average rate of growth/decline. Leveraging these key data elements provide a good starting point for analyzing regional economies and determining where business and industry opportunities may exist. Ultimately, before any economic development strategy is undertaken, these critical data elements should be considered and further evaluated in the community and/or region. Table 13.3 provides a sample of regional economic data at the 2-digit NAICS level.

Occupation Data

Just as industry data is necessary for understanding how the local economy operates, knowing the occupational structure of the economy is just as important. Historically, little weight was given to the workforce structures in an economy. Workers were viewed as labor "inputs" to a production or service process. In recent decades scholars and practitioners have become increasingly interested in the workforce and occupational structure of local and regional employment. From an economic development standpoint, knowing what occupation (and skills) talent reside in an economy can help to more narrowly target and solicit prospective employers, or assist current employers with expansion opportunities. In a 2012 corporate executive survey conducted by Area Development, 89.4% of the more than 200 respondents (mostly CEOs and senior positions) rated skilled labor availability as "very important" or "important" among site selection criteria.[2]

As described in the "Types of Data" section, occupation data is classified by a coding system known as Standard Occupation Classification system, or SOC. Much in the same way industry data is reported and used, occupation data can also be constructed in a similar fashion, though a few addition key data points can also be calculated. The more prevalent types of occupation data utilized by economic development and regional planning practitioners are:

- Employment: historical, current and future occupation projections provide a complimentary perspective to industry. Since industries are responsible for employing occupations, industry sector growth signals demand for new occupational entrants to the industry. Inasmuch, knowing what occupations coincide with industry needs can enhance the capabilities of existing companies or be used as a tool to recruit new companies.
- Percent change: much like the industry data perspective this can provide a deeper understanding of the types of occupations that are growing or declining. Occupations in decline can become targets and leverage points for demonstrating available workforces to compatible industries.
- Location quotient: signifies concentration of specific occupations, when compared to a larger geographic context (e.g. state or nation).

Table 13.3 *Sample Region Data and Analysis*

		Employment			*Change*		*Earnings*			*Competitiveness*		
NAICS code	*Industry description*	*2005 Jobs*	*2010 Jobs*	*2015 Jobs*	*2010–2015 Projected change*	*2010–2015 % Change*	*Total earnings*	*2010 Total EPW*	*2010 National LQ*	*Nat growth effect*	*Expected change*	*Competitive effect*
11	Agriculture, forestry, fishing, and hunting	2,884	3,025	2,971	(54)	(2%)	$67,850,750	$22,430	0.26	229	(97)	43
21	Mining, quarrying, and oil and gas extraction	424	655	774	119	18%	$45,519,880	$69,496	0.14	50	119	0
22	Utilities	1,182	1,214	1,162	(52)	(4%)	$106,924,264	$88,076	0.61	92	16	(69)
23	Construction	34,060	27,715	30,384	2,669	10%	$1,564,982,905	$56,467	0.93	2,102	2,475	194
31–33	Manufacturing	56,574	42,037	38,623	(3,414)	(8%)	$2,578,423,469	$61,337	1.04	3,188	(1,729)	(1,686)
42	Wholesale trade	18,500	17,229	17,780	551	3%	$1,262,230,998	$73,262	0.85	1,307	786	(234)
44–45	Retail trade	61,284	52,486	54,887	2,401	5%	$1,624,494,186	$30,951	0.90	3,981	1,457	945
48–49	Transportation and warehousing	15,423	15,133	16,080	947	6%	$741,834,793	$49,021	0.74	1,148	867	81
51	Information	12,218	11,527	12,230	703	6%	$777,035,070	$67,410	1.06	874	428	275
52	Finance and insurance	32,215	34,235	38,099	3,864	11%	$2,675,533,720	$78,152	1.10	2,596	4,026	(162)
53	Real estate and rental and leasing	24,271	23,738	26,934	3,196	13%	$577,616,754	$24,333	0.96	1,800	3,713	(517)
54	Professional, scientific, and technical services	35,656	37,540	43,203	5,663	15%	$2,282,469,540	$60,801	0.96	2,847	5,062	601
55	Management of companies and enterprises	8,913	9,312	10,204	892	10%	$1,032,011,712	$110,826	1.44	706	489	403
56	Administrative and support and waste management and remediation services	31,717	27,742	30,509	2,767	10%	$860,002,000	$31,000	0.82	2,104	3,105	(337)
61	Educational services	25,399	28,438	32,409	3,971	14%	$1,297,739,692	$45,634	2.09	2,157	3,594	377
62	Health care and social assistance	80,597	87,055	94,674	7,619	9%	$4,163,927,705	$47,831	1.37	6,603	11,054	(3,435)
71	Arts, entertainment, and recreation	13,313	13,964	15,644	1,680	12%	$276,738,552	$19,818	1.11	1,059	1,806	(127)
72	Accommodation and food services	44,726	43,582	46,131	2,549	6%	$825,660,990	$18,945	1.10	3,305	3,571	(1,021)
81	Other services (except public administration)	25,828	25,262	26,701	1,439	6%	$839,683,618	$33,239	0.85	1,916	2,027	(588)
90	Government	72,701	69,932	71,601	1,669	2%	$5,342,175,412	$76,391	0.87	5,304	3,350	(1,681)
	Total	597,885	571,820	611,002	39,182	7%	$28,942,856,010	$50,615		43,369	46,121	(6,938)

Occupational LQs vary by region due to the industry makeup of a region. For example, Silicon Valley has a significant concentration of computer engineers and computer programmers that support the burgeoning information technology industry sector.

- Demand (new and replacement jobs): one of the most fundamental, yet cornerstone concepts in economics is demand. If gaps exist between supply and demand, the economic system is considered to be in a state of disequilibrium and resources are deemed misallocated. The same concept applies to occupational demand. Simply stated, economic growth creates new jobs while existing workers quit, retire, or otherwise move on, which all contribute toward creating demand for occupations. If demand for occupations exceeds the supply of competent workers, then business costs rise as companies seek skilled labor and production is stunted. Understanding the future demand for both new and replacement jobs, especially those in critical industry occupations can assist in developing the proper talent supply pipelines (e.g. education, workforce training, etc.) to support economic development.

Data Sources

The data and analysis components described above can be collected and structured through multiple data sources, organizations and private companies. The list of sources below is not exhaustive, though many of the sites and organizations provide good starting points for economic research.

- Federal sources: Many federal agencies report specific data on the economy. Some data is focused on national trends, while others are focused on regional and local industries. Below are a few sources and the types of data that can be collected from each:
  - Census Bureau: reports data on households, poverty, income, commuting patterns (i.e. where people travel to work) and local industry.
  - Bureau of Labor Statistics: reports data on occupations, employment, inflation and prices.
  - Bureau of Economic Analysis: reports data on trade, industry output and input–output multipliers.
- Regional/state: Many sources of local-level data can be obtained through regional and state organizations. The following are specific resources that can be used to meet specific data needs:
  - State labor market information (LMI) agencies: Each state has an LMI agency that gathers specific industry, occupation, education and other data to provide to state-level leaders. Many of these agencies can be contacted directly for special data requests by local economic developers.
  - Utility companies: Many utility companies maintain regional datasets on the areas they serve. As part of community outreach and economic development, the utility companies provide a great resource for obtaining needed data.
  - Economic Development Corporations (EDCs) and Area Development Districts (ADDs): As part of their operations, EDCs and ADDs often conduct research and surveys on their local areas. The information and data collected can provide good insight into the types of economic activity taking place in a local economy.
  - Workforce Investment Boards (WIBs) and Workforce Development Councils (WDCs): Similar to EDCs and ADDs, the workforce boards also collect and maintain regional data on the workforce and often times on industries.
- Local: Depending on location, many local area chambers of commerce collect and retain data on their respective economies. Knowing the resources available through the Chamber of Commerce can provide strong quantitative and qualitative understanding.

- Universities: Two federal government agencies, specifically the Department of Commerce and the Department of Labor, provide funding to specifically designated universities to develop data and analysis tools to be made available to the public. Familiarizing oneself with the types of data and tools available through university-led research can add additional perspective to economic activity. The following two sites are among the more prevalent sources:
  - Indiana University: STATS America (2014): STATS America provides profiles of U.S. states, counties and metro areas, including industry and occupation cluster profiles. The site also includes tools for regional investment planning and mapping innovation.
  - Harvard Business School: Institute for Strategy and Competitiveness (2014): provides tools for assisting in cluster development and understanding clustering advantages and performance throughout the US. Specific tools include the Cluster Mapping Project, as well as several research papers on various economic development strategy topics.
- Private sector sources: several firms provide subscription-based data and analysis tools to various public organizations and private consulting companies. Each tool is typically used for specific economic development purposes and some can be compatible with other systems (public and private data sources). A few of the more popular private data and analysis systems are:
  - Economic Modeling Specialists Inc (EMSI, 2014): a web-based subscription tool, which contains detailed industry, occupation, demographic, education, and economic impact (input–output) modeling tools. The data is available at the zip-code level and is updated four times a year.
  - Dunn and Bradstreet (D&B): a popular source of company-level information across multiple geographies (Dun and Bradstreet 2014).
  - Conway Data (2014): provides industry- and company-level data, specifically new investment construction, expansions and relocations—valuable insight for economic developers involved in site selection and other development activities.
  - IMPLAN (IMpact analysis for PLANning 2014): provides an input–output impact modeling tool used to conduct local and regional level impact analysis. The tool is often used by economic developers and consultants to determine both the value an industry or organization brings (or would bring) to a local economy as well as incentive packages a local area would be willing to offer a company.
  - Environmental Systems Research Institute, Inc. (ESRI 2014): develops software known as ArcGIS, an integrated mapping tool containing economic, household and other data. The tool can integrate geographic-specific data from most any source.

Industry Type (Basic vs. Nonbasic)

The simplest way to think about a community's economic base is to think of a community's industries as being composed of two fundamentally different types: (1) one sells to non-residents and thereby brings outside dollars to the community, (2) while the other sells to residents and thereby intercepts monies already in circulation. The former industries are called "basic," while the latter are called "nonbasic." A community can grow by adding either basic or nonbasic industry, but without basic industry (i.e. industry that is predominately export oriented and brings in outside income) the community cannot exist. Note that historically when one thought of basic industry it usually meant extractive activities such as agriculture, timber, and mining or finished-product manufacturing, such as, food processing, textiles, and wood products. Nowadays, the list will often include the non-labor incomes generated by retired persons or from tourism. The key, from the standpoint of the community/regional economy, is to obtain enough outside income needed to purchase goods and services to sustain economic activity.

Basic Industries

Basic industries are made up of local businesses largely reliant on the economy outside the local area. For example, Boeing builds large commercial aircraft that they in turn sell to airline companies. Boeing does not sell their products to local residents and households, which makes the company largely reliant on the external marketplace to export their goods. Typically, manufacturing and national resource-based (i.e. mining, logging, etc.) firms tend to be considered basic industries because they export their final product to economies outside their respective local economies.

Economic base theory supports the development of basic industry sectors, as the driver of economic expansion and prosperity. In other words, a local economy cannot continue to grow and develop if new, outside money is not flowing in through export activity. Nonbasic sectors, though important in supporting internal functions of the economy, will be developed as demand for local-level services increase through economic expansion. In some cases, nonbasic sectors can help spur further growth in basic industry, with import substitution strategies (replacing products or services obtained externally with those provided locally, especially in the case that these new products or services can also be sold externally). See the case at the end of the chapter for more details as well as the section on input–output analysis.

The theory also emphasizes the importance of diversification within the basic industries. The primary assumption in this approach is a diverse economic base sector can help to insulate a local economy from economic downturns because hopefully outside markets will not be as heavily affected.

Nonbasic Industries

Nonbasic industries are those local businesses that primarily rely on local economic factors to grow. In other words, these businesses tend to circulate existing money within the economy. For example, a local restaurant primarily sells its goods and services to local households and perhaps businesses. Since the restaurant's clientele are local, the prepared products are also consumed locally.

The nonbasic industry sector is very important within a local economy because the nonbasic businesses provide services and products required by local households, residents, and businesses. Furthermore, the sector helps to prevent leakage

BOX 13.2 SOME LIMITATIONS OF ECONOMIC BASE THEORY

Economic base theory best explains near-term growth and decline in area economies (cyclical change). It cannot really explain structural change. More extensive work, in the form of econometric modeling, would be needed to gauge longer-term changes. Malizia and Feser (1999: 51) explain its limitations this way:

> As a basis for understanding the reality of local economic development, however, economic base theory offers limited useful insights. In the form of a quantitative model, it can be applied for impact analysis and for making predictions of economic growth only as long as the structure of the local economy does not appreciably change. For long-term analysis of economic development, economic base theory has limited application because key export sectors and local economic structure change over time, often significantly.

The Editors

of money from the local economy, allowing further increases in economic multiplier effects. Note that there is more crossover between basic and nonbasic industries, with the reality being that some enterprises sell to both a local and external market.

Introduction to Input–Output Modeling

Rather than just using the arbitrary division of an economy into two sectors (basic and nonbasic), input–output modeling (I–O) can be used to gain more insight into the economy (Malizia and Feser 1999). This serves to expand the usefulness of economic base approaches. The input–output model uses a matrix representation of a region's economy to predict the effect of changes in one industry on other industries, consumers, government and foreign suppliers to the economy. It predominantly focuses on business spending components of the nonbasic sector, tracking the "intricate web of production linkages among different industries in the area, and reveals the extent to which the natures of these linkages ultimately lead to more or less income for each unit of final sales of area goods and services" (Bendavid-Val 1991: 87). The model considers inter-industry relationships in the regional economy, depicting how the output of one industry goes to another industry, where it serves as an input. This thereby makes one industry dependent on another, both as a customer of output and a supplier of inputs. In addition the model also takes into account additional inputs, such as workforce and other various capital requirements. Effects are calculated based on the availability of inputs, cost of input providers and output consumers. These calculated effects are called regional purchase coefficients (RPCs) and represent the proportion of regional demand for inputs fulfilled from regional production. In other words, a region containing a large, diverse economy with many suppliers will typically have high industry RPCs. As a result, when production in an export-oriented industry increases, the demand for regional input supply will go up, creating additional economic benefits within the region as dollars filter through supply chains—a.k.a. the multiplier effect. The multiplier effects calculate the added value to the regional economy from an initial injection of money and resulting "rounds" of spending of that money for goods and services (i.e. inputs) until the money dissipates through "leakage" to economies outside the region. Leakage in a regional economy is inevitable; however, a diversified and well-integrated regional economy can have less leakage than a narrow, fragmented regional economy.

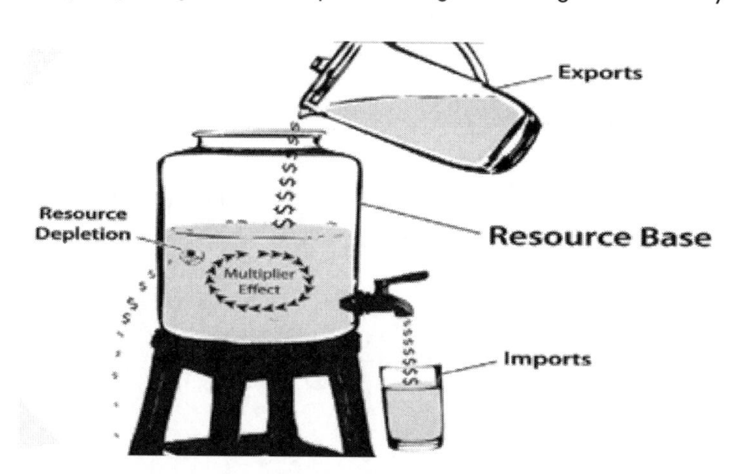

Figure 13.2 *Community Economic Development Leaky Faucet*

Another way to view input–output modeling is as a tub. At one end of the tub you have a faucet that would represent dollars flowing into the economy due to local companies exporting their products, or outside payments to local residents (i.e. social security, disability, etc.). At the other end of the tub you have a drain that represents money leaving the economy and being spent on imports. As a rule of thumb, the more an economy can increase export (e.g. turn up the faucet) or reduce the amount of imports (e.g. plug the drain or leakages) the faster the economy will grow (e.g. increase the level of water in the tub) due to multiplier effects.

Multiplier Effects

When evaluating the effectiveness of industry investment, or the impact of an industry, multiplier effects aide in gauging the level of ripple (or spillover) effects the industry's operation has in a given economy. Impacts are generally broken into three component effects: direct, indirect, and induced. Direct effects are created directly in the industry being measured or invested in. Indirect effects pertain to the upstream supply-chain effects resulting from the direct effect. In other words, if a business is expanding and hiring 100 new workers, that business will require more inputs from their suppliers, resulting in increased productivity needs for the supplying industries (i.e. the suppliers will have to expand or hire more workers). Induced effects pertain to the increases in wage and salary spending by both directly and indirectly affected industries. If the businesses must hire more workers, those businesses will have to increase their overall payroll budget. As workers collect wages, they spend on goods and services in the local economy such as supermarkets, restaurants, etc.

Typical multiplier effects are categorized into three groups: sales, jobs and earnings.

Sales Multipliers

Sales multipliers show how "deeply rooted" an industry is in your region—for example, a highly developed cluster will have a high sales multiplier because every dollar fed into the cluster from the outside has a high ripple effect, propagating through the regional economy for some time before it leaks out. One dollar of sales coming into a highly developed automotive manufacturing cluster, for example, might have a ripple effect of 2.8 (i.e. that $1 led to a total of $2.80 in regional sales). Industries and clusters with very low multipliers are usually owned outside of the region (so the profit is lost immediately) and also buy mostly from outside the region (a "shallow root system").

Jobs Multipliers

A jobs multiplier indicates how important an industry is in regional job creation. A jobs multiplier of 3, for example, would mean that for every job created by that industry, two other jobs would be created in other industries (for a total of three jobs).

Jobs multipliers are easily misinterpreted—jobs multipliers of 17 or higher are sometimes seen—

BOX 13.3 INDUSTRY CLUSTERS AND SUPPLY CHAINS

A case study of Dayton, Ohio's 14-county regional approach to economic development is discussed in the context of identifying clusters of economic activity, supply chains serving these industries, and related job skill needs. See the following link for the study: www.economicmodeling.com/2013/01/07/emsi-for-economic-developers-part-2-analyzing-industry-clusters-supply-chains-talent-gaps but a high jobs multiplier for a set of one or more industries in an added-jobs scenario does *not* necessarily mean that attracting businesses in those industries to the region is the best or most viable option for regional economic growth. Other considerations such as social, economic, and environmental feasibility must also be taken into account.

Jobs multipliers are primarily tied to the type of industries in the scenario—industries with a high sales-to-labor ratio typically have a high jobs multiplier, and vice versa. For example, a nuclear power plant might have only 20 workers, but "behind" each of those workers there are millions of dollars of equipment costs and millions of dollars of electricity being generated. Thus, if we bring 20 more nuclear power jobs into the region, that would involve a huge amount of investment flooding into the region (to build another nuclear power plant or double the size of the current one) and millions of dollars in new sales and profits.

Some of that money would go to the employees' high salaries, while other money would go to local construction companies, real estate, janitorial services, etc. The overall jobs multiplier would be impressive—each new job in nuclear power might support 14 other jobs scattered throughout the rest of the economy (i.e. a jobs multiplier of 15). However, the effort required to attract 20 jobs in nuclear power (with the entire necessary infrastructure) is substantially more than to attract 20 jobs in an industry with a lower jobs multiplier.

Earnings Multipliers

An earnings multiplier of 1.5 means that for every dollar of earnings generated by a new scenario, a total of $1.50 (the original $1 direct effect, plus the additional $0.50 indirect and induced effects) is paid out in wages, salaries, and other compensation throughout your economy. This is important for understanding how a given scenario will affect quality of jobs within an area. A scenario whose ripple effect—indirect and induced effects—bring two dozen lawyers and accountants into your region would have a much higher earnings multiplier than a scenario that brings two dozen food service workers. Note that understanding the degree and types of multiplier effects is critical in understanding the scale and scope of any economic development activity. Specific considerations for each of the multiplier effects lead to additional economic, socioeconomic, and tax base gains.

Conclusion

Understanding the local economy is critical to assessing the overall community and economic development situation and choosing the best policy options. Practicing community and economic development without a detailed picture of the local economic and demographic situation is like flying an airplane in a cloud without radar or any navigational tools—the mountain range and disaster could be just ahead and nobody has a clue about the unforeseen disaster they are rapidly approaching. Socrates' famous philosophical quote "Know Thyself" is equally applicable in community and economic development, particularly in terms of analyzing the local economy.

CASE STUDY: PLUGGING THE "LEAKS" IN A COMMUNITY'S ECONOMY—THE CASE OF MARTHA'S VINEYARD

Martha's Vineyard is a small island community with 15,000 permanent residents located off the coast of Massachusetts. It is known for its beauty and as a resort town with a seasonal population.

This case provides a different approach to analyzing a community's economy, one that focuses on growth from within communities using the strategy of import substitution. While export-based industries are vital in an economy, as described in the chapter and evidenced by development, considering ways to "plug the leaks" in a local economy are important too. The basic idea is to identify products and services that are brought into a community from other places and find ways to produce them in the local economy. This isn't a new idea and has long been considered a way to help grow local enterprise as well as keep more money circulating in the local economy. As Shuman and Hoffer (2007) describe in their study, "Leakage Analysis of the Martha's Vineyard Economy: Increasing Prosperity through Greater Self-Reliance," import-substitution strategies do not imply stopping export-based business development. Rather, as Jane Jacobs noted in her 1969 classic, *The Economy of Cities*, import substitution helps further develop export-based businesses. Instead of focusing extensively on export-based industry recruitment or development, this approach centers on developing a variety of small businesses emanating initially from local markets. The intent is that some will become exporters. It can be an excellent strategy for building reliance within local economies, producing benefits for a given region because local ownership is more deeply connected to its host community and also because of rounds of spending in a local economy (the multiplier effect). This is defined as the longer money circulates in a community's economy, the more income, wealth, and jobs it creates; it is a basic concept in community economics highlighting the importance of maximizing the numbers of dollars being spent locally and minimizing their "leakage" (Shuman and Hoffer 2007: 5).

Plugging these "leaks" in an economy can generate the following results:

- Policy makers have a clearer vision of how to allocate scarce public resources for economic development.
- Existing small-business proprietors have a better sense of promising opportunities for expansion, and entrepreneurs see the most profitable markets for start-ups.
- Local banks, lenders, and investors can better calibrate their allocations of commercial capital.
- Consumers have a better appreciation of the potential payoffs of buying more goods and services locally.
- Foundation, nonprofits, and grass-roots groups have a clearer sense of whom to mobilize for community action.

(Shuman and Hoffer 2007: 13)

Martha's Vineyard is a particularly challenging case to show application of these approaches because it is so small and rural in the sense of its isolation. This small resort town has a small permanent population of 15,000 and a large number of seasonal visitors. It is difficult, for example, to apply economic base analysis to small communities, because data are easier to obtain at the county, multi-regional, or national levels. However, after conducting the analysis, Shuman and Hoffer found in their study that even in this small, seasonally influenced economy, there are 13 sectors where

> import-replacement seems especially plausible given the assets and goals of the island: growing local food, producing local electricity, manufacturing local biofuels, building affordable housing, increasing overall demand during off-season months, creating cottage-industry-scale manufacturing, expanding various local services (for business, health, and finance), starting a telecommunications utility, creating new institutions for local pension fund reinvestment, stimulating more local procurement by government (and school entities), and creating local purchasing pools that can reduce costs of all kinds of inputs to production
>
> (23–27).

It was also found that with a small 10% shift in expenditure patterns just by the permanent residents in these sectors, off-peak employment should increase by 6%, and "probably much more if other seasonal residents, visitors, and workers on the island began to shift their consumption patterns as well" (26). This study provides evidence to help support some of the "Local" movements emerging in the US as well as other countries across the globe, focusing on ways residents can contribute to their own economy. For more information on these approaches, see Chapter 6 in this book.

For more details on findings in the Martha's Vineyard case, see the full study at: www.islandplan.org/planning/livelihood.html.

The Editors

Keywords

Economic development, industry analysis, local economy analysis, economic impact leakage and economic opportunity, industry composition.

Review Questions

1 Can you describe how changes to location quotients are linked to shift share?
2 What are export-based industries and why are they important in a local economy?
3 What are economic multipliers and how are they used in community and economic analysis?
4 Why is it important to assess the local economy for community and economic development decisions and policy making?
5 What key data elements provide a good starting point for analyzing regional economies and determining where business and industry opportunities may exist?

Notes

1 2000 Standard Occupational Classification (SOC) User Guide.
2 www.areadevelopment.com/laborEducation/November2012/skilled-labor-availability-training-programs-1677109.shtml

Bibliography

Bendavid-Val, A. (1991). *Regional and Local Economic Analysis for Practitioners*. New York: Praeger.

Center for Community and Economic Development, University of Wisconsin Extension. (2014). Local and Regional Analysis. Accessed February 24 from: http://fyi.uwex.edu/downtown-market-analysis/understanding-the-market/local-regional-economic-analysis

Conway Data. (2014). Accessed July 15 from: www.conway.com

Dun and Bradstreet. (2014). Accessed July 10 from: www.dnb.com

Economic Modeling. (2014). Accessed July 15 from: www.economicmodeling.com

Environmental Systems Research Institute, Inc. (ESRI). (2014). Accessed July 15 from: www.esri.com

Harvard Business School. (2014). Institute for Strategy and Competitiveness. Accessed August 1, 2014 from: www.isc.hbs.edu

Holzheimer, T. (2006). Benchmarking the Creative Class at the Local Levil. *Economic Development Journal*, 5(3), 34–39.

IMPLAN. (2014). Accessed July 10 from: www.implan.org

Jacobs, J. (1969). *The Economy of Cities*. New York: Random House.

Malizia, E.E. and Feser, E. (1999). *Understanding Local Economic Development*. New Brunswick, New Jersey: Center for Urban Policy Research.

Shuman, M. H. and Hoffer, D. (2007). Leakage Analysis of the Martha's Vineyard Economy: Increasing Prosperity through Greater Self-Reliance. Silver Springs, Maryland: Training and Development Corporation.

Stats America, Indiana University. (2014). Accessed August 5, 2014 from: www.statsamerica.org

Connections

Explore economic impact analysis via the University of Minnesota's Extension service site at: www.extension.umn.edu/community/economic-impact-analysis

Find out more about Gap Analysis by reading the following article: http: //pods.dasnr.okstate.edu/docushare/dsweb/Get/Document-1631/F-917 web.pdf

Delve deeper with the excellent "How to" guide for Community Economic Analysis by Hustedde, Shafer, and Pulver at: http: //community-wealth.org/sites/clone.community-wealth.org/files/downloads/tool-cmty-econ-manual.pdf

Find sources of data at the following sites:
County Business Patterns. www.census.gov
U.S. Bureau of Economic Analysis. www.commerce.gov/egov.html
U.S. Bureau of Labor Statistics. www.bls.gov
U.S. Census Bureau. http: //factfinder.census.gov
UK National Statistics: www.statistics.gov.uk/hub/index.html
Discover a world of data at the UN site: http: //data.un.org

PART III

Programming Techniques and Strategies

14 Human Capital and Workforce Development

Monieca West

Overview

Communities and states must focus on raising the skill levels of their resident workers or risk becoming noncompetitive in the global economy and losing jobs to other states and countries. This chapter discusses the kind of human capital development program that states and communities should develop to assure they will have sufficient numbers of trained workers. It provides examples of state and local initiatives for reaching this goal.

Introduction

As the workforces of the US and other developed countries compete for jobs in an international arena, policy and practice must be revolutionized to meet the demands of this new environment. A public school system designed for the rural, agricultural economy of the 1800s must now prepare students to succeed in a very different twenty-first-century life. Colleges and universities must become more responsive to their role in economic prosperity. Immigration policies that are politically driven must be revamped to permit the flow of needed talent into the country, especially in light of expected shortages due to the aging and retiring U.S. workforce. Federal and state workforce development programs that struggle to keep up with the emerging needs of business and industry in this ultra-competitive environment must be streamlined with policies that easily adapt to changing conditions. Perhaps most important, some observers believe that the US as a society has not demonstrated an understanding of the value of lifelong learning nor a willingness to demand rigor and excellence of all students at all points along the educational system. All of these issues must be addressed if the economic prosperity Americans have taken for granted is to continue.

The changing economic realities of the twenty-first century cannot be predicted with certainty, but communities can prepare with a variety of tools and a flexible workforce is among the most essential.

> It is not the strongest of the species that will survive, or the most intelligent, but the ones most responsive to change.
>
> (Charles Darwin)

Primary and Secondary Education

Attention to early education is the most cost-effective strategy to increase the number of skilled workers in the workforce. Primary and secondary education provides students with strong foundations in reading, writing, mathematics, reasoning, and computer skills. It also paves the way for success in higher education and placement in employment (Judy and D'Amico 1997).

Community leaders often mistakenly focus workforce development efforts too narrowly, thinking only in terms of customized training for industrial recruitment or business retention. In reality, holistic workforce development begins in earnest at the secondary level with career exploration, academic preparation for postsecondary education, or technical preparation for entry into apprenticeship programs or job placement. Numerous reports and studies affirm that academic achievement at the high school level is extremely important in preparation for the high-skilled, high-wage, high-demand jobs of the new economy. This is equally true for the "vocational" student that traditionally moved through the education system focusing almost exclusively on trades-based studies, with less emphasis on academic subjects such as math and communications skills that are now required across all levels and types of employment.

To make significant progress, public education can no longer engage in "business as usual." America must consider new approaches to public education commensurate with that required of industry as it faces the challenges of the new economy. The standard approach to K-12 education has only marginally changed since the late 1800s. After decades of "educational reform" and immense increases in funding, far too many high school graduates still leave public education with inadequate skills for the workplace, or require remedial classes to raise their level of academic proficiency for college. High schools must do a better job of preparing students for either entry into the workforce or postsecondary education, as both are equally important to a competitive workforce.

Unfortunately, education reform has been slow to nonexistent, which results in U.S. students lagging behind their peers in countries that compete with American businesses for the best and brightest talent, the currency of the new economy. To demonstrate the significance of this brain drain, consider that the 25% of the Chinese population with the highest IQs is greater than the total population of the US. This means that China has more honors students than the US has children, which certainly positions China to compete for high-skill and high-wages jobs.

Because K-12 education is the foundation of all future learning, a comprehensive workforce development strategy should include support and advocacy for excellence in education, beginning with K-12 and extending through two- and four-year postsecondary institutions. The issues surrounding K-12 reform are very difficult and often controversial but community leaders and community developers must find ways to facilitate a rational public discussion of the issue. Support for educational reform must be included if workforce development efforts are to be effective.

Postsecondary Education

If a community or state is to increase its competitive standing nationally and internationally, more graduating high school seniors and working adults must start or return to college. Postsecondary education is a crucial way that working adults can acquire the skills and credentials necessary to succeed in current and emerging business and industry. Unfortunately, since working adults have full-time jobs and family responsibilities, they often lack the time, money, and flexibility of schedule to fit into traditional higher education models. Recent studies report that working adults get very little financial aid from federal or state sources. Working adults who hold full-time jobs are typically able to attend school on a less than half-time basis, which renders them ineligible for most aid. The problem is not just the absence of financing but, on the part of most institutions of higher education, the lack of programs and schedules developed to accommodate working adults. Degree attainment and other credential requirements often seem too daunting for a working adult attending school

BOX 14.1 A PRESSING GLOBAL ISSUE: YOUTH UNEMPLOYMENT

Throughout the world, high unemployment rates among youth are a severe issue. In some countries, rates of unemployment are over 50%. The Organization for Economic Co-operation and Development's recent report highlights the difficulties of high unemployment. The following excerpt explains:

> The global financial crisis has reinforced the message that more must be done to provide youth with the skills and help they need to get a better start in the labour market and progress in their career. Sharp increases in youth unemployment and underemployment have built upon long-standing structural obstacles that are preventing many youth in both OECD countries and emerging economies from making a successful transition from school to work. Not all youth face the same difficulties in gaining access to productive and rewarding jobs, and the extent of these difficulties varies across countries. Nevertheless, in all countries, there is a core group of youth facing various combinations of high and persistent unemployment, poor quality jobs when they do find work and a high risk of social exclusion. In the context of rapid population ageing, successful engagement of this group in the labour market is crucial not only for improving their own employment prospects and well-being, but also for strengthening overall economic growth, equality and social cohesion. In many countries, the immediate short-term challenge is to tackle a sharp increase in youth unemployment. In April 2013, the youth unemployment rate was close to 60% in Greece and Spain. For the OECD as a whole it was stuck at 16.5%, up from 12.1% just prior to the crisis, and two-and-a-half times the unemployment rate for those aged 25 and over.
>
> (*OECD Action Plan for Youth*, 2013, available at: www.oecd.org/els/emp/Youth-Action-Plan.pdf)

For additional information on youth development, see UNESCO's work in youth development and global youth indicators at: www.unesco.org/new/en/social-and-human-sciences/themes/youth/development-and-violence-prevention. A text box on UNESCO's youth development program is provided in Chapter 10.

The Editors

part-time, who may have been out of school for several years and may require some level of basic skills remediation. The pathway connecting education with improved jobs is obstructed for many adults. The states do not effectively provide the financial aid, student support, and basic skills remediation essential so that working adult students can advance in their education and career.

Yet, the ability to compete effectively in the new economy is directly related to a community's or state's ability to increase the educational attainment of all its citizens, from those with minimal job skills to those seeking the highest in academic achievement. Numerous national studies have confirmed the direct correlation between academic attainment and earnings potential. The benefits of postsecondary education accrue not only to individuals, but to families, communities, states, and to the nation as a whole. For individual citizens, it is well documented in U.S. Census Bureau data and other sources that personal and family "lifetime" earnings increase directly with higher education levels. The Bureau also documents that quality-of-life factors, such as improved health for individuals and families, are dramatically enhanced by higher levels of education. As quality-of-life issues are improved for the citizens of a state, the need for other state programs directed toward health issues and correctional activities is reduced. Policy makers increasingly recognize that improving educational attainment is a key ingredient in their efforts to strengthen economic competitiveness and enhance quality of life.

The role and contributions of community colleges in workforce development initiatives cannot be overstated. These institutions are uniquely positioned as the bridge between high school and university studies, yet are flexible enough to offer classes responsive to general community needs. Community colleges provide developmental, remedial classes to high school graduates without college-level academic preparation and provide continuing education for non-degree-seeking students. These various approaches to education are important to the overall level of workforce quality, making a strong community college a valuable local asset. Because community colleges are deeply embedded in the local community, they are naturally attuned to the needs of local business and industry. Typically, community colleges are the institution of choice for nontraditional students above the age of 25 who also happen to make up the majority of the local workforce. The U.S. Bureau of Labor Statistics has noted that one-third of the fastest growing occupations nationally are those that require certificates or associate degrees provided by community colleges. Community colleges are extremely important in producing students to fill critical shortages in occupations such as nursing; providing industry certifications such as Oracle and Cisco; offering services that help students transition from secondary to postsecondary education or from postsecondary into the workplace; and providing direct services to local employers and organizations.

Virginia offers a model for comprehensive workforce development. Through its 23 colleges, the Virginia Community College System extends workforce development services and transitional programs into the community, directly serving businesses, employees, and special populations. Programs focus on instructional curricula that prepare incumbent, upcoming and displaced employees for jobs in current and emerging occupations. In partnership with the office of Workforce Development Services, Virginia's community colleges house the state's Workforce Development Service Centers that provide a single point of entry for job seekers and employers.

Secondary to Postsecondary Career Pathways

One of the most common barriers to educational attainment is the difficulty encountered by students as they progress through the separate secondary and postsecondary systems. Rare is the occasion

BOX 14.2 VERMONT WORKS FOR WOMEN

This Vermont-based nonprofit has created programs aimed at developing the human capital potential and workforce skills of women. The idea is to assist those in poverty or transition (nearly one-third of the state's single mothers live in poverty) with training and capacity development for long-term economic independence. These programs include skill development for careers that pay a living wage (defined as rising above poverty and able to provide a living for themselves and their families. A few of their Step Up programs include: Step Up to Law Enforcement for Women—in partnership with local and regional law enforcement agencies, this training program prepares for careers in police and correctional facilities.

Building Homes, Building Lives is designed to help incarcerated women focus on building of modular homes that in turn provide affordable housing in communities throughout Vermont. This program prepares women for employment in construction trades, provides valuable job readiness skill development, and a resume or portfolio of work to show prospective employers when transitioning back into communities.

(R. Phillips, B. Seifer, & E. Antczak, *Sustainable Communities, Creating a Durable Local Economy*. London: Routledge, pp. 61–62.)

Gather more information about Vermont Works for Women at: http: //vtworksforwomen.org when high schools and colleges develop occupational programs of study with curricula that are aligned. Aligned systems allow for articulation agreements; that is, concurrent and dual enrollment programs designed to allow high school students to acquire college credit prior to graduation. To facilitate this alignment, the Department of Education has established 16 career clusters with 81 separate career pathways.

Career clusters provide a framework for grouping occupations according to common knowledge and skills. Using sequences of coursework required at secondary and postsecondary levels, a career pathway provides a student with a plan of study that leads to specific occupational titles found in his or her career cluster. Rather than training for a specific job, this allows students to acquire skills that are transferable across related occupational fields. For example, the health science career cluster includes pathways for therapeutic services, diagnostic services, health informatics, support services, and biotechnology research and development. A student wishing to become a nurse would develop a plan of study in the therapeutic services pathway, which would include foundational skills that would be transferable to the other pathways in the health sciences career cluster if desired at a later time. This plan of study would map core, technical, and elective coursework beginning with the ninth grade through graduation. If the high school and the postsecondary institution have aligned coursework using the career cluster framework, the student could also map courses required for the nursing degree and identify opportunities to take college-level work while in high school. This would save tuition dollars and reducing the time between high school and employment.

Virginia has enhanced its career pathways initiatives with career coaches—community college employees who spend most of their time on high school campuses, where they serve as resource specialists for career planning and connect students with businesses and community colleges. To further support the career pathways initiative, Virginia is piloting the "Middle College" concept which will provide an opportunity for high school dropouts between the ages of 18 and 24 to attain a GED and enroll in college course work that will enhance basic workforce skills as well as count toward a certificate or associate's degree and attainment of a workforce readiness certificate.

Virginia has also established the Workforce Development Academy to train workforce development practitioners and advance the workforce development profession. Professional development opportunities include a graduate Certificate in Workforce Development and a noncredit certificate for the completion of the Workforce Development Professional Competencies course.

Career pathways to improve adult postsecondary attainment initiatives can be designed for secondary to postsecondary transitions. However, career pathways are also needed to enable adult individuals to advance, particularly in high-wage, high-growth careers, and to meet the long-term demands of the new economy. A framework of study must be developed to help an adult student start at the beginning of the career path, if necessary, then move along the path to a certificate or degree. Steps along the path could include basic skills development, earning a GED, learning how to integrate education into work, how to be successful with higher level responsibilities, and how to manage wages. The pathway is not a single program, but a collective framework of multiple programs and services to serve the client at all points along the path.

The Southern Growth Policies Board (SGPB 2002), in its report titled *The Mercedes and the Magnolia*, looked at how the South is preparing for the new economy and offered recommendations that stress the need for graduates to understand and prepare for future career advancement, rather than simply meeting the requirements of today. These recommendations include: creating a seamless workforce system to maximize the client (worker) control over the outcomes; identifying and developing underutilized sources of workers and talents; and creating a self-directed workforce with the attitudes, learning habits, and decision tools necessary for making wise career choices throughout life.

Workforce development has often been aimed at one career path, which is much too narrow to apply in today's economy. Institutions must now focus on the entire client, not just the skill the client wants to learn. This new focus will help clients raise the standard of living for themselves as well as future members of their families. Too often in the past, workforce training programs relied on someone else to focus on the personal and social development needs of the client, which resulted in confusion as to who was responsible.

Therefore, workforce development must now be an entire approach organized with an economic development focus applied to the current and future training of clients. At the same time, it must provide lifelong learning, a wide range of employment support services, and attention to character development and career awareness. Thus, a much broader scope of organizations and individuals must be involved in the design and delivery of the new workforce development program.

This concept has been developed in many states, in many different forms, and is now a reality in the State of Arkansas. The "Arkansas Career Pathways Initiative" (CPI) is a comprehensive project designed to move low-income adults who are eligible for Temporary Assistance for Needy Families (TANF) into self-sufficiency. CPI provides a series of connected or sequential education courses with an internship or on-the-job experience, as well as enhanced student and academic support services. This combination of structured learning creates achievable steppingstones for career advancement of adult workers and increases the quality of workers at all levels. As such, career pathways directly link investments made in education with economic advancement.

CPI pathways start with employability skills and adult education, then progress to a certificate of proficiency, a technical certificate, an associate degree and, finally, a baccalaureate degree. Along the way, students are taught to apply knowledge gained in the classroom to local jobs, and to gain work experience through internships, on-the-job training, and shadowing services. These real-world experiences allow students to make connections between education and jobs in a given industry or sector. To increase chances of success, intensive student services such as career assessment, advising, tutoring, job search skills, and job placement assistance are provided, in addition to more flexible class schedules needed by working adults. In some cases extraordinary support, such as assistance with transportation and child care, is provided.

CPI is a partnership between the state Departments of Higher Education, Workforce Services, and Human Services; the Arkansas Transitional Employment Board; the Arkansas Association of Two Year Colleges; and the Southern Good Faith Fund. CPI programs have been funded for all two-year colleges in the state.

Developing a Demand-Driven Workforce Development System

Designing an effective workforce development system requires a new thought process. Traditionally, workforce training has focused on job training—developing the specific skills necessary for a particular job as well as the human interaction and leadership skills required for the team approaches of the last decade. Unfortunately, this approach may result in limited training and preparation, designed to move individuals into entry-level jobs without consideration of long-term skill development, thereby increasing the number of working poor. Job training must be differentiated from workforce development, as the former is but a single strategy and too narrow for a comprehensive workforce development plan.

> The world we have created is a product of our thinking; it cannot be changed without changing our thinking.
>
> (Albert Einstein)

The ideal workforce development approach is a demand-driven system of both short- and long-term programs and policies which integrate workers, employers, educators, community and economic developers, and government policies into one seamless system. The system must be coordinated as well as flexible and responsive to changing economic conditions. It must include strategies for incumbent workers, the unemployed and the underemployed, special needs populations, mature workers, and the future workforce. This will require ongoing collaboration among employers, trade associations and labor organizations, chambers of commerce, economic developers, public and private secondary and postsecondary institutions, and community-based nonprofit organizations.

Workforce development strategies for the twenty-first century should be grounded in the community development process and support the economic development needs of the community. An effective demand-driven workforce development system will both bridge the traditionally separate policy domains of education, labor, and economic development, and address both demand- and supply-side issues (Porter 2002). The demand side includes workforce development policies and programs that respond to employer needs in finding and retaining qualified workers, and increasing their skills so that the employer can be more competitive. This can include employer practices that increase on-the-job learning and getting employees to work in teams. On the supply side, workforce development systems support short- and long-term economic priorities by producing increased skills among current and future workers. These systems help people find and keep jobs and advance in employment, with programs for out of school youth, those already working, those seeking employment, and special populations such as women, minorities, and the disabled.

What sets demand-driven systems apart from other approaches is the focus on lifelong learning and advancement. To achieve this, systemic changes will be required in education policy to improve achievement levels of all students, to offer alternative pathways which facilitate lifelong learning, and to create more flexible credentials and course delivery methods that allow working adults to attain education in small increments over extended periods of time (Porter 2002).

The community development process detailed in this book should be the basis for a workforce development plan because it should be part of a larger overall strategic plan for the community. Because most communities have an established workforce development entity, it may be that a review, redirection, or consolidation of efforts is in order rather than starting from scratch. Common elements of most effective programs include programs that are employer/employee centered, maintain strong networks of diverse partnerships between government- and community-based organizations, contain employers who value training for their employees, and serve a network of clusters in the industry, government, and educational sectors.

In addition to the community development process, the following are points to consider when developing a demand-driven workforce development system:

- Establish a vision for the workforce network supported by an action plan with key indicator targets and performance measures. The vision and action plan should contain labor market policies that both support local and regional economic development goals for job creation and economic growth, and contain financial strategies and incentives that support public and private sector investment in skills development.
- Ensure that the vision and subsequent decision making are data- and fact-driven to provide an analysis of the labor market including current economic conditions, major forces, and local trends. Routinely evaluate current programs funded under the Workforce Investment Act and other public programs using data to

measure progress toward goals and to modify and adjust strategies as conditions warrant. Use data and creativity to replace bureaucracy with innovation.

- Involve all stakeholders in visioning and planning, especially if agencies may be eliminated, reduced, or combined. Develop strategic linkages among stakeholder groups including service providers (secondary and postsecondary educators, state agencies, private companies); clients (employers, unions, current workers, those in the pipeline); government agencies (policy makers, state and federal funding sources); and community-based organizations and individual citizens. In this way, the system will be characterized by responsiveness to economic needs, continuous improvement and results-based accountability. Because industry representatives are the best source of information for layoff aversion strategies, as well as of advance knowledge related to preparing for emerging jobs, they are critical team members in developing workforce strategies.
- Give strong consideration to regional programs since today's mobile workforce is not necessarily restricted to a specific municipality or county. Because economies are regional in nature, develop workforce development strategies that are also regionally focused. Develop industry-led regional skills alliances to address areas of skills shortages across the region, in order to provide training for industry clusters rather than individual companies.
- Design incumbent worker training programs so that recipient firms are required to develop long-term, work-based training plans before financial assistance is provided. Firms should be required to consider overall training needs rather than providing quick fixes for immediate needs.
- Build a seamless learning system that focuses on K-16 or K-20 systems that produce graduates with marketable skills. To achieve this, education must be viewed as one system rather than separate secondary and postsecondary silos. It also involves advocating for overall high school reform and implementing increased linkages between high school and college.
- Make sure higher education understands the economy and advises students of current and emerging job opportunities so that education is aligned with available employment opportunities. Increase resources of community colleges so that services can be expanded for students pursuing a four-year degree, those seeking technical certifications for beginning employment, and incumbent workers needing to acquire new skills to keep abreast of changes in the workplace. This will help people raise their standards of living through increased educational attainment.
- Create strategies for certifying knowledge and skills that are recognized by employers though gained outside the formal education system, and build pathways for continuing education in the informal and formal learning systems for adults and youth including literacy programs, apprenticeship, vocational training, youth development, technology-enabled learning, and alternative education programs.

Federal Resources

The federal government has demonstrated its interest in workforce development for many years through several legislative acts. The Workforce Investment Act of 1998 (WIA) provides the framework and direction for the bulk of federal funds for workforce development. The Carl D. Perkins Career and Technical Education Act of 2006 provides significant funding for career and technical education at both secondary and postsecondary levels. Other legislation relevant to workforce development includes the Higher Education Act, the previously mentioned Temporary Assistance for Needy Families, the Vocational Rehabilitation Act, and the Adult Education and Family Literacy Act.

BOX 14.3 THE MISSISSIPPI BAND OF CHOCTAW INDIANS

The Mississippi Band of Choctaw Indians (MBCI) provides a case study for the development of local workforce skill strategies to address high rates of unemployment, poverty, and the lack of employment opportunities in a rural area. The MBCI has approached skill development by using a method of self-reliance and self-determination as well as participating with other organizations for collaborative partnerships. The results have been remarkable, with poverty and unemployment rates dropping significantly since the early 1970s from highs of 80% to a low of 2% in 2007. The focus for workforce development is on technology-intensive manufacturing as well as the hospitality industry, aimed at upgrading the skills of workers, while also training those without skills. The emphasis is also on tribally owned and managed enterprises to foster self-reliance (Phillips, 2009: 155).

(R. Phillips (2009). The Choctaw Tribe of Mississippi. In F. Froy, S. Giguere, & A. Hofer (eds.). *Designing Local Skills Strategies*. Paris: OECD, pp. 155–173 (DOI : 10.1787/9789264066649–6-en).)

Department of Labor—Employment and Training Administration

The Workforce Investment Act is administered by the U.S. Department of Labor's Employment and Training Administration (ETA). The ETA is responsible for federal government job training and worker dislocation programs, federal grants to states for public employment service programs, and unemployment insurance benefits. The ETA provides the majority of its workforce development support through the Office of Workforce Investment, which has operations that focus on one-stop operations, business relations, and services to adults, youth, and workforce systems.

The WIA reformed and consolidated federal job training programs and created a new, comprehensive workforce investment system intended to be customer-focused to help Americans more effectively plan and manage their careers. It authorized the establishment of state workforce investment boards and state workforce development plans. The purpose of the law was to increase the employment, retention, occupational skill attainment, and earnings of participants.

Department of Education—Office of Vocational and Adult Education

Funds provided by the Carl D. Perkins Career and Technical Education Act are administered by the Department of Education's Office of Vocational and Adult Education (OVAE). The OVAE oversees programs that prepare students for postsecondary education and careers through high school programs and career and technical education. It also provides opportunities for adults to increase literacy skills, ensures equal access and opportunities for special populations, and promotes the use of technology for access to and delivery of educational instruction.

Perkins focuses on integrating academic and technical study for career and technical education students, strengthening the connections between secondary and postsecondary education, and preparing students for nontraditional occupations. In addition, it focuses on serving the needs of special populations, improving professional development opportunities for educators and administrators, and promoting partnerships between education and industry.

Federal Perkins funds are awarded annually to a state agency designated by its governor. Typically, it is a department of workforce services, a department of education or, in some cases, a department of higher education. Once the award is received by the state, the recipient agency determines the amount of funds to be provided to secondary and postsecondary institutions.

State and Local Resources

State Agencies

The delivery of workforce development services at the state level is structured in various ways and often involves several agencies including workforce services, economic development, employment security, general education, higher education, and the workforce investment board. Several states, such as California, have created a cabinet-level interagency team to coordinate all economic development activities including workforce development.

State-level services are provided for employees through occupational and labor market information; skills assessment; job matching and job banks; and special services for dislocated workers, veterans, the disabled, and other special need populations. Services provided to employers include assistance in finding employees; information related to grants, training incentives, tax credits, and state and regional employment data and analysis. Additional services are provided related to job skills profiling and assessment, employee skills gap identification, and related training.

To better compete in the new global economy, several states have developed strategic plans specifically for workforce development. The State of Oregon plan includes increased funding for prekindergarten through postsecondary education; a Cluster Investment Fund targeted to demands of employers and clusters of businesses in high-demand areas; and a "Skills Up Oregon" fund to both upgrade the skills of unemployed and lower-wage workers, and to increase GED and professional certificates among high school dropouts. The plan also includes targeted education investments such as career pathways, the creation of work readiness certificates, and health care training. The program also focuses on increased math and science skills, engineering and manufacturing research and development, and apprenticeship programs.

Florida has developed initiatives for "First Jobs/First Wages," which targets youth and adults entering the workforce for the first time; "Better Jobs/Better Wages," which helps underemployed workers improve skills to advance to higher wages, and "High Skills/High Wages." The latter acts as a catalyst among industry, economic development organizations, and training providers, in order to identify job skills that are critical to business retention, expansion, and recruitment activities.

Customized training is an integral element of workforce development and should be provided to meet the needs of a specific business rather than those of individuals. In most states, this type of training is provided through its economic development agency. For example, in North Carolina, free customized training is offered to businesses in targeted business sectors that create a certain threshold of new jobs annually. Instruction and training are furnished by local colleges, a state network of training specialists, and third-party vendors. The program also provides both extensive continuing education through community colleges for the existing workforce, and a special retraining program for programs with high technical costs.

State and Local Workforce Investment Boards

The Workforce Investment Act established state workforce investment boards as well as local workforce investment boards. The state boards must include the governor and state legislators; representatives of business and labor; chief local elected officials; relevant state agency heads; and representatives of organizations with expertise in workforce and youth activities. The majority of the board must be representatives of business. The board advises the governor on developing the statewide workforce investment system, the statewide employment statistics system, performance measures, and a system for allocations. It also develops a state strategic plan which describes the workforce development activities to be undertaken in the state. The plan also describes how the state will enact WIA requirements including those related to special populations, welfare recipients, veterans, and those with other barriers to employment.

Local workforce investment areas are designated by the governor, as well as a local board responsible for planning and overseeing the local program. Like the state board, the local board must also have a majority of business members and include education providers, labor organizations, community-based organizations, economic development agencies, and each of the one-stop partners. A youth council is also established as a subgroup of the local partnership. It is comprised of local board members, youth service organization representatives, local public housing authorities, parents, youth, and other organizations as deemed appropriate locally.

Under the WIA, each local area must establish a one-stop delivery system through which core employment-related services are provided by local one-stop partners, and through which access to other related federal programs is provided. Each local area must have at least one one-stop service center which may be supplemented by a network of affiliated sites.

The law designates certain partners that are required to participate in the one-stop system including those authorized under the following federal legislation: adult, dislocated worker, youth, and vocational rehabilitation activities under the WIA; postsecondary Carl Perkins Act funding; and welfare to work grants. Under the Older Americans Act; veterans' employment and training programs; community services block grants; Housing and Urban Development-administered employment and training programs; Trade Adjustment Assistance and NAFTA–TAA (North American Free Trade–Transitional Adjustment Assistance); and unemployment insurance and state employment services required by the Wagner-Peyser Act.

The local board either selects the operator of a one-stop center through a competitive process, or may designate a consortium of no fewer than three one-stop partners to operate the center. An eligible operator may be a public, private, or nonprofit entity or a consortium of such; a postsecondary educational institution; the employment service organization authorized under the Wagner-Peyser Act; another governmental agency; or other organizations.

The cornerstone of the new workforce investment system is the one-stop service delivery, which unifies numerous training, education, and employment programs into a single system in each community for youth, adults, and dislocated workers. The goal is for services from a variety of government programs to be delivered in a seamless manner including customer intake, case management, and job development and placement services. Access is also facilitated to support services that reduce barriers to employment such as transportation providers, child care services, nonprofit agencies, and other human service providers. Required core services for adult and dislocated workers include: job search and placement assistance; career counseling; labor market information related to job vacancies; skills training for in-demand occupations; skills assessment; information on available services and programs; case management and follow-up services to assist in job retention; development of individual employment plans; and short-term vocational services.

Using the Workforce Investment Board as its base, the State of Maryland has made dramatic change at all levels of its workforce development system. In so doing, it has created a model, demand-driven workforce development system with partnerships among the three "E" cornerstones: education, employment, and economic development. The system includes all stages of education and training, from K-16 education through retirees. While geared toward meeting the needs of business, the system also provides support services for workers, such as day care and transportation, and is aligned with the economic development goals of the state.

The Maryland model encourages the following:

- increased cooperation and collaboration among state agencies to eliminate duplication, reduce costs, unify planning, and maximize resources and organized activities around industry clusters;
- multiple initiatives that support existing Maryland businesses with retention, growth and development of the workforce including a statewide Web-based career management and job marketing system;
- refocused postsecondary workforce development initiatives to include an industry sector approach, customized coursework, and degree resources targeted to high-demand industry sectors.

Conclusion

Communities must develop effective approaches to human capital development for fostering an agile workforce in the throes of continual change and uncertainty, characteristics of the new workplace. The U.S. Department of Labor estimates that today's learners will have 10 to 14 different job assignments by age 38. Former Secretary of Education, Richard Riley, predicts that the 10 jobs most likely to be in demand in 2010 will not have existed in 2004, and that the US must prepare students for jobs that don't yet exist, using technologies that haven't yet been invented in order to solve problems not yet recognized. This churning makes the workplace very unpredictable for employers, employees, and policy makers. Employees must be equipped not only with industry-specific technical know-how, but have both the ability to create, analyze, and transform information that helps solve real-world problems, and to work and communicate effectively with others. Both formal and informal learning must become lifelong in nature, as opposed to stopping after a degree or certificate is attained or employment found. Employers must understand that investing in and facilitating employee training affects productivity and, therefore, determines the company's competitive position. Policy makers must provide a demand-driven workforce development system that is flexible, responsive and innovative—more than just "business as usual."

CASE STUDY: COLLABORATING TO DEVELOP A HIGH-TECH WORKFORCE IN TULSA, OKLAHOMA

One of the biggest challenges many communities face today is training and retraining the workforce for high-skilled jobs in technology-based industries. No one program or organization can accomplish such a major workforce development effort—its takes a coordinated, community-wide effort. As such, workforce development is an exercise in community development.

Sleezer et al. (2004) studied how Tulsa, Oklahoma succeeded in creating an advanced workforce development program with participation from key organizations in both the public and private sectors. By 1990, Tulsa had begun to attract a significant number of telecommunications companies. A community assessment revealed that there was a shortage of skilled workers for this industry in the area and no workforce development program in place to train more. Business, political and academic leaders began working together to address the issue and concluded that the region needed a master of science degree with an emphasis in telecommunications.

Implementation of this recommendation proved to be a challenge. After some initial cooperation, the branches of the two major state universities located in Tulsa could not agree on how to move forward collaboratively to create a joint program. Each university decided to proceed independently, which would mean fewer resources for each program. In 1999, a change in leadership at one university provided a spark to rekindle the cooperative effort. With input and assistance from business and political leaders, the two universities worked together to create the Center for Excellence in Information Technology and Telecommunications. Since that time, Tulsa has been cited in several studies as having a strong concentration of information technology workers.

Sleezer et al. describe Tulsa's workforce development effort as an exercise in community development, as well. Business, political, and academic leaders had somewhat different motives and goals going into the process. For example, faculty members emphasized providing a high-quality curriculum, while business leaders emphasized meeting their firms' educational needs. Through mutual trust and good-faith negotiations, the parties resolved their differences and achieved success. Contributing to this success was the fact that the parties involved in the negotiations had a history of working together on other committees and community efforts. In other words, there was a preexisting degree of social capital or cohesion in Tulsa which helped the community plan and act together effectively.

Keywords

Demand-driven workforce development, demand driven, career pathways, postsecondary education, new economy, workforce investment boards.

Review Questions

1. Why should workforce development strategies be integrated into primary and secondary education?
2. What are career pathways and career clusters and how are they related?
3. What is a demand-driven workforce development program and what are its advantages?
4. Can you identify an innovative human capital development program from another country? What is different from U.S. approaches?
5. In your own community, what are some challenges facing workforce development?

Bibliography

Arkansas Department of Higher Education. Online. Available at: <www.arpathways.com/> (accessed December 11, 2007).

California Economic Development Partnership. Online. Available at: <www.labor.ca.gov/cedp/default.htm> (accessed December 11, 2007).

California Labor Federation AFL-CIO. Workforce and Economic Development Strategies. High Road Partnerships. Online. Available at: <www.calaborfed.org/workforce/dev_strat/index.html> (accessed December 11, 2007).

Career Clusters Institute. Online. Available HTTP: <www.careerclusters.org/> (accessed 11 December 2007).

Carl D. Perkins Career and Technical Education Act. Online. Available at: <www.ed.gov/policy/sectech/leg/perkins/index.html> (accessed December 11, 2007).

Commission on the Skills of the American Workforce. Online. Available at: <www.skillscommission.org/staff.htm> (accessed December 11, 2007).

Council for Adult and Experiential Learning. Online. Available at: <www.cael.org/> (accessed December 11, 2007).

Council for a New Economy Workforce. Online. Available at: <www.southern.org/cnew/cnew.shtml> (accessed December 11, 2007).

Employment and Training Administration. Online. Available at: <www.doleta.gov> (accessed December 11, 2007).

Global Workforce in Transition. Online. Available at: <www.gwit.us/> (accessed 11 December, 2007).

Gordon, E.E. (2005) *The 2010 Meltdown: Solving the Impending Jobs Crisis*, Westport, CT: Praeger Publishers.

Government of South Australia. (2005) *Better Skills. Better Work. Better State: A Strategy for the Development of South Australia's Workforce to 2010*. Online. Available at: <www.dfeest.sa.gov.au/dfeest/files/links/wds2005.pdf> (accessed December 11, 2007).

Indiana Strategic Skills Initiative. Online. Available at: <www.in.gov/dwd/employus/ssi.html> (accessed December 11, 2007).

Jobs for the Future. Online. Available at: <www.jff.org/> (accessed December 11, 2007).

Judy, R.W. and D'Amico, C.G.L. (1997) *Workforce 2020: Work and Workers in the 21st Century*, Indianapolis, IN: Hudson Institute.

Maryland Department of Labor, Licensing and Regulation. Governor's Workforce Investment Board. Online. Available at: <www.mdworkforce.com/> (accessed December 11, 2007).

National Association of State Workforce Agencies. Online. Available at: <www.workforceatm.org/> (accessed December 11, 2007).

National Center on Education and the Economy. (2003) *Toward a National Workforce Education and Training Policy*. Online. Available at: <www.ncee.org> (accessed December 11, 2007).

National Council for Workforce Education. Online. Available at: <www.ncwe.org> (accessed December 11, 2007).

National Governors Association. Center for Best Practices. Online. Available at: <www.nga.org> (accessed December 11, 2007).

North Carolina Department of Commerce. Online. Available at: <www.nccommerce.com/en/WorkforceServices> (accessed December 11, 2007).

OECD Action Plan for Youth (2013). Online. Available at: <www.oecd.org/els/emp/Youth-Action-Plan.pdf>

Office of Vocational and Adult Education. Online. Available at: <www.ed.gov/about/offices/list/ovae/index.html> (accessed December 11, 2007).

Phillips, R. (2009). The Choctaw Tribe of Mississippi. In F. Froy, S. Giguere, & A. Hofer (eds.). *Designing Local Skills Strategies*. Paris: OECD, pp. 155–173 (DOI : 10.1787/9789264066649–6-en).

Porter, M.E. (2002) *Workforce Development in the Global Economy*. Online. Available at: <www.gwit.us/global.asp> (accessed December 11, 2007).

Progressive Policy Institute. (2004) *Economic Development Strategies for the New Economy*. Online. Available at: <www.ppionline.org/> (accessed December 11, 2007).

Progressive Policy Institute. New Economy Index. Online. Available at: <www.neweconomyindex.org/index.html> (accessed December 11, 2007).

Public/Private Ventures. Online. Available at: <www.ppv.org/ppv/workforce_development/workforce_development.asp> (accessed December 11, 2007).

SGPB (Southern Growth Policies Board) (2002) *The Mercedes and the Magnolia: Preparing the Southern Workforce for the New Economy*. Online. Available at: <www.southern.org> (accessed December 11, 2007).

Sleezer, M., Gularte, M.A., Waldner, L., and Cook, J. (2004) "Business and Higher Education Partner to Develop a High Skilled Workforce: A Case Study," *Performance Improvement Quarterly*, 17(2): 65–81.

University of Virginia Workforce Development Academy. Online. Available at: <www.workforcedevelopmentacademy.info/> (accessed December 11, 2007).

U.S. Bureau of Labor Statistics. Online. Available at: <www.bls.gov/> (accessed December 11, 2007).

U.S. Census Bureau. Online. Available at: <www.census.gov/> (accessed December 11, 2007).

U.S. Department of Education. Online. Available at: <www.ed.gov/> (accessed December 11, 2007).

Virginia Community College System Workforce Development Services. Online. Available HTTP: <www.wdscommunity.vccs.edu> (accessed December 11, 2007).

Workforce Florida. Online. Available at: <www.workforceflorida.com> (accessed December 11, 2007).

Workforce Investment Act. Online. Available at: <www.doleta.gov/usworkforce/wia/act.cfm> (accessed December 11, 2007).

Workforce Oregon. Online. Available at: <http: //governor.oregon.gov/Gov/workforce/workforce_vision.shtml> (accessed December 11, 2007).

Workforce Strategy Center (2001) *Workforce Development: Issues and Opportunities*, Brooklyn, NY: Workforce Strategy Center.

WorkforceUSA.net. Online. Available at: <www.workforceusa.net/index.php> (accessed December 11, 2007).

WorkKeys. Online. Available at: <www.act.org/workkeys/index.html> (accessed December 11, 2007).

Connections

Sample plans of study in all career clusters are available from Career Clusters Institute at www.careerclusters.org

See Rutgers' Center for Women and Work's Innovative Training and Workforce Development Research for resources at: http: //smlr.rutgers.edu/CWW/workforce-development-programs

15 Marketing the Community[1]

Robert H. Pittman

Overview

The competition to attract new investment and jobs into communities is fierce. There are thousands of communities in the US competing for business expansion and relocation projects annually. In today's economy, the competition for new investment is increasingly global. Communities must also market themselves to firms already located in their areas (business retention) and to new business start-ups. Marketing is a critical component of a successful community and economic development program. This chapter defines marketing in the community and economic development context, explains the components of a successful industry marketing program, and discusses the key elements in the marketing plan.

Introduction

Communities market themselves to a variety of entities: new firms, existing firms, nonprofit organizations, tourists, new residents, restaurant chains, etc. While the principles behind marketing to these various audiences are similar, the specifics of marketing programs vary according to the target audience. The focus of this chapter will not be on marketing to tourists (the subject of Chapter 18). Nor will it be on attracting retail stores, restaurants, or new residents. Instead, it will focus on the marketing of communities in a specific economic development context; that of attracting new firms and retaining and expanding existing ones. In most communities these companies—manufacturing operations, call centers, service firms, etc.—form the basis of the local economy, which in turn supports retail stores, restaurants, and new residents.

Marketing Definitions

There are many textbook and lay definitions of marketing. However, what does marketing mean in the specific context of economic development? Let us start with a broad definition of marketing and narrow it down to the economic development field.

General marketing: "The performance of business activities that direct the flow of goods and services from producer to consumer" (American Marketing Association). This is a broad, consumer product definition of marketing.

Societal marketing: "To determine the needs, wants and interests of target markets and to deliver the desired satisfactions more effectively and efficiently than competitors in a way that preserves or enhances the consumer's and the society's well being" (Kotler 1991: 26). This is getting closer to a definition of marketing in the economic development context. In order to attract new investment that will create jobs and benefit the community at large, communities are trying to meet the needs and wants of their target markets. A company or organization needs a good location for business. This, in turn, is a complex and multifaceted concept including potentially hundreds of factors affecting business operations.

Economic development marketing can be thought of as creating an image in the minds of key company executives who rarely make expansion decisions; staying in contact with them so that when the time comes to act, they consider a particular community; and, finally, ensuring that their business location needs are fully satisfied.

For purposes of this chapter, economic development marketing can be defined in this way: Creating an image is a key to successful community and economic development marketing. Of course, actively pursuing firms with mailings, visits, phone calls, etc. is also critical to a successful economic development campaign. Some observers might refer to this part as "sales" instead of "marketing." However these activities are defined, both will be covered in this chapter.

Components of Marketing

Community marketing can be divided into four broad steps:

- Define the product and message.
- Identify the audience.
- Distribute the message and create awareness.
- Satisfy the needs and wants of the customer.

Define the Product and Message

It would be remiss to discuss community marketing without covering two of the key elements on which a marketing program must be based: community SWOT analysis (strengths, weaknesses, opportunities and threats) and strategic visioning and planning. SWOT analysis (or community assessment) is covered in Chapter 9, while strategic planning is covered in Chapter 6. However, it is worthwhile to review these tools in the context of a marketing program.

A SWOT analysis identifies the economic development strengths and weaknesses of a community. Which community strengths—such as highly productive and trained labor force, good utility cost and service, and supportive local government—will attract industry? Conversely, which weaknesses—such as inadequate transportation infrastructure, lack of available industrial site and buildings, and permitting difficulties—will hurt the recruitment of new businesses, retention and expansion of existing community, and new business start-ups? All the basic site selection factors should be covered but there will also be community-specific issues that will surface in the analysis such as factions quarrelling over which direction the community should go. Opportunities might include economic development potential from an excellent community college while threats might include the potential closure of a local plant.

It should be obvious why a SWOT analysis is such a critical component of community marketing. If an audit of strengths and weaknesses hasn't been conducted, how can a product (the community) be defined and made attractive to its potential audiences (types of industries)? Many communities claim they are a great place to "live, work and play." However, what specific set of assets does a community have relative to the thousands of others also vying for new investment? Without a SWOT analysis, this question cannot be answered, the product cannot be defined, nor can an effective marketing image and campaign be created. A SWOT analysis also identifies the weaknesses and threats that need to be addressed in the short and long term.

The SWOT analysis gives the community a foundation on which to base effective community strategic planning and visioning. Once community leaders and citizens compare their economic development strengths and weaknesses with those of other communities, they can more effectively chart a direction for future growth and development. So many communities want to attract the next automobile assembly plant or company headquarters, yet have a completely unrealistic assessment of their chances for attracting such large projects. Perhaps they should focus instead on smaller advanced manufacturing companies or customer support centers. It is great to dream about a future in which the community is growing and attracting high-paying jobs, but if a dream is unrealistic then it will likely never happen.

Another component of strategic planning and visioning involves what the community wants to be in the future. Does it want to focus on attracting manufacturing and service firms, or does it want to become a "bedroom" community and thus rely more on tourism and retail and jobs in surrounding communities to support its residents? Again, it is easy to understand why strategic planning and visioning is critical to an effective marketing campaign. Without this element based on SWOT analysis a community does not even know to whom it wants to market. Imagine that a person wakes up one day and decides to get in their car and go somewhere. However, they are not sure where to go or why so they drive around aimlessly and return two hours later, happy they made good time on the highway. Their neighbors and family would certainly question this person's sanity. Community marketing without a SWOT analysis and strategic plan is just as pointless.

Marketing Image

A fundamental part of defining the product and marketing message is to create an effective marketing image. A community marketing image can be defined as: "The sum of beliefs, ideas and impressions that people (residents, target audience, outside public, etc.) have of a place" (Kotler et al. 1993: 141). Another term often used for image is the "market brand" of a community. The image or brand of a community or place can profoundly affect companies' location decisions. As an example, consider North Dakota. The state has an image in many people's minds of being too cold, remote, and snowy for effective business operations. In reality, North Dakota boasts a number of leading national and international companies such as Great Plains Software (acquired by Microsoft) and the Melroe Company (producer of Bobcat construction equipment). The state has worked to counteract this negative image by running advertisements featuring testimonials from North Dakota executives. In magazines and other publications, these testimonials tout the state as a profitable business location with productive labor and few down days due to weather.

As another example, consider the image boost Alabama enjoyed when Mercedes-Benz (now DaimlerChrysler) decided in the early 1990s to locate its first North American assembly plant in the state. That decision sent a message around the world that a leading global company found Alabama to be a good location for its operations. Any negative stereotypes in the minds of corporate executives (e.g. that Alabama was a poor, agricultural state with a lack of skilled labor) were refuted by that one location decision.

A state's or community's image often determines whether it is on the initial facility location search list or not. Once the short list of locations has been selected, it also often has an intangible but strong effect on the final location decision. Most of the time, when the decision is down to the final two or three communities, the profit and cost profiles of the final communities are similar. The final decision is affected by a number of intangible factors including image. For example, a company executive might be afraid to recommend a location with a negative image for fear his management and board of directors would think he was making a mistake.

A marketing image should be:

- valid;
- believable;
- simple;
- appealing;

- distinctive (not just another great place to live, work and play); and
- related to the target audience.

If the marketing image does not meet these criteria, it can do more harm than good to a community.

Slogans and Logos

Slogans and logos can be used to communicate a community's image and rise above the marketing clutter of hundreds or thousands of other communities. Slogans are briefly stated ideas, themes, or "catchphrases" used to convey a community's or an area's image (see examples in Box 15.1). Logos are graphic or pictorial images used to help convey a community's image. They can provide a visual unifying theme for marketing materials such as brochures; letterheads; and promotional giveaways such as key chains, pens, coasters, etc. Again, themes and logos must clearly reflect a community's carefully crafted marketing image. It has been said that we live in a "sound bite" world today and slogans and logos can be effective in instantly communicating an image.

A slogan can also effectively communicate a community's geographic location to corporate executives worldwide. For example, two hours north of Dallas, the southern Oklahoma City of Ardmore uses the slogan "Dallas is Coming Our Way." This slogan successfully communicates two facts. One is that some businesses are choosing Ardmore instead of Dallas as a location. Another is that Ardmore is close to Dallas, because corporate executives use the better-known Texan city as a point of reference.

Identify the Audience

After the community image and message have been crafted, the next step is to identify the audience for the development marketing campaign. That market is certainly broad but, again, this chapter focuses on marketing to companies. This audience includes:

- outside companies, site selection consultants, industrial real estate companies, and others involved in corporate expansion (recruitment);
- lead-generating economic development organizations (state and regional economic development organizations, utilities, etc.);
- existing businesses already in the community (business retention and expansion); and
- entrepreneurs (new business start-ups).

This group can be referred to as an *external* marketing audience: those organizations and individuals that can help a community attract investment and create jobs. There is another marketing audience that can be referred to as *internal*:

- community stakeholders (elected officials, board members, sponsors, etc.);
- media (newspapers, TV, radio); and
- the general public.

BOX 15.1 EXAMPLES OF ECONOMIC DEVELOPMENT MARKETING SLOGANS

What image do these slogans create in your mind?
Mississippi: Yeah, We Can Do That!
Michigan: The Future is Now
Seattle: Leading Center of the Pacific Northwest—the Alternative to California
Fairfax County, VA: The Nation's Second Most Important Address
Atlanta, Georgia: Center of the New South
Henderson County, NC: Smart Business Location with a Metropolitan Blue Ridge Lifestyle

Community and economic development organizations need to communicate with this internal audience so that they may continue to secure strong cooperation and support in the overall development effort. Many economic development marketing programs have died premature deaths because not enough attention was given to ensuring that a strong community coalition stayed together to support the overall effort. As discussed in Chapter 1, this is one area where community development and economic development have a strong overlap.

The external development marketing audience is still quite large and successful and communities identify their prime prospects by drilling down to specific sectors and organizations in the marketplace. This is critical because communities have limited marketing budgets and cannot spread their resources too thinly by marketing to a broad audience that may not even be interested in their message. The challenge is to identify those industries and companies that would be most likely to view a community as a potential location match. In order to do this, communities need to segment their markets.

Community economic development market segments include:

- *Sector/industry.* Manufacturing, service, professional, wholesale/distribution, etc. Within each sector (e.g. manufacturing), which industries would make the best targets (e.g. auto parts, plastics, electronics, etc.)?
- *Geography.* For a community in rural Arkansas, for example, it would probably be more fruitful to try to recruit companies in higher-cost locations such as Chicago, Dallas, or another urban area rather than similar rural areas nearby.
- Type of company/organization. Smaller communities may be better off targeting smaller privately owned firms. Larger communities can target these as well, but also may want to target larger publicly traded companies. Again, it is all about the best match with the community.

Substantial effort should be given to the target industry analysis because this often makes the difference between a successful and unsuccessful marketing campaign. Target industry selection criteria include:

- a match between the location needs of the industries and the strengths/weaknesses of the community (from the SWOT analysis);
- historical and potential growth rates of industries;
- skill levels and wage rates of industries;
- diversification potential of the industries for the local economy;
- other community-specific considerations such as environmental friendliness, industry image, etc.

Once the target industries are selected, the next step is to identify companies in those industries that will make likely targets for recruiting. The community should look for companies that are most likely to expand and that also meet the above mentioned criteria of geography, company size, and ownership.

The target industry and company selection process must be based on sound research. Large amounts of data on growth rates, location criteria, and patterns, etc. should be collected and analyzed in order to select the best targets. Many communities hire professional consultants or rely on local universities to assist with this complicated procedure. If the community undertakes the analysis itself, there are numerous sources of information on industries and companies that are available through government and private organizations. So much of this is now only a few clicks away on the internet.

Distribute Message and Create Awareness

After creating the message and identifying the target audience, the third basic step in community marketing is to distribute the message and create awareness. There are different means that a community can use to distribute its marketing message and promote itself. The mix of marketing elements should be chosen very carefully and with much research in order to make the best use of limited marketing dollars. Ways to promote and market a community for economic development include:

- *Advertising.* Media choices include television, radio, and print. Types of print media for economic development advertising include national business publications such as the *Wall Street Journal* or *Business Week*. However, advertising in these national publications is quite expensive and probably not feasible for smaller communities. Other print media include industry trade publications oriented toward industries such as aerospace or plastics and site selection/development magazines such as *Expansion Management*, *Area Development*, and *Site Selection*.

  Industry trade publications are read by executives within specific industries and thus can be used to reach a community's target industries. Site selection and development magazines target corporate executives that are typically involved in corporate expansion decisions but cover a broad range of industries. There are other niche print publications, such as local business journals and chamber of commerce magazines, which communities can explore as potential advertising options.
- *Direct mail.* We are all regularly bombarded with "junk mail" and 99% of it winds up unopened in the trash. However, many communities operate successful direct mail campaigns aimed specifically at the industries and companies they have carefully selected as described above. Research shows that repeated mailings to a select audience has a much higher response rate than general blanket mailings. If a community sends out 1,000 targeted mailings, gets ten company prospects, and lands one new company, the benefit will be tremendous even though the batting average is low.
- *Email.* Many communities use email to market themselves. Some email messages are direct "sales pitches" while others are electronic copies of newsletters, announcements, etc. Targeted email can be effective but random mass e-mailings are usually treated as just more spam.
- *Trade shows.* Many communities successfully market themselves to their target industries at trade shows. There can be hundreds of executives from a target industry there, so attendance can be a very cost-effective way to reach the target audience. Many communities buy or share booths at the trade show and stock them with marketing materials. Others just "walk the floor" and look for opportunities to meet executives from key companies face to face.
- *Personal contact.* This is one of the best ways to market a community but it is often difficult and expensive to get a personal contact opportunity with a corporate executive. Trade shows are one way to do this, as is the telemarketing follow-up which is part of the direct mail campaign discussed above. Another personal contact method is to visit prospects but this is expensive and usually done after a prospect has expressed interest in a community.
- *Networking.* Many leads come to communities through networking with other economic development officials and agencies. Networking contacts should include the state economic development agency, utility companies, real estate brokers and developers, site selection consultants, existing employers in the community, railroads, and other organizations that may know of expanding companies.
- *Public relations.* While all of the activities above can be considered public relations, to get their message out to their target audience and general public, successful communities have even more proactive programs at their disposal. Local, regional, and state news channels (television, radio, and print) should

be regularly sent press releases and otherwise contacted to get the community's name in print and publicize significant events.

- *Websites.* The internet is now a primary way to gather information and site selection data is no exception. It is almost taken for granted today that a community will have a website with relevant economic development information. Not having one sends a negative message to potential investors. In addition to containing basic economic and demographic information about a community, a website can contain testimonials from local executives, preferably in audio or video format; community pictures; and interactive video "virtual tours" of the community.

Good marketing materials are essential to effectively promote a community. Types of marketing materials include:

- *Brochures.* One of the best formats is a six-to-eight-page four-color professional brochure highlighting the economic development assets and livability of the community. However, these can be expensive so many communities opt for "desktop" versions.
- *Postcards and other brief mailings.* As discussed above, these brief reminder pieces sent out quarterly or semiannually help keep a community in the minds of target company executives.
- *Community profiles.* These are more in-depth fact books about communities. Often 15 to 20 or more pages in length, they are really more valuable as a follow-up tool once an executive's interest in a community has been piqued. Brochures or other shorter pieces should be used to create the initial interest.
- *Audiovisual presentations.* This category includes CDs, DVDs, computer presentations and the like (websites are covered above). They can be expensive to develop and are being replaced in many instances by websites that can do the same thing over the internet at the click of a mouse. However, these presentations can be effective as a downstream selling tool once initial interest and specific contacts have been established. Mass mailings of these media are almost never effective.
- *Promotional items.* This category includes T-shirts, pens, tote bags, and many, many other creative items. As mentioned above, these promotional gifts can help with community branding and imaging.
- *Newsletters.* Short newsletters mailed to prospects can be effective marketing tools. However, newsletters mailed to the *external* audience should not contain purely local information such as the date of the next city council meeting. Instead, newsletters sent to the external audience should briefly highlight recent positive developments, company locations, and expansions in the community.

BOX 15.2 SUGGESTIONS FOR EFFECTIVE ECONOMIC DEVELOPMENT WEBSITES

- Don't bury the economic development website in a municipal, chamber or other website. At most, locate the economic development page one link away from the home page.
- Orient your economic development website toward external users (prospects). Don't make users wade through chamber banquet announcements.
- Keep the information up to date. Old data gives a bad impression and may contribute to eliminating your community from consideration.
- Liven it up with graphics, pictures, and videos; however, keep it accessible via lower-speed connections.
- In marketing materials, encourage people to visit the website.

Satisfying the Needs and Wants of the Customer

The final basic step in marketing the community is satisfying the needs and wants of the customer. To his definitions of marketing, Kotler (1991) adds: "Marketing...calls upon everyone in the organization to think (like) and serve the customer" (p. xxii). This is a critical but often overlooked distinction in community marketing. Development organizations must keep their focus on attracting and satisfying the needs and wants of the companies, tourists, restaurants, nonprofits, or other groups they are trying to attract. When making a location decision, most companies refer to a "must have" (or needs) list of key location drivers for the particular project. These might include a skilled labor force, low tax rates, low transportation costs, high-quality utilities, or any number of factors. In addition to the needs list, they usually have a "want" list that might include a low cost of living, good education, recreational opportunities, etc. It should be stressed that all company location projects are unique. There is no "universal" list of top location factors. Obviously the location criteria for a steel mill are different from those of a corporate research and development center or nonprofit organization.

Attracting companies and new investment often requires communities to be flexible and make accommodations for companies such as improving water and sewer infrastructure, helping with labor training, or granting incentives. Of course it is up to the community to decide which accommodations to make, if any, based on the economic benefits of new corporate investment and on the desires and values of local residents.

The Marketing Plan

A marketing plan is the roadmap communities follow for successful promotion. It should be a written document widely agreed upon by all elements of the community. The marketing plan should contain:

- a mission statement for the marketing plan and a vision statement for the community;
- a situation analysis (SWOT analysis summary);
- a description of the target audience (industries and companies);
- marketing goals and objectives;
- strategic action items to achieve each objective;
- budget and resource requirements;
- clearly defined staff requirements and positions;
- clearly defined responsibilities of participating organizations and stakeholders.

Marketing goals are what a community or other entity has set out to accomplish. For example, a major goal of a marketing plan could be to "Attract new business and industry to the county through an aggressive targeted marketing campaign." Marketing objectives generally describe what needs to be done to achieve the goal. An example of an objective could be to "Convince prospects with an active project to visit the county." Strategic action items are specific, measurable activities designed to accomplish an objective. An example could be to "Invite corporate prospects as expense-paid guests to enjoy the county's Fall Apple Festival and local quail hunting."

Marketing plans can be 10 or 100 pages long, depending on the community and level of detail desired. Too little detail causes the marketing plan to be viewed as just another general statement of goals and objectives; too much detail can discourage local stakeholders from reading the plan and buying into it. Detailed marketing plans are best and they can include an executive summary for the general public's benefit.

Marketing plans should be dynamic. No one can predict with certainty how the various elements of the plan will work in practice. As the community learns about the elements of the marketing program that work best, and as external conditions such as the economic situation change, the marketing plan should be revisited and modified at least yearly.

Conclusion

Community marketing is a broad topic. This chapter has focused on economic development marketing and specifically on recruiting new firms to a community. However, the marketing elements discussed in this chapter have broad applicability throughout the field of community and economic development. They are particularly relevant to business retention and expansion and new business start-ups. Research shows that most new jobs in communities are created by retention and expansion of existing companies and business start-ups as opposed to recruitment of new companies. Marketing plans should include goals, objectives, and strategic action steps for all of these important economic development activities.

Even the smallest, most isolated communities can achieve marketing success through applying the principles described in this chapter. For a metropolitan area, marketing success may be attracting a new Fortune 500 company with 1,000 local employees. For a small rural community, marketing success may be as basic as helping a start-up company with five employees or attracting a McDonald's Restaurant. The key point to remember is that manna rarely falls from heaven. Success is more likely to come to communities that develop solid marketing programs and stay the course.

BOX 15.3 SOME *SUCCESS* FACTORS FOR DEVELOPMENT MARKETING

- Unify your community vision and strategic plan.
- Understand the product you are marketing.
- Know which audience wants the product.
- Understand the target audience.
- Write a marketing plan including:
  - how to get the marketing message to the target audience; and
  - adequate budget, resources, and staff.
- Have committed parties to stay the course to ensure that:
  - board members, elected officials, the private sector, and other citizens understand the plan and process; and
  - public and private concerns cooperate and form partnerships.

BOX 15.4 SOME *FAILURE* FACTORS FOR DEVELOPMENT MARKETING

- *Ideal community syndrome.* "This is the best place in the world to live, work and play. Why wouldn't anybody in his right mind want to put a facility here?" The fallacy is that there are thousands of "ideal communities" competing for a relatively few number of corporate investments. Often, residents cannot or chose not to view their community objectively from an outside investment standpoint. This is one reason a SWOT analysis is a key part of the marketing process.
- *If you build it they will come.* This relates to the Ideal Community Syndrome above. If a community just develops a site or industrial building and waits for the phone to ring, the chances are slim that they will be successful. In addition to developing the product, it must be marketed.
- *If it ain't broke, don't fix it.* "We've been doing fine for years so why should we change things now?" What if "doing fine" means that the local textile mill has not laid off yet? If communities do not actively assess their economic well-being and industrial base on a regular basis (as in Chapter 11) nor have development plans to move forward, they will inevitably move backwards.

- *Throwing the baby out with the bathwater.* "We didn't meet our goal of recruiting new jobs last year so we must change the program or quit wasting money." Marketing success is dependent on a number of external influences beyond the control of the community such as national economic conditions and corporate spending cycles. While success goals should be set, programs should not be solely judged on "hitting the numbers." Real economic development success comes from establishing a good marketing plan, evaluating it regularly, making midcourse corrections as necessary, and staying the course.

CASE STUDY: BUFFALO NIAGARA REGION

Located on Lake Erie in far Western New York State, Buffalo originally thrived as a Great Lakes port and manufacturing center. In the late twentieth century, as shipping routes and technologies changed and U.S. manufacturing growth abated, Buffalo's economic development fortunes declined. In 1999, Buffalo Niagara Enterprise (BNE) was founded. A regional economic development marketing organization, its mission was to market the eight-county area and attract new corporate investment.

In the minds of some corporate executives, Buffalo is well known for being featured in the video highlights on weather reports. With its high average snowfall, Western New York, gets a lot of "lake effect" snow and storms can be quite intense. To change their image from a place known for harsh weather to a place recognized as a good business location, BNE undertook a systematic, well-funded marketing campaign widely supported by the region's stakeholders. Their marketing program included:

- Advertising to modify the region's image. BNE became a National Public Radio sponsor using the slogan "Buffalo Niagara: Home of Snow-White Winters." This is a good example of directly tackling a perceived marketing weakness and turning it into a positive. The phrase "snow-white winters" evokes a more pleasing image of winter in the Buffalo region than blizzard footage shown on weather reports. It also appeals to winter sports enthusiasts.
- Target industry research. BNE spent a year carefully identifying industries that might be well suited to the economic development strengths and weaknesses of the region and are growing on a national basis. The target industries identified included primary life sciences, agribusiness, and back-office operations. BNE developed a list of 600 companies in these industries for direct marketing activities.
- Call trips to prospects. To personally explain the benefits of locating business operations in the Buffalo Niagara region, BNE began making trips to visit high-level executives in these target companies. Taking advantage of its location, BNE paid visits to numerous Canadian companies, touting the Buffalo Niagara region as a good place for Canadian companies to establish U.S. operations.

While it is sometimes difficult to directly link marketing activity with recruiting success, there is no doubt that the Buffalo Niagara region has been successful in attracting corporate investment. For BNE's 2006–2007 fiscal year, it announced 21 new or expanded projects worth more than $1.8 billion in private sector investment, 1,052 new jobs, and 1,337 retained jobs. Companies that created new jobs in the region included Astronics Corporation, 100 jobs; Multisorb Technologies, 95 jobs; and Citicorp, 500 jobs.

Sources: author discussions with BNE, the BNE website (www.buffaloniagara.org), and Glynn (2007).

Keywords

Marketing, recruitment, business retention and expansion, business retention, business expansion, image, market brand, slogan, target industry, marketing plan.

Review Questions

1 How is community and economic development marketing different from general or "societal" marketing?
2 What are the four major components of community marketing?
3 What are a community's external and internal marketing audiences?
4 What are some economic development market segments?
5 What are some key components of a community's marketing plan?

Note

1 Acknowledgement: The author wishes to express appreciation to David Kolzow, Rhonda Phillips, and Jennifer Tanner. Together, we have developed and taught many of the concepts in this chapter. However, the author assumes responsibility for any errors or omissions.

Bibliography

American Marketing Association. Online. Available at: <www.marketingpower.com> (accessed December 5, 2007).

Finkle, J.A. (2002) *Introduction to Economic Development*, Washington, DC: International Economic Development Council.

Glynn, M. (2007) "Buffalo Niagara Enterprise Returns Its Focus to Marketing," *Knight Ridder Tribune Business News*, 4 Oct.

Koepke, R.L. (ed.) (1996) *Practicing Economic Development*, Rosemont, IL: American Economic Development Council.

Kotler, P. (1991) *Marketing Management*, 3rd ed., Englewood Cliffs, NJ: Prentice Hall.

Kotler, P., Haider, D.H., and Rein, I. (1993) *Marketing Places*, New York: Free Press.

Shively, R. (2004) *Economic Development for Small Communities*, Washington, DC: National Center for Small Communities.

Connections

Learn more about economic development marketing with this resource guide provided by the U.S. Department of Housing and Urban Development (HUD) at: www.hud.gov/offices/cpd/economicdevelopment/toolkit/edt_manual.pdf

Explore marketing tools online, provided by The Municipal Research and Services Center (MRSC) a private, nonprofit organization based in Seattle, Washington at: www.mrsc.org/subjects/econ/ed-mark.aspx

Find a marketing plan template to use, provided by the Louisiana Community Network at: www.louisianaeconomicdevelopment.com/assets/LED/docs/LCN/Module-10-Marketing-Plan-Template.pdf

16 Retaining and Expanding Existing Community Businesses

Robert H. Pittman and Richard T. Roberts

Overview

While the recruitment of a new business into a community usually garners front-page headlines, the retention or expansion of an existing business rarely does. Communities tend to take their existing businesses for granted, which can be a big mistake. It is a lot easier to create 25, 50, or 500 jobs through the retention and expansion of existing industry than to spend the time and money it takes to recruit new industry. Plus, the benefits of generating business and enterprise locally can be large, including fostering a greater sense of connection to the community, and being vested in its success and long-term future.

Introduction

When people hear the term "economic development," they often think about recruiting new businesses into their community or region, sometimes visualizing large projects such as automobile assembly plants, high-tech research facilities, or corporate headquarters. These types of projects make headlines and can transform a regional economy almost overnight. Because of the publicity and attention these large projects command, many communities make the mistake of pursuing industrial recruitment, overlooking the small and medium-size businesses that create most new jobs. According to the Organization for Economic Cooperation and Development (OECD), small and medium-size enterprises (fewer than 250 employees) account for 95% of the firms and over half of private sector employment in most OECD countries (OECD 2013).

As discussed in Chapter 1, the three legs of the economic development stool are recruiting new businesses, retaining and expanding existing businesses in the community, and facilitating new business start-ups. Research has shown that retention and expansion and new business start-ups account for the majority of new jobs created in most areas. Consider these findings through the years:

- In pioneering research into job creation in the United States, David Birch found that 60–80% of all new jobs created in most areas come from retention and expansion of existing businesses and new small businesses (Birch 1987).
- Data compiled by the consulting firm Arthur D. Little revealed that between 31 and 72% of jobs in six states surveyed were created by business retention and expansion alone (Phillips 1996).
- *Site Selection* magazine found that 57% of manufacturing capacity increases in the US in 2006 (50 or more jobs or at least 20,000

square feet) were expansions of existing facilities (*Site Selection* 2007).

- *Southern Business and Development* magazine reported that, in manufacturing and nonmanufacturing companies, 55% of all new projects in 2005 that created at least 200 jobs in the South, were expansions of existing facilities (*Southern Business and Development* 2006).

Competition for a limited number of projects is very fierce and the odds of a given community landing one are low. It has been estimated that there are over 15,000 economic development organizations in the US actively recruiting new businesses while only about two to three thousand industry moves happen per year. To increase their chances of success, smart communities figure out the types of industries and size of businesses most suitable for their area and concentrate on recruiting those projects.

While this is certainly a better approach than pursuing large "trophy" projects, many of these communities still put too much emphasis on recruiting new industry. In the heyday of postwar manufacturing in the US, this was referred to as "smokestack chasing." Recruiting new firms, even small- to medium-size businesses, is extremely expensive and time-consuming. To stand out from the dozens or hundreds of competitors vying for the same new facility, communities must market themselves. They must advertise, attend trade shows, call on companies in distant cities or countries, and engage in a host of other expensive activities.

Companies don't simply knock on doors and ask to locate in a community. Many communities and citizens do not understand how intensely competitive it is to recruit new businesses, and fall prey to what can be referred to as the perfect community syndrome, to wit: "My community is a

BOX 16.1 JOB GROWTH IN US DRIVEN ENTIRELY BY START-UPS, ACCORDING TO KAUFFMAN FOUNDATION STUDY

Although conventional wisdom suggests that the annual net job gain at existing companies is positive, the fact is that net job growth in the U.S. economy occurs only through start-up firms, a new report from the Ewing Marion Kauffman Foundation (www.kauffman.org) finds. Based on the U.S. Census Bureau's business dynamics statistics, the report, *The Importance of Startups in Job Creation and Job Destruction*, found that both on average and for all but seven years between 1977 and 2005, existing firms were net job destroyers, losing a combined one million jobs per year. In contrast, during their first year new firms added an average of three million jobs. The report also found that while job growth patterns at both start-ups and existing firms were pro-cyclical, there was much more variance in job growth patterns at existing firms. Indeed, during recessionary years job creation at start-ups remained relatively stable, while net job losses at existing firms were highly sensitive to the business cycle. And it's not just net job creation that start-ups dominate. Although older firms lose more jobs than they create, the gross flows decline as firms age. On average, one-year-old firms create nearly one million jobs, while 10-year-old firms generate only 300,000. In other words, the notion that firms bulk up as they age is not supported by data. Because start-ups that develop organically are the principal driver of job growth in the economy, job-creation policies aimed at luring larger, established employers inevitably will fail, said the report's author, Tim Kane. Such city and state policies are doomed not only because they are zero-sum but because they are based on unrealistic employment growth models, added Kane. "These findings imply that America should be thinking differently about the standard employment policy paradigm," said Robert E. Litan, Kauffman Foundation vice president of research and policy. Policy makers tend to focus on changes in the national or state unemployment rate, or on layoffs by existing companies. But the data from this report suggest that growth would be best boosted by supporting start-up firms.

Source: Ewing Marion Kauffman Foundation Press Release 7/07/10.
http: //bit.ly/byg7Kh, as cited in Phillips et al. (2013: 32).

great place to live, work and play. I've raised a family here, operated a successful business, and enjoyed the quality of life. Why wouldn't anyone in their right mind want to locate a new corporate facility here?" There are two major problems with this attitude. First, citizens often do not see the shortcomings of their community from a business location standpoint; and, second, there are thousands of self-perceived "perfect" communities competing for every project.

If recruiting new companies—large or small—is difficult and competitive, especially in the era of globalization, what should communities do to create jobs, increase incomes, and improve the quality of life? The answer is to adopt a balanced approach to local economic development. It is ironic that when a new firm moves into town it makes the front page of the local newspaper, even if only a few jobs are created. However, if a business already located in and contributing to the community announces an expansion that creates hundreds of new jobs, the announcement often is buried in the back of the report. In short, many citizens take their existing industries for granted. They do not realize that local companies, especially branch facilities of national firms, regularly consider other communities for their expansions. They may even decide to relocate or consolidate all operations in a different area. Often, one community's loss is another's gain. What elected official or citizen wants to lose an existing business and/or its expansion to another community?

Even thriving local economies can be subject to sudden downturns from unforeseen events such as the closure of a manufacturing plant, call center, or other local facility. In such cases there is a downward economic multiplier effect. The closure of a local facility has a ripple effect throughout the economy. Primary job and income losses translate into less consumer and business spending in an area, a loss that hurts the retail industry and local consumer and business service industries. A loss of 100 primary jobs can translate into a total job loss of 200 or more.

Recovering from such a shock to the local economy can take years. Surely no one in a community wants such an economic shock to occur, regardless of the strength of the local economy. However, in today's extremely competitive and fluid global economy, economic dislocation is becoming more common. Industries are moving offshore; product lines and therefore production facilities are rapidly changing; and mergers and acquisitions are occurring more frequently. If there is agreement that these economic shocks are undesirable, then communities should do what they can to prevent them in order to maintain and grow local industries. Fortunately, most economic developers understand the importance of business retention and expansion. As shown in Figure 16.1, a survey by the Southern Economic Development Council, a professional association of economic developers in 17 southern states, indicated that 95% of the members were engaged in some sort business retention and expansion activities (SEDC 2006).

Companies may decide to relocate or expand away from where they are currently located because of a variety of external factors that are completely outside the control of the community such as mergers, acquisitions, moving offshore, changing product mix, etc. However, factors that are within the control of the local community often enter into a company's decision to relocate or expand elsewhere. Some examples include:

- inadequate services and infrastructure (water/sewer, electricity, roads, telecommunications, etc.);
- poor business climate (e.g. the community's permitting and regulatory policies and procedures don't support business);
- inadequate labor force;
- lack of available sites and buildings;
- poor education system and workforce training;
- low amenities and resulting quality of life.

Seemingly insignificant issues that the community can easily address can contribute to a business's decision to relocate or expand in a different area (see Box 16.2).

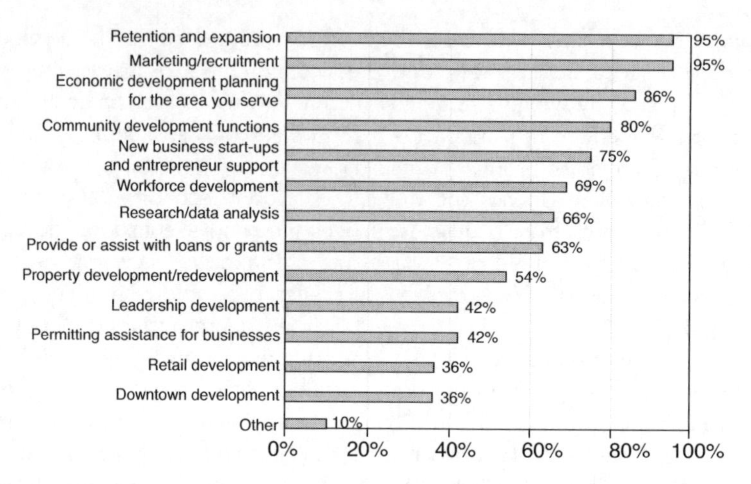

Figure 16.1 *Economic Development Activities Performed by SEDC Members*

BOX 16.2 CAN POTHOLES CAUSE LOCAL BUSINESSES TO RELOCATE?

In the course of doing an economic development assessment for a town in the South Central US, a consultant interviewed the manager of a large plant, one that employed almost 500 persons, about the strengths and weaknesses of the community as a business location. The manager did not point to any major problems—the labor force was suitable, the community's location on a major highway was good, and the local regulations were not onerous. As the manager escorted the consultant out the front door after the session, one of the company's trucks pulling into the plant hit a large pothole, shaking the truck and potentially damaging the cargo. The plant manager's smile turned into a frown as he said "I've been trying to get the City to fix that pothole for over a year, and nobody does anything about it." That pothole, an insignificant issue to the City, was a sign to the manager of the City's lack of support and concern for local businesses. If the City won't fix the pothole, how can it be trusted to upgrade water and sewer infrastructure and improve local roads to support the growth of local businesses?

Factors like the pothole repair issue can influence business location decisions. Imagine—as often happens—an executive visiting the community to screen it as a potential location hearing about this event. What impression would he take away concerning the City's support for local businesses? Rarely does a company move to a community without first talking to local executives about their experiences in that community. The local executive with the pothole problem would certainly not give the City a glowing reference.

Source: Pittman and Harris (2007).

What Does Business Retention and Expansion Encompass?

Business retention and expansion (BRE) covers a wide spectrum of activities but a general definition is to work with existing businesses in an area to increase the likelihood that they will remain and expand there instead of relocate or expand elsewhere. BRE programs should be tailored to the specific needs of a community—one size does not fit all. However, there are several components that usually are included in BRE programs. They include business visitations and surveys to identify problems and issues; troubleshooting community-related problems that companies might have; and industry recognition/appreciation programs.

Regular visits to businesses and interviews with executives are important components of a BRE program. While BRE programs should be tailored to fit a community's unique situation, they should generally include the following components:

- Interviews and surveys with local businesses to detect trends, problems, or issues. It is preferable to conduct personal interviews with key executives from local businesses. This demonstrates the value the community puts on these businesses and facilitates better communication and information gathering. Surveys can also be used to gather data and they are relatively easy to create and administer, especially if they are internet-based. Often a survey followed by a personal interview is the best approach. Interviews and surveys give local businesses a feeling that the community cares, and if there is a problem an executive will likely identify it in the interview or survey. It is critical, however, that any problems identified be addressed immediately or the business will not fill out a survey or consent to an interview again. All local businesses should be visited and/or surveyed at least once a year. This can be a very large undertaking, especially in larger communities. Economic development agencies often solicit the assistance of board members and volunteers to help conduct the interviews and surveys.
- Industry appreciation functions such as annual banquets, etc. These typically provide a casual setting in which to interact with local businesses and provide an excellent opportunity for businesses to network and discuss issues of mutual interest.
- Initiatives to improve the general business climate. This might include streamlining permitting processes, making government agencies customer-friendly, etc.
- Sponsorship of meetings and forums to help companies deal with issues of concern to them. One example would be roundtable discussions with local operations or human resource managers. These work most effectively when the host provides talking points to get the discussion started, possibly over breakfast or lunch. However, there should be plenty of time for unstructured discussion to let issues bubble to the surface.
- An area industry directory listing all businesses in a community including manufacturing, service, retail, etc. These directories can help businesses find local products and services they need to operate effectively. Information for each business should include key executives, contact information, a brief description of the product or service, and the North American Industry Classification System (NAICS) number.
- Labor and wage surveys to help local businesses gauge their pay scales.

Some BRE programs even offer business consulting services. These services may include assistance in identifying export markets for local products or sponsoring informative sessions about market trends and specific topics of interest. One of the most valuable BRE services is helping local companies attain "lean and agile" production status to help them remain competitive in the global economy (see Case Study on p. 276).

Benefits and Advantages of BRE

As a part of the overall community and economic development program, the benefits and advantages of business retention and expansion are significant.

- *BRE is a cost-effective way to create jobs and keep local economies strong.* Existing firms are already located in the community. There is no need to incur travel expense, spend staff time, and mount large marketing campaigns as required when recruiting new firms. Economic development, just as any sales-related activity, is influenced by personal relationships. Companies want to feel comfortable that local officials and citizens support local businesses. Personal relationships and trust are important. Economic developers, committee members and many other local stakeholders can more easily establish these relationships with local managers than with out-of-town firms. They have the opportunity to do so because they live and work in close proximity.

Furthermore, when new companies move in, communities incur new expenses. Often new infrastructure must be developed, site preparation may be necessary and cash incentives may be granted. The expansion of an existing business in its current location is usually a less expensive proposition.

BOX 16.3 STRENGTHENING COMMUNITIES WITH LOCALLY OWNED BUSINESSES

Having locally owned businesses can generate numerous benefits for a community. In addition to the local serving nonbasic businesses (as described in Chapter 13), there is much opportunity to support development of enterprises serving not only local but other markets as well. Recent studies and resulting benefits are described as:

> The impact of a dollar spent at local businesses has a far greater impact than money spent at chains, big box, or formula retail. Local businesses yield two to four times the multiplier benefit (reiterative spending in an economy) as compared to nonlocal businesses. Local businesses have higher multipliers because they spend more locally. In other words, local management uses local services, advertise locally, and enjoy profits locally.
>
> (Phillips 2014)

An illustration is that of the locally owned community franchise, the Green Bay Packers (it's the only one with this form of ownership in the National Football League). Organized as a nonprofit with residents of Wisconsin serving as primary shareholders, the Green Bay Packers are a critical source of wealth and economic multipliers for Green Bay and will be around for many years—they can't leave (Schuman 2000).

Here's an additional reason to consider a locally owned business development strategy: strong local economic ownership improves a community's prosperity because this type of enterprises can help support transition to a more sustainable economy that simultaneously is locally and globally oriented (Phillips 2014).

The following provides a brief overview of strategies for encouraging local ownership of business:

1 Improve substitution start-ups where new, locally owned business is started to produce a product or a service which major local employers must presently import from out of state. Major employers stand to benefit from these enterprises through the provision of a ready, convenient source of supply which would reduce their need to carry excess inventory. New, locally owned enterprises would receive a temporary, sheltered, local market which can assist a company in its early start-up stages.
2 Conventional, entrepreneurial start-ups where local entrepreneurs proceed to organize a locally owned business on the strength of a new product or service idea designed for a variety of markets—local, national, and international. Likely sources of new business ideas could include university research and development centers, oriented toward encouraging local ownership.
3 Conversions of retiring owner businesses where existing healthy, local businesses find no likely or desirable conventional "outside" buyer and where local, "internal" management or management/ employee groups move to purchase the firm themselves (Phillips et al. 2013: 51).

Various mechanisms exist to encourage locally owned businesses. In addition to traditional ownership, Employee Stock Ownership Programs (ESOPs) can be used.

Sources

Phillips, R., Seifer, B. & Antczak, E. (2013). *Sustainable Communities, Creating a Durable Local Economy.* London: Routledge.
Phillips, R. (2014). Building Community Well-Being across Sectors: "For Benefit" Community Business. In S.J. Lee, Y. Kim and R. Phillips, *Community Wellbeing: Conceptions and Applications.* Dordrecht: Springer, forthcoming.
Shuman, M.H. (2000). *Going Local, Creating Self-Reliant Communities in a Global Age.* New York: Routledge.

- *Relationships with local executives can be leveraged to create even more jobs.* As noted above, key components of a good BRE program are regular visits and contacts with local businesses and executives. Through these visits, relationships and trust are developed. Local executives who feel supported and appreciated think of themselves as members of the economic development team. There are many examples of communities and local executives working together to entice companies that supply local firms, or with whom they have other business relationships, to move to the community. This strengthens the supply-chain relationship, makes both firms more competitive, and helps the local economy.

  Many communities leverage their relationships with local firms to "move up the food chain" and gain access to decision makers at parent company headquarters located in different cities or countries. If a company already has a successful branch operation in a community, they know that community is a good place for business and may be more inclined to expand existing facilities or locate new facilities there.
- *BRE can facilitate the recruitment of new industry.* Despite the difficulty and expense of recruiting new firms, it still is a critical component of the overall economic development process for most communities. Very often, local business executives are a community's best recruiters. In almost all cases, before a company makes a decision to locate a new facility in a community, representatives from that company or their consultants will talk at length with executives of local firms about operating a business there. A local executive who feels the community supports his business and has benefited from a BRE program will make a much better salesperson for the community. On the other hand, a disgruntled executive (such as the one with the pothole problem in the example above), will not be a good community advocate. All too often, unfortunately the latter is the case.

Companies make location decisions very carefully based on a myriad of factors—some are quantifiable (wage rates, transportation costs, etc.) and some are not as easily quantifiable (local work ethic, labor/management relations, etc.). Facilities are long-term investments for companies. The last thing they want to do is close them prematurely after a few years because of problems with the local business climate or global economic events. A strong BRE program sends a signal to

BOX 16.4 EARLY WARNING SIGNS OF A POSSIBLE BUSINESS RELOCATION, DOWNSIZING OR CLOSURE

- Declining sales and employment: Are they having issues with cost of production, struggling with supply-chain issues, or is employment declining just because their processes are becoming more automated?
- Obsolete buildings and/or equipment: Why aren't they spending money on the facility to maintain buildings and equipment?
- Ownership change—merger, buyout, acquisition: What plans do the new owners have for the local facility?
- Older industries and product lines: Do they need help with advancing their technology or will they require a complete retooling due to obsolete products?
- Excess production capacity: If they have not gone through a lean initiative (see Case Study), allowing them to produce more efficiently, why are they utilizing less of their production capacity?
- Complaints about labor supply, local business climate, infrastructure and services, etc. Have their complaints fallen on deaf ears, or is someone communicating the concerns to the right people?
- Expansions in other locations: Why aren't they choosing to expand in the local community? Are they landlocked or did they receive incentives to start a new facility elsewhere?

companies making location decisions, that a community will help support their profitability as conditions change in the future. With global markets and competition, product lifetimes are shorter and facilities become obsolete sooner. The willingness of a community to "partner" with local firms and help them meet their changing resource needs as products and markets change, increases the likelihood that facilities can be retooled to produce different products or services and avoid closing down. It is important for anyone involved in the economic development process to question what existing businesses would have to say when asked about the community. If "I don't know" is the answer, the community is likely in need of a better BRE program.

BRE Is a Team Effort

Successful BRE programs are usually team initiatives. It is not just the job of the chamber of commerce or economic development professional to operate the program. Issues affecting local businesses are varied and complex, often requiring special expertise. There are many specialty areas—such as utilities, permitting and regulation, government services, transportation, etc.—that may require attention from local government agencies or other private organizations.

BRE is most effective when undertaken by a team composed of representatives from economic development organizations, local government agencies, educational institutions, and other fields. Knowledgeable specialists from utilities, transportation departments, etc. should be on the team as necessary to help resolve technical issues. The multidisciplinary BRE team should work together and coordinate communications and activities. All members should assist in BRE activities, such as business visitations and surveys, and the team leaders should know whom to contact to handle a specific situation. The team leaders should always make sure that any problems or concerns reach the appropriate experts on the BRE team. This is often where the ball is dropped and, as a result, the issues and concerns fall on deaf ears.

How Successful Are BRE Programs?

As shown in Figure 16.1 above, the vast majority of economic development professionals report that their organizations are involved in some facet of business retention and expansion. Most of them would undoubtedly offer stories and anecdotes on the success of BRE programs if asked. In a survey of economic developers, Morse and Ha (1997) gathered information on the benefits and best practices of BRE programs. They found that 30% of the plans adopted in BRE written reports had been substantially or completely implemented and another 35% of the plans were being actively pursued. Among the organizations reporting implementation or pursuit thereof, the most common programs included (percent reporting on each program in parentheses):

- publicizing the area's strengths (72%);
- sharing information on business programs (66%);
- continuous BRE program (64%);
- labor training (53%);
- informing politicians of concerns (51%).

The survey respondents reported that the following factors, in their opinions, facilitated success in BRE programs:

- adhering closely to the BRE written plan;
- sharing the results with the community;
- assuring adequate funding and personnel for the program;
- specifying responsibility for program implementation;
- concentrating on key aspects of the BRE program and implementation rather than dilute the effort with too many efforts and initiatives.

Because of the many benefits and high return on the investment of resources and time, a BRE program should be one of the initial and primary functions in a community's overall economic development program.

BOX 16.5 HOW TO CONDUCT BRE INTERVIEWS AND SURVEYS

In regard to their operations and potential for relocation, downsizing, or expansion, obtaining sensitive information from local companies requires a diplomatic approach. This is why it is important to establish a relationship of trust with local executives. It is best to obtain BRE information in a personal interview or meeting as opposed to a mail survey. However, a survey can be used to gather routine statistical data before a personal interview, and to gather information from a broader cross section of local companies.

A good BRE program should track the health of local companies. Sample questions could be: How much are employment and revenues at the local facility fluctuating year after year? How much are they fluctuating at corporate headquarters, if applicable? In either or both cases, are they growing or declining? To avoid asking repetitive questions and to show they have done their homework, the BRE team should do thorough research into a local firm before initiating discussions with them. A good BRE interview is a conversation, not an interrogation. When setting up the interview, the BRE team should request 30–45 minutes of the interviewee's time and confirm that this is acceptable. The interview should never go over time unless the interviewee prolongs it.

The interview is composed of three parts: opening, main body, and closing. The opening is the icebreaker, in which the interviewer introduces himself, explains the purpose of the visit, and establishes rapport with the interviewee. Always assure the interviewee that the information shared will be kept confidential within the BRE team and ask permission to take notes. The main body of the interview is the information gathering portion. Rather than the interviewer simply reading questions and jotting down answers, the interview should be conversational. The interviewer should know the questions so well that he can ask them at the appropriate place in the interview (not always necessarily in the same order) without having to read them. Often, interviewees have one or two key issues they want to "get off their chest." If the interviewee seems anxious to get on to the next question, let that person lead the discussion and get to what is really important to them. In the closing, the interviewer thanks the interviewee for their time, once again gives assurances of confidentiality, and asks them to call if the BRE team or local economic development agency can ever be of assistance.

Below are sample BRE interview and survey questions. While each situation is unique, these are the types of questions that will uncover clues on company issues and potential future actions.

- What are the current levels of employment and production at this facility? How do these compare with previous levels?
- Is your facility adequate for current and anticipated future operations?
- How old is this facility and how long has it been in operation?
- What products do you make? Do you anticipate phasing out this/these products or adding new ones in the near future?
- What/where are your principal market areas? Where do you ship your product or provide your service?
- Is the company locally owned or do you have a parent company (who and where)?
- What do you consider to be the advantages of doing business in this area (labor, transportation, business climate, utilities, etc.)? The disadvantages?
- Are there local issues or problems affecting the success of your operations? How can we help resolve them?
- Do you anticipate any downsizing, closure, or expansion of this facility in the near future? Why and when?
- Do you have any suppliers or customers that might consider locating in our community?
- If a new company were looking to relocate to the area and asked you for information on the business climate in the community, what would you tell them?
- May we speak to or visit your corporate headquarters to express our appreciation for the economic contribution you make? Whom should we contact?
- Are there any other issues that you would like to discuss? Feel free to contact us anytime if issues arise.

Some of these questions may be sensitive and must be asked in a diplomatic way with the assurance of confidentiality. Some companies will be comfortable answering all the questions, while others will decline to answer some in spite of confidentiality. Any information obtained will be useful. For more guidance on appropriate questions for BRE interviews, see Sell and Leistritz (1997) and Canada (1999).

CASE STUDY: THE CENTER FOR CONTINUOUS IMPROVEMENT

In the course of conducting an economic development assessment in the Athens, Georgia region, a consultant identified a powerful example of a successful business retention and expansion program. During an interview at a local plant that employed several hundred persons, the manager stated that corporate headquarters had almost decided two years ago to shut down the facility due to low productivity. He went on to say that, if not for the Center for Continuous Improvement's help with increasing productivity, the axe would have fallen.

Hearing such a strong testimonial, the consultant investigated and discovered that the Center for Continuous Improvement was founded at the Athens Technical Institute in response to the needs of several local companies. These needs were discovered in the course of BRE interviews. The main symptom was that productivity was lagging at many of these facilities. In both manufacturing and service sectors, these companies' managers knew they needed to improve productivity and quality but lacked the knowledge and/or resources to implement an improvement program by themselves. At the suggestion of Athens Tech, several local companies agreed to combine resources and establish the Center for Continuous Improvement. The Center would work jointly with the companies to improve productivity and quality.

The Center became a research, training and implementation resource for Total Quality Management (TQM), also called "lean" operations. Lean can be described as the minimization of waste while adding value to the product or process. Lean distinguishes between *value-adding* activities that transform information and materials into products and services that the customer wants, and *non-value-adding* activities that consume resources but don't directly contribute to the product or service. In today's global economy, lean operations are key to remaining competitive and profitable (Stamm and Pittman 2007).

Each member company working with the Center was able to accomplish the following:

- Instill team culture throughout the facility.
- Create a vision and values.
- Perform cycle time studies resulting in improved quality, less waste, and higher productivity.
- Implement ergonomic practices to reduce injuries and cycle time.
- Improve customer service.
- Improve purchasing and material management techniques.

TQM or lean transformation is not just a one-time event; it is an ongoing process. The Center continues to serve as a focal point for research and training in the field. Member companies help each other, sharing ideas and visiting each other's facilities. This example demonstrates that productivity and quality can be dramatically improved in local companies by adopting TQM or lean operations, increasing the likelihood that companies will remain competitive, stay, and grow in a community. The revolution in management and production techniques required to achieve this can take place using local resources. Through BRE programs, economic development agencies can be a catalyst to make this happen. For more information on this case study, see Ford and Pittman (1997).

Keywords

Business retention and expansion, business interviews and surveys, business climate, industry directories, labor and wage surveys, early warning signs.

Review Questions

1 What are some common reasons companies decide not to expand in their present location or to relocate to another community?
2 What are the key components of BRE programs?
3 What are the advantages of BRE programs?
4 What are some key success factors in BRE programs?

5 What are some early warning signs that a company may be considering closing or relocating away from a community?
6 What are some important questions to ask in a BRE interview?

Bibliography

Birch, D. (1987) *Job Creation in America*, New York: The Free Press.

Canada, E. (1999) "Rocketing Out of the Twilight Zone: Gaining Strategic Insights from Business Retention," *Economic Development Review*, 16(2): 15–19.

Ford, S. and Pittman, R. (1997) "Economic Development through Quality Improvement: A Case Study in Northeast Georgia," *Economic Development Review*, 15(1): 33–35.

Morse, G. and Ha, I. (1997) "How Successful Are Business Retention and Expansion Implementation Efforts?" *Economic Development Review*, 15(1): 8–12.

Organization for Economic Cooperation and Development. (2002) *OECD Small and Medium Enterprise Outlook*, Paris: OECD.

Organization for Economic Cooperation and Development. (2013) *Financing SMEs and Entrepreneurs 2013: An OECD Scoreboard, Final Report*, Paris: Centre for Entrepreneurship, SMEs and Local Development.

Phillips, P.D. (1996) "Business Retention and Expansion: Theory and an Example in Practice," *Economic Development Review*, 14(3): 19–24.

Phillips, R., Seifer, B., and Antczak, E. (2013) *Sustainable Communities, Creating a Durable Local Economy*, London: Routledge.

Pittman, R. and Harris, M. (2007) "The Role of Utilities in Business Retention and Expansion," *Management Quarterly*, 48(1): 14–29.

Sell, R.S. and Leistritz, F.L. (1997) "Asking the Right Questions in Business Retention and Expansion Surveys," *Economic Development Review*, 15(1): 14–18.

Site Selection Online. (2007) *New Corporate Facilities and Expansion*. Online. Available at: <www.siteselection.com/issues/2007/mar/cover/pg02.htm> (accessed November 1, 2007).

Southern Business and Development. (2006) Available at: <www.sb-d.com> (accessed November 1, 2007).

Southern Economic Development Council (SEDC). (2006) *2006 Member Survey*. Online. Available at: <www.sedc.org> (accessed December 11, 2007).

Stamm, D.J. and Pittman, R. (2007) "Lean is a Competitive Imperative," *Business Expansion Journal*, March. Online. Available at: <www.bxjonline.com> (accessed December 11, 2007).

Connections

Explore strategies for increasing high-growth enterprises in your community with information from the Organization for Economic Co-operation and Development at: www.tem.fi/files/28938/High-Growth_Enterprises_What_Governments_Can_Do_to_Make_a_Difference.pdf

Learn more about employee-owned business structures with information from The Employee Stock Ownership Program Association at: www.esop associatiom.org

Support retention and expansion of minority-owned businesses in your community by connecting with resources at the Minority Business Development Agency at: www.mbda.gov

Go deeper with a guide provided by Entergy Arkansas for business retention and expansion: www.entergy-arkansas.com/content/economic_development/docs/Business_Retention_Expansion_Guidebook.pdf

Find the fundamentals of how to set up visitation programs, courtesy of North Dakota State University and Mississippi State University's Extension programs: www.ag.ndsu.edu/pubs/agecon/market/cd1605.pdf

17 Entrepreneurship as a Community Development Strategy

John Gruidl, Brock Stout, and Deborah Markley

Overview

Support and encouragement of entrepreneurship within communities is one of the most effective development strategies. This chapter provides fundamentals for implementing a strategy of supporting entrepreneurs and creating an entrepreneurial ecosystem for growing these businesses. It is not about how to start a business; rather, it is about how to create an ecosystem that nurtures new business startups within communities. The chapter describes possible actions that communities can take to bolster entrepreneurship and provides resources and examples of communities that are succeeding with this strategy.

Introduction

A profound change in community economic development strategy over the past decade has been the emergence of entrepreneurship. Now, as never before, community developers recognize that entrepreneurship is critical to the vitality of the local economy. This change in strategy is due to several factors. A primary reason is the impact of globalization in driving many manufacturing jobs to overseas locations and, thus, reducing the effectiveness of using industrial recruitment as a strategy.

Another factor leading to the rise of an entrepreneurship strategy is the evidence that entrepreneurs are driving economic growth and job creation throughout the world. For example, the National Commission on Entrepreneurship reports that small entrepreneurs are responsible for 67 per cent of inventions and 95 per cent of radical innovations since World War II (National Commission on Entrepreneurship 2001). On the international level, studies demonstrate the close connection between entrepreneurship and economic development. According to the Global Entrepreneurship Monitor Project, there is not a single country with a high level of entrepreneurship that is coping with low economic growth (Reynolds et al. 2000).

Furthermore, entrepreneurship has also succeeded in revitalizing many communities. For example, in Fairfield, Iowa, support from the Fairfield Entrepreneurs Association has sparked equity investments of more than $250 million in more than 50 start-up companies since 1990, generating more than 3,000 high-paying jobs (Chojnowski 2006). Another example is Littleton, Colorado which pioneered the concept of "economic gardening." The City of Littleton provides entrepreneurs, many in high tech enterprises, with extensive market information, either free or at low cost, and arranges networking opportunities with other entrepreneurs, trade associations, universities and think tanks for the latest innovations. Littleton created twice as many jobs in its first seven years of "economic gardening" as were created in the previous fourteen years (Markley, Macke, and Luther 2005). It is clear from these examples that an entrepreneurial strategy can lead to significant increases in new businesses, jobs, and private investment in a community. And this phenomenon of community or regional-supported entrepreneurship is not limited to the United States. Even in greater China, for example, some areas are much more entrepreneurial than others (Li, Young & Tang, 2010).

The goal of this chapter is to provide the fundamentals for implementing a community-based strategy of supporting entrepreneurs and creating a vibrant ecosystem for growing these businesses. Although this chapter is about new enterprises, it is not intended to provide a primer on how to start a business or prepare a business plan. Instead, the chapter is written from the perspective of the community developer or leader who seeks to expand the number of enterprise start-ups in the community and to create a culture of entrepreneurship among community residents. The main premise of the chapter is that communities can, over time, improve their entrepreneurial ecosystem by focusing on the needs and wants of entrepreneurs.

We use the term "ecosystem" because, just as in a biological system, entrepreneurs operate best when they have access to the resources that they need and operate in an environment that is supportive. Furthermore, to develop sustainable entrepreneurship in a community, it is rarely sufficient to change just one element in the ecosystem. Instead, community developers and local leaders must change several elements simultaneously. This is reinforced by the examples of communities such as Fairfield, Boulder (Colorado), Silicon Valley, and Iceland where a number of ecosystem elements are strong and have evolved more or less simultaneously. In this chapter, we define the various elements of a strong ecosystem and provide examples of how communities have improved their ecosystems with beneficial results.

BOX 17.1 TWO IDEAS ABOUT ENTREPRENEURSHIP

There are two popular, competing ideas about entrepreneurship. Joseph Schumpeter believed that entrepreneurship required the existence of new information to develop new offerings, while Israel Kirzner believed that only differential access to information was necessary to identify opportunities (Dahlstrom, 2013). In Schumpeter's view, entrepreneurship involved 'creative destruction,' a process where old, inferior offerings are destroyed and replaced by new, improved offerings. Kirzner, on the other hand, saw entrepreneurship as much more incremental in nature, producing improvements on existing offerings rather than radical replacements of current offerings.

The scope of entrepreneurship is wide. It can come from large businesses, new or small businesses, nonprofit organizations, and government. Business entrepreneurship is oriented toward encouraging for-profit business start-ups and growing existing small businesses, while social entrepreneurship is oriented toward meeting community needs. Both of these types of entrepreneurship are important for community development because the community can influence them and the community benefits directly from them (Dahlstrom, 2013).

Timothy R. Dahlstrom
Faculty Associate, School of Public Affairs, Arizona State University, USA

Source

Dahlstrom, T. (2013). *The Role of Business Counselors in the Human Capital Resource Acquisition of Entrepreneurs.* Ph.D. thesis, Arizona State University. Retrieved from ProQuest Digital Dissertations. (3559606).

Who Is an Entrepreneur Anyway?

We begin by considering the word "entrepreneur" which brings to mind different images for different people. Some people think of high-tech wizards like Steve Jobs of Apple Computer or Bill Gates of Microsoft. Others may think of a local quilt-maker running a home-based business or a "mom and pop" store struggling to survive in the downtown.

A useful definition of an entrepreneur is a person "who perceives new opportunities and creates and grows ventures around such opportunities" (Markley, Macke and Luther, p. 35). This definition reminds us to focus on the person, not the venture itself. Only about one in 10 American adults is currently an entrepreneur; that is, actively engaged in the process of starting an enterprise (Markley, Macke and Luther). Bill Koch's story in the sidebar provides an example of how a successful entrepreneur can have an enormous benefit to a community.

It is a common notion that any and all business owners are entrepreneurs. But this is not the case. The characteristics of an entrepreneurial business are innovation, growth, and a high degree of risk. Not all businesses have these characteristics. For example, business owners who are focused on managing their current enterprise and are not seeking new opportunities or growth are better categorized as business managers, rather than entrepreneurs.

The same applies to a person who decides to buy an existing franchise or retail store. While managing an existing business requires business skills and willingness to take some risk, it is not necessarily concerned with innovation and growth. Thus, it is not very entrepreneurial. However, a business owner, either new or existing, can become more entrepreneurial by seeing an opportunity, implementing a new idea, and taking on more risk in an on-going effort to grow the business.

BOX 17.2 MEET AN ENTREPRENEUR

Bill Koch, Santa Claus, Indiana

Bill Koch started working for his father's company in the tiny town of Santa Claus, Indiana after returning from overseas duty with the U.S. Navy during World War II. His father had built a small amusement park which he had named Santa Claus Land. At that point, the Santa Claus project wasn't much more than a mail-order house and the bare bones of a theme park. It had a Mother Goose train, some storybook characters and a toy shop with elves. It was America's first theme park, opening nine years before Disneyland.

Koch wasn't impressed by his father's Santa Claus undertaking. "I thought it was sort of a folly," he recalled. "I didn't think it would go anywhere." However, he started working in mail-order. The company had a catalog that offered Christmas gifts and unusual items that couldn't be found at stores.

"We decorated the carton with things that showed it came from Santa Claus, Indiana. It was unique. It went well for a few years, then it fizzled out. When stock came in and sold out, we couldn't get more," said Koch.

At one point the family decided to stock the warehouse with Erector Sets and Gilbert trains, expecting to make a small fortune—but lost a big fortune. "We didn't sell any because they could go to the local store and get them," said Koch. "We sold all of those items at about 35 cents on the dollar. That was the end of that."

As the mail-order part of the business was failing, Koch turned his attention to the theme park. He expanded Santa Claus Land to include a toy shop, a gift shop, a restaurant and exhibits. Rides were added and advertising got the word out. "People kept coming here by the thousands," said Koch. "There were no parking lots. The highway would be parked with cars for a couple of miles back in every direction. We would have as many as 10 state police directing traffic to get people in and out."

It was hard to make ends meet, much less grow the business. In 1955, Koch proposed an admission charge of 50 cents for adults, with kids free. "Every member of the family tried to talk me out of charging. I was so doggone stubborn I wasn't going to give in," he said. It paid off. Visitors paid up. The revenue enabled the park to grow and add new attractions.

Santa Claus Land grew to include more shops, shows, a wax museum and rides. The site also served as a community center for meetings and meals. People lined up for chicken and turkey dinners, and exotic fare such as Baked Alaska. The four initials for the dining room were F.F.F.F., which stood for Famous For Fine Food. With the growth, new problems – or new opportunities, as an entrepreneur would say – arose. There weren't any motels nearby and the cabins and campgrounds at Lincoln State Park, five miles away, often were full. Koch saw it as an opportunity. In 1958, he opened a 150-site campground, Lake Rudolph Campsites, across the lake from the theme park.

Koch didn't stop there. He built a gated subdivision called Christmas Lake Village with lakes, tennis courts and a golf course. He built a town hall, a medical center and a bank for the community that grew from 37 to 2,000 residents. He served on 27 boards at one time. He was instrumental in getting a major interstate highway that changed the face and the economy of Southern Indiana.

He had a passion for making the community better, a sense of where the opportunities were, and the courage to take calculated risks. He networked and acquired the skills he needed to take an evolving dream and grow it into successful businesses that also improved the town.

Although he died in 2001, his son Will serves as President of Holiday World that today encompasses the family entrepreneurial business. Holiday World has more than 1 million visitors each year and employs more than 1,000 people in the summer. Bill Koch stands as a great example of how a creative, spirited entrepreneur can not only build successful enterprises, but can also serve the community by contributing to its economy and quality of life.

Source: Holiday World and Splashing Safari website:
http://www.holidayworld.com/news/holiday-worlds-patriarch-leaves-legacy

The Entrepreneurial Talent Pool

Rather than debate who is and isn't an entrepreneur, it is more useful to consider a community as having a pool of entrepreneurial talent (Markley, Macke, and Luther). This idea of entrepreneurial talent recognizes that not all entrepreneurs are the same. It also recognizes that there are many people who are interested or active in starting businesses or launching new markets for their current business. Entrepreneurial talent ranges from the stay-at-home mom who is researching a new home business idea to the high-growth entrepreneur who is considering an offer from a venture capitalist to a current business owner who would like to expand into a web-based store.

Table 17.1 shows the different types of entrepreneurial talent. The four main categories of entrepreneurial talent are potential (those who may become entrepreneurs), existing business owners (some of whom may be innovating into new services, products, or markets), business entrepreneurs (including those with growth and even high growth potential), and social entrepreneurs (those creating enterprises for social betterment). Entrepreneurs in all these categories are people who community leaders want to identify and nurture.

It is important to recognize that the needs of entrepreneurs are not all the same. For example, aspiring and start-up entrepreneurs need someone who can help them move from ideas to a solid game plan. They need help making sure all the pieces are in place. Is the management team strong? Is there capital to start the venture? Are markets clearly identified and strategies for tapping them tested? These needs can be met with moral support, networking and mentoring, business counseling, and entrepreneurship training.

Table 17.1 *The Entrepreneurial Talent Pool*

Types	*Characteristics*
Potential – Aspiring	Actively considering going into business Researching a business idea Motivated toward a life change May include youth (under age 25) some of whom are still in school
Potential – Start-ups	In the process of starting a business May or may not have a good plan May or may not have the necessary skill
Business Owners – Survival	Struggling to make enough income Often stressed and reluctant to seek help Don't have the time or energy to see new markets or opportunities
Business Owners – Lifestyle	Successful and well established Not actively seeking to grow or change the business model
Business Owners – Re-Starts	Have been in business before with limited success In the process of starting another business Determined to succeed this time Willing to seek outside help
Entrepreneurs – Growth-Oriented	Successful in business Have a growth orientation and drive Seeking to be more competitive, actively seeking new ideas Actively seeking new markets, processes, products, services
Entrepreneurs – Serial	History of creating and growing more than one business Don't like to mange existing businesses, often sell business when it is up and running Generally on the lookout for new ideas and opportunities
Entrepreneurs – Entrepreneurial Growth Companies	Experiencing rapid growth in employment or sales Reaching new markets, developing new products and services Innovative and dynamic leadership and workforce Sometimes referred to as "gazelles"
Social Entrepreneurs	Seeking new, innovative solutions to social problems Must follow solid business practices, including effective financing, marketing, and product development (just as business entrepreneurs do)

Source: Developed by the authors drawing from Markley, Macke, and Luther, Chapter 4.

Lifestyle business owners might include a small retail store or a family medical practice. They may not desire to grow the business at this time. Often the motivations of lifestyle entrepreneurs can change unexpectedly – when a son or daughter returns to town and wants to be involved in the business or when the business owner is faced with an opportunity such as the potential to expand into an adjacent storefront on Main Street or to take the business "online." Business owners who are motivated to grow their businesses need many of the same support services as start-up entrepreneurs – networking opportunities, training to build their skill sets, one-on-one assistance with specific business issues, such as creating a new website, developing e-commerce tools or tapping new markets.

Since growth-oriented entrepreneurs are often interested in developing a new product or cultivating a new market, they may need knowledge of capital sources, assistance with marketing or expanding production. They may need to expand the management team to encompass new skills. As their success has grown, their support needs have changed from broad forms of general assistance to very specific, targeted business information needs. To effectively support these growth-oriented entrepreneurs, support services should focus on customized assistance, often of a technical and specialized nature.

As part of the discussion about entrepreneurial talent, let us look at social entrepreneurs. Social entrepreneurs create programs and resources that benefit our communities and our lives. They develop children's museums, organize a new chapter of Big Brother/Big Sisters, provide public health care, and build new playgrounds and parks. Often, this work is done through a non-profit organization or informal community or neighborhood association. Social entrepreneurs need skills in planning their enterprise, marketing their product or service, earning revenues or obtaining funding to keep the organization financially solvent, and creating value. As with business entrepreneurs, they perceive and act upon opportunities.

BOX 17.3 CHANGE AGENTS: SOCIAL ENTREPRENEURS

Entrepreneurs can operate in any sector, whether private, public or nonprofit or nongovernmental. Tackling social issues and challenges with new and innovative ways to effect positive change needs collaborative effort across sectors. Some consider that lines across sectors have become blurred, with a social entrepreneurship emerging to help address complex social issues. Heerad Sabeti, of the Fourth Sector Network, provides a description of these activities at the cross section:

a The convergence of organizations toward a new landscape – a critical mass of organizations within the three sectors has been evolving, or converging, toward a fundamentally new organizational landscape that integrates social purposes with business methods;
b The emergence of hybrid organizations – pioneering organizations have emerged with new models for addressing societal challenges that blend attributes and strategies from all sectors. They are creating hybrid organizations that transcend the usual sectoral boundaries and that resist easy classification within the three traditional sectors (Sabeti, 2009, 2).

Whichever sector is home to a social entrepreneur, it is evident that these individuals are seeking innovative solutions for effecting positive change. Ashoka describes them this way,

> "just as entrepreneurs change the face of business, social entrepreneurs act as the change agents for society, seizing opportunities others miss and improving systems, inventing new approaches, and creating solutions to change society for the better. While a business entrepreneur might create entirely new industries, a social entrepreneur comes up with new solutions to social problems and then implements them on a large scale".
>
> (Ashoka, 2013)

In other words, these are individuals and their organizations involved in non-market entrepreneurship, or pursuing opportunities not solely for profit maximization. This may encompass such diverse entrepreneurial activities as public sector entrepreneurship, policy entrepreneurship, social enterprise, nonprofit entrepreneurship, and philanthropic enterprise (Shockley, Frank, & Stough, 2008). More communities are recognizing the value of supporting social entrepreneurship, as it contributes to resiliency and durability; for an example of a community that is highly committed to this, explore Burlington, Vermont's programs, policies, and outcomes (Phillips, Seifer, and Antczak, 2013).

The Editors

Creating an Entrepreneurial Ecosystem

The entrepreneurial talent in a community does not exist in isolation, but instead is connected to many people and organizations in the community, from the local banker to the Internet provider to the local business counselor. This external environment has a major impact on the prospects for the entrepreneurs. We use the term "entrepreneurial eco-system" to describe an environment that is nurturing for emerging businesses. A biological ecosystem is a system of elements such as air, water, soil, nitrogen, light, and other organisms that are interconnected and interact together. Some ecosystems in nature are healthier than others. In a similar way, elements and actors in human communities can interact to provide the growth that entrepreneurs need.

Creating the necessary ecosystem is a huge challenge, but the payoff can also be huge: an economically viable oasis community in a desert of industrial hollowing out. In Entrepreneur Magazine's list of the top ten entrepreneurial ecosystems, four are outside the U.S. (Davis, 2013). Tel Aviv is second behind only Silicon Valley. Also included are London (seven), Toronto (eight), and Vancouver (nine), suggesting the world-wide importance of entrepreneurship.

The entrepreneurial ecosystem requires many things, but can be understood through four categories of elements: Culture (sunlight), Human Capital (fertile soil), Support Systems (rain), and Momentum (surrounding plants that are sprouting and growing). Eleven important conditions are described here as part of the four ecosystem elements, offered with the caveat that even though not all conditions can be changed by community developers, awareness of the conditions is crucial to an entrepreneurship development initiative.

Human Capital is represented by fertile soil

Human capital refers to the skills, abilities, education, and experience of those living in the community.

1 Entrepreneurial skills

One vital skill set is the ability of individuals to start ventures and to marshal others in the community in the *bricolage* process. Bricolage refers to the entrepreneur's ability to create something out of resources on hand. True entrepreneurs don't complain or give up when they don't have available resources. They use the available means and build something out of those means. Conversely, a community that has vast resources for entrepreneurs but no entrepreneurial people to take advantage of those resources, or skills to assemble them into a new enterprise, will not succeed.

2 Presence of skilled workers

Once people muster up enough courage to start innovative companies, they must be able to hire workers to code the software, operate the machines, generate press publicity, train other employees, and produce more ideas for future improvement. Potential entrepreneurs usually consider worker availability before deciding if a community would suit their potential business. A region with an aggressive workforce development policy will make entrepreneurs feel more positive about their likelihood of success. Russian entrepreneurial levels were down because business owners in the Soviet era were "often deemed criminals for making a profit" (Aidis, Estrin, and Mickiewicz, 2007, p. 5). Yet Moscow's entrepreneurial ecosystem is now viewed as strong (Davis, 2013), as a talented workforce is available.

Culture is represented by sunlight

Culture is "a set of attitudes, beliefs, customs, mores, values and practices which are common to or shared by any group" (Throsby, 2001, p. 4). It applies not just to a nation, but also to communities. Each community has its own set of norms, attitudes, and practices regarding entrepreneurship. Building an entrepreneurial culture in the region should accompany efforts such as expanding technical and consulting assistance to entrepreneurs (Gibb, 1999).

3 Perceived desirability of being an entrepreneur
Social norms can encourage entrepreneurial motivation. In your community, is entrepreneurship thought of as a "cool" career? This perception affects all age groups, but could be more critical to startup rates among younger people. Thus altering this perception is particularly useful for encouraging youth entrepreneurship.

4 Community support
Just as a company's workers need to feel they have support from management in order to try new things, entrepreneurs need to feel their communities will support them. Do community leaders have their backs? Ambitious entrepreneurs require less support than most people, but if the goal is to increase the percentage of people trying out business ideas, encouraging the perception of support is beneficial. Communities can communicate their support for entrepreneurs and potential entrepreneurs and encourage openness and generosity with time and advice, through mentoring and networking initiatives.[1] In successful communities, there is a widespread belief that cooperation and building synergy will help everyone.

5 Risk and failure tolerance
Conventional thinking says that entrepreneurs are people who love to take risks, but contrary to this common conception, entrepreneurs may in fact be more risk averse than non-entrepreneurs (Xu and Ruef, 2004). Because fear of the consequences of failure discourages action, communities that encourage a little risk are more likely to increase the number of participants in the ecosystem. Communicating toleration of failure could help, but training people to ignore failure intolerance might be just as important. Venkataraman and Sarasvathy (2008) offer a solution: a community should be "empathetic towards the entrepreneurial process and evolve from treating setbacks as 'failures' to one that considers them to be 'options' 'experiments' and 'learning opportunities' and move away from a language of 'failure', 'biases', and 'risk.'" (p. 13). Of course, this is a major shift in mental models for many communities and requires a long, sustained effort. One step may be for community leaders themselves to talk about their initiatives in entrepreneurship-building as experiments, not guaranteed to succeed. However, even if the initiatives do not immediately succeed, the leaders should reassure residents that the community will learn from them and continue on the path toward a strong ecosystem.

Support System is represented by the nourishing rains

An effective support system for entrepreneurs has many players, including investors, local governments, universities, and service providers. The players have different roles within the system, for example, banks and investors supply financial capital, universities create new technologies that might be commercialized, service providers provide business counseling, and schools offer entrepreneurial training. Each community must develop its own set of program offerings based on the needs of its entrepreneurial talent pool. Among the key offerings to consider are entrepreneurial education, access to capital, and access to networks (Pages, 2005).

6 Networking support

Fogel (2001) asserts that new entrepreneurs spend almost half of their time networking as they seek support and motivation, counseling, access to opportunities, resources, and other needed information. Research finds that "the diversity of individuals' relationships is strongly correlated with the economic development of communities" (Eagle, et al, 2010, p. 1029). Providing opportunities for creative people to network is critical, but the networking programs need to be substantive, not just chats with the same people that already know each other. Feld (2012) cites some good examples of efforts: "hackathons, new tech meetups, open coffee clubs, startup weekends, and accelerators," and open office hours from successful entrepreneurs and financiers.

7 Positive regulatory environment

Of course balance is important, as protection of creditors and enforcing of contracts is necessary. Moreover, governments can compensate for lack of necessary institutions in transition economies and turbulent environments (Smallbone, et al, 2010). But too many rules and procedures can de-motivate entrepreneurs (van Stel, et al, 2007). Entrepreneurs often find it difficult to navigate through the various regulations and requirements. Also they can become overwhelmed in trying to determine which agency to contact for technical assistance on a specific issue. Communities can help with a one-stop shop that provides assistance with complying with regulations, as well as a directory of various service providers, such as the local Small Business Development Center.

8 Access to customers

If a person opens a retail store, will that store have enough potential customers to enter and browse? Likewise, manufacturers will need access to freight transportation, web-commerce startups will need reliable internet access, and public relation firms will need proximity to companies requiring their services. Access to transportation and other infrastructure can bolster both the courage to start and the likelihood of success. Smart entrepreneurs consider this issue in their plans. They might also consider the economic health of the region in deciding if enough customers will be able to afford their offerings. Some communities that cannot provide sufficient customer access bring in broadband and ecommerce infrastructure and training.

Momentum is represented by surrounding plants that are healthy and vibrant

Momentum creates the feeling that "people can succeed around here." Once a community has some entrepreneurial movement, energy will pull other potential entrepreneurs into startup mode. Just as organisms in an ecosystem encourage other organisms, successful companies help buoy up potential entrepreneurs.

9 Perception of employer enterprise birth rates

Startups that make the leap into hiring staff are different from "solopreneur" efforts. Potential entrepreneurs are encouraged by seeing the presence of companies growing large enough to create employment. An obvious action by communities is to publicize the growth of companies and their hiring of staff.

10 Business churn: calculation of birth and death rates

How fast are new companies being created and shuttered? Is the net calculation positive or negative? The answer is also based on perception, but perception is what encourages or hinders entrepreneurial motivation. Promoting positive stories and data on business churn might have an impact on that perception.

11 Positive role models

Some communities have more entrepreneurs than others, and that "proximity effect" has an impact.

Entrepreneurial motivation is benefited by the presence of other entrepreneurs (Giannetti & Simonov, 2004). This is a huge challenge because it requires momentum, as someone needs to succeed first in order to act as a role model for others. Shining a light on those who do exist might be a place to start: communities can publicize the successes they do have. Just one shining example can trigger momentum, such as Erik Hersman in Kenya. Hersman is a pioneer in technology, technology, and humanitarianism who developed iHub, an Innovation Hub in Nairobi, an open space for technologists, investors, and tech companies.[2]

In summary, an entrepreneurial ecosystem requires four elements, human capital, culture, support system, and momentum to function optimally. Community developers should recognize that in a dynamic economy, churn will occur. Choosing to support only the most promising ideas and enterprises is never a successful strategy. Leaders should encourage everyone to participate in the ecosystem. The more participants, the more winners, and the better off everyone will be.

Communities such as Fairfield, Iowa, a small town of only 10,000 people which is profiled in the sidebar, show that it is possible to develop a vibrant ecosystem and that the results in economic growth and quality of life can be staggering. To create an entrepreneurial ecosystem, communities must demonstrate their own willingness to experiment to find the best approaches that work locally.

BOX 17.4 THE BOULDER THESIS

Boulder, Colorado is a city that has become famous for the remarkableness of its entrepreneurial ecosystem. Brad Feld, successful venture capitalist and evangelist for Boulder's accomplishments, has published his account in a book entitled "Startup Communities: Building an Entrepreneurial Ecosystem in Your City."

The Boulder story could be summarized by an evolution through stages. First, five different companies were successful, and they spawned experienced executives in their respective industries, creating five separate and parallel entrepreneurial ecosystems in biotech, technology, natural foods, clean technology, and a lifestyles and sustainability industry. This created some entrepreneurial momentum before the entrepreneurial culture began swelling. Next, some of those entrepreneurs founded additional companies in the five industries.

The effort then gained momentum as serial entrepreneurs wanted to stay, and local investors started realizing they had good investment options at home. Finally, TechStars, an accelerator, multiplied startups numbers. Because of the existing momentum of these forces, Boulder's experience may not be completely replicable in other towns, which is why we recommend that communities experiment to discover what kind of efforts will fit their needs.

Feld coined the term "Boulder Thesis," which he sees as a "framework for creating a vibrant, long-term startup community." The four tenets of this thesis are that:

"Entrepreneurs must lead the startup community."
Civic leaders can be helpful, but people who have experienced the entrepreneurial journey are better at providing guidance to those beginning the journey. Leaders are not hierarchal heads, but those who emerge organically. Committed non-entrepreneurs are feeders, and they make an important contribution. The feeder group includes investors, governments, universities, service providers, and anyone else who adds value to the ecosystem.

"The leaders must have a long-term commitment."
Building the ecosystem takes about 20 years. Leaders can't focus on entrepreneurship for a few years then become distracted by something else when the economy starts dragging. Anyone throwing in support should recognize how long the commitment they are making is.

"The startup community must be inclusive of anyone who wants to participate in it."
Leaders can't choose which entrepreneurs will be successful, so they should encourage everyone to participate.

"The startup community must have continual activities that engage the entire entrepreneurial stack."
These activities need to be substantive, not just talk, for all involved, such as investors, educators, business leaders, and mentors. Some activities are more effective than others, so experimenting is important. Governments can be supportive, but should not get in the way.

Source

Feld, B. (2012). *Startup Communities: Building an Entrepreneurial Eco-system in Your City*. New Jersey: John Wiley & Sons.

CASE STUDY: FAIRFIELD, IOWA: AN ENTREPRENEURIAL SUCCESS STORY

In 1989, a group of entrepreneurial businesspeople formed the Fairfield Entrepreneurs Association (FEA). The FEA was designed to increase the success rate of start-up companies and nurture companies in the second stage of development; that is, after they had generated over $500,000 in annual revenues. The FEA supports entrepreneurs through recognition and awards, acceleration of second stage companies, and mentoring and networking activities for entrepreneurs. Fairfield is characterized by extensive sharing of information on the "how-to's" of business start-up financing and marketing. It has developed a pool of shared wisdom and experience and a culture of guarded openness about business ideas. The result is great synergy among entrepreneurs, overlapping, and sometimes copy-cat business models.

For example, an entrepreneur named Earl Kappan, after several other start-up ventures, started a company that became known as "Books Are Fun." The company sold best-selling hardcover books via book fairs in schools, hospitals, and businesses across the United States. With financing from an outside investor, the company grew rapidly and developed a network of sales representatives across the country. A synergy occurred when local author Marci Shimoff proposed a collaborative venture to Kappan to explore the potential for distributing a book through Books Are Fun. The book was *Chicken Soup for the Mother's Soul*. Books Are Fun tested the book and started buying the title in lots of tens of thousands, propelling the book to the top of best seller lists.

Other Fairfield residents developed Chicken Soup topics that went on to become *Chicken Soup for the Pet-Lover's Soul, Chicken Soup for the Gardener's Soul, Chicken Soup for the Veteran's Soul,* and more. Approximately 13 Chicken Soup authors live, or formerly lived, in Fairfield. Another resident designed covers for the Chicken Soup series and has become a sought-after book jacket designer. Another Fairfield resident is a leader in self-publishing enterprises. The original enterprise Books are Fun also did well. When it grew to more than 500 employees and $400 million in annual revenue, it was purchased by Reader's Digest for $380 million.

There are many more examples of success in Fairfield among financial services, e-commerce, telecommunications, and art-based businesses. Burt Chojnowski, President of the Fairfield Entrepreneurial Association, attributes much of the success to the networking and sharing among businesses. Chojnowski states that "having access to this specialized knowledge, especially about funding opportunities, was clearly a competitive advantage for entrepreneurs and improved the financial literacy and sophistication of Fairfield entrepreneurs" (p.3–4).

The FEA, in conjunction with other Fairfield organizations, offers additional support to entrepreneurs. Entrepreneurial boot camps and an up-to-date library are available, including specialized information for "art-preneurs," social entrepreneurs, and "food-preneurs." Entrepreneurship training has been expanded to include workshops for youth entrepreneurs. The results have been amazing for a small town. Since 1990, equity investments of more than $250 million have been made in more than 500 start-up companies, generating more than 3,000 jobs. This entrepreneurial ecosystem with strong organization and culture has produced striking results.

Source: Chojnowski (2006).

Community Options to Improve the Ecosystem

There is no magic bullet, no single set of actions that a community should adopt to improve its entrepreneurial ecosystem. As we have emphasized, much depends on the entrepreneurial talent and the current support system and culture within the community. However, in this section, we provide some ideas to spur action. Since communities are at different stages in developing their ecosystem, we provide two sets of options. The first set consists of actions for communities in the starting phase. The actions are, generally speaking, not very expensive, yet can have an immediate impact.

Actions for Getting Started

1 Create Legitimacy through Conversation and Storytelling. A good first step is to start a conversation with the community before you begin any intervention. Build a vision for an entrepreneurial ecosystem. Warn people that discovering the initiatives that fit your particular community will require experimentation, and that means some initiatives will probably fail. Some individual ventures may also fail, but "now we allow failure here, so go ahead and try." Also, seek support for the belief that hitting chuckholes won't stop the gradual growth of the ecosystem. This will start the culture in the right direction, and preempt the naysayers. The content of the conversation can also eventually become a message for evangelizing the community to the outside.

2 Form an Entrepreneurship team. If there is little current support for entrepreneurs in your community, you can start the ball rolling by forming a team of leaders and interested citizens to lead the entrepreneurial effort. It may be that an existing organization, such as the local economic development group, has an interest in entrepreneurship and could be the umbrella organization for the new team. Once the team has come together, one of its first actions would be to examine what the community is currently doing to foster entrepreneurship.[3]

3 Talk to Entrepreneurs. Yes, this is an obvious one, but it is critical to build your initiative on the needs and wants of entrepreneurs. Conversations with entrepreneurs may reveal common needs and concerns and clarify specific actions for the community team to take.

4 Create Awareness of Entrepreneurs. Create a focus on entrepreneurs. This might include raising the awareness level of community residents and leaders about the role of entrepreneurship within the community. Going a bit further, a community might identify entrepreneurs and provide periodic recognition for their contributions to the community. An Entrepreneur of the Year award presented at the Chamber of Commerce banquet would be a specific example. This recognition is important because it helps residents to see the importance of entrepreneurs. Gradually the culture can shift to one that is supportive of people who start new businesses.

5 Connect to Business Services. Take stock of your current access to appropriate business services (e.g., legal, marketing, production, financial, accounting). Access to the right services is important. Remember, if these services are not available within the local community, as is sometimes the case in rural areas, they can be accessed over long distances using today's technology.

6 Networking and Mentoring. Entrepreneurs at all levels of venture formation benefit from a network of peers. Entrepreneurial networks can be formal or informal. An example of a formal network would be monthly forums sponsored by the chamber of commerce that offer an opportunity for entrepreneurs to meet their peers and share information about service providers, markets, or frustrations about doing business. A network is a great place for an

aspiring entrepreneur to get moral support as he speaks with others who have already traveled the path to a successful business.

However, networking does not have to be formal. Entrepreneurs will network whenever they come together in one place. An informal breakfast meeting for young entrepreneurs, a friendly Friday happy hour for entrepreneurs on Main Street can provide opportunities for entrepreneurs to develop connections and identify the resources and support they need from their peers. In many cases, simply providing a venue for entrepreneurs to come together is all the work the community needs to do – the entrepreneurs often take it from there.

Mentoring programs can be effective in strategically linking an experienced entrepreneur with an aspiring or start-up entrepreneur. Mentoring can happen organically. For example, experienced Hispanic restaurant owners in Hendersonville, North Carolina "adopted" new immigrants who were interested in starting restaurants or catering businesses to help them learn the ins and outs of the sector. Mentoring programs can also be established by creating a pool of experienced entrepreneurs who are willing to work with new entrepreneurs in sectors where they have expertise (Markley, Macke and Luther).[4] Low-population regions without breadth of potential mentors might be able to establish virtual mentoring by recruiting mentors from outside the region who can use internet and telephone to communicate with local potential entrepreneurs.

More Advanced Options to Improve the Ecosystem of Your Community

Once the basic elements of an entrepreneurial ecosystem are in place, a community can consider a number of advanced activities to further energize entrepreneurs. Remember, more advanced support doesn't mean that things should become more complicated for the entrepreneur. Massive directories and complicated pathways for entrepreneurs to access support can be counterproductive. We urge communities at this level to create some kind of simple organization (probably using existing organizations) to ensure that entrepreneurial support efforts are understandable, easy to access and seamless.

1 Entrepreneurship Training and Business Counseling Programs

Aspiring and start-up entrepreneurs can often benefit from participation in training and counseling programs, either one-on-one or with other entrepreneurs. Small Business Development Centers (SBDC) are an important resource in providing both training and counseling opportunities to aspiring and start-up entrepreneurs. Furthermore, a number of well-tested "how to" training programs take entrepreneurs through the process of starting their own businesses. FastTrac (http://fasttrac.org/) and NxLeveL (http://www.nxlevel.org/) are examples of programs that provide an excellent curriculum for entrepreneurs who are in the early stage of building their ventures. Community developers and leaders need to be closely linked with the Small Business Development Center and other regional providers of training and counseling.

Counseling programs may be more appropriate for entrepreneurs who have already developed a business plan but need assistance with specific aspects of the business. For example, an entrepreneur might need assistance in accessing export markets or in understanding the licensing requirements for operating a commercial kitchen. These types of questions are best addressed through the services of a business counselor who works one-on-one with the entrepreneur. Community developers need to know enough about an entrepreneur and her skills so that you can steer her toward the training program or business counselor that can best meet her needs.

2 Youth Entrepreneurship

How can young people even learn about business opportunities? Networking with entrepreneurs and business owners is an excellent way to open the door to entrepreneurship. A mentoring or apprenticeship program which matches a business owner to a young person in a one-on-one mentoring relationship gives the potential entrepreneur the chance to see how business is created in the real world. This, at least in part, may provide the young person with the insight and encouragement for choosing the entrepreneurial path. A first step in getting a mentoring program going in your community might be to invite community entrepreneurs into the classroom. Entrepreneurs who tell their stories in the schools are providing the role models these young people need to envision an entrepreneurial future for themselves.

How can youth acquire the skills that they need to become successful business entrepreneurs? Traditionally, K-12 education has focused on preparing students to be good workers, rather than successful entrepreneurs. However, this is beginning to change as more schools are offering activities, such as Junior Achievement, that provide youth with the knowledge and practice of business. Some schools have gone even further by incorporating notable entrepreneurship training programs in the curriculum. A successful example is the Rural Entrepreneurship through Action Learning (REAL) program which is now being used in 43 states and foreign countries. In the REAL program, youth create viable businesses that generate on average 2.2 jobs and sometimes move out to a permanent site in the community (Corporation for Enterprise Development.)

Another excellent youth training example is the CEO program, (Creating Entrepreneurial Opportunities, http://www.effinghamceo.com/) in Effingham, Illinois. CEO was developed by business leaders in the community to train the next generation of business leaders. They funded the program for three years and hired a dynamic individual to design and deliver the training in the local school.

The effort was designed to be a "stretch" for the community. The program conformed to school regulations for curriculum but was designed by entrepreneurs. Inclusion by juniors and seniors is by application, and has become quite competitive. The program soon spread to neighboring counties, and CEO is now partnering with schools in other regions to help them create similar systems.

Youth entrepreneurship camps are another approach to bring youth together with experienced businesspeople for intensive training and guidance in starting a new business. For example, the youth entrepreneurship camp sponsored by the University of Wisconsin Small Business Development Center includes 40 hours of training in real-world business skills, team building, leadership development, financial management, verbal communication, and business etiquette. Participants also learn how to successfully negotiate for business materials, set goals, and recognize real business opportunities (University of Wisconsin Small Business Development Center.)

3 Provide Customized Assistance (to the full range of local entrepreneurs)

This might involve entrepreneurial coaches to work one-on-one with aspiring or start-up entrepreneurs. It might involve the city and library providing extensive and specialized market information to growth-oriented entrepreneurs, as the City of Littleton, Colorado does. This customized help may require significant investment by local community organizations.

4 Create "Angel" Investment Networks

Communities can build on current financing resources by creating area-based "angel" investment networks and pathways to more traditional venture capital resources (which may be external to the community). Sooner or later, growing ventures need more sophisticated forms of capital including access to equity capital. As entrepreneurial deals emerge and grow, the ability to help these ventures meet their capital needs is critical to keeping these businesses within the community.

5 Design a High-Capacity Organization Dedicated to Supporting Entrepreneurs

These entrepreneurial support organizations, such as the Fairfield Entrepreneurial Association, are rooted in communities and provide a comprehensive and sophisticated package of support that energizes start-up entrepreneurs and develops entrepreneurial growth companies.

6 Start an Accelerator

An accelerator is very different from an incubator (smaller communities with fewer mentors might be better off starting an incubator). An accelerator raises a small fund, then brings together three groups: mentors (executives from established firms, etc.), high-potential entrepreneurs, and investors. The program is of limited duration, such as 90 days. The three groups work together to create knock-out companies. The accelerator invests seed money, then venture capitalists invest larger sums. For communities short on resources, an accelerator can be sponsored by a university, using the school facilities and resources. Often people will borrow large sums of money to realize their dreams of owning a business, but face bankruptcy when they learn the business is not needed by the market. Accelerators help potential entrepreneurs to test ideas before funding and launching.

Keep in mind that entrepreneurs need better networks, not simply more programs. It is vital to establish relationships with entrepreneurs and to be responsive to their needs. Adding more complexity and more layers of service is not going to lead to success. Instead, having a clear pathway to available services and helping to establish networks will bring your community closer to having a vibrant entrepreneurial ecosystem.

The references at the end of this chapter provide useful resources as you take these next steps in your community.

Conclusion

Communities are recognizing that entrepreneurship is an important strategy of community economic development. However, as has been seen in this chapter, entrepreneurs are not all the same. Communities must be flexible and responsive to meet the varying needs of entrepreneurs. It is essential that support be tailored to fit the needs and wants of entrepreneurs, rather than based on satisfying an external agency or funding source.

An entrepreneurial ecosystem involves four major components: human capital, culture, support system, and momentum. While fully developing an ecosystem requires a long-term commitment, there are many options for communities. To get started, communities can design networking and mentoring opportunities for entrepreneurs. Indeed, other entrepreneurs may be the most valuable source of information and ideas! At more advanced levels, communities can arrange customized assistance and counseling, provide entrepreneurs with training opportunities, establish "angel" investment networks, create an accelerator, and encourage entrepreneurial curricula in local schools.

Entrepreneurship has always been the engine that brings new vitality to places in the U.S. and around the world. Yet it has been largely overlooked as a development strategy. Community developers will be surprised at the potential of local entrepreneurs. As the community learns to support and honor entrepreneurship, remarkable and surprising things can happen.

Keywords

Entrepreneur, entrepreneurial ecosystem, social entrepreneur, youth entrepreneurs, potential entrepreneurs, business owners, lifestyle entrepreneurs, innovation, small business development centers, economic gardening, micro-enterprise, microenterprise, the Boulder thesis.

Review Questions

1 How would you define "entrepreneur?"
2 What are the different types of entrepreneurs?
3 What are the needs of these different types of entrepreneurs that the community can provide to support them?
4 How can a community discover current entrepreneurial strengths?
5 How can communities allow potential entrepreneurs to feel safe to try starting a new venture?
6 Why is networking so important to entrepreneurs?
7 What is a social entrepreneur?
8 Briefly explain what we mean by an "entrepreneurial ecosystem".

Notes

1 TechStars has published a useful manifesto for mentors (See Cohen at http://www.davidgcohen.com/2011/08/28/the-mentor-manifesto)
2 For more information on Hersman who is a TED Senior Fellow, see, for example, http://www.ted.com/speakers/erik_hersman.html. For information on his venture fund in Africa, see Clayton, 2012.
3 A useful tool from the RUPRI Center for Rural Entrepreneurship enables you to rate your community's support for entrepreneurship and get a sense of what your community is doing and what you might do next. It is available at: http://www.energizingentrepreneurs.org/site/images/research/tp/et/et4.pdf
4 A good resource in building an entrepreneurial network is the guide, Hello, My Business Name Is ... A Guide to Building Entrepreneurial Networks in North Carolina, available at http://www.entreworks.net/Download/HelloMyBusinessName.pdf

Bibliography

Ashoka. (2013). "What is a Social Entrepreneur?" www.ashoka.org/social_entrepreneur (Accessed July 17, 2013).

Aidis, R. A., Estrin, S., & Mickiewicz, T. (2007). "Institutions and Entrepreneurship Development in Russia: A Comparative Perspective", Journal of Business Venturing, 23(6), 656–672.

Chojnowski, B. (2006) "Open-Source Rural Entrepreneurship Development," *Rural Research Report*, Illinois Institute for Rural Affairs, Western Illinois University, 17 (2). http://www.iira.org/pubs/publications/IIRA_RRR_660.pdf (Accessed October 29, 2013).

Cohen, D. (2011). "The Mentor Manifesto". http://www.davidgcohen.com/2011/08/28/the-mentor-manifesto (Accessed October 29, 2013)

Clayton, N. (2012). " Savannah fund brings silicon valley-style accelerator to africa. *Wall Street Journal*", http://blogs.wsj.com/tech-europe/2012/06/07/savannah-fund-brings-silicon-valley-style-accelerator-to-africa/ (Accessed October 31, 2013).

Corporation for Enterprise Development, *Real Entrepreneurial Education*. Online. http://cfed.org/programs/youth_entrepreneurship/national_resources/rural_entrepreneurship_through_action_learning/ (Accessed October 31, 2013).

Davis, K. (2013). Entrepreneur magazine. http://www.entrepreneur.com/dbimages/article/the-worlds-20-hottest-startup-scenes.jpg. (Accessed October 29, 2013)

Eagle, N., Macy M., and Claxton R. (2010). "Network Diversity and Economic Development," *Science* 328.5981: 1029–1031.

Feld, B. (2012). *Startup Communities: Building an Entrepreneurial Ecosystem in Your City*, Hoboken, New Jersey: Wiley.

Fogel, G. (2001). "An Analysis of Entrepreneurial Environment and Enterprise Development in Hungary," *Journal of Small Business Management* 39.1: 103–109.

Giannetti, M. and Simonov, A. (2004). "On the Determinants of Entrepreneurial Activity: Social Norms, Economic Environment and Individual Characteristics." *Swedish Economic Policy Review* 11.2: 269–313.

Gibb, A. S. (1999), "Creating an Entrepreneurial Culture in Support of SMEs", Small Enterprise Development, 10(4), 27–34. doi: http://dx.doi.org/10.3362/0957-1329.1999.040 (Accessed October 10, 2013).

Holiday World and Splashing Safari (2013). http://www.holidayworld.com/news/holiday-worlds-patriarch-leaves-legacy (Downloaded September 19, 2013)

Li, J., Young, M. N., and Tang, G. (2010). "The Development of Entrepreneurship in Chinese Communities: An Organizational Symbiosis Perspective", Asia Pacific Journal of Management, 29(2), 367–385. doi: 10.1007/s10490-010-9192-x

Markley, D., Dabson, K. and Macke, D. (2006). *Energizing an Entrepreneurial Economy: A Guide for CountyLeaders*. http://www.energizingentrepreneurs.org/site/images/research/cp/prwp/prwp1.pdf (Accessed September 13, 2013)

Markley, D., Macke, D., and Luther, V.B. (2005) *Energizing Entrepreneurs: Charting a Course for Rural Communities.* RUPRI Center for Rural Entrepreneurship and Heartland Center for Leadership Development. http://www.energizingentrepreneurs.org/site/images/research/e2/e2book.pdf (Accessed October 29, 2013).

National Commission on Entrepreneurship (2001). *High Growth Companies: Mapping America's Entrepreneurial Landscape.* http://papers.ssrn.com/sol3/papers.cfm?abstract_id=1260389 (Downloaded September 19, 2013)

Nielsen, Kristian. *Bringing the Person and Environment together in Explaining Successful Entrepreneurship: A Multidisciplinary and Quantitative Study.* Aalborg, Denmark: Aalborg University.

Pages, E. (2005). "Creating Systems for Entrepreneurial Support", Economic Development America, Winter, Economic Development Administration: 4–6. http://www.entreworks.net/Download/EDAWinter2005Final.pdf (Accessed October 31, 2013).

Pages, E. (2006) Hello, My Business Name Is ... A Guide to Building Entrepreneurial Networks in North Carolina. http://www.entreworks.net/whatsnew/06/HelloMyBusinessName.pdf (Downloaded September 23, 2013).

Phillips, R., Seifer, B., and Antczak, E. (2013). *Sustainable Communities: Creating a Durable Local Economy.* Oxan, England: Earthscan.

Reynolds, P.D., M. Hay, W.D. Bygrave, S.M. Camp, and E. Autio (2000). *GEM 2000 Global Executive Report.* http://www.gemconsortium.org/docs/download/2408 (Accessed October 29, 2013).

Rural Policy Research Institute (RUPRI) Center for Rural Entrepreneurship. (2006) "Energizing Your Economy Through Entrepreneurship," *Assessment Series: Understanding Entrepreneurial Talent.* http://www.energizingentrepreneurs.org/site/images/research/tp/at/at3e.pdf (Accessed October 31, 2013).

Sabeti, H. (2009). *The Emerging Fourth Sector*, Washington, D.C.: The Aspen Institute.

Shockley, G., Frank, P., & Stough, R. (2008). *Non-Market Entrepreneurship: Interdisciplinary Approaches.* Cheltenham, England: Edward Elgar Publishers.

Smallbone, D., Welter, F., Voytovich, A., & Egorov, I. (2010). "Government and Entrepreneurship in Transition Economies: The Case of Small Firms in Business Services in Ukraine", *The Service Industries Journal, 30*(5), 655-670.

Smilor, R.W. (2001) *Daring Visionaries: How Entrepreneurs Build Companies, Inspire Allegiance, and Create Wealth*, Cincinnati, OH: Adams Media Corporation.

Throsby, D. (2001). *Economics and Culture.* Cambridge, England: Press Syndicate of the University of Cambridge.

University of Wisconsin Small Business Development Center, http://bus.wisc.edu/cped/sbdc/program-topics/special-programs/youth-entrepreneur-camp (Downloaded September 23, 2013).

van Stel, A., Storey, D. J., & Thurik, A. R. (2007). "The Effect of Business Regulations on Nascent and Young Business Entrepreneurship", *Small Business Economics, 28*(2–3), 171–186.

Venkataraman, S., & Sarasvathy, S. (2008). "Fabric of Regional Entrepreneurship: Creating the multiplier effect", White paper invited by the World entrepreneurship Forum, September, Lyon, France.

Xu, H., and M. Ruef (2004). "The Myth of the Risk-Tolerant Entrepreneur." *Strategic Organization* 2(4): 331–355.

Connections

Play a game! "Build a socially conscious business," start the game at: http://www.pbs.org/opb/thenewheroes/engage/business.html.

Read a book. *Rippling: How Social Entrepreneurs Spread Innovation Throughout the World* by Beverly Schwartz, Jossey-Bass, 2012.

Watch an inspiring video. Entrepreneur Erik Hersman explains how technology is saving lives in Africa in this TED talk: http://www.ted.com/talks/erik_hersman_on_reporting_crisis_via_texting.html

Explore these sites:

www.energizingentrepreneurs.org The RUPRI Center for Rural Entrepreneurship has abundant resources and guides to help rural communities develop a strong community-based entrepreneurial program.

www.Ashoka.org – a leading nonprofit with worldwide operations to encourage and support social entrepreneurship

http://caseatduke.org/ – Center for Advancement of Social Entrepreneurship, a research center with a variety of resources and tools for social impact via entrepreneurship

18 Arts, Culture, and Community Development

Rhonda Phillips

Overview

Arts and culture, expressed as activities, venues, heritage, planning, participation, or any other manner, matter to communities. It adds elements of interest and a way to express a culture, its history, and its future. Arts and culture can be used as a means for social action and for community economic development through tourism, for example. This chapter provides a range of considerations, including cultural concerns as well as approaches to tourism-based development as a community development strategy.

Introduction

Even "beyond the aesthetic appeal" (Aquino et al., 2013), arts and culture are a powerful factor addressing social and economic needs of society (Brennan et al., 2009; Florida, 2002; Foster, 2009; Markusen & Schrock, 2006; Zukin, 2010). Expressed in artistic form, culture can serve as a basis on which to renew or revitalize communities, helping advance quality of life and encouraging redevelopment and revitalization (Grodach, 2011; Polèse 2011; Pratt, 2009). The use of arts and culture as a basis for community economic development and revitalization is not a new endeavor.

During the late 19th century, arts and cultural efforts such as the City Beautiful Movement aimed at beautifying communities and increasing quality of life for residents, which included many community-based arts programs (Aquino et al., 2013; Korza, Brown, & Dreeszen, 2007; Phillips, 2004). Communities fostering arts and culture as industries can gain both economic and social benefits for their communities. For example, cultural industries can help spur job growth, turn ordinary cities into destination cities, create interconnections between arts and business, revitalize urban areas, attract skilled workers, and create spin-off and supporting businesses (Creative City Network of Canada, 2005; Currid, 2007; Florida, 2002; Grodach, 2011; Markusen & Schrock, 2006; Paddison, 1993; Pratt, 2009; Stern & Seifert, 2010). Furthermore, arts spaces not only promote cities and neighborhood revitalization, they aid in boosting tourism in addition to enhancing arts consumption.

Arts and Social Action

Artists have long expressed views on social and other conditions impacting societies through the years. Arts, as a form of expression, can be powerful tools for eliciting change: for example, music or visual arts as ways to convey messages and encouragement (McGrath & Brennan 2011). For example, could not the country music genre

BOX 18.1 INDIGENOUS COMMUNITY DEVELOPMENT

Indigenous design and planning is informed by an emerging paradigm using a culturally responsive and value-based approach to community development. As generations of people have successively lived over time in the same place, they have evolved unique world-views. Adherence to values such as stewardship and land tenure temper the immediacy of exploitative practices and reactionary planning. Leadership strives to balance the immediacy of action (short term) with a comprehensive vision (long term). And in the lifetime of an individual, it is not unusual that their extended family consisted not only of oneself but three generations before and three generations after (known as the seven generational framework).

Communities have the additional need to make their projects culturally viable. Their contributions have too often been dismissed as inconsequential to the evolution of "great" building traditions and settlements. Often relegated to anthropology and the study of quaint vernacular traditions, accomplishments in indigenous architecture and planning have been consigned to anonymity and obscurity. Today, one of the greatest challenges that tribes face is to see their populations, especially young people, shift away from their cultural traditions and toward urbanization. There is a heightened urgency to develop community environments suitable for retaining their cultural identity.

An example of design and planning approaches that are informed by indigenous theory, practice, and research is the Zuni Pueblo, the first Main Street project awarded to an Indian tribe in the US. The Main Street program is a vital community revitalization program of the National Trust for Historic Preservation. This groundbreaking work is the outcome of graduate coursework offered by the University of New Mexico's Community & Regional Planning program in conjunction with the Indigenous Design + Planning Institute. Working closely with the newly established Pueblo of Zuni Main Street Board, the students divided into their degree concentration areas (community development, physical planning, and natural resources planning) to address the compelling issues of the Pueblo's planning and development along State Highway 53. This roadway not only provides access onto and through this community but it bisects prehistoric traditional village areas from more contemporary settlements. Zuni is a world destination for both its cultural patrimony and its renowned traditional artisans.

The project brought into focus the unique aspects of human settlements and the role of placemaking in sustaining cultural identity. It brought forth challenges on the role of economic development and local empowerment. It introduced students to the evolution of tribal sovereign authority and protocol, especially as applied to local governance and infrastructure development.

Tribally based activities are intended to educate and empower faculty, students, practitioners, and community leaders to establish culturally responsive community development. Increasing awareness of indigenous approaches to community development will help reclaim abilities to shape and mold communities for the betterment of everyone.

Theodore (Ted) Jojola, Ph.D.,
Director, Indigenous Design + Planning Institute,
School of Architecture and Planning,
University of New Mexico, USA
http: //idpi.unm.edu

often be considered lamenting the woes of life ("defeatist"), while ska and reggae could be seen as a "defiant" genre? The Theater of the Oppressed started in the 1970s by Brazilian artist activist Augusto Boal serves as a participatory theater with the ideas of democratic and cooperative interaction. Here is a description of this remarkable form of art:

> It is a rehearsal theater designed for people who want to learn ways of fighting back against oppression in their daily lives…(it is) "Forum Theater," for example, the actors begin with a dramatic situation from everyday life and try to find solutions—parents trying to help a child on drugs, a neighbor who is being evicted from his home, and individual confronting racial or gender discrimination, or simply a student in a new community who is shy and has difficulty making friends. Audience members are urged to intervene by stopping the action, coming on stage to replace actors, and enacting their own ideas. Bridging the separation between actor (the one who acts) and spectator (the one who observes but is not permitted to intervene in the theatrical situation), the Theater of the Oppressed is practiced by "spect-actors" who have the opportunity to both act and observe, and who engage in self-empowering processes of dialogue that help foster critical thinking. The theatrical act is thus experienced as conscious intervention, as a rehearsal for social action rooted in a collective analysis of shared problems.
>
> (Brecht Forum 2014: 1)

The roots of this type of theater is found in the work of Paulo Freire, an educator who promoted the following: (1) to see the situation lived by the participants; (2) to analyze the root causes of the situation, including both internal and external sources of oppression; (3) to explore group solutions to these problems, and (4) to act to change the situation following the precepts of social justice (Brecht Forum 2014).

BOX 18.2 ARTS AND CULTURE IN RURAL DEVELOPMENT

Examples from County Leitrim, Ireland

Leitrim County (population 31,798) is the smallest and most rural county in the northwest of Ireland. Both the county and the wider region are recognized as home to a very vibrant and creative arts and culture community. Throughout the 1980s and 1990s, a significant number of artists and craft workers—many from other parts of Ireland, the UK, and European countries—settled and established small enterprises in the county. Over time, a number of key local organisations and institutions have emerged, driven by those working within the sector and through support from national and European development funding. This includes in particular the Leitrim Design House (http: //intoleitrim.com), the Dock Arts Centre (www.thedock.ie, the Leitrim Arts Office (www.leitrimarts.ie) in Carrick-on-Shannon, and the Leitrim Sculpture Centre (www.leitrimsculpturecentre.ie) in Manorhamilton. These local organisations, which have evolved from within the creative community, have been critical in sustaining the area's creative sector over time. Each offers a distinctive dimension of support, which in combination has enhanced the 'creative capacity' of the local community and economy. Other official bodies, such as the local community and rural development organisations, local government, and the county enterprise board, are all actively involved in linking with these organisations and providing programmes of support through national and European sources of funding (e.g. LEADER Rural Development Programme and PEACE III).[1]

While there is a considerable involvement and interest among local people in rural areas in developing and applying their creative talents, there can be major capacity challenges as well as social hurdles with this kind of work. In the latter case, limited opportunities to socialize and network with others can lead to feelings of isolation. Economically, creative expression, application, and technology can be highly competitive arenas, particularly where distinctiveness and innovation are much sought after. While artists and craft workers are heavily invested in the creative dimensions of their work, translating this to a sustainable livelihood requires other forms of support and capacity, such as technical and business knowledge. The following case outlines two community-based organisations offering distinctive contributions to the sector and which have recently collaborated to invigorate the rural cultural economy.

Case Study: Leitrim Sculpture Centre and Design House

Leitrim Sculpture Centre is a major focal point for visual and material art practices regionally, nationally, and internationally. It was established in the 1980s by a community of local artists. In recent years the centre has undergone substantial investment and redesign which makes it a major resource, space, and supportive environment for artists, craft workers, and the local community. Located in a small town in the north of the county, the centre incorporates a wide range of spaces, equipment, and rooms devoted to metalwork, glass, ceramics, printmaking, experimental art exhibitions, and artist studios. It offers a hugely significant community dimension through its educational and outreach work, public art space, and as a source of paid and voluntary employment. The Centre is the focal point for a community of practice based principally on 'craft knowing', based on a combination of aesthetic, kinaesthetic, and embodied knowledge (Amin and Roberts 2008).[2] Funded principally from the Arts Council of Ireland, the Centre is firmly grounded in experimentation and creative exploration. It offers between 25 to 50 workshops per year delivered by experts in various specialisms, and draws both novice and advanced participants. It also attracts artist residencies from other parts of Ireland and internationally who use the facilities or provide training. In addition to workshops, the Centre enables the linkage with local experts, who can offer technical support within their area of specialism, for creative workers looking to explore new techniques and practices. It hosts various master classes, presents at local schools, and hosts symposia and additional projects that explore the local landscape and alternative and sustainable models of arts practice. The Centre has a very strong presence in the town and region and offers an inclusive space for the local population. The Centre offers space that can be hired at low cost by local groups in the community. The manager summarises the attraction of the centre for a range of people:

> We have such diverse technologies of work and kinds of spaces and kinds of human interactions that there are many ways in which a person might see a function for themselves in such a place ... it's becoming an important place for people to think about realigning their careers and getting back into work.

In the south of the county, the Leitrim Design House was formed in the mid-1990s in response to the types of rural challenges identified earlier. It is the longest established centre and retail outlet in Ireland. As well as being a retail outlet for high-quality arts and craft work, it directly supports craft makers through training initiatives tailored to the sector, in such areas as branding, presentation, packaging, pricing, and showcasing. It has forged relationships with significant national experts in the field of product development, who are involved in delivering various training workshops and programmes. Through mentoring, advisory work, and some small financial supports in particular areas, this type of work is critical to up-skilling in the sector. The retail space offers the possibility for makers to showcase their work and get initial feedback. One of the contributing aspects to the visibility and sustainability of the Design House is its location and linkage with the Dock Arts Centre, an integrated centre for the arts comprising large theater space, galleries, artist studios, and arts education room. Since it would be unable to rely solely on retail activity, the Design House acts as the booking service for the venue.

Both the Sculpture Centre and Design House joined together for the first time in 2013 to make a LEADER rural development funding application to Leitrim Development Company, which administers the programme locally. The successful bid resulted in a joint initiative known as 'Live Craft,' aimed at revitalizing the economic position of craft in the county. It was recognized that a new impetus was needed to support and sustain the local community of artists and craft workers, especially in a difficult economic climate. The 'Live Craft' initiative combined the strengths of both organisations, who provided instruction and supports across a range of different material and visual craft practices, and product development and business skills, such as portfolio development and product design. The tailor-made training initiative was able to draw in around 70 existing creative individuals and businesses as well as new entrants without experience in creative work, some of whom were experiencing unemployment. In addition to newly developed business plans and portfolios, a new network of creative workers was established in the county who meet on a regular basis and provide confidence and support to members.

Brian McGrath,
Lecturer/Programme Director, MA in Community Development,
National University of Ireland, Galway, Ireland

Linking Arts and Culture to Tourism

Community-based tourism development is defined as efforts by communities or neighborhoods to develop and manage their assets for tourism opportunities (Matarrita-Cascante et al., 2010; Murphy, 1988). This type of approach integrates community involvement or citizen/resident participation, and is very consistent with the principles of community development practice (see Chapter 7 for a listing of these principles). Integrating arts and culture (as well as heritage inherent in these resources), community-based tourism can serve as foundation to help foster community economic development outcomes.

Ready, Set, Go: Starting the Cultural-Based Planning Process

Community participation is a crucial factor for the long-term viability of arts- and cultural-based development approaches, including tourism. In an effort to counter friction resulting from development's negative impacts, many researchers are suggesting that communities should plan their evolution more systematically, thereby taking into account residents' attitudes and perceptions about its growth at the outset (Reid et al. 2004: 624). The significance of community involvement is to provide a voice for those involved in or impacted by tourism, to make sound decisions making use of local knowledge, and to reduce possible conflicts between tourists and members of the host community. The emphasis in community planning models on the need for coordination and collaboration among all sectors involved leads to the need for planning. Simply put, planning provides the opportunity to envision what a community wants and how to get there. Without it, there is no direction to achieve desirable outcomes—the community will have to accept whatever comes its way. All planning processes, whatever the focus, always begin with an inventory or research phase and cycles to an evaluation or monitoring of outcomes phase. Note that the process is continuous and is never "over" because change is the only constant! Further, planning is not an end in itself but rather a necessary means for achieving outcomes that people value in their communities (Bunnell 2002). The following process is recommended to start a community on its way to achieving tourism and cultural development:

1 *What do we have?* Inventory assets (people; organizations; cultural/heritage, natural, financial, and built resources) and contexts (political, economic, social, environmental) of the community. This is the research phase and can include a variety of sources and tools such as surveys, focus groups, asset mapping, etc.
2 *What do we want?* At this point, the all-important vision as a guide to seeing what could happen is crafted by stakeholders—those in the community that have an interest in helping achieve a more desirable future. Belief is a powerful tool and can inspire a community to achieve remarkable outcomes. The vision should be bold enough to inspire and realistic enough to attain.

3 *How do we get there?* This stage is about developing the plan that is a guide with specifics for achieving the vision and includes goal statements and actions. Most importantly, it selects the strategies or approaches desired—will an approach such as National Trust for Historic Preservation's Main Street program or a theme-based approach be used? Historic preservation is a powerful tool that, when combined with community development approaches, can help revitalize historic downtowns as well as neighborhood corridors (Smith 2007). What trends should be looked at and which markets are sought? It also identifies which organizations or groups of collaborators will tackle the tasks and action items. Collaborative efforts typically work best, but in some cases it takes a "champion" to start the efforts, and others will join in later.
4 *What have we done, and what do we need to do now?* Monitoring is critical to see if the above steps are working; if not, then adjustments and revisions are needed. Because the nature of this process is continuous, it provides feedback for refining ongoing activities as well as starting new initiatives.

Getting from Here to There: Selecting Approaches

One of the most critical components of selecting development approaches is determining who is going to carry it forward. Many communities develop a collaborative arrangement with both public and private entities; others focus on one organization to serve as the umbrella agency for the community's efforts. Here are examples of approaches that a community can use to implement development. Some like to use a combination or create an entirely unique approach with the bottom line being: *whatever works and reflects the desires of the community!*

Main Street Approach

In 1980, the National Trust for Historic Preservation established the Main Street Program to focus on traditional downtown revitalization. It transcends historic preservation and includes community and economic development, infrastructure, and marketing elements. The approach includes not only aspects of encouraging heritage and cultural preservation but also tourism in the form of visitors to shop and spend in downtown, but also more broad-scale community development outcomes. The Program is successful, with an average return of $35 reinvested for every $1 spent on revitalization. While not all can be a designated Main Street community, there is much to be learned from looking at their development strategy, the Main Street Four-Point Approach™:

1 *Organization* involves getting everyone working toward the same goal and assembling the appropriate human and financial resources to implement a Main Street revitalization program. A governing board and standing committees make up the fundamental organizational structure of the volunteer-driven program. Volunteers are coordinated and supported by a paid program director as well. This structure not only divides the workload and clearly delineates responsibilities, but also builds consensus and cooperation among the various stakeholders.
2 *Promotion* sells a positive image of the commercial district and encourages consumers and investors to live, work, shop, play, and invest in the Main Street district. By marketing a district's unique characteristics to residents, investors, business owners, and visitors, an effective promotional strategy forges a positive image through advertising, retail promotional activity, special events, and marketing campaigns carried out by local volunteers. These activities improve consumer and investor

confidence in the district and encourage commercial activity and investment in the area.

3 *Design* means getting Main Street into top physical shape. Capitalizing on its best assets—such as historic buildings and pedestrian-oriented streets—is just part of the story. An inviting atmosphere, created through attractive window displays, parking areas, building improvements, street furniture, signs, sidewalks, street lights, and landscaping, conveys a positive visual message about the commercial district and what it has to offer. Design activities also include instilling good maintenance practices in the commercial district, enhancing the physical appearance of the commercial district by rehabilitating historic buildings, encouraging appropriate new construction, developing sensitive design management systems, and long-term planning.

4 *Economic Restructuring* strengthens a community's existing economic assets while expanding and diversifying its economic base. The Main Street program helps sharpen the competitiveness of existing business owners and recruits compatible new businesses and new economic uses to build a commercial district that responds to today's consumers' needs. Converting unused or underused commercial space into economically productive property also helps boost the profitability of the district.

(National Trust for Historic Preservation 2014: 47)

Many of the Main Street Programs have been met with success, and often tourism is a vital component of that success. The Program offers many sources of information and help to communities and should be a first stop for a community embarking on revitalization efforts. The Program has worked with nearly 2000 communities since its inception.

Heritage Tourism

Whether it is ethnic activities or styles of architecture, using heritage as a basis for fostering community development can be very rewarding. Reclaiming history in communities can serve as a:

> foundation for directing future growth and development of communities. By understanding the history of the physical patterns in the land and built environment, communities can evaluate how to preserve desirable quality of life attributes, as well as opportunities for enhancement … reclaiming history can be a powerful and catalyzing force providing numerous positive impacts (including) building social capital, enhancing community identity and sustaining the environment.
>
> (Phillips and Stein 2011: 1–2)

Heritage tourism has grown exponentially with fascinating dimensions and consequences (Chhabra 2010). Preserving heritage and tourism have not always been congruent ideas, but in the recent past it has become one of the most popular forms of tourism with heritage travelers typically staying longer and spending more than any other type of tourist. The benefits of this approach are numerous including new opportunities for preserving and conserving an area's heritage while giving the visitor a learning and enriching experience. It can begin with using the community's built heritage, such as the case with Cape May, New Jersey, or Eureka Springs, Arkansas. Other communities use their ethnicity to develop their approach, as with Solvang, California, with a population of just over 5,000 and over 2 million visitors per year. How did they do it? Solvang had a rich Danish history and parlayed this into becoming the "Danish Capital of America" complete with festivals and other special events (Phillips 2002). This Danish culture mecca for tourists encourages new buildings and rehabilitations to use ethnic style architecture for the built environment as well.

Natural/Recreational Tourism

Many communities or regions have a bounty of natural resources that lend themselves as a basis for tourism. The U.S. national park system and the individual states' park systems are major destinations for natural and recreational tourists each year. However, at the community level this type of tourism can yield benefits as well. While the scale may be different, the appeal is still high. For example, are there unique landscapes or even transportation features with unique recreational opportunities such as a canal or railroad corridor? The Rails to Trails program has become extremely popular and can attract numerous recreational tourists. Combining trails with venues at community locations is a successful approach. Three small towns in south Florida recently did just this—by combining efforts and evoking the help of the U.S. Army Corps of Engineers, a walking/biking trail has been constructed around the levy of a large lake that borders the towns. Visitors have increased to nearly 300,000 on the trail and represent an opportunity to visit the towns when venues are offered.

BOX 18.3 ELEMENTS FOR ATTRACTING HERITAGE TOURISTS

Historical and Archaeological Resources

How many of these historic resources are in your community?

- museums;
- historic properties listed on the National Register of Historic Places or otherwise designated landmarks;
- historic neighborhoods, districts, or even entire towns or villages;
- depots, county courthouses, or other buildings that have historic significance because of architectural or engineering features, people associated with them, or contribution to historic events;
- bridges and barns, battlefields or parks;
- fountains, sculptures, or monuments.

Ethnic/Cultural Resources

In addition to historic and archaeological sites, consider the ethnic or cultural amenities of your community by exploring the traditions and indigenous products of:

- artists;
- craftspeople;
- folklorists;
- other entertainers such as singers or storytellers;
- galleries, theaters;
- ethnic restaurants or centers;
- special events like re-enactments or ethnic fairs; and
- farming and commercial fishing and other traditional lifestyles.

Source

National Trust for Historic Preservation (1999) Getting Started. *How to Succeed in Heritage Tourism*, Washington, DC: National Trust, p. 22.

BOX 18.4 TOURISM DESTINATION PLANNING

A few examples of destination planning cases as the Hopi in Arizona, Bouctouche Bay in New Brunswick, Canada, and Sanibel Island in Florida. The Hopi in Arizona aimed to achieve a balance between economic development through tourism planning and the integrity of their culture. They created their own code of ethics that imposed several limitations on tourists such as restricting the recording and/or publication of tourist observations, providing guidelines regarding appropriate tourist clothing and behavior, and respect for community restricted areas and events. In their effort to strive toward cooperative solutions to issues related to tourism development, the community has been considered successful.

The Bouctouche Bay community in New Brunswick was considered successful in their tourism planning because it considered community participation as integral to their development process. The Bouctouche Bay Ecotourism Project was considered successful worldwide in its efforts to achieve public–private partnerships with its emphasis on resource protection and tourism interpretation. This project promoted educational workshops that involved entrepreneurial/community groups and included leadership by professionals as well as field trips and the sharing of ideas for the future (Gunn and Var 2002: 277). The local stakeholders, community groups, government, industry, and aboriginals were all encouraged to participate in the planning process so that a shared vision and plan could be created.

Deepak Chhabra, Ph.D.
Associate Professor
Arizona State University, USA

Sources

Chhabra, D. and Phillips, R. (2009). Tourism-Based Development. In R. Phillips and R. Pittman, *Introduction to Community Development*, London: Routledge.
Gunn, C. and Var, T. (2002). *Tourism Planning*, 4th ed. London: Taylor & Francis.

Note

Additional information about tourism marketing can be found in *Sustainable Marketing of Cultural and Heritage Tourism* by Deepak Chhabra (London: Routledge, 2010).

Popular Culture Approaches

Popular culture runs the gamut from visual arts and music to filmmaking. Using these elements as a basis for building a tourism-based community development strategy is speculative yet very exciting. The entertainment industry can play an important role in stimulating development outcomes. This approach is not without risks though as some communities find the intensity of energy and resources required may be too high to sustain a "popular" appeal. Others find it has worked very well. There are three major popular culture strategies that have been used with success, as described in *Concept Marketing* (Phillips 2002).

Arts-based

Some view the arts as a powerful catalyst for rebuilding *all* aspects of a community, not just the economic sphere, in terms of venues to attract tourists. So the benefits of this strategy can be far-reaching beyond tourism alone. The number and types of tourist venues based on art are tremendously diverse. Some communities use themselves as the palette for the venue, painting murals on the walls of their buildings or incorporating public art on a major scale into the community—Stuebenville, Ohio; Toppenish, Washington; and Loveland, Colorado, are such examples. Care must be taken with this approach so that the commercialization or theming aspect does not threaten community ambience. In addition to the community as a visual arts palette, others attract artists to live and display their work, often in arts districts including converted warehouses or other obsolete industrial facilities. This strategy is not limited to only large urban areas. Small communities, such as Bellows Falls, Vermont (population 3,700), have used their downtown structures to develop artist live-work space, attracting visitors to the galleries for shows and to shop. The following considerations are offered to communities interested in pursuing an arts-based strategy:

- General support for the Arts. Citizens and local government officials need to recognize that a healthy arts presence is a vital part of community infrastructure and is important in terms of community development. Participation approaches in community decision making should be used to further build support.
- Seek out untapped resources. Local governments may have more resources than direct funding that can be used to support arts-based businesses and other activities which in turn attract visitors to the community. Examples include rent-free facilities from a variety of sources such as school classrooms and auditoriums, commercial warehouses, conference centers, or vacant retail spaces.
- Integrate the support of arts with community development benefits. Whenever possible, the community should strive to link benefits with arts-based activities. For example, artisans could participate in programs such as placing their art in public venues.
- Maximize resources through community sharing. The centralization of facilities and resources is a significant factor in the success of arts-based programs. A centralized facility, such as a production studio, gallery, office, or retail space, can be used by numerous groups to provide cost savings.
- Adopt a flexible approach to arts support. All artists are different and need different kinds of support and assistance. Business management assistance to arts entrepreneurs is usually a critical need in communities.

(Phillips 2004)

Building an arts-based development strategy will result in increased tourism, particularly for shopping in galleries and visiting during special venues such as public art exhibits.

Music

Music has long been used as a basis for attracting tourists. One only has to look as far as Nashville or New Orleans to see the widespread impact that music and related activities can have on an area. Branson, Missouri, is often cited as an example of how entertainment and particularly the music side of the business can be used as a basis for building a tourist-dependent economy. One strategy that seems to work particularly well is when ethnic or heritage music is coupled with a tourism development strategy. Mountain View, Arkansas, with a population fewer than 3,000, found that its remote location in the Ozark Mountains has not prohibited it from becoming a tourist destination because of its music heritage. The "Folk Music Capital," Mountain View holds the Arkansas Folk Festival as well as offering other venues and special events. Eunice, Louisiana, is now known as the "Prairie Cajun Capital" because of its Zydeco and Cajun music heritage that is kept alive and well with special venues and a regular television and radio show, the Rendez-Vous Des Cajuns—the Cajun version of the Grand Ole Opry. Eunice attracts many visitors to its town and to its renovated Liberty Theater as a spectacular venue for music offerings. Its visitors are not limited to the region—the British Broadcasting Company has visited several times to film the shows which has prompted many British visitors to journey over to see the live performances. An approach by Playing for Change, a nonprofit dedicated to music education, has pioneered the way for connecting musicians around the world with youth for developing music skills. The ability to build community via music is evident, with inspiring results.

Filmmaking

While filmmaking and television production can be an attractive strategy for economic development outcomes, it can have mixed results as a tourism venue. Typically after filming, the community attracts tourists for a period of time to see the venues. However, the attraction can fade after time if the community does not offer complementary venues. Winterset, Iowa, the location of the film *Madison County*, experienced a great influx of tourists for several years after the film was shot. While the initial crowds have gone, the community has incorporated several other venues to attract tourists over the long term such as the Covered Bridges Festival since the county is home to six remaining bridges. If a community can continue to attract periodic filming activities, then the tourist interest stays steadier.

BOX 18.5 CREATIVE PLACEMAKING

Creative placemaking is a new way of thinking about how to make places better with and through the arts. Placemakers have known for more than a century that the arts can help improve quality of life and enhance local economies. The traditional way is to think about doing projects and expecting good things to follow. This leads people to focus on such things as building performing arts centers, creating cultural districts, or organizing crafts festivals. Creative placemaking focuses on people first; bringing diverse people together and developing sustainable partnerships for action. The projects then follow.

Ann Markusen and Anne Gadwa, who coined the term Creative Placemaking in 2010 in a white paper for the National Endowment for the Arts, described it this way:

> In Creative Placemaking, partners from public, private, nonprofit, and community sectors strategically shape the physical and social character of a neighborhood, town, city, or region around arts and cultural activities. Creative placemaking animates public and private spaces, rejuvenates structures and streetscapes, improves local business viability and public safety, and brings diverse people together to celebrate, inspire, and be inspired.
>
> In turn, these creative locales foster entrepreneurs and cultural industries that generate jobs and income, spin off new products and services, and attract and retain unrelated businesses and skilled workers. Together, Creative Placemaking's livability and economic development outcomes have the potential to radically change the future of American towns and cities.
>
> (Creative Placemaking, Ann Markusen, and Anne Gadwa, 2010)

Why people first? After all, projects are easy to see, easy to measure, and are what people talk about when they describe a place as being 'artsy'. But community and economic development isn't as much about what's on the ground as about how people connect to what's there. A new performing arts center in a poor part of town may be successful in bringing wealthy people to its theaters and halls. However, if those people just go to the performing arts center (PAC) and leave without spending money anywhere nearby, that PAC is not achieving the kind of economic impact that community leaders imagined. (And because the community probably spent a lot of time and money to build the PAC, it might be costly to the people it was designed to help.) When *diverse groups of people work well together*, they are more likely to develop a wider range of projects that together would have broader (and maybe even deeper) impacts than the single jump-starter project.

The National Consortium for Creative Placemaking developed a model that builds on the pioneering work of Gadwa and Nicodemus. In the NCCP model, creative placemaking is a set of strategies that:

- improves the quality of life in communities for as many people as is reasonable;
- enhances economic opportunity in communities for as many people as is reasonable;
- promotes a healthier climate for individual and collective expression;
- maximizes the use of existing assets in the community (for example, by connecting them in new ways);
- responds to the distinct qualities of place in the community.

For more on this model, see *Creative Placemaking: Integrating Community, Cultural and Economic Development* (at: www.artsbuildcommunities.com).

This model is rooted in sustainability, asset-based community development and cultural competency. It challenges creative placemakers to be cost-effective, build on the best qualities of their community, and to be sensitive to the diverse points of view and concerns in their community.

The traditional model of using arts in urban planning is to focus on big cultural institutions, districts, and attracting artists. This can lead to unchecked gentrification that can push out artists and make neighborhoods unaffordable to low- and moderate-income residents. Smaller arts organizations might benefit from the increased attention, or they might be now forced to compete against new institutions and initiatives that have more support from public officials. What is good for the arts is not necessarily always good for communities.

The key to thinking about creative placemaking is that it is not about a particular set of strategies; it is about a motivation and an approach to addressing problems within communities. There is a lot of overlap between cultural planning and Creative Placemaking. Both focus on the arts (and cultural activities). Recent cultural plans, like Creative Placemaking plans, talk about how arts and culture help to benefit communities. Cultural planning is planning to help the arts and culture industry—museums, historic sites, performing arts groups, etc. Strategies in these plans are focused on enhancing arts and culture for its own sake. Cultural plans can and should play an important role in advocating for the interests of arts and culture. Creative placemaking plans should focus on how arts and culture can help address issues that prevent communities from being better places to live, work and visit.

While a cultural plan may mention that connection to arts and culture can help students do better in school, a Creative Placemaking plan should offer strategies for connecting schools and arts organizations to—as a team—develop programming and other strategies that serve their mutual missions. For example, the Creative Placemaking plan might identify the types of arts and culture programs that could be created or expanded in specific schools, as well as identifying the types of learning or personal development outcomes team members would like to see from these programs. (Of course, to do all this, the team should include or partner with school officials.)

In short: cultural plans are about promoting cultural development, mostly by helping organizations and creative professionals. Creative placemaking is mostly about resolving issues facing communities through and with arts and culture.

While many people enjoy the arts and see it as a tool for economic development and revitalization, it is difficult to do creative placemaking well. Too many people cling to certain myths about arts and culture: 'The "arts" happen only in cities or in artist colonies.' 'Arts are decoration or a luxury.' 'You need a big performing arts center or arts district.' 'It can't happen here because there are not enough creative people.'

NCCP developed a model to address these issues and develop a core group of culturally competent leaders to develop and sustain creative placemaking efforts in their communities. Community coaching pairs a trained coach with a group of 10 to 30 community stakeholders from a district, city or region. The coach works with the group to help them build a plan for creative placemaking, as well as a team of leaders. Team members do more than develop projects and priorities; they explore difficult social and economic issues that affect their community's quality of life.

In Louisiana in 2013, ten cultural districts, cities and regions engaged in community coaching through the Louisiana Creative Communities Initiative program. While many of the community teams started out focused on doing arts projects, or arts as a way to attract tourists, several wound up pursuing bigger strategies. One town's team developed strategies to engage the arts to address long-standing divisions between African-American and White communities. Another town's team decided to build better connections with a neighboring community that was separated by a river. A team of five parishes (the Louisiana term for counties) decided to use arts and culture to help the area build a new identity and create more economic opportunities.

Leonardo Vazquez, AICP, PP,
Executive Director, National Consortium for Creative Placemaking, USA

Note

Adapted from "Building teams and making plans through community coaching: A guide for coaches in the Louisiana Creative Communities Initiative," by Leonardo Vazquez, AICP/PP.

Sources

Markusen, A. and Gadwa, A. 2010. *Creative Placemaking.* A white paper for the National Endowment for the Arts' Mayors' Institute on City Design. http: //kresge.org/sites/default/files/NEA-Creative-placemaking.pdf

Vazquez, L. 2013. *Building teams and making plans through community coaching: A guide for coaches in the Louisiana Creative Communities Initiative.* Training guide prepared for the State of Louisiana Office of Cultural Development.

Vazquez, L. 2013. *Creative Placemaking: Integrating Community, Cultural and Economic Development.* White paper. www.artsbuildcommunities.com/cpmodel

Corporate Culture Approaches

Corporate culture has long fascinated Americans. Fast growing tourist venues are factory tours and corporate museums, with currently over 300 such attractions in the US and more expected. When Binney and Smith Inc.'s Crayola Factory museum opened in Easton, Pennsylvania, over 300,000 visitors came the first year. Take a look at corporate communities such as Hershey, Pennsylvania—"Chocolate Town, USA"—or Austin, Minnesota—"Spam Town, USA"—to see the appeal. These venues are mostly private sector initiatives, but when combined with integrated community efforts as a development strategy, they can be very effective. The drawbacks are the risks of tying a community's image to one corporation, the issue of who has control over or influence on major decisions affecting the venues, and the resulting community outcomes. There is also the question of authenticity which may conflict with historic preservation and cultural dimensions of communities. Yet the benefits include bringing in external revenues and visitors that may otherwise not have ventured to the community. This strategy would have to be approached cautiously and with concern for the community's desires.

Surrealistic Approaches

This category is in a world of its own. This type of strategy is of an incongruous nature that defies or exceeds common expectations. It emerges when a community does not have inherent natural, cultural, ethnic, or built resources to use as the basis for their tourism development approach. So they create it with energy and imagination. In other instances, a community may already have the genesis of a resource and then take it to a different level altogether. The results can be astoundingly successful in some cases or a flop. It takes a high measure of courage (and sometimes audaciousness) to embark on this strategy, but the flip side is that the rewards can be very high if done well. There are three elements to consider in this strategy:

1. *Shock value*: Is the approach completely unnatural to the point that visitors are not attracted? Or is it workable because of the delight of the visitors when encountering the unexpected?
2. *Scale:* This concerns the ability to develop the created environment (the venues) to a plausible level so there is enough there to draw visitors. One venue, depending upon its size, may not do it. Combining venues and carrying the theme throughout can work better.
3. *Scope:* Communities must find the strategy or approach that incorporates activities and venues appealing to a market. In other words, it needs to be targeted to an audience.

(Phillips 2002)

A case in point: tiny Cassadaga, Florida, has found its niche in its surrealistic strategy by building its reputation as a center for providing palm readings and other mystic services. Its shock value is seeing an entire community that revolves around these services in a rather removed location; its scale is appropriate as most businesses in town have some elements of the theme; and its scope is focused on tourists who enjoy a dip into the mystical side of life. Perhaps the best case to illustrate this approach though is Helen, Georgia. Located in the north Georgia mountains, this small community developed itself as an alpine village with strict design standards, venues, and special events year round. Transforming itself from a dying town with only six businesses open to now over 200 and, at its peak, over a million visitors per year, Helen is a great story of how creativity can lead to a total transformation of an economy. This is not the approach for a community concerned with authenticity!

Conclusion: Bringing It All Together

Arts, cultural and tourism-based development represents an excellent opportunity for communities to capture benefits of one of the most dynamic industries in the world. This chapter has presented reasons why these strategies are appropriate. Identifying, designing, and implementing a development process is complex, multifaceted, and requires a large measure of energy, resources, and commitment. It necessitates that communities go beyond marketing efforts to consider all facets—from planning and project and program management to gauging outcomes and adjusting strategies as needs and desires change through time. The stakes are high because these types of development have the potential to dramatically change a community in many ways and aspects. These changes can be negative or positive; with the correct approach and well-designed strategies, the results can be tremendously beneficial.

CASE STUDY: AGRITOURISM: CONNECTING FARMS, COMMUNITIES, AND CONSUMERS

Many farms in Vermont and throughout the US are struggling to keep their land in farming and their families employed on their farms while sustainably managing resources for the long term. Suburban sprawl and second home development have led to increases in land values and property taxes, threatening working farms and available farmland. At the same time, tourism and culinary industries are expanding, providing opportunities for farms interested in diversifying and sharing their products, heritage, and landscapes directly with consumers. In Vermont, tourism is the first or second revenue-generating source and agricultural operations are in the top ten. The economic impacts of blending tourism and agriculture have significant potential for supporting the working landscape while educating consumers about the importance of farming.

Agritourism can be defined as "a commercial enterprise on a working farm or ranch conducted for the enjoyment, education, and/or active involvement of the visitor, generating supplemental income for the farm or ranch" (Chase 2008). Agritourism can take many forms including overnight farm stays, retail sales, hay rides, corn mazes, pick-your-own operations, and use of woodlands on farms for hunting, hiking, horseback riding, and other activities. There may be educational components including programs for schoolchildren, as well as exhibits and demonstrations tailored to specific visitor groups. Agritourism enterprises allow farms to diversify their core operations, add jobs for family members and others, and keep land in production while preserving scenic vistas, maintaining farming traditions, and educating non-farmers about the importance of agriculture to a community's economic base, quality of life, history, and culture. In essence, agritourism is providing educational and authentic agricultural experiences that enhance direct marketing of farm products and improve public support for agriculture.

Agritourism is growing rapidly in Vermont and other areas of the US; however, the industry remains underdeveloped in many states, lacking technical assistance support, infrastructure, and networking opportunities to ensure best practices. To address this need, University of Vermont Extension specialists collaborated with other Northeastern states to develop agritourism training modules. With funding from a Sustainable Agriculture Research and Educations (SARE) grant and additional resources, 19 workshops were held in ten states in the Northeast over the course of a year. More than 750 farmers participated. The workshops were followed by technical assistance available to farmers to help them follow through on plans to adopt agritourism best practices.

A survey immediately after the workshops indicated that 97% of respondents increased their knowledge of income-generating opportunities for agritourism, 92% planned to assess their business to determine where improvements or new ventures were needed, and 84% planned to implement improvements or new ventures. A web survey one year after the workshops found that 81% of respondents had assessed their business to determine where improvements or new ventures were needed while 64% had implemented improvements or new ventures. To evaluate contributions to farm viability, increases in farm profitability as well as increases in quality of life of farmers were measured. Of farmers responding to the survey, 64% reported a positive impact on profitability and 79% reported increases in their quality of life as a result of changes made based on information received through workshops and technical assistance (Chase et al. 2013). Trevin Farms of Sudbury, Vermont is one of the farms benefiting from the Extension program on agritourism. Visitors to Trevin Farms can learn about making goat cheese from the dairy goats on the farm, and they can enjoy a breakfast made from local ingredients when they stay overnight at the farm's bed and breakfast. With assistance from Extension, Trevin Farms reports, "We were able to implement our business plan and reach our goal of increased profitability. Revenue was more than double our expectations."

Beth Kennett, farmer and innkeeper at Liberty Hill Farm in Rochester regularly participates in workshops and programs offered through Extension. Liberty Hill Farm's dairy business is complemented by a farm stay bed and breakfast and maple syrup production. With three generations of the Kennett family living on the farm, diversification is key for providing stability when milk prices fluctuate. Beth explains, "I am constantly learning and adapting my business. There are few opportunities other than Extension for business development relevant to agritourism, and the schedule of life on the farm precludes me from finding other sources of continuing education." One workshop stands out for Beth as being particularly helpful: Website Tools for Farms, a hands-on workshop held in a computer lab for farms interested in agritourism and direct sales. The workshop was partially funded by a grant from the National e-Commerce Extension Initiative. According to Beth, "I discovered that my website was 'pretty' but absolutely dysfunctional. Search engines were not able to read it and, therefore, not many visitors were finding my website and making reservations at the farm stay." Beth worked with a web service provider found through Extension resources and quickly experienced a dramatic increase in her farm stay business reservations. Beth states, "You don't know what you don't know. I am so thankful that Extension offered this workshop and I attended. It made all the difference in our financial viability last year. Without the changes made, our farm might not have survived the downturn in the economy."

Lisa Chase, Ph.D.
Director, Vermont Tourism Research Center
University of Vermont, USA

Sources

"Agritourism," by L.C. Chase, 2008, In *Encyclopedia of Rural America: The Land and People*, 2nd ed., G.A. Goreham (Ed.) pp. 70–74, Grey House Publishing, Millerton, NY. Copyright 2008 by Grey House Publishing.

Chase, L.C., B. Amsden, and D. Kuehn. 2013. Measuring quality of life: A case study of agritourism in the Northeast. *Journal of Extension* 51(1): 1FEA3. www.joe.org/joe/2013february/a3.php

For more information:
University of Vermont Extension and other agricultural service providers continue to support farms interested in education, tourism, and direct sales. Visit www.uvm.edu/tourismresearch/agritourism/ for resources and research for farms and service providers working with agritourism and culinary tourism.

Keywords

Arts-based development, heritage and cultural resources, eco-tourism, community concept and place marketing, social action via arts.

Review Questions

1 Why is it important to include arts and culture in community development?
2 Describe a situation you have observed in a community where culture was not recognized or honored properly. What could be done to remedy this, from a community development perspective?
3 Select a community with which you are familiar, and outline how one or more of the approaches in this chapter could be applied.
4 Why has ecotourism gained interest?
5 Discuss at least one positive benefit of historic preservation, and how it integrates into community development approaches.

Notes

1 Launched in 1991, LEADER (Liaison Entre Actions pour le Development d'lEconomie Rurale) is a Rural Development Programme part-funded by the European Union. It is administered by local companies also known as Local Action Groups (LAGs) who distribute grants and other supports to projects within their areas. The PEACE III Programme is the part-EU-funded EU Programme for Peace and Reconciliation in Northern Ireland and the Border Region of Ireland.
2 Amin, A. and J. Roberts (2008). Knowing in action: beyond communities of practice. *Research Policy*, 37, 2: 353–69.

Bibliography

Aquino, J., Phillips, R., & Sung, H. (2013). Tourism, culture, and the creative industries: Reviving distressed neighborhoods with arts-based community tourism, *Tourism Culture and Communications*, 12(1): 5–18. DOI: http: //dx.doi.org/10.3727/109830412X13542041184658

Brecht Forum. (2014). Accessed January 5, 2014, at: http: //brechtforum.org/abouttop

Brennan, M.A., Flint, C.G., & Luloff, A.E. (2009). Bringing together local culture and rural development: Findings from Ireland, Pennsylvania and Alaska. *Sociologia Ruralis*, *49*(1), 97–112.

Bunnell, G. (2002). *Making Places Special, Stories of Real Places Made Better by Planning*, Chicago, IL: American Planning Association.

Chhabra, D. (2010). *Sustainable Marketing of Cultural and Heritage Tourism*. London: Routledge.

Creative City Network of Canada. (2005). Making the case for culture: Culture as an economic engine. Retrieved from www.creativecity.ca/making-the-case/culture-economic-engine.pdf

Currid, E. (2007). How art and culture happen in New York. *Journal of the American Planning Association*, *73*(4), 454–467.

Fennell, D. (1999). *Ecotourism: An Introduction*, London: Routledge.

Florida, R. L. (2002). *The Rise of the Creative Class: And How It's Transforming Work, Leisure, Community and Everyday Life*, New York: Basic Books.

Foster, D. (2009). The value of the arts and creativity. *Cultural Trends*, *18*(3), 257–261.

Fyall, A. and Garrod, B. (1998) Heritage tourism: at what price? *Managing Leisure*, 3: 213–228.

Grodach, C. (2011). Art spaces in community and economic development: Connections to neighborhoods, artists, and the cultural economy. *Journal of Planning Education and Research*, *31*(1), 74.

Holden, A. (2000). *Environment and Tourism*, London: Routledge.

Inskeep, E. (1998) *Tourism Planning. An Integrated and Sustainable Development Approach*, New York: John Wiley and Sons.

Korza, P., Brown, M., & Dreeszen, C. (2007). *Fundamentals of Arts Management*. Amherst, MA: University of Massachusetts.

McGrath, B. & Brennan, M.A. (2011). Tradition, cultures and communities: exploring the potentials of music and the arts for community development in Appalachia. *Community Development* 42(3), 332–350.

Matarrita-Cascante, D., Brennan, M.A., & Luloff, A. (2010). Community agency and sustainable tourism development: The case of La Fortuna, Costa Rica. *Journal of Sustainable Tourism, 18*(6), 735–756.

Murphy, P. (1988). Community driven tourism planning. *Tourism Management*, *9*(2), 96–104.

National Trust for Historic Preservation. (1999). *Getting Started, How to Succeed in Heritage Tourism*, Washington, DC: National Trust.

National Trust for Historic Preservation. (2013). The Main Street Four-Point Approach™ to commercial district revitalization. Accessed January 13, 2014, at: www.mainstreet.org/content.aspx?page=47 andsection=2

National Trust for Historic Preservation. (2014). Four-Point Approach. Accessed May 25, 2014 at: www.preservationnation.org/main-street/the-approach/

Paddison, R. (1993). City marketing, image reconstruction and urban regeneration. *Urban Studies, 30*(2), 339–349.

Phillips, R. (2002). *Concept Marketing for Communities: Capitalizing on Underutilized Resources to Generate Growth and Development*, Westport, CT: Praeger Publishers.

Phillips, R. (2004). Artful business: Using the arts for community economic development. *Community Development Journal, 39*(2), 112.

Phillips, R. & Stein, J. (2011). An indicator framework for linking historic preservation and community economic development. *Social Indicators Research*, 113: 1–15, DOI 10.1007/s11205-011-9833-6

Polèse, M. (2011). The arts and local economic development: Can a strong arts presence uplift local economies? A study of 135 Canadian cities. *Urban Studies*, Advance online publication. DOI: 10.1177/0042098011422574

Pratt, A.C. (2009). Urban regeneration: From the arts 'feel good' factor to the cultural economy: A case study of Hoxton, London. *Urban Studies, 46*(5&6), 1041–1061.

Reid, D., Mair, H., & George, W. (2004). Community Tourism Planning of Self-Assessment Instrument. *Annals of Tourism Research*, 31(3): 623–639.

Smith, K.L. (2007). Historic preservation meets community development. *Communities & Banking, 18*(3), 13–15.

Stern, M.J. & Seifert, S.C. (2010). Cultural clusters: The implications of cultural assets agglomeration for neighborhood revitalization. *Journal of Planning Education and Research, 29*(3), 262–279.

Strom, E. (1999). Let's put on a show! Performing arts and urban revitalization in Newark, New Jersey. *Journal of Urban Affairs, 21*(4), 423–435.

Zukin, S. (2010). *Naked City: The Death and Life of Authentic Urban Places*. New York, NY: Oxford University Press.

Connections

Americans for the Arts. Online. Available at: <www.americansforthearts.org/issues/comdev/index.asp> Great resource! Books such as the "Field Guide" list arts organizations throughout the US. Also see the economic impact calculator feature—"Arts and Economic Prosperity Calculator"—under the community development section.

Community Arts Network. Online. Available at: <www.communityarts.net/links> See the "Links for Arts and Community Development" for examples of projects and activities throughout the US.

See Playing for Change for inspiring stories of connecting musicians around the world via music, for community development outcomes at: <www.playingforchange.com>

Downside Up is an inspiring film about how art can change a community and serve as the basis for revitalization. The National Endowment for the Arts has partnered with New Day Films to bring this documentary to each state by loaning the film for viewing in your community. To obtain the film, go to: <www.downsideupthemovie.com>

Art and Social Justice Education is a book offering inspiration and tools for educators to craft critical, meaningful, and transformative arts education curriculum and arts integration projects. See more information and a companion website at: <www.routledge.com/cw/quinn>

Incubating the Arts: Establishing a Program to Help Artists and Arts Organizations Become Viable Businesses by Ellen Gerl is a very helpful book. Published by the National Business Incubation Association and available online at: <www.nbia.org>

International Downtown Association. Good resources by topic for members. Online. Available at: <www.ida-downtown.org>

National Endowment for the Arts. Online. Available at: < http: //arts.endow.gov/> Various links, resources, and information on grants.

National Main Street Program of the National Trust for Historic Preservation. See "Success Stories" as sometimes these are themed in their redevelopment strategies. Online. Available at: <www.mainstreet.org>

Travel and Tourism Research Association (TTRA). Numerous publications for travel and tourism information. Online. Available at: <www.ttra.com/>

World Tourism Organization (WTO). Resources for promoting tourism-based development. Online. Available at: <www.unwto.org/index.php>

19 Housing and Community Development

Rhonda Phillips

Overview

Housing is the most prevalent land use in communities. It is essential for supporting residents and without a strong component of housing, communities cannot be sustained for any length of time. This chapter provides an overview of housing, and its connections to community development through the years as well as newer approaches including ecovillages and cohousing.

Introduction

In most communities, housing plays a dominant role in creating the sense of place and providing for basic needs to sustain well-being. A variety of types of housing exist in many areas, from government-provided public housing to privately owned individual and group units. Depending on policy, the residential built environment can look very different across communities, and countries. The level of home ownership can vary widely, having implications for community planning and development outcomes as well. Recent economic downturns have widely influenced the ability of residents to purchase homes, with many losing homes in foreclosure. Affordability, and the ability to qualify for home ownership, has intensified. Aging populations place new demands on existing housing supply while questions of density continue. All these factors influence community development and its impact on community well-being and quality of life.

Approaches to community development are reflected in housing infrastructure. In the US, long-standing policies since the 1940s have encouraged single-family home ownership, provided by private developers. The resulting sprawling patterns of development evidenced throughout the country are in many ways a result of these housing policies. Dependence on automobile transportation has led to a variety of issues for cities, their suburbs, and surrounding regions for decades. Development predominately driven by both consumer demand and private sector supply looks very different from that guided by public policy throughout the countries of Western Europe and other regions of the world. For example, in Taiwan, nearly half of the country's eight million residents live in public housing. Many of these developments are massive and very high density; the Kin Ming Estate project consists of ten blocks housing 22,000 people.

At the same time, the need for housing units in the US was so great after World War II that suburban development seemed the best option. Other results of this era of housing policy included expansion of federal mortgage backing, leading to an explosion of growth in the mortgage industry, as well as urban renewal programs. Results of the urban renewal programs were very mixed, ranging from build up of successful commercial areas in most major cities to slum clearance occurring in communities, predominately in areas of minority populations. Resulting displacement of neighborhoods and concurrent rise of public housing in urban cores led to social issues that had not been evidenced at that scale prior. Community development responses to these social issues, focusing predominately on housing provision, proliferated in the 1960s with establishment of community development corporations (CDCs). The number of CDCs focusing on housing in the US was in the thousands at that time and continuing into future decades. There has been more consolidation of CDCs and related types of nonprofits due to issues in the economy; however many of these groups have long worked to counter the "effects of a swinging housing market, bad loans, and vacant properties" (NACEDA 2010: 1).

A Bit of History

The history of community development is intricately tied to housing, given a legacy of housing work in the 1800s and early 1900s. Jane Addams may be one of the best known advocates and reformers, both for housing and other aspects of society such as women's rights. She visited the first settlement house in the world, Toynbee Hall in London in 1887. Inspired, she returned to Chicago to establish Hull House in 1889 and to push for tenement code reform. Her work in social reform led was recognized with a Nobel Peace Prize in 1931. Jacob Riis in the noted work, *How the Other Half Lives: Studies among the Tenements of New York* (1890), was a photojournalist who called attention to deplorable housing conditions. He proposed model tenements as part of the answer to the overwhelming conditions. Later, Clarence Stein, a planner, promoted the idea that design was essential to helping create communities that are more equitable. Stein's work was dedicated to "regeneration of public life by making communities that would serve truly public needs," including those serving the underrepresented (Bose et al. 2014: 1). Good design for safe and quality housing was a major focus of his work, as well as others pushing for reform and new approaches as reflected in New Town planning, the Garden Cities concepts, and supportive urban policy (Bose et al. 2014). These reformers paved the way for subsequent generations to continue to push for progress on housing conditions, supply, and policy reforms.

Both unintended consequences of government policies and lack of response by society led to the need for a nonprofit response to issues created in the social context. Community development, with this strong origin in advocacy from prior generations, responded. Advocates pushed for reform and voice for those underrepresented, and along with the civil rights movement of the 1950s and 1960s, attention began to be called to housing discrimination. Paul Davidoff, a civil rights activist and community planner, pushed strongly on the issue of housing discrimination, influencing others to follow suit and address inequities.

Housing Types

Housing reflects demographic and social trends, as well as government policy and private sector market influences. Demographics have a major influence on housing development throughout the world. For example, in some countries the preference for single-family homes is clear while others prefer smaller units in multi-family buildings. Aging populations also influence housing markets, including types of units offered along with services and amenities to appeal to various market segments. Income has a direct impact on housing types and choices, and can be evidenced in the physical characteristics of communities.

Social trends often correlate with demographic changes, although sometimes a reflection of government policy as well. China, for example, has long encouraged a one child per family policy, and although that policy is now being relaxed, it has influenced the type and scale of housing. Lifestyle preferences with some desiring less commuting times or a more active urban environment have influenced housing supply as well. High-density developments, neo-traditional designs, and newer approaches have resulted in response. The New Urbanist approach, started in the US in the 1980s centers on high-density, pedestrian-oriented lifestyles with housing reflecting these urban design imperatives. In many ways, this approach tries to recapture what existed prior to rapid suburbanization, and still exists in many places throughout older countries in the world.

There are widely varying types of housing and it is difficult to suggest a typology. The nature and structure of housing will reflect market forces, government policies, demographic and social trends. Table 19.1 provides a short overview of various types of housing.

Table 19.1 *Housing Typology*

Conventional
Single-family, detached
Constructed housing
Single-family, attached
Row house
Town house
Multifamily
Low-rise (1–4 stories)
Medium-rise (5–9 stories)
High-rise (10 or more stories)
Factory-built housing
Manufactured
Modular
Mobile homes & trailers
Group quarters
Commercial Dormitories & rooming houses
Hotels & motels
Organizational
Fraternities & sororities
Seminaries & convents
Shelters
Institutional
Boarding schools
Military bases
Jails

Source: J. Macedo, 2009. Housing and Community Development. In R. Phillips and R. Pittman, *Introduction to Community Development*, London: Routledge.

BOX 19.1 HOUSING POLICY INNOVATION IN EUROPE: PUBLIC HOUSING FOR YOUNGER GENERATIONS AND THE ROLE OF RODA—PARENTS IN ACTION, A COMMUNITY INITIATIVE IN ZAGREB, CROATIA

The Welfare Innovations at the Local Level in Favor of Social Cohesion project (WILCO) is a three-year European Union-funded project (2010–2013) which compared 77 cases of socially vulnerable populations in two cities from ten countries aimed at the young unemployed, women and migrants. Approximately 16% of the populations in European countries are considered socially vulnerable. The project was coordinated by ten European Universities and two European research networks. The goal of the project was to understand how these innovations affect social inequalities, favor social cohesion and can be transferred to and implemented in other settings. The project had three specific goals:

1 To identify innovative practices in European cities and the factors that make them emerge and spread.
2 To set them against the context of current social problems and urban policies.
3 To make recommendations how to encourage local social innovation.

This spotlight focuses on one case—how one community-based initiative's innovation created change for a socially vulnerable population, youth. Housing prices during the studied period have increased everywhere, leading to problems of affordability for vulnerable people superficially younger generations. Over time, reductions in state-funded public support in Europe and a decentralized housing policy have led to decreased resources available to support local housing policies and programs for socially vulnerable populations.

In Zagreb, Croatia, it was well documented that there was a lack of affordable housing for younger generations due to the inability to have a controlled market and the inability to access housing loans. A community-based civic initiative, RODA—Parents in Action, sought to assist the socially vulnerable younger families to obtain decent permanent housing. RODA, by promoting and advocating the rights of children has impacted changes in public policies for children, women and families (Evers et al. 2014: 38). RODA challenged the norm that familial well-being from safe housing to education was strictly a private matter and perceived as the responsibility of parents in traditional societies and their extended families. This expansion of the initiative's goal came about due to the increase in the number of young families with children living in Zagreb, a large city, without the support of parents and close relatives.

In a survey by a WILCO project member of civil society organizations, RODA documented how their members live in terrible housing conditions: a four-member family of young generations live in a flat of 26 m^2. Affordability is a crucial problem for young families and it is a serious obstacle for them having more children, www.wilcoproject.eu/book/chapters/about-this-book, p.42, n. 5). Additionally, permanent housing is required for kindergarten-age children to attend school. Lack of affordable housing was now becoming both a public health and education issue for Croatia's city-dwelling younger families.

The specific challenge to this crisis was the lack of coordination of governmental programs serving this socially vulnerable population. The local public housing authority was not tasked with finding affordable housing for younger generations, but rather, those who were defined as economically poor. Policy makers recognized that it was time for innovation and paid attention to this civic initiative. The innovative idea gained traction and was placed on the political agenda and came before the city council. Various stakeholders from politicians to administrators to civic organizations participated in the public debate. The innovation was to blur the line between those who were socially vulnerable with those who were economically vulnerable to solve the housing crisis. Tenants had to have an income of at least 30% of the average income in Zagreb. Tenants would make a down payment and sign a contract that they would pay timely rent and keep the premises in good condition. Tenants in public housing are often not willing to pay rent and other costs related to housing such electricity, gas, heating, water, and the communal fee, etc. These services were subsidized by the welfare state.

The innovation was found to be economically efficient and sustainable given the terms for public rental. Families with kindergarten-age children were able to register and send their children to school. Family social cohesion most likely will improve but these results were not captured in the study. An unintended benefit was the how enthusiastically public housing officials embraced the program. Through this innovation administrators had a "sense of a social investment in a program with very viable returns in the near future." In other words, public administrators who never saw or imagined a break in the cycle of economic vulnerability were able to imagine a groundbreaking shift being a part of breaking cyclical social vulnerability that leads people onto a path of economic poverty. "This innovation put the issues of planning and cooperation of different local stakeholders on the agenda of local social policy" (Evers et al. 2014: 44).

RODA's success is an identifiable social policy innovation that can be replicated across Europe, the context of the project. The RODA project is an innovation which also can shine a beacon of light on public and private actors worldwide looking for planning and policy solutions for similar housing crisis facing young families.

Patsy Kraeger, Ph.D.
Faculty Associate, Arizona State University, USA

Resources

Evers, A., Ewert, B., and Brandsen, T. (eds.) (2014) *Social Innovations for Social Cohesion*, WILCO Consortium.
www.wilcoproject.eu/the-project/
www.roda.hr/article/print/about-the-roda-association

Numerous issues and considerations impact housing, whether it is local codes and planning regulations for the location; mix, type, and density of housing; or affordability. The latter looms large for many communities. Affordability across incomes and locations continues to be an area of interaction and focus for many CDCs and communities. Some local governments control through

BOX 19.2 HOUSING AFFORDABILITY

Affordability is a crucial element of housing delivery, whether the government or nonprofit sector is involved in giving families access to housing, or whether families, regardless of income, are left to the market's devices. What does affordable mean? In the US, the government definition of "affordable" is that the expenditures in rent and utilities consume no more than 30% of a household's pretax income. Although this percentage has been used as a rule of thumb by consumers as well as financial institutions to estimate how much housing one can afford, there is no scientific explanation for the 30% figure.

Housing affordability can be achieved by different means. It can be made more affordable through innovative design, less stringent building and land development standards, lower financing costs, or improvements in other areas such as education and employment. These strategies will indirectly result in making shelter accessible to people by way of increasing their skills, marketability, and consequently, their income levels.

A number of organizations compute indices to measure the ability of typical families to afford a home. Some of these indices are:

- Affordability Index, calculated by the National Association of Realtors (NAR), measures the ability of a household earning the median family income to qualify for a conventional loan covering 80% of the median existing single-family home price in its area.
- Housing Opportunity Index, calculated quarterly by the National Association of Home Builders (NAHB), computes the median family income and the percentage of all homes sold during the quarter which a family earning the median income could afford.

Affordable housing is a crucial element in any comprehensive plan or improvement program, be it local or regional. In an attempt to promote a better quality of life for the entire population, regardless of income level, community planning must address the issue of housing and relate it to all other basic components of a plan such as health care, education, and employment. Programs based on a comprehensive examination of current conditions and realistic projections have a better chance to succeed than those dictated by policies whose foremost interest is creating a final product, whether it suits the target population or not.

Delivery of Affordable Housing

Different approaches have been taken in the development of affordable housing programs. Two of the most traditional approaches are the supply-side and demand-side approaches. In basic terms, supply-side solutions are public housing and the incorporation of options with the private providers, while demand-side solutions consist of improved financing or other subsidies and incentives such as negative income tax payments or rent certificates.

- Supply-side approach: the main characteristic of supply-side assistance programs is that the government builds or subsidizes new housing to be occupied by those who meet the established criteria. Supply-side policies generally increase housing consumption.
- Demand-side approach: the main characteristic of demand-side assistance programs is that existing housing stock is occupied by families who need shelter and receive government assistance. The government gives vouchers to be used as income supplements by those who meet the established criteria to rent existing housing. This approach has been successfully used by nonprofit organizations.

Government programs established by supply-side policies, namely new public housing, are generally more expensive than those devised by demand-side policies, or cash payments. Demand-side policies tend to increase housing prices. The danger of surplus demand is that increasing housing affordability for all income levels lays the groundwork for all home prices to increase. Equilibrium can be eventually achieved because home builders and developers would recognize advantageous opportunities in building more, thus increasing the supply and subsequently driving prices down.

Source: Joseli Macedo (2009), Housing and Community Development, in R. Phillips and R. Pittman, *Introduction to Community Development*, London: Routledge.

affordability policies, or rent control. A good example of enhancing affordability through policy rather than rent control is found in Burlington, Vermont. The City of Burlington founded the Champlain Housing Trust in 1984 to provide affordable housing to low- and moderate-income residents. In 2008 this organization received the World Habitat Award at UN-HABITAT's global celebration of World Habitat Day. It recognizes the Trust's permanent affordable housing programs as innovative, sustainable and transferable (Phillips et al. 2013).

Approaches to Housing

There are numerous approaches to housing, from public sector, to private sector, to nonprofit or civic sector initiatives. In some cases, joint partnerships across sectors work well to provide housing. In many communities, nonprofits or the civic sector lead efforts, and because of the implications for community development, the remainder of this section will focus on a few examples of this type approach—cohousing and ecovillages.

Throughout the world, the idea of cohousing is gaining interest. Cohousing is essentially collaborative housing with residents managing their neighborhoods. It may be initiated directly by groups of residents, local government, or a nonprofit promoting housing development. Real estate ownership may be individually titled units, similar to condominiums or a housing cooperative structure with common space maintained by community residents. Financing is easier to obtain in the US when units are individually titled, versus in a cooperative model. The United States cohousing organization CohoUSing.org confirmed confirms more than 120 active cohousing communities in the country as of 2013 (Coho*US*ing, 2013). At least six of the communities are designated specifically to people over the age of 55. In 1991, Davis, California became the first site in the US to have a cohousing project.

While cohousing serves limited populations, it represents a community development approach that attempts to meet several needs. Cohousing projects include a process for incorporating resident voice reflecting a basic tenet of community development, the desire for democratic participation. This participatory process is key to success of community development projects, and particularly for housing. It is explained below, along with other characteristics of cohousing according to the organization, Coho*US*ing:

- *Participatory process.* Future residents participate in the design of the community so that it meets their needs. Some cohousing communities are initiated or driven by a developer. In those cases, if the developer brings the future resident group into a process late in the planning, the residents will have less input into the design. A well-designed, pedestrian-oriented community without significant resident participation in the planning may be "cohousing-inspired," but it is not a cohousing community.
- *Neighborhood design.* The physical layout and orientation of the buildings (the site plan) encourage a sense of community. For example, the private residences are clustered on the site, leaving more shared open space. The dwellings typically face each other across a pedestrian street or courtyard, with cars parked on the periphery. Often, the front doorway of every home affords a view of the common house. What far outweighs any specifics, however, is the intention to create a strong sense of community, with design as one of the facilitators.
- *Common facilities.* Common facilities are designed for daily use, are an integral part of the community, and are always supplemental to the private residences. The common house typically includes a common kitchen, dining area, sitting area, children's playroom and laundry, and also may contain a workshop, library, exercise room, crafts room and/or one or two guest rooms. Except on very tight urban sites, cohousing communities often have playground equipment, lawns and gardens as well. Since the buildings are clustered, larger

sites may retain several or many acres of undeveloped shared open space.

- *Resident management.* Residents manage their own cohousing communities, and also perform much of the work required to maintain the property. They participate in the preparation of common meals and meet regularly to solve problems and develop policies for the community.
- *Non-hierarchical structure and decision making.* Leadership roles naturally exist in cohousing communities, however no one person (or persons) has authority over others. As people join the group, each person takes on one or more roles consistent with his or her skills, abilities or interests. Most cohousing groups make all of their decisions by consensus, and, although many groups have a policy for voting if the group cannot reach consensus after a number of attempts, it is rarely or never necessary to resort to voting.
- *No shared community economy.* The community is not a source of income for its members. Occasionally, a cohousing community will pay one of its residents to do a specific (usually time-limited) task, but more typically the work will be considered that member's contribution to the shared responsibilities (Coho*US*ing 2013: 1).

Cohousing locations can be built from undeveloped land or can be conversion of old industrial buildings or other underutilized sites. The sites may be privately owned or financed by a cooperative of residents or the nonprofit sector. Some communities are retrofitting existing properties to develop cohousing projects. There are several benefits of cohousing projects, which include the following:

- reduced cost of living due to shared resources;
- social capital—economics, care, education, and culture;
- sustainability;
- low crime due to community awareness and support; and
- affordability.

Cohousing is experiencing rising popularity worldwide. It first began in Denmark in 1972 outside of Copenhagen when 27 families developed a cohousing project. As of 2010, there were 700 cohousing projects in this small country (McCament and Durrett 2011). The idea of cohousing spread throughout Scandinavia which now has the largest number of cohousing projects worldwide. Cohousing in the United Kingdom began in the 1990s. As of late 2013, there are 14 existing cohousing communities with an additional 40 in the planning or building stages. Other countries are following suit, with more projects planned in New Zealand, Australia, Spain, France and beyond.

The ideas and inspiration to be more sustainable has resulted in another form of cohousing, ecovillages. Global Ecovillage Network defines it as "an intentional or traditional community using local participatory processes to holistically integrate ecological, economic, social, and cultural dimensions of sustainability in order to regenerate social and natural environments" (2013: 1). From Scotland to South Africa, ecovillages are springing up from within communities, often by collaborative efforts of residents. Ecovillages may include economic sharing of resources and with the emphasis on green housing and

BOX 19.3 BURLINGTON, VT, COHOUSING EAST VILLAGE

The City of Burlington, Vermont, helped develop the Burlington Cohousing East Village, a 32-home development offering shared common spaces, with community gardens, outdoor areas and shared laundry facilities. In 2010 this project received the Home Depot Foundation Award for Sustainable Community Development as an exemplar for best practices in housing, natural resources, land use and development.

sustainability measures, they differ from cohousing. There are numerous ecovillages springing up across the globe, including as best estimate at least 500 in the US (Kirby 2003). Part of the reason that both cohousing and ecovillages are gaining in popularity is connected to the lack of community and sense of belonging that results from sprawling patterns of development. Houses that are widely separated and isolated tend to not foster opportunities for getting to know neighbors and building relationships in the community.

Conclusion

The types of housing vary widely among communities, reflective of demographics, social trends, economics, policies, and market forces. Community development organizations have long played a role in housing, and continue to tackle challenges in the current environment. Resident or civic-sector-led initiatives have increased recently, including with such innovative approaches as ecovillages incorporating sustainability concepts and affordability. Cohousing is another approach that is rapidly gaining popularity as a way to connect with community.

CASE STUDY: SPRINGHILL COHOUSING, STROUD, ENGLAND (AS TOLD BY RESIDENT MAX COMFORT)

Springhill cohousing in Stroud, Gloucestershire is a cohousing project in the UK, based on the original Danish principles. It was the brainchild of David Michael, whose offer for the site—2 acres on a south-slope and within 5 minutes walk of the High Street—was accepted in the summer of 2000. He then went about gathering a group around him who all became Directors and shareholders in the limited company that was to own the site. By September that year, when Contracts were exchanged, some 15 households had signed up, each paying £5,000 for 5,000 £1 shares.

At this point, architects were engaged to work with the group in laying out the site and designing the homes—three-, four-, and five-bedroom houses and one- and two-bedroom flats. During this process it was decided to add 12 studio units to the mix. Once the site layout was agreed—positioning the Common House in relation to everything else was crucial—everyone chose their plot on a "first come first served" basis and paid their plot fee. In the case of the five-bed houses this was £36,000, for one-bedroom flats it came to £18,000. The money thus raised paid back David's deposit of £150,000 and completed the purchase of the site (£550,000) as well as establishing a fund to cover professional fees and admin costs.

While more households joined the group over the winter of 2000–2001 (28 by end of March) a planning application was submitted to Stroud District Council. It had the full support of Stroud Town Council and the Chamber of Trade, who saw this as another feather in the town's cap. However, although the planning officers recommended approval, a group of Councillors decided to oppose the application on the grounds that we "were going to eat together, would be bulk purchasing and dragging our furniture across the park at night"! The application was thrown out and the officers then had the task of dreaming up "genuine" reasons for refusal. The group appealed and simultaneously reapplied, and in September 2001 we had our permission.

Meanwhile we worked closely with Architype in designing each individual house and flat type. Once a generic design had been agreed, individual households were able to customize their homes, resulting in big variations in layout and accommodation in all the homes.

At the same time we realized that we couldn't get "self-build"-type mortgages on leasehold homes (ours were all to be on 999 year leases, with each household having a share and being a Director of the freehold-owning company, a legal device that is well tried and tested), so David began negotiations with Triodos Bank for a commercial loan to cover the build cost—estimated then at £3.4M on the basis of £70/ft^2. Despite their excellent reputation, we found the Triodos negotiations cumbersome and eventually turned to the Co-op Bank, who could not have been more helpful. A facility was agreed and once we drew down the first monies, we all started sharing the interest payments—those who were to be living in the biggest homes paid the most interest—around £300/month.

Finding a contractor was a challenge. What we hadn't realized, and were not told by our cost consultants, was that the housing market was overheated and builders were really only interested in nice flat sites, traditional noddy-box houses and a conventional developer client (ex-accountants in smart suits!). Naturally, a difficult site with unconventional homes (we were probably the first triple-glazed housing development in Gloucestershire) and a "dead weirdy" client, was not very attractive to contractors intent on making as much hay as they could while the boom still shone. We went through partnering, design-build and, after a frustrating year during which we almost gave up (silly quotes of £6.7M and £7.2M from two builders) we settled on a Management contract (cost-plus) with a local builder, John Hudson. The contract suffered delays and significant price hikes (from our somewhat naïve £70/ft² to £120/ft²) but, eventually, in September 2003, the first phase of choosers moved in; by the Spring of 2005, the project was complete, with the Common House built last, contrary to our original intentions. Final build cost was around £4.2m.

Now, three years on, it all seems very normal to us but we have regular—and increasing—groups of visitors on our open days, most intent on replicating what we have achieved.

A few stats:

- 34 units on two acres; 6 x five-bed houses, 6 x four-bed houses, 8 x three-bed houses, 4 x two-bed flats, 4 x one-bed flats and 12 studio units, some of which have been amalgamated to form maisonettes.
- 75 people approx., including 25 children, ranging in age from 3 days to 70-year-olds and over.
- A mix of single men and women, couples without children, single-parent families and families with up to four children.
- One dog and 14 cats (7 per acre quota).
- Parking is to one side of the site (one per household in our planning approval) with the rest of the site fully "pedestrianised".
- All houses and flats are timber frame with 150mm Warmcell insulation, and triple glazed. All houses have photovoltaic tiles which were monitored by the DTI, who gave us a £321,000 grant.
- We have SUDS (sustainable urban drainage system) which worked perfectly in the dreadful floods of July 2007.
- We eat together (if we wish) every Wednesday, Thursday and Friday evening, with a "pot-luck" on Saturday. We all have to cook at least once a month. Meals are vegetarian despite a majority of meat and fish eaters (it's easier that way) and cost us £2.60.
- All homes are completely self-contained and there is no compulsion to "join in"; however, the Common House is treated as an extension of our living rooms and we have 24/7 access to it.
- We are required to do 20 hours community work a year, consisting of deep-cleaning the Common House, looking after the boundaries, and other maintenance tasks.
- We have a very low turnover (four units in three years), with most movements occurring within the community; only one home made it onto the open market. If we wish to sell, we have to give 28 days' notice to the freehold-owning company of our intention to do so at a certain price (which we choose) and they have the right to put forward people from the waiting list to make an offer, which we are not obliged to accept, however.
- When homes change hands, they do so at around 15% to 20% premium over similar properties in Stroud. By being our own developer, we have gained from the notional 20% developer's profit and the value of our homes significantly exceeds the build cost.

Lessons learned:

- Pick the professional team very carefully: we experienced difficulties in that some of the professionals didn't respect us and this, in our experience, led to less than full service in some cases.
- Don't beat the professionals down too much on their fees—you need them to be loyal enough to come out at weekends.

- Use an architectural practice to do the design but consider a seasoned surveyor to run the project on site. Don't use internal project managers but do use an external one.
- Don't individualise the homes—it's too much for the professionals to cope with and probably puts the price up.
- Get good advice on realistic build costs before you begin detailed planning.
- Spend time on growing the group—it doesn't work to do the build first and tip the people in at the end, despite what the Americans will tell you.
- Get the "storming" bit of forming, storming and norming over as soon as possible.
- Expect endless meetings and having to make major decisions in a hurry with very little information.
- It needs a strong, bloody-minded and very, very determined individual or small group to get cohousing going in this country, but there are now individuals and companies out there who are experienced in cohousing and are supporting would-be cohousers, working closely with developers, local authorities, carefully selected professionals and land-owners.

Source: Max Comfort, Springhill, December 12, 2008 (www.cohousing.org.uk/springhill-cohousing).

Keywords

Housing, cohousing, affordability, planning, public housing, ecovillages, nonprofit organizations.

Review Questions

1 What is the role of government in housing?
2 What is the role of nonprofits and the private sector in housing?
3 Why are the connections between community development and housing strong?
4 How do current trends in housing, such as cohousing and ecovillages, impact community development outcomes?

Bibliography

Bose, M., Horrigan, P., Doble, C., and Shipp, S. (2014). *Community Matters: Service-Learning in Engaged Design and Planning.* London: Routledge.

Coho*US*ing. What Is Cohousing? November 22, 2013, www.cohousing.org

Global Ecovillage Network. What is an Ecovillage? November 22, 2013, http: //gen.ecovillage.org/index.php/ecovillages/whatisanecovillage.html

Kirby, A. (2003). Redefining social and environmental relations at the ecovillage at Ithaca: A case study. *Journal of Environmental Psychology* 23(3): 323–332.

McCament, K. and Durrett, C. (2011). *Creating Cohousing, Building Sustainable Communities.* Gabrieola Island, British Columbia: New Society Publishers.

Macedo, J. (2009). Housing and Community Development. In R. Phillips and R. Pittman, *Introduction to Community Development,* London: Routledge pp. 249–65.

National Alliance of Community Economic Development Associations. (2010). "Rising Above: Community Economic Development in a Changing Landscape." Washington, DC: NACEDA.

Phillips, R., Seifer, B., and Antczak, E. (2013). *Sustainable Communities: Creating a Durable Local Economy.* Oxon., England: Routledge.

Riis, J. A. (1890). *How the Other Half Lives: Studies among the Tenements of New York.* New York: Scribner's Sons.

Connections

Read the papers of activist Paul Davidoff at the Cornell Library: http://rmc.library.cornell.edu/EAD/htmldocs/RMM04250.html

Visit a museum to learn more about the rich history of Jane Addams' advocacy work in housing: www.uic.edu/jaddams/hull/hull_house.html

See where ecovillages are located the world over at: http: //gen.ecovillage.org/index.php/ecovillages/worldmap.html

Learn about approaches to sustainable community development as related to housing with a multitude of case studies available from the Canada Housing and Mortgage Corporation (Canada's National Housing agency) whose goal of providing these cases is to promote community livability. Providing examples of best practices in design and development, there are a variety of ideas and tools discussed at: www.cmhc-schl.gc.ca/en/inpr/su/sucopl/

Check to see if a U.S. community has enough affordable housing with this interactive map: www.urban.org/housingaffordability/?utm_source=iContact&utm_medium=email&utm_campaign=UI%20Update&utm_content=Mar+2014+-+1st+Thursday

Explore data sources:

Public Housing:
http: //portal.hud.gov/hudportal/HUD?src=/topics/rental_assistance/phprog

Community Planning and Development:
http: //portal.hud.gov/hudportal/HUD?src=/program_offices/comm_planning

Homeownership and Rental Vacancy Information:
www.census.gov/housing/hvs

American Housing Survey:
Most comprehensive national housing survey in the United States www.census.gov/housing/ahs/

2011 United States Housing Profile:
www.census.gov/prod/2013pubs/ahs11-1.pdf

20 Neighborhood Planning for Community Development and Revitalization

Kenneth M. Reardon

Overview

Efforts to encourage local residents, institutional leaders, and elected and appointed officials to work together to improve the quality of life for those living in economically distressed communities has been an important historic concern for American city planners. Our nation's increasingly uneven patterns of metropolitan development that has widened the income, wealth, and power gap separating rich and poor communities in our society has led to a renewed interest among city planners and other design professions in resident-led neighborhood stabilization and transformation. This chapter provides an introduction to various forms of participatory neighborhood planning and development which have, in recent years, enabled community-based development organizations serving our poorest communities to plan and implement increasingly complex and transformative economic and community development projects.

Introduction

Interest in participatory approaches to neighborhood planning has skyrocketed among city residents, professional planners, elected officials, and urban scholars during the past two decades. This renewed interest in resident-led planning is the result of a number of powerful economic, social, and political trends affecting our nation's major metropolitan regions. The increasingly uneven pattern of development characterizing many of our metropolitan areas has led to a disturbing expansion in the number of economically distressed neighborhoods where the quality of life is often shockingly low. The failure of Urban Renewal and other centrally conceived revitalization strategies to address the critical economic and social problems confronting these neighborhoods has undermined public confidence in and support for top-down urban regeneration efforts. Federal cuts in intergovernmental assistance to local governments have increased the burden local, county, and state government must shoulder for economic and community development. These budget cuts have forced increasing numbers of cash-strapped villages, towns, and cities, hurt by our most recent recession, to transfer responsibility for these programs to local nonprofit agencies and community-based organizations. The total quality management movement, which stresses the importance of continuous improvement in the quality of service delivery to an increasingly diverse citizenry, has encouraged municipal planning directors and city managers to emphasize more participatory approaches to governance. Community development and planning professionals have also been encouraged to adopt more collaborative forms of practice due to increasing pressure from cultural identity groups, including African-Americans, Latinos, Asians, and Native Americans, seeking a greater voice in public policy decisions affecting their communities. Finally, public and private funders of urban revitalization are increasingly mandating active participation of local stakeholders at every stage of the planning, design, and development process, further reinforcing the movement toward participatory neighborhood planning and development (Peterman 2000).

The Rising Tide

Evidence of the growing popularity of participatory neighborhood planning is widespread. There has been an explosion in the number of community-based organizations, particularly community development corporations (CDCs), involved in resident-led planning, design, and development (Brophy and Shabecoff 2001). The number and variety of municipal government planning departments that have established active neighborhood planning units is also impressive. Cities as diverse as New York, Washington, DC, Savannah, Austin, Chicago, Portland, Los Angeles, and San Diego have created specialized units to help neighborhood residents design and implement revitalization strategies aimed at improving their quality of life. Within the last ten years, the Annie E. Casey Foundation, the nation's largest family foundation, has partnered with the American Planning Association to sponsor two national symposiums to establish principles of good practice and alternative models for collaborative neighborhood planning. In recent years, growing numbers of colleges and universities have established ambitious community/university partnership programs and centers to support collaborative approaches to urban problem solving, planning, and development. Those involved in these efforts, along with scholars who have documented and evaluated this work, have contributed to important practice-oriented publications such as *City Limits*, *Shelterforce*, *The Neighborhood Works*, and *Progressive Planning* as well as many new academic journals and books including: the *Journal of the Community Development Society;* the *Michigan Journal of Community Service Learning;* Healey's *Collaborative Planning: Shaping Places in Fragmented Societies*; Rohe and Gates' *Planning with Neighborhoods*; Jones' *Neighborhood Planning: A Citizen's Guide to Practice*; William Peterman's *Neighborhood Planning and Community-Based Development*; Forester's *The Deliberative Practitioner*; and Rubin's *Renewing Hope within Neighborhoods of Despair*.

Another indication of the growing importance of these endeavors is the number of regional and national foundations that have established ongoing programs to support resident-led neighborhood planning. Among these foundations are: The Ford Foundation, the Rockefeller Brothers Foundation, the Annie E. Casey Foundation, the MacArthur Foundation, the Surdna Foundation and the Wachovia Foundation. A final indication of the increasing importance of this work is the growing number of graduate students concentrating and/or specializing in affordable housing, economic development, and community development, in order to prepare to work as neighborhood planners.

Emerging Principles

While individual professionals may adopt somewhat different approaches to the practice of participatory neighborhood planning, an emerging set of principles of good practice have emerged in recent years. Most practitioners are committed to a model of participatory neighborhood planning that seeks to:

- Improve the overall quality of life enjoyed by poor and working class families by adopting a *place-based approach* to community development, one that emphasizes the importance of a healthy, safe, and nurturing environment characterized by high-quality public services, living wage jobs, affordable housing, and supportive local institutions.
- Involve the active participation of a broad cross section of local residents, businesspeople, institutional leaders, and elected officials as decision makers at every step of a *resident-led* research, planning, design, and development process.
- Connect previously uninvolved and often ignored residents, including youth, seniors, immigrants, ethnic/racial/religious minorities, individuals with disabilities and others, with participatory neighborhood planning processes. The involvement of previously marginalized residents *produces plans* that address the most critical issues confronting the community and *expands political support* for such plans thereby increasing their chances of adoption and implementation.
- Recognize the need to complement traditional "bricks and mortar" approaches to community revitalization by implementing *comprehensive revitalization* strategies that invest in high-quality educational, health care, public safety, job training and placement, and business development programs in order to rebuild the social capital base of economically challenged communities.
- Pursue an *asset- rather than a deficit-based* approach to neighborhood planning that builds upon the significant knowledge, skills, resources and commitment that every established community, regardless of its economic status, possesses.
- View successful neighborhood planning as a *developmental process* in which momentum generated by implementing modest, near-term projects enables local residents and their allies to build the necessary support to undertake more ambitious and complex long-term projects.
- Identify and cultivate local residents' organizing, research, planning, and development knowledge and skills to enhance the *organizational capacity* of local community-based development organizations.
- Recruit *strategic public and private investment partners* willing to commit long-term resources to support projects designed to restore the health and vitality of the community.
- Engage local residents in *municipal, regional, state, national, and international public interest campaigns* that promote both redistributive economic and community development policies and participatory decision making.
- Embrace a *reflective approach to professional practice* that systematically engages participants in critical reflection upon their past practices in order to reframe the planning problems being addressed and expand the solutions being considered, so as to maximize the positive outcomes of the neighborhood planning process.

A Short History of Participatory Neighborhood Planning

Participatory neighborhood planning has a long but often overlooked place in city planning history. Patrick J. Geddes, the Scottish botanist turned planner who made seminal contributions to early planning theory, methods, and practice, successfully encouraged residents of several of Edinburgh's poorest neighborhoods to undertake a series of slum improvement campaigns in the 1890s with the help of his students. Geddes' revitalization efforts, which he referred to as "conservative surgery," featured the systematic collection and analysis of data describing local conditions within their regional context. They also featured active resident participation in the formulation of innovative yet workable solutions to local problems, and the aggressive mobilization of local citizens and sympathetic supporters to implement key elements of cooperatively developed plans. Through his conservative surgery approach to community development, local residents transformed trash-filled lots into vest-pocket parks, converted dilapidated buildings into worker and student housing cooperatives, and enhanced the appearance and functionality of local public markets. Geddes encouraged local residents and officials to undertake increasingly challenging urban regeneration projects and to document their efforts by organizing exhibitions, adult education courses, summer school programs and extension activities at his Outlook Tower. This building was viewed by many as the first sociological laboratory dedicated to nurturing civics—the science of community-building and town planning (Welter 2002). American city planners were introduced to Geddes participatory approach to neighborhood planning and development through Lewis Mumford's articles, especially those appearing in his highly influential the "Sky Line" columns that appeared in the *New Yorker* Magazine for more than 30 years.

In the years immediately following the first national planning conference held in Washington, DC in 1909, Charles Mulford Robinson wrote a series of influential newspaper articles highlighting the importance of inspired city planning and design in the creation of vibrant, sustainable, and just cities. These articles urged residents of the nation's rapidly expanding industrial cities to initiate a wide range of physical improvement schemes at the neighborhood and municipal levels of government to restore the environmental health and enhance the aesthetic appearance of their residential communities (Robinson 1899). Robinson's campaign, which came to be known as the "City Beautiful Movement," was enthusiastically embraced by local garden clubs, women's organizations, religious institutions, and business associations. These and other organizations came together in hundreds of American cities to encourage local officials to improve basic sanitation services, establish building construction and maintenance codes, implement urban design standards, install public art, and expand public playgrounds and parks to protect the health and enhance the quality of life enjoyed by local residents. Among Robinson's many publications was his *Third Ward Catechism* (McKelvey 1948), which contained his philosophy of neighborliness and civic improvement through cooperative action.

In the 1920s, an interdisciplinary team of social scientists and design professionals under the direction of Thomas Adams created the Regional Plan for New York and its Environs. This plan was sponsored by the business-led Regional Plan Association and focused considerable attention on the design and construction of healthy residential neighborhoods in cities such as New York that were struggling to cope with intense urbanization pressures. Clarence Perry offered his Neighborhood Unit concept and designs based upon lessons learned from Forest Hills, Queens, the highly successful residential community conceived as a model working class development by the Russell Sage Foundation (Heiman 1988). Perry believed that a significant portion of the New York region's future growth could be accommodated in well-designed residential communities of 10,000 to 15,000. He envisioned their organization around common open spaces and community school centers that would serve as focal points for local civic and social life. In Perry's scheme, commercial services would be restricted to the periphery of medium-density residential communities that offered a mix of rental and homeownership housing. Traffic would be minimized to encourage pedestrian activity by laying out most of the proposed streets as cul-de-sacs following the land's natural topography. The scheme also limited the number of streets that continued through the length and width of the planned developments. In 1928, Clarence Stein and Henry Wright designed Radburn, a planned community near Paterson, New Jersey organized around pedestrian-oriented neighborhoods reflecting many of Perry's ideas (Stein 1957).

While a number of exciting neighborhood planning initiatives were advanced within the New Deal programs of the Roosevelt Administration, these were soon curtailed due to the manpower and supply needs of the nation's war production effort. As growing numbers of industrial workers crowded into northern cities to participate in wartime manufacturing and nationwide rationing prioritized the needs of the troops over those of the state-side civilian population, the housing, community facilities, and infrastructure within neighborhoods where war-related production facilities were located suffered. Following the war, returning servicemen and women quickly came to understand the impact that years of intense use and deferred maintenance had inflicted on their former urban neighborhoods. Eager to secure employment with the growing number of firms seeking lower-cost greenfield locations in rapidly expanding suburban communities, young workers soon found it possible to purchase new homes in those communities with assistance provided by their GI benefits. The movement of firms and people from older central city neighborhoods soon resulted in falling inner city property values and rising vacancy rates. These conditions made it increasingly difficult to finance the improvement of residential and commercial buildings in our major cities. As economic, social, and physical conditions in our urban neighborhoods deteriorated, local business leaders, organized through the Urban Land Institute, successfully lobbied for the passage of the Taft, Ellender, and Wagner Housing Act of 1949.

Between 1949 and 1973, more than 2,000 local communities secured federal funding to conduct large-scale clearance, infrastructure improvement, and residential, commercial, and civic development under the Federal Urban Renewal Program. Local communities initiated their Urban Renewal Programs by establishing Local Renewal Agencies (LRAs), charged with identifying "blighted areas" where occupancy rates and property values were falling. A detailed plan would then be prepared, often with the assistance of outside consultants, to demolish seriously deteriorated structures within the target area, install state-of-the-art infrastructure, construct needed public buildings and community facilities, reduce the selling prices of the remaining vacant land to stimulate private investment, and provide public insurance to those lenders willing to finance improvements within these redevelopment districts.

To support this dramatic expansion in the role of the federal government in urban real estate markets, advocates of the Taft, Ellender, and Wagner Housing Act of 1949 based their arguments on the need to improve the quality and availability of affordable housing for poor and working-class families. However, the impact of the federal Urban Renewal Program on these families was devastating. More than 600,000 housing units, most of which were affordable to those living on modest incomes, were destroyed through the Urban Renewal Program. Of the 100,000 units of new housing constructed through this program, only 12,000 were affordable to those of modest means. During the program's first five years, displaced renters, including the poor, elderly, and individuals with disabilities, received no relocation assistance. In addition, the majority of the families displaced by the activities of Local Renewal Agencies moved into substandard housing for which they paid higher rents. Furthermore, six out of ten families displaced by the Urban Renewal Program were African-Americans or Latinos. While the program produced millions of square feet of attractive retail, commercial, and performing arts space, millions of poor and working-class families of color suffered as a result of the program's clearance activities (Anderson 1966).

The widespread pain and suffering that the federal Urban Renewal Program inflicted on hundreds of poor and working-class communities of color throughout the United States produced widespread protests. Mel King, a community educator and settlement house worker in Boston's South End, organized low-income residents to occupy a parking lot where their former homes had been demolished by the Boston Redevelopment Authority. This protest continued until community leaders were placed on the BRA's citizen advisory board. The success of Boston's "Tent City" occupation encouraged other communities to stand up to the federally funded bulldozers targeting their neighborhoods (King 1981). In Lower Manhattan, the residents of the Cooper Square neighborhood, led by advocacy planner Walter Thabit, successfully opposed a clearance-oriented renewal plan proposed by Robert Moses. For more than 40 years, Cooper Square residents and leaders have doggedly and successfully pursued the implementation of their own preservation-oriented approach to community revitalization (Thabit 1961). With the support of progressive planner, Chester Hartman, residents of San Francisco's Yerba Buena community waged an inspired but unfortunately unsuccessful campaign to save this long-term immigrant enclave from federally funded clearance (Hartman 1974). Along with the growing influence of the Civil Rights movement, these oppositional planning efforts led to a significant shift in federal policies toward distressed urban neighborhoods. The passage of the Economic Opportunity Act of 1964 funded the creation of nonprofit organizing, planning, and development organizations. These Community Action Agencies used local resources to leverage outside funds to implement educational, job training, small business development, health care, and legal services designed to improve the quality of life for the urban poor. The Act used the term "maximum feasible participation" to describe the central role that the poor would play in determining future economic and community development priorities for these federally funded community transformation organizations (Moynihan 1970).

The emphasis on community-based and resident-led planning and development, in both the Equal Opportunity Act of 1964 and the subsequent Demonstration Cities and Metropolitan Development Act of 1966, prompted leaders of Cleveland's Hough Area and Brooklyn, New York's Bedford-Stuyvesant neighborhood to create multipurpose economic development organizations in the mid-1960s. These organizations sought to attract public and private investment needed to support the revitalization plans formulated by the poor and their allies. In the late 1970s and 1980s, as the Carter, Reagan, and Bush administrations sought to balance the federal budget by reducing spending on a wide range of urban development programs, the number of community development corporations exploded. Seeking to promote sustainable forms of development in our nation's low-income urban and rural communities, growing numbers of CDCs undertook an increasingly variety of affordable housing, job training, small business assistance, youth development, and public safety programs. By 2005, the National Congress for Community Economic Development (NCCED) estimated that the number of professionally staffed community development corporations exceeded 4,600 (NCCED 2006: 4). Many of these organizations were engaged in various forms of participatory neighborhood planning. In increasing numbers of low-income communities, the efforts of CDCs and other community-based nonprofits were encouraged by municipal planning agencies that collaborated with them to initiate Comprehensive Community Initiatives (CCIs). The financial support, leadership training, and technical assistance required by these efforts were typically provided by a national network of financial intermediaries that arose in the early 1970s including the Local Initiatives Support Corporation, Enterprise Community Partners, Neighborhood Reinvestment Corporation, and Seedco. In many communities, the efforts of these institutions to advance the goals of participatory neighborhood planning was reinforced by networks of civic-minded architects, landscape architects, and urban planners working with Community Design Centers sponsored by local professional associations or nearby universities.

In recent years, one of the most important sources of organizational support for participatory neighborhood planning and development in severely distressed neighborhoods has been local religious congregations and their national denominations. The Catholic Campaign for Human Development, the Methodist Race and Justice Fund, and the Presbyterian Self-Development of the Peoples Fund are examples of church supported community organizing, planning, and development programs. One of the most rapidly expanding national community development organizations is the National Christian Community Development Association founded by John M. Perkins that attracts more than 5,000 delegates to its annual conference. The growing importance of religious institutions to local economic and community development was highlighted in 2001 when the Bush Administration established the White House Office for Faith-Based Organizations and Community Initiatives.

Types of Neighborhood Plans

Today, most mid- and large-sized cities have several neighborhood planners, often organized into a specialized unit within their planning departments, supporting resident-led neighborhood revitalization efforts. They are responsible for assisting local residents and leaders in designing and implementing comprehensive strategies that enhance the unique quality of life available in their residential neighborhoods. These planners tend to work with local residents to pursue one of the following six types of neighborhood planning strategies based upon the environmental, economic, and social conditions confronting their communities:

- Growth management strategies: These plans are developed by neighborhoods seeking to encourage growth while protecting long-term residents, businesses, and institutions from displacement.
- Preservation strategies: These plans are developed by local residents eager to protect historically, culturally, and aesthetically significant places and structures. These sites are central to the identity of local communities but they can be easily threatened or lost through neglect, insensitive reuse, or demolition.
- Stabilization strategies: These plans are formulated by those committed to reducing and/or eliminating the out-migration of people, businesses, institutions, and capital from a community, which, if unabated, will undermine its long-term stability.
- Revitalization strategies: Through a comprehensive transformation strategy featuring investments aimed at improving the community's physical fabric and rebuilding its social capital base, these plans are designed by residents seeking to restore the former vitality of an area in severe decline.
- Post-disaster recovery strategies: These plans are pursued by local leaders eager to rebuild neighborhoods following devastation caused by a significant natural disaster such as an earthquake, tornado, hurricane, flood, or fire.
- Master plan strategies: These plans relate to the design and construction of new, often mixed-use, communities at former urban "brownfield" sites or exurban "greenfield" locations.

In a dynamic, rapidly changing region, residents of a given neighborhood may undertake several of the above-mentioned types of neighborhood planning over time as environmental, demographic, economic, and social conditions affecting their communities, cities, and regions change.

Steps in the Neighborhood Planning Process

While there are many types of neighborhood plans, most tend to be produced through a process that reflects the following seven steps: steering committee formation; basic data collection and analysis; visioning and goal-setting; action planning; plan presentation, review, and adoption;

implementation; and monitoring, evaluation, and modification. The following section provides a brief description of the major activities and deliverables produced at each of these steps in the neighborhood planning process (Jones 1990).

Steering Committee Formation

Following the decision by local residents, community leaders, and municipal officials to prepare a neighborhood plan, steps must be taken to identify the community's major stakeholder groups. One-on-one meetings are then scheduled with representatives of these groups to inform them about the pending neighborhood planning process; elicit their views regarding the plan's overall improvement goals; and invite each stakeholder group to identify one or more individuals to serve on the steering committee for the planning process. This body serves a number of critical functions, including: legitimizing the overall effort; serving as advocates defending the initiative from both internal and external challenges; designing a planning process uniquely suited to the civic, social and cultural history of the community; and mobilizing others, especially those who have previously been uninvolved in community affairs, to become active participants in the effort.

Working with the steering committee, neighborhood planners draft a preliminary scope of services, timetable, and budget for the process. When this work has been completed, an ambitious media campaign is designed and implemented to inform all those living, working, and serving in the target area about the goals, objectives, activities, and opportunities for participation in the upcoming neighborhood planning process. Among the typical elements of such a media campaign are ads in local newspapers; articles in community newspapers and publications; appearances by steering committee members on local radio stations and cable television programs; storefront posters, sidewalk tables, church bulletin notices, and pulpit announcements; flyers sent home with local school children; establishment of a project website and Facebook page; and a kickoff press conference featuring steering committee members, councilpersons, and the mayor. The use of a project logo, common graphic layout, and motto, such as "Northside Turning the Corner" or "Southside Blooming," helps establish a strong identity for the project in the consciousness of local residents, leaders, and officials (Bowes 2001).

Data Collection and Analysis

Patrick Geddes encouraged citizens and planners to "survey before plan." Following this advice, neighborhood planners tend to collect and analyze a wide range of data to determine the environmental, economic, and social assets and challenges confronting the community in which they are working. The typical neighborhood plan is based upon the collection and analysis of the following types of data:

- a detailed social history of the community highlighting its past successes in overcoming local social divisions in order to solve critical community problems;
- a systematic review of past public and private surveys, studies, reports, and plans for the area;
- a longitudinal analysis of population, education, employment, income, poverty, and housing trends using U.S. Census data;
- a study of the land features, water resources, and other natural assets as well as current and future environmental threats;
- an investigation of current land use patterns, building conditions, site maintenance levels, infrastructure, and community facilities quality;
- a review of the zoning history and inventory of current land use issues facing the community;
- a survey of local businesses and their economic and community development concerns and planning priorities;

- a survey of residents' perceptions of existing conditions, desired development directions, and future neighborhood improvement priorities;
- a parallel survey of the opinions of local "movers and shakers" regarding existing conditions, desired development directions, and future neighborhood improvement priorities.

Under optimal conditions, residents would participate in the development of the instruments used to gather these data, actively engage in the collection of this information in the field, and collaborate with professional planners in interpreting the meaning and determining the implications of such analyses for local planning and policy making. Following the completion of the above-mentioned data collection and analysis activities, residents are typically invited to revisit their data analysis using the Stanford Research Institute's "SWOT analysis" technique (SRI International 2007). When the major findings from each dataset have been categorized according to this system—that is, whether they represent a present strength (s), a present weakness (w), a future opportunity (o), or a future threat (t)—local residents are asked to organize these observations into major themes. Bernie Jones in his volume on neighborhood planning recommends the use of the "PARK System" in preparing a preliminary planning response to conditions highlighted in the SWOT analysis. Residents and planners can work together using this system to determine which of the following planning treatments or "interventions" each important neighborhood feature should receive.

- Preserve (P)—Take steps to save a valued neighborhood characteristic;
- Add (A)—Pursue opportunities to add or expand a highly valued neighborhood feature;
- Remove (R)—Undertake actions to remove a particularly offensive local feature; and
- Keep Out (K)—Initiate policies, programs, and actions to keep a particularly noxious external threat from undermining the local quality of life.

Visioning and Goal-Setting

Several times during the steering committee formation and data collection and analysis phases of the neighborhood planning process, representatives of local stakeholder groups as well as individual citizens are asked to describe their version of an "ideal" or "improved" neighborhood. Over time, a series of alternative future development options emerge for community stakeholders to review. Neighborhood planners typically work with steering committee members to prepare a draft community profile, SWOT analysis, and alternative future development scenarios report for local stakeholders and their allies to consider. These materials are distributed via a newsletter and/or website a week or so before a community-wide meeting to evaluate these documents. During what is often referred to as a Neighborhood Summit, local stakeholders review and comment on the historical, archival, and research documents; environmental conditions surveys; building/site/infrastructure analysis; business surveys; resident and official interviews; and other data collected by the neighborhood planning team. They also examine and revise the SWOT analysis that has been prepared by the steering committee and their neighborhood planners based upon these data.

Having completed these tasks, they review, and occasionally amend and/or expand, the alternative future development scenarios that have emerged during the neighborhood planning process. The later portion of the Summit typically focuses on an analysis of the pros and cons of each alternative development scenario and selection of a preferred path forward. Having determined the kind of community they will work together to create (the end stage), those attending the Summit proceed to craft an overall development goal to guide their future activities. They conclude their Summit activities by identifying the major development objectives that will enable them to achieve this goal. Among the most common development objectives featured in such plans are those that seek to improve public safety; restore the urban environment; expand local employment and entrepreneurial opportunities; provide access to quality affordable housing; serve the specialized educational, health and transportation needs of youth and senior citizens; ensure the provision of a full range of basic municipal and social services; and encourage ongoing citizen mobilization for economic and community development. Before residents, businesspeople, private sector funders, institutional leaders, and elected officials leave the Neighborhood Summit, they are challenged to join an action planning team focused on one of the plan's major programmatic elements (i.e. public safety, affordable housing, etc.). Each team is charged with the responsibility for formulating a set of immediate-term (year one), short-term (years two and three), and long-term (years four and five) improvement projects to achieve one of the plan's major development objectives.

Action Planning

For several months after the Summit, a significant number of neighborhood planning participants meet in (issue specific) groups to formulate developmentally based action plans to achieve one of the major redevelopment objectives articulated at the Summit. With the assistance of neighborhood planners and supportive allies from the local human services, foundation, and local and state government communities, these groups typically begin their work by brainstorming the longest possible lists of plausible economic and community development proposals to help the neighborhood gradually accomplish its development objectives.

Following what is often referred to as a "blue sky" session, in which any and every proposal is seriously considered, the action planning teams often invite representatives of sympathetic public and private funding agencies to provide initial input regarding proposals. These professionals rate the community's initial wish lists of possible projects as very likely, somewhat likely, and unlikely to receive political and financial support from local economic and community development policy makers and regional funders. These economic and community development professionals are also encouraged to add their own suggestions to the community's list of possible improvement projects. With input from these supportive development professionals, members of each action team make a preliminary attempt at prioritizing their project list using the following template for guidance.

For Year One, they are encouraged to identify one high-priority project that could be implemented by the neighborhood's existing volunteer base with little or no outside funding or technical assistance. They are further encouraged to identify

Table 20.1 *Proposed Action Plan*

Neighborhood housing improvement plan

Time frame	*Immediate term*	*Short term*	*Long term*
Year	Year 1	Years 2–3	Years 4–5
Number of projects	One project	Two projects	Three projects
Volunteer base requirements	Current	Slightly expanded	Significantly larger
Approximate costs	$0—in-kind	$ 50,000 each	$200k–$300k each
Needed technical assistance	None	Modest/short term	Significant/ongoing
Projects	1.	2.	4.
Projects		3.	5.
Projects			6.

two high-priority projects for Years 2 and 3. These could be implemented with what they hope would be a slightly expanded volunteer base; $50,000 or less in local and outside funds; and a small amount of high-quality technical assistance. Finally, they are encouraged to identify three high-priority projects that could be implemented in Years 4 and 5 of the neighborhood implementation planning process through the efforts of a significantly expanded volunteer base; $200,000 to $300,000 in local and outside funds; and a significant amount of high-quality technical assistance. Table 20.1 is a preliminary planning tool that can be used to assist neighborhood action teams in identifying a tentative list of projects for the broader community to consider.

Once each action planning team has prepared an initial list of development projects for their assigned policy/planning area, these are reviewed by other process participants and finally adopted. When this has been completed, members of each action team and their neighborhood planners consult the economic and community development literature to identify both principles of good practice and models of program excellence. Armed with this knowledge, they develop a specific action plan that includes the following information for each one of their proposed development projects:

- name;
- detailed description;
- rationale;
- major implementation steps;
- lead organization/agency;
- supporting institutions;
- approximate costs; and
- potential funding sources.

Sources of quality technical assistance include:

- location requirements;
- comparable model programs; and
- design requirements/guidelines.

Plan presentation, review, and adoption: Once the action plans have been written and approved by those participating in the neighborhood planning process, a complete draft of the neighborhood plan is prepared under supervision of the steering committee. Copies of the plan are then made available for review at the town or city hall, public libraries, community centers, schools, and senior citizen centers. In addition, a PDF version of the plan is typically posted on either the local municipal and/or dedicated neighborhood planning website. If possible, prior to a final neighborhood meeting held to publicly discuss, review, and vote on the plan, an executive summary of the document is prepared and distributed to each household, business, religious institution, nonprofit organization, and government office within the community for review. Following a formal vote on the plan by local stakeholders, the document is usually forwarded to the neighborhood's city council representative. It is then typically sent by the neighborhoods' council representative to the full legislative body with a formal resolution asking the local government to adopt the plan through an ordinance, thereby amending it to the community's existing master and/or comprehensive plan. In reviewing and approving neighborhood plans, the specific process that local, county, and state governments follow is somewhat different in each state. In general, neighborhood plans are officially approved through the following steps:

- Local residents are publicly informed that a particular neighborhood plan is being considered for approval.
- The city planning commission is given the opportunity to review the plan; hold one or more public hearings to elicit resident input; and vote on the document.
- The city council is then asked to review the plan, go over the planning commission's resolution in favor of its adoption, and hold one or more public hearings to elicit additional resident input. Before voting on the plan, they

generally ask municipal planners to complete both an environmental impact statement and a historic preservation report to assess the plan's likely impact on the local environment and its historic and cultural resource base. If these reports show little or no negative ecological or historical impacts, the city council is likely to approve the document.

- Depending on the form of local government, the mayor may have an opportunity to complete an independent review of the document. If he/she has the right to veto the plan, the document may be returned to the city council for reconsideration.
- Once the neighborhood plan has been approved by the local city council and mayor, adjacent municipalities, and/or the county where the plan was generated may have the opportunity to review and comment on the likely impact of the plan on the metropolitan region.
- Assuming a positive review by the surrounding municipalities and/or the county where the plan was generated, the document is then sent to the secretary of state's and/or attorney general's office to determine if it meets basic state standards.
- If state officials believe the plan meets state standards, it is then returned to the city clerk, who files the document as an official amendment to the master plan.

The legal authority of officially approved neighborhood plans differs by state. In some states, such as New York, local governments are bound to consider neighborhood and master plans when making significant physical planning decisions. However, they are not strictly bound to follow the specific policies contained within officially approved neighborhood and master plans.

Plan implementation: Following the adoption of the plan by municipal officials, local residents and stakeholders who have participated in the planning process mobilize their neighbors, local businesspeople, institutional leaders, and elected officials to carry out the immediate-term projects contained in the plan that not require either extensive technical assistance or significant levels of outside funding. As success is achieved on these modest self-help projects, efforts are subsequently made to identify nonprofit organizations or government agencies which may be willing to serve as the lead agency for the plan's more ambitious development projects contained. Such projects require significant organizing, research, planning, and fundraising activities and, in communities where networks of highly effective community-based organizations remain, it is preferable that agencies with deep programmatic expertise in a given policy area undertake related development projects. While this may be preferable, a special effort must also be made to coordinate each agency's development projects in order to capture their synergistic benefits.

In communities where the disinvestment process has proceeded unabated for a long time, there may either be a single or no existing nonprofit agency capable of carrying out the more ambitious development projects contained in the neighborhood plan. In the former situation, efforts may be needed to assist community-based organizations in carrying out the neighborhood's more ambitious development agenda. In the latter case, local institutions may need to work together to establish a community development corporation to carry out the proposed development agenda. Whether implementation occurs through a single community-based development organization or a number of different nonprofits, a varied funding base should be developed to implement the neighborhood plan. Such a multipronged development strategy will reduce the likelihood for possible disruption in the plan's implementation due to changing politics, policies, or personnel within a single funding agency.

Monitoring, evaluation, and revision of neighborhood plans: Like the U.S. Constitution, neighborhood plans are living documents designed to provide local civic leaders and elected officials with general policy and program development guidance. Economic and political conditions, as well as local, regional, state, and federal policy contexts, can and often do change rapidly requiring participants in neighborhood planning processes to reevaluate their planning activities on a regular basis. In many such processes, members of the steering committee are brought together on a quarterly basis to evaluate the extent to which the neighborhood action plan is being effectively carried out and having the desired impact. In light of changing conditions, modifications are routinely made in the neighborhood action plan to better achieve the plan's overall development goals and objectives. For example, a new federal grant program may become available, which may enable the neighborhood to expand one of its programmed activities or to move it ahead within the neighborhood's overall neighborhood planning timetable. Likewise, the decision of a nearby CDC to undertake one of the plan's most challenging projects contained may enable the neighborhood to reevaluate its role in the project from that of developer to monitoring agency.

The Global Movement toward Participatory Neighborhood Planning

The historic argument is that broad-based participation can only be achieved if a community is willing to accept significant delays and additional costs during the implementation phase of the planning process. However, this assertion is contradicted by a growing list of successful economic and community development projects that have been carried out using highly participatory planning methods. Increasing numbers of planners and designers, along with the elected officials with whom they work, have come to appreciate the important contribution that active participation by residents, business owners, and institutional leaders can make. At each step of the planning and design process, these collaborations can both improve the quality of specific development proposals and broaden their political base of support. Due to the growing body of participatory neighborhood planning projects being undertaken by community development corporations; municipal, county, and regional planning agencies; and private planning and design firms in the US, planners and designers in other parts of the world have been inspired to undertake similar efforts. Such work is being strongly encouraged by the European Union and the United Nations Research Institute for Social Development. As a result, U.S. planners and planning scholars have an important new venue in which to share their participatory planning and design ideas, methods, and practices. For instance, those who belong to one of the nine disciplinary associations organized on a regional basis throughout the world by the Global Planning Educators Association Network (GPEAN) now have regular meetings and publish comparative research related to, among other topics, citizen participation (Afshar and Pizzoli 2001).

Conclusion: Challenges on the Horizon

Looking toward the future, there are numerous challenges confronting those committed to participatory neighborhood planning. First, there is an urgent need to further refine the core curriculum within professional planning programs to provide graduate planning students with a stronger grounding in urban ethnography, participatory action research methods, and community organizing theory and methods. Second, opportunities for neighborhood activists, institutional leaders, practicing planners, and elected officials to acquire basic training in the theory, methods, and practice of participatory neighborhood planning must be expanded. Third, there is a critical need for local citizen leaders and neighborhood planners engaged in bottom-up planning and development efforts to come together, on a more regular basis, to discuss local, state, and federal policies to promote more equitably and participatory forms of economic and community development. Fourth, specific steps must be taken to strengthen the citizen participation requirements of such new federal programs such as HUD's Choice Neighborhood Program which appears to accept forms of resident engagement from the lower rungs of Arnstein's Ladder of Citizen Participation. Finally, there is a need to encourage planning scholars to undertake more systematic evaluations of various participatory neighborhood planning models, taking place in contrasting contexts, in order to provide planners with more empirically grounded practice advice and direction.

BOX 20.1 CROWDSOURCING

An exciting new citizen engagement technique that uses the internet and social media to elicit insightful analysis, innovative solutions, and, in some cases, financial resources from the broadest possible base of interested individuals to support the community-building, problem-solving, and neighborhood transformation efforts of neighborhood residents and leaders (see Daren C. Brabham 2013 for more information).

CASE STUDY: PROMOTING CITIZEN PARTICIPATION IN NEIGHBORHOOD PLANNING

Overcoming resident skepticism regarding the possibilities for change is often the single most important challenge confronting those involved in participatory neighborhood planning. Citizens living in economically distressed communities have witnessed local employers moving away, soaring unemployment and poverty rates, local retail stores closing, credit sources evaporating, municipal services deteriorating, and long-term residents departing. They have also listened to several generations of elected officials and appointed administrators promising, as did Herbert Hoover, that prosperity was "just around the corner." This widening gap between existing conditions and promised improvement has left many residents of low-income communities highly skeptical of government-sponsored or -endorsed community renewal efforts (Schorr 1997). The following section describes how a small group of academic planners from Cornell University and their students, working in a once-vibrant resort community in New York State's Catskills Mountains Region, succeeded in involving a large and representative cross section of local residents and stakeholders from the Village and Town of Liberty, New York, in a highly participatory process to create and implement a comprehensive economic development strategy for their community.

In 2003, representatives of a local family foundation asked Cornell University's Department of City and Regional Planning to assist residents, business owners, human service providers, and officials from the Village and Town of Liberty in formulating a comprehensive community economic development plan. While the community had worked hard to create a sense of optimism and momentum regarding its future, this work had fallen on the shoulders of a very small number of volunteers. While developing the plan, the Cornell faculty and students realized their need to design a process capable of identifying and recruiting new volunteers committed to assisting in its implementation. Therefore, the planners decided to pursue a highly participatory "bottom-up, bottom-sideways" planning approach encourage previously uninvolved individuals and institutions to become actively engaged in the planning process and subsequent community development efforts. To do so, they used the following methods:

One-on-One Meetings

Working with the foundation staff, and elected and appointed officials from the Village and Town of Liberty, the planners identified more than 40 civic leaders representing the community's major stakeholder groups. The planners subsequently contacted each of these individuals to elicit their perceptions regarding existing community conditions, the need for a comprehensive economic development plan, their proposed goals and objectives for such a plan, and their willingness to participate in the process. Special effort was made to reach out to area youth and to members of the Latino, Hassidic, and Muslim communities that had recently migrated to the community. Despite their numbers, these groups had not previously participated in local community planning and development activities.

Steering Committee
Following a series of more than 40 one-on-one interviews, 25 local leaders representing a broad cross section of the community were invited to serve on the Steering Committee for the Liberty Economic Action Project (LEAP). The primary functions of the Steering Committee were to assist the planners in formulating a basic research design for the project; encourage local residents, business leaders, and elected officials to participate in the process; serve as spokespeople for the effort; and defend the undertaking from outside criticism and attack.

Community Media Campaign
Members of the steering committee subsequently worked with project planners to devise an ambitious local media campaign to inform residents about the project and encourage them to actively contribute to the effort. The local media campaign featured a kickoff press conference; a weekly news update which appeared in the community's weekly newspaper; a weekly discussion of the planning process on a popular radio talk show; storefront posters; weekend informational tables and sandwich boards set up along Main Street; regular updates presented at Liberty Volunteer Fire Department meetings; bulletin and pulpit announcements at area churches, synagogues, and mosques; and notices posted on a community bulletin board located in the heart of the Village's downtown.

Social History Project
One of the planners' first activities focused on the completion of a series of oral history interviews with long-time residents and officials who had helped lead successful community development projects in the past. The stories of these community renewal efforts were collected to remind Village and Town residents of their history of working together to overcome critical economic, social, and political problems confronting their community. These interviews were also helpful in identifying effective leaders who might be recruited to participate in the newly initiated comprehensive community renewal program. A "Short and Glorious History

of Liberty's Community Improvement Legacy," that chronicled local residents' efforts to restore and preserve the town's original public buildings, among other tales, was distributed throughout the community. It became a highly sought after document that residents eager to learn more about their community's rich social change history copied and distributed.

Door-to-Door Campaign

The week following the project's kickoff press conference, 25 students and six steering committee members spent three days in the pouring rain visiting every house, business, and institution in the Village and the Town. They informed local stakeholders about the economic development planning process that was under way, elicited their ideas regarding the Village and Town's future, and invited them to an initial community planning meeting to be held at the Liberty Volunteer Fire Department headquarters. During the course of a single weekend, more than 1,200 informational brochures were distributed describing the goals, objectives, and desired outcomes of the economic development planning process.

Community Mapping

Approximately 60 local residents came to the first meeting of the Liberty Economic Action Project. Following a very brief overview of the goals and objectives of the planning process provided by a Village Councilman, residents were asked to form six-person teams for the purpose of sharing their knowledge of the community. Each team was seated at a round table on which there was a very large base map of the Village and Town and a set of colored markers. They were invited to use their purple markers to identify their community's most significant subareas (i.e. communities of interest or neighborhoods); green markers to locate important community resources or assets; red markers to identify areas of concern; and orange markers to identify untapped resources available to advance their economic development efforts.

Within minutes, the noise level in the room rose as residents introduced themselves to each other and began to share their in-depth knowledge of the community with those seated at their table. After approximately 30 minutes, a spokesperson from each table was asked to share the highlights of his or her team's mapping exercise with the entire assembly. Attendees were astounded by the vast knowledge and deep insights they and their neighbors possessed of their community; they were also blown away by the number and quality of untapped assets and resources for economic and community development which this initial exercise had surfaced.

Camera Exercise

At the end of the first community meeting, residents were asked to assist the steering committee in documenting the community's most important attributes. Each meeting attendee, regardless of age, was given a simple disposable camera with 27 exposures that they were asked to use to identify nine each of Liberty's most important assets, serious problems, and underappreciated/untapped resources. Along with the camera, each volunteer was given a simple log book to provide captions for each of their photos. During the week following the preliminary meeting, more than 40 Liberty residents were seen taking photos of the community's many natural areas, historic buildings, residential neighborhoods, community facilities, and most colorful residents. As they did so, other residents asked what they were doing. Following a brief explanation, they invited these individuals to join the planning process by attending the next meeting.

Shoe Box Planning

Those attending the second community planning meeting were again seated at round tables in groups of six, where they were given 50 of the recently resident generated photos to sort into one of four shoe boxes. Following Stanford Research International's (SRI)'s strategic planning tool, affectionately referred to as the "SWOT" Analysis, residents were asked to place photos depicting positive community traits in the box marked "S" for current strengths; negative community attributes in the box marked "W" for current weaknesses; potential opportunities in the box marked "O" for future opportunities; and possible threats in the box marked "T" for future threats. Following their initial sorting of these images, the teams worked together to group photos within each box according to overarching themes. For examples, images of historic structures placed in the "Current Strengths" box might be organized under the theme, "Strong Building Stock," while photos of illegal dumpsites from the "Current Weaknesses" box might be placed under the theme, "Environmental Degradation." This meeting ended with the teams working together to integrate their preliminary assessments of existing conditions and future development possibilities.

Spike Lee Exercise

As local stakeholders worked together to develop a consensus regarding Liberty and its future, middle school students were asked to share their perspectives on the community and its future through an art activity referred to as Spike Lee's "The Good, the Bad, and It's Gotta Change Now, Baby" Exercise. On a Friday afternoon, teams of Cornell planning students visited the Liberty Middle School where they worked with 50 12 to 14 year olds on a mural project. Following a mid-afternoon snack of pizza and soda, each middle schooler was given a 30 x 40 inch piece of newsprint to share their sense of the "Best Liberty had to offer" young people, the "most serious problem affecting Liberty youth" and the one improvement project they would like to see take place if they were "Mayor for a day." Within minutes, the room quieted down as the participating youth sketched their images of the city and proposals for improving the quality of urban life. As the students did so, university volunteers took their photographs and collected basic biographical information to create informational plaques similar to those traditionally featured in museum exhibitions.

Several weeks after this activity, the students' work was displayed as part of an interactive community-building exhibit organized by the steering committee with the assistance of student planners and volunteers from the Liberty Historical Museum. Each student's work was neatly matted, framed, and accompanied by a small biographical plaque with their name, age, career aspiration, family profile, and picture. More than 200 Liberty residents visited this installation gaining a better understanding of how their community's young people view their Village and Town. Many of those attending this event had not participated in the community's past planning activities admitting they were drawn into the process by their children's participation.

Community-Building Exhibition

A third community planning meeting was organized to involve residents in a final review and analysis of the existing conditions and future projections data generated by the planning students. This meeting was preceded by a week-long interactive planning exhibit at the Liberty Museum sponsored by the steering committee and the Museum's Board of Directors. The students presented the data organized by theme through a series of interactive installations designed to elicit additional resident input for the planning process.

Among the interactive installations was a board where local residents could place pins to identify locations within the community to which they regularly drove. They were then asked to connect the location of their home, workplace, child's day care center, or older children's school with rubber bands.

Over the course of the week, the community's most heavily traveled areas became apparent, enabling residents to think about the following question posed by the planners: "Are there ways we could simplify and improve your life by relocating certain community facilities (i.e. day care centers, schools, retail stores, the "Y," etc.)?" Another exhibition gave people the opportunity to mount a fake pulpit, wear a Burger King crown, and as "King for a Day," pronounce the single most important improvement they would make in Liberty if given the opportunity. Each resident's photo was taken as "King" and their proposals were documented in the exhibit. Local residents visiting the Museum were both amused by the photos and impressed by the dozens of innovative ideas proposed by their neighbors.

Envision Liberty Week

One of the most enjoyable and productive activities that the university planners undertook was "Envision Liberty Week," which occurred two-thirds of the way through the planning process. It was advertised by a banner hung across Main Street featuring an enormous pair of Armani sunglasses accompanied by the slogan, "Envision Liberty: Making a Great Community Even Better, October 15–22." The event began with the week-long planning exhibition at the Liberty Museum, already described. On Friday evening, local volunteers and student planners cleaned, relit, and reopened the Village's Art Deco era theater.

More than 80 residents came to an event entitled, "Liberty Today: A Community on the Move" in which the student planners discussed with residents the planning and policy-making implications of their recently completed research activities. The evening ended with a dialogue on the strengths and weaknesses of four alternative development scenarios and a vote on their preferred option. The following morning, more than a hundred residents appeared for the second part of the program entitled, "Guided Visualization: Crafting Liberty's Future."

As people gathered at the theater, they were organized into 15-person groups and assigned a facilitator, recorder, and image-maker (a.k.a. artist). These groups were then asked to follow their facilitator to one of the many vacant storefronts on Liberty's Main Street, which had been cleaned out, equipped with chairs, and supplied with ample amounts of newsprint, markers, tape, and a CD player. A total of eight groups spent the remainder of the morning engaged in a rather unusual "Envisioning Activity" in which they were asked to get as comfortable as they could in their chairs and to close their eyes while listening to the jazz music of Miles Davis. Each facilitator then asked participants to imagine that they were home alone (no kids, parents, spouse, or pets), sitting in their favorite room, resting in their favorite chair, drinking their favorite New York State wine, and falling into a very deep and restful sleep.

They were then asked to imagine the years flying by, just as they did for another historic Upstate New Yorker—Rip Van Winkle. They passed through 2004, 2005, 2006, 2007... until it was 2018. They were then asked to imagine that Liberty, New York had, through local resident and official effort, become everything they every wanted it to be! They were invited to imagine themselves stepping out the front door of their home and, with their mind's eye serving as a virtual video camera, recording the most exciting aspects of the "New and Improved" Liberty.

Next, they were asked to take a few minutes with their eyes still closed to remember the most powerful, transformed, and uplifting aspect of the New Liberty community. At the count of three, they were asked to wake up and share their most powerful image of the New Liberty community with their group. As they shared their imagined visions of a New Liberty, the artist who accompanied the group to their storefront created a large image of the described scene in what we referred to as an "idea bubble." Following these reports, each participant was given a set of five green dots and a skull and crossbones sticker. Finally, they were invited to tour the images that now covered the walls of their storefront using the green dots or skull and cross bones stickers, to approve the visions they found most compelling and to disapprove the project they would fight the hardest to oppose.

After this exercise, the groups reconvened at the former Liberty Art Cinema where they briefly shared a quick summary of their ideas taking a few extra minutes to describe their most popular proposal. Following these reports, all of the participants were given a second set of green dots and skull and crossbones stickers to share their views on the entire set of proposals developed by the assembly. Envision Liberty Week ended with the identification of five action areas residents and officials viewed as critical to the community's future economic development. Among these were community organization, small business assistance, quality affordable housing, workforce training and development, youth leadership, and urban design. Before the meeting was adjourned, each participant was asked to both identify which of these community development areas they were most interested in and to set a date to meet with those sharing these interests.

Action Teams

In the months following Envision Liberty Week, more than 100 Liberty residents met on several occasions to identify a set of immediate-, short-, and long-term development projects to be undertaken by the community in the above-mentioned program areas. While the five action planning teams worked in somewhat different ways, they all tended to follow a similar process. Each team began their work by inviting members to "brainstorm" the longest list of possible development projects within their individual program area without consideration of feasibility. Having done so, each team identified two to three local funders with considerable development experience in their area to share their "quick and dirty" evaluation of: (a) their sense of the long-term value of each potential project and (b) their assessment of the feasibility of funding such a project in the current local and regional context. Following these two steps, each group was asked to produce a list of six to eight high-priority projects that could be undertaken during the next five years. When this list was completed, two or three residents, along with a student planner, were assigned to research the best practice literature related to this project as well as model projects successfully undertaken by communities similar to Liberty. For several weeks, these teams worked on a series of three to five page briefing papers summarizing their best practices and model projects research. Then, prior to determining the final list of projects to be pursued in each program area and creating phases for each one, these briefing papers were sent to each member of the action team.

Final Community Forum

When this work was completed, another community-wide meeting was held in which residents and officials from the Village and Town were given the opportunity to discuss and adopt a final list of project proposals. Community participation was bolstered by the meeting taking place at the corporate headquarters of the sponsoring foundation and because it was attended by that institution's founder, a highly respected international business leader and philanthropist and the then newly appointed President of Cornell University.

In the year following the completion of the Liberty Economic Action Plan, the Village and Town worked with their allies to accomplish many things. With the help of the local foundation, the two municipalities came together to establish the Liberty Community Development Corporation, to mobilize public and private resources to implement its very ambitious economic and community development agenda. With $100,000 it had raised, and the assistance of Cornell Extension Service, this group implemented an after-school business internship program for "at risk" high school students. Youth participating in the process worked together to: secure land for the construction of a skateboard park; complete a design for the project; and raise funds to construct the initial improvements needed to establish that facility.

The Village, with the assistance of Cornell Historic Preservation Planning students, faculty, and alumni, succeeded in cleaning and stabilizing a Mondrian-inspired commercial building that had long stood vacant at the corner of a prominent downtown intersection. This resulted in the purchase of the building by an outside developer who adapted it for use as an antique furniture and vintage clothing store. Again with the assistance of Cornell students and faculty, the Village and Town completed plans for downtown traffic and transportation improvements and the revitalization of the Village's major public park. The collaboration also produced a new subdivision proposal for the nearby hamlet of Swan Lake, based upon behavioral guidelines suggested in the Torah, to meet the needs of the area's rapidly growing Hassidic community. Finally, the local development corporation, with the help of historic preservation students, faculty and alumni from Cornell University stabilized and retrofitted a former cabaret structure that was part of a long-abandoned hotel property for use as a much-needed community meeting and cultural space.

Keywords

Participatory neighborhood planning, community-based development organizations, resident-led planning and design, empowerment planning, asset-based community development, participatory action research.

Review Questions

1 Explain how an asset-based approach to community planning and development differs from a deficit-based approach to neighborhood transformation?
2 What is a SWOT analysis and how can it used in neighborhood planning?
3 Describe a successful process of citizen participation in neighborhood planning and development decision making that you participated in or observed?
4 Identify one technique or approach described in this chapter that would be applicable to a situation in your own neighborhood.
5 When would you use as Spike Lee or Shoebox exercise?

Bibliography

Afshar, F. and Pizzoli, K. (2001) "Editors' Introduction," *Journal of Planning Education and Research*, 20(3): 277–280.

Anderson, M. (1966) "The Federal Bulldozer," in J.Q. Wilson (ed.) *Urban Renewal: The Record and the Controversy*, Cambridge, MA: MIT Press, pp. 491–508.

Arnstein, Sherry. (1969) "A Ladder of Citizen Participation," *Journal of the American Institute of Planners* 35(4): 216–224.

Bowes, J. (2001) *A Guide to Neighborhood Planning*, Ithaca, NY: City of Ithaca Department of Planning.

Brabham, Daren C. (2013) *Crowdsourcing*, Cambridge, MA: MIT Press.

Brophy, P. and Shabecoff, A. (2001) *A Guide to Careers in Community Development*, Washington, DC: Island Press.

Hartman, C. (1974) *Yerba Buena: Land Grab and Community Resistance in San Francisco*, San Francisco, CA: Glide Publications.

Heiman, M. (1988) *The Quiet Evolution: Power, Planning, and Profits in New York State*, New York: Praeger Publishers, pp. 30–97.

Jones, B. (1990) *Neighborhood Planning: A Guide to Citizens and Planners*, Chicago: Planners Press, pp. 1–38.

King, M. (1981) *Chains of Change: Struggles for Black Community Development*, Boston, MA: South End Press, pp. 111–118.

Kubisch, Ann, P. Brown, R. Chaskin, J. Hirota, M. Joseph, H. Richman, and M. Roberts. (1997) *Voices from the Field: Learning from the Early work of Comprehensive Community Initiatives*, Washington, DC: Aspen Institute.

McKelvey, B. (1948) "A Rochester Bookshelf," *Rochester History*, 10(4): 7–13.

Moynihan, D.P. (1970) *Maximum Feasible Misunderstanding*, New York: Free Press, pp. 75–101.

National Congress for Community Economic Development (NCCED) (2006) *Reaching New Heights: Trends and Achievements of Community-Based Development Organizations, 5th National Community Development Census*, Washington, DC: NCCED.

Perkins, John M. (2007) *Beyond Charity: The Call to Christian Community Development*, Grand Rapids: Michigan: Baker Books.

Peterman, W. (2000) *Neighborhood Planning and Community-Based Development: The Potential and Limits of Grassroots Action*, Thousand Oaks, CA: Sage Publications, p.1–32.

Robinson, C.M. (1899) "Improvement in City Life: Aesthetic Progress," *Atlantic Monthly*, 83 (June): 771–185.

Schorr, L.B. (1997) *Common Purpose: Strengthening Families and Neighborhoods to Rebuild America*, New York: Anchor Books, pp. i–xxviii.

SRI International (2007) *Timeline of SRI International Innovations*, Palo Alto, CA: SRI International.

Stein, C. (1957) *Toward New Towns for America*, Cambridge, MA: MIT Press, pp. 37–64.

Stiftel, Bruce and Vanessa Watson (2004) "Introduction: Building Global Integration in Planning Scholarship," in *Dialogues in Urban and Regional Planning, Volume 1*, edited by Bruce Stiftel and Vanessa Watson, New York, New York: Taylor & Francis, pp. 1–14.

Thabit, W. (1961) *An Alternative Plan for Cooper Square*, New York: Cooper Square Community Development Committee and Businessmen's Association.

Welter, V. (2002) *Biopolis: Patrick Geddes and the City of Life*, Cambridge, MA: MIT Press.

Connections

Learn about the many multifaceted dimensions of neighborhood planning with the Municipal Art Society of New York's very useful *Livable Neighborhoods Toolkit* at: http://mas.org/urbanplanning/community/training-assistance

See a sample of a neighborhood planning toolkit provided by the City of Phoenix at: http://phoenix.gov/webcms/groups/internet/@inter/@dept/@nsd/documents/web_content/nsd_neighborhood_toolkit.pdf

Gaining Ground: www.youtube.com/watch?v=D8PiIRE6TQg—A presentation describing the origins, evolution, and accomplishments of the Dudley Street Neighborhood Initiative (Boston, MA).

Some Kinda Funny Porto Rican: www.youtube.com/watch?v=SzCbt4eR5M4—An overview of the award-winning documentary on the destruction of the Cape Verdean community of Foxpoint by well-intentioned city planners (Providence, RI).

New Community Corporation: www.youtube.com/watch?v=ByqKG1y7H0E—A short history of one of the nation's oldest and most accomplished Community Development Corporations (Newark, NJ).

Ron Shiffman Wins Jane Jacobs Award: www.youtube.com/watch?v=_g4HX_szAnQ—An interview with Ron Shiffman one of the founders of the Bedford-Stuyvesant Restoration Corporation and long-time Director of the Pratt Institute Center for Community and Environmental Development (Brooklyn, NY).

What is neighborhood planning?: www.youtube.com/watch?v=L6imxsbN33s—An entertaining introduction to visual tools for engaging local residents in participatory neighborhood planning (Bristol, UK).

Enterprise Community Partners Post-Katrina Planning: www.youtube.com/watch?v=_w2eBkKM-2A—An interview with Michelle Whetten, Director of Gulf Coast Programs for Enterprise Community Partners who describes her intermediary's efforts to support resident-led recovery planning and development in Post-Katrina, New Orleans (New Orleans, LA).

Yorktown Survival Plan: www.youtube.com/watch?v=Lyc-66wVWMw—An inspired example of a community-based and resident-driven neighborhood sustainability plan (Philadelphia, PA).

21 Measuring Progress

Rhonda Phillips and Robert H. Pittman

Overview

Measurement and evaluation of community development progress is not only challenging, it is essential. Communities must be able to demonstrate the value and outcomes of their activities in order to be accountable to residents, secure funding, and to assess the efficacy of their programs. Increasingly, concerns with community well-being and happiness are emerging along with the need to measure and assess progress in these dimensions. Community indicators can be used to evaluate the progress of communities and community development organizations. Communities face many needs and opportunities and must allocate limited funds and human resources as efficiently as possible to successfully achieve their goals across these areas. Additionally, best practices and benchmarking are valuable tools in community decisions on development program structure, operations, and follow-up modifications.

Introduction

What is evaluation? Simply put, it is a way to figure out the importance, value, or impact of something. There are numerous ways to "figure it out" and numerous "things" that may need to be evaluated. So that it is clear what is being evaluated and which approaches will be utilized, evaluation is typically conducted in a methodical manner with a defined process or approach. Thus, evaluation can be defined as a systematic determination of the value or quality of a process, program, policy, strategy, system, and/or product or service including a focus on personnel (Davidson 2005).

Getting what a community wants in the future requires evaluation. Past performances can be reviewed to estimate future outcomes but, more importantly, evaluation should be included in the continuous cycle of program and policy development and implementation. Evaluation is not a one-time effort; it should be ongoing and periodic. Evaluation helps communities to develop, evolve, and improve in a constantly changing environment. Every time something new is tried—be it a policy, strategy, program, process, or system—its value must be considered (Davidson 2005). In community development, evaluation is particularly critical because citizens' quality of life is affected by such policies, programs, strategies, etc. If the impact and outcomes have not been soundly evaluated, can it be said that one approach is better than another or has a more positive influence?

While all communities are unique, many share common problems and issues. These may be addressed by previously developed strategies and solutions. To avoid wasting resources and "reinventing the wheel" when confronting an issue, a community should first conduct research into such proven best practices solutions. Benchmarking, or measuring one or more aspects of a community or program against its counterparts, is also a useful way to measure progress and provides additional perspectives on community indicators. This chapter discusses how indicators, best practices, and benchmarking can be used to assess community development progress.

BOX 21.1 COMMUNITY OUTCOME MEASUREMENT PRACTICES

Accurately measuring and documenting outcomes from community development interventions and processes has become almost essential in working with funding agencies as well as being expected by local businesses and other investors. In the past, many agencies focused on number of jobs created or retained and amount of private investment after the development process had occurred. Neither of these approaches was especially useful for local management because the links to the intervention process were tenuous and affected by a host of unrelated external factors.

Much of the thinking about measurement has now shifted to a broader set of measures based on an underlying conceptual foundation such as the Community Capitals. The Community Capitals Framework (CCF) "offers a way to analyze community and economic development efforts from a systems perspective by identifying the assets in each capital (stock), the types of capital invested (flow), the interaction among the capitals, and the resulting impacts across capitals" (Emery and Flora, 2006: 19). Using this approach allows managers to see how various sectors of the community are changing in response to the development strategies. An approach like Community Capitals also provides more immediate feedback that allows more precise adjustments to strategies as the development process unfolds. A recent special issue of *Community Development* was devoted to "Innovative Measurement and Evaluation of Community Development Practices," (44: 5, 2013). It contains examples of effective approaches used not only in the US but in other countries as well.

Effective outcome measurement usually contains several key components. A clearly documented *vision* for the community must exist to communicate to participants the directions of the development process. Without this vision, it is difficult to find agreed-upon measures that monitor progress.

The community vision is accompanied by a clear set of *goals* that ultimately will lead to achieving the visions. The goals should be specific, measurable, attainable, realistic, and timely to maximize their effectiveness in guiding the development process. In addition, they should affect and engage a broad section of community groups with diverse interests.

The vision and goals, then, lead to *strategies* and *targets*. This may be the area where communities have the most difficulties and where an effective measurement system can provide considerable assistance. It is often difficult to see precisely how a strategy will lead to a specific goal unless targets and other operational measures are included in the development framework.

Some development efforts have created a community dashboard with numerous measures to help monitor the development process. While an effective approach, the dashboard can be difficult to monitor and determine how specific measures lead to a desired community-based outcome. Nevertheless, it provides broad insight into the community development process.

The final measurement system is an important step but also essential to successful development is a feedback loop that links changes in key measures to resulting changes in policy approaches or initiatives. This loop can be difficult to create and over time may be hard to maintain. Nevertheless, it is one of the most important components in the development approach and is key to success.

Norman Walzer and Andy Blanke
Center for Governmental Studies, Northern Illinois University, USA

Sources

Blanke, A.S. and Walzer, N. (2013) Measuring Community Development: What Have We Learned? *Community Development*, 44: 5, 534–550 (DOI: 10.1080/15575330.2013.852595).

Emery, E. and Flora, C. (2006) Spiraling-Up: Mapping Community Transformation with Community Capitals Framework. *Community Development* 37(1): 19–35.

Walzer, N. and Cordes, S. (2011) Innovative Approaches to Community Change. *Community Development* 43: 1, 2–11.

Community Indicators

Given the importance of evaluation, contrasted with the complexity and barriers to conducting it, what should a community or community development organization do? Among evaluation techniques in the field of community/economic development planning, the use of community indicators is reemerging. These indicators were first used over 100 years ago but their new application is more beneficial and useful across a spectrum of domains impacting communities (Phillips 2003).

When used as a *system*, indicators hold much promise as an evaluation tool. What makes community indicators any different from other measures of community development such as job growth or changes in per capita income? The key is developing an integrative approach, to consider the impacts of development not only in terms of *economic* but its *social* and *environmental* dimensions. A community indicators system reflects collective values, providing a more powerful evaluative tool than simply considering the economics of change and growth. When properly integrated into the early stages of comprehensive community or regional planning, community indicators hold the potential to go beyond mere activity reports because they can be used systematically, making it easier to gauge impacts and evaluate successes (Phillips 2005). Furthermore, these indicators incorporate frameworks of performance and a full spectrum of process outcomes, both of which facilitate evaluation and decision making. The functions of indicators are listed in Table 21.1.

Just what are community indicators? Essentially, they are bits of information that combine to provide a picture of what is happening in a local system. They provide insight into the direction a community is taking; whether it's improving or declining, moving forward or backward, increasing or decreasing. Combining indicators creates a measuring system to provide clear and accurate information about past trends, current realities, and future direction in order to aid decision making. Community indicators can also be thought of as a report card of community well-being. It is important to note that these systems generate much data. It is the analysis of these data that can be used in the decision-making and policy/program improvement processes.

A well-developed indicator system provides a way to look at social, economic, and social phenomena and shows comprehensive pictures about communities, regions, states or provinces, and even countries. A familiar indicator is that of the gross national product (GNP), providing more understanding of economics and economic market valuation. There are obvious limitations to the GNP for measuring nonmarket activities and more esoteric domains such as quality of life. These lead to other measures attempting to reflect social and environmental well-being of society in more objective ways (Phillips et al. 2013). GNP does capture the full dimension of community. Social indicators and systems including economic and environmental have been developed that more fully express community well-being. Indeed, the idea that GNP cannot do justice to the entire picture of a community or country is reflected in ideas behind sustainable development and the need to measure impacts and outcomes in better ways. For example, the Canadian Index of Well-Being (2014) provides a composite indicator system to gauge situations across the following domains:

- community vitality
- democratic engagement
- education
- environment
- healthy populations
- leisure and culture
- living standards
- time use.

Table 21.1 *Functions of Indicators*

Key concept	*Functions*
Finding	• Revealing core concerns • Identifying information gaps • Clarifying opportunities • Information about past to present
Measuring	• Tracking progress toward achieving result • Evaluating performance
Monitoring	• Monitoring collaboration between citizens, experts and decision makers • Producing a feedback system for decision maker • Identifying emerging threats to community • Early warning system
Setting	• Setting community's priorities • Predicting quantifiable thresholds • Suggesting feasible goals • Implementing choices underlain by clear goals
Changing	• Shifting attention to particular area • Keeping track the progress in new dimensions of human responsibility and concern • The ability to changes in process and policy
Reflecting	• Providing a broader perspective • Sharing of decision-making power via better information, communication and dialogue • Increasing public accountability

Sources: Reprinted from Phillips et al. (2013), State-level Applications: Developing a Policy Support and Public Awareness Indicator Project. In J. Sirgy, R. Phillips, & D. Rahtz (Eds.) *Community Quality-of-Life Indicators: Best Cases VI*, Dordrecht, Netherlands: Springer.

BOX 21.2 PERSPECTIVES ON INDICATORS RESEARCH

My association with indicators research was closely related to the Conservative government regime in the late 1980s and early 1990s. Under the neo-liberal ideology of the government and its advocate of a strong auditing culture, a range of studies were commissioned by different government departments to identify indicators to measure their specific policy concerns (see Wong, 2003; 2006). Most of the official studies were commissioned to provide policy inputs to a particular government department, hence their nature tended to be short term and ad hoc. This means the scope for fundamental theoretical and methodological analysis was highly constrained. The nature of this wave of urban indicators research in Britain, both policy-related and pure academic studies, tended to be strongly grounded in an empirical approach. My work with colleagues at the Centre for Urban and Regional Development Studies of University of Newcastle upon Tyne at the time led to the publication of a number of official reports for the then Department of the Environment (Coombes, Raybould, and Wong, 1992; Coombes, Raybould, Wong and Openshaw, 1995), Department of Employment (Coombes, Wong and Raybould, 1993a) and Scottish Homes (Coombes, Wong, and Raybould, 1993b). We made some significant methodological contribution to shape subsequent research development in the area by demonstrating the importance of:

1 clarifying the concepts to be measured and deriving a classification of domains/issues to guide the selection of indicators;
2 identifying a clear and transparent set of assessment criteria to evaluate the quality of the data and the interpretation of indicators;
3 exploring the advantages and limitations of different weighting schemes to inform policy makers;

4 developing a four-step methodological procedure—working from general to specific—to guide indicators research (see Coombes and Wong, 1994; Wong, 2006);
5 using innovative use of data sources from directories and unofficial sources (see Wong, 2001); and
6 differentiating indicators of individual/household level from those measuring the wider neighbourhood and local environment.

Since the 2000s, several research studies were carried out to continue the development of innovative methodologies to improve the technical and statistical measures of indicators that have a spatial dimension. This was achieved not only by experimenting with robust statistical methods, but also by improving the understanding of policy concepts; linking the analysis to different socio-political contexts and visual interpretation; and engaging key stakeholders in the process. Research funded by the UK Economic and Social Research Council on local economic development (Wong, 2001; 2002) showed the importance of user perspectives and contextual interpretation of indicators; demonstrated the value of using principal component analysis to develop composite indices; empirically tested the importance of different factors for local economic development; and showed the visualisation of indicator values via mapping analysis. A research project for the Department for Communities and Local Government and the Royal Town Planning Institute demonstrated the value of bundling indicators to yield meaningful policy intelligence and the value of adopting a collaborative, reflexive and double loop learning approach on monitoring complex spatial policies (see Wong et al, 2008; Wong and Watkins, 2009; Rae and Wong, 2012).

Cecilia Wong,
Professor of Spatial Planning and Director of Centre for Urban Policy Studies,
University of Manchester, UK

References

Coombes, M., Raybould, S., and Wong, C. (1992) *Developing Indicators to Assess the Potential for Urban Regeneration*, London, HMSO.
Coombes, M., Raybould, S., Wong, C., and Openshaw S. (1995) *The 1991 Deprivation Index: A Review of Approaches*, London, HMSO.
Coombes, M. and Wong, C. (1994) Methodological steps in the development of multi-variate indexes for urban and regional policy analysis, *Environment and Planning A*, 26(8), 1297–1316.
Coombes, M., Wong, C., and Raybould, S. (1993a) *Indicators of Disadvantage and the Selection of Areas for Regeneration*, Scottish Homes Research Report No. 26, Edinburgh, Scottish Homes.
Coombes, M., Wong, C., and Raybould, S. (1993b) *Local Environment Index: Infrastructural Resources Dimension*, a final report to the Department of Employment, Newcastle upon Tyne, Centre for Urban and Regional Development Studies, University of Newcastle upon Tyne.
Rae, A. and Wong, C. (2012) Monitoring spatial planning policies: toward an analytical, adaptive and spatial approach to a 'wicked problem', *Environment and Planning B: Planning and Design*, 39(5): 880–896.
Wong, C. (2001) The relationship between quality of life and local economic development: an empirical study of local authority areas in England, *Cities*, 18(1): 25–32.
Wong, C. (2002) Developing indicators to inform local economic development in England, *Urban Studies*, 39 (10): 1833–1863.
Wong, C. (2003) Indicators at the crossroads: ideas, methods and applications, *Town Planning Review*, 74 (3): 253–279.
Wong, C. (2006) *Quantitative Indicators for Urban and Regional Planning: the Interplay of Policy and Methods*, Royal Town Planning Institute Library Book Series, Routledge, London.
Wong, C. and Watkins, C. (2009) Conceptualising spatial planning outcomes: toward an integrative measurement framework, *Town Planning Review*, 80 (4/5): 481–516.
Wong, C., Rae, A., Baker, M., Hincks, S., Kingston, R., Watkins, C., and Ferrari, E. (2008) *Measuring the Outcomes of Spatial Planning in England*, the Royal Town Planning Institute and the Department for Communities and Local Government, RTPI.

There are four common frameworks used for developing and implementing community indicators systems in the US: (1) quality of life; (2) performance evaluation; (3) healthy communities; and (4) sustainability (Phillips 2003). A summary of each type is presented below.

Quality of Life

Joseph P. Riley, Jr. has served as Mayor of Charleston, South Carolina for over 20 years and is noted for providing excellent civic support and direction to make this historic city very livable. He's quoted as saying, "what you need to do is make sure that place in which you live, and your people live, is as nice as the places they would dream about visiting" (cited in Jackson 2012: xii). This embodies the concept of quality of life which can be thought of as being reflective of the values existing in a community (Budruk and Phillips 2011). One method to gauge a host community's quality of life is by measuring and aggregating the quality-of-life indicators of various stakeholders.

Quality of life is reflective of the values that exist in a community. Indicators can be used to promote a particular set of values by making clear that residents' quality of life is of vital importance. If agreement can be reached, the advantage of this type of system is its strong potential to stimulate all types of community outcomes, not the least of which is evaluating progress toward common goals. The disadvantage is that measuring quality of life is a political process. What defines "the good life" differs among individuals, groups, and institutions (Phillips 2003).

One of the most notable examples of this quality-of-life framework in the US is the Jacksonville Community Council Inc. project. Many U.S. community indicators projects are based on this model, which started in 1974 and has become a part of the Floridian city's ongoing evaluation and decision making (Swain 2002). It attempts to integrate indicators into overall planning activities while monitoring for consistency with the comprehensive plan and other plans. The system has ten indicator categories and annual quality-of-life reports and indexes are released on each annually. Numerous examples exist now, given the interest in integrating quality-of-life approaches in community planning and development.

Performance Evaluation

This type of indicator system is mostly managed by state or local governments as a way to gauge the outcomes achieved by their activities. It is very beneficial as an evaluative technique because it provides reports on progress and outcomes, usually annually or semiannually. It is typically part of the annual budgeting process so that adjustments can be made for priority areas.

Healthy Communities

This approach is gaining popularity as it attempts to cultivate a sense of shared responsibility for community health and well-being. It focuses on indicators that reflect health care in the phases of life that often do not show on typical economic indicators concerned with working adults: prenatal, early childhood, and youth. Healthy Communities also prioritizes education and other human development facets including social concerns.

An example is Hampton, Virginia's "Healthy Families Partnership." By focusing on healthy children and families, Hampton has garnered desirable community and economic development outcomes (Hampton, VA, Healthy Families Partnership 2007). Some of the indicators in their system have provided remarkable evidence of progress on a variety of challenging issues. Hampton was one of the first U.S. cities to develop an indicator set integrated with its community economic development approach, and since that time, numerous cities and regions have incorporated healthy living and well-being measures into their approaches.

Sustainability

Community indicators systems can provide the mechanism for monitoring progress toward balanced or sustainable development because they provide information for considering the impacts of development, not only in economic terms but in social and environmental dimensions. The concept of sustainable development includes such characteristics as broad citizen participation, ongoing assessment, and a guiding vision. Indicators are consistent with these principles. The difficulty with the approach is to fully integrate the use of indicators into overall community planning so that sustainability can be a reality instead of rhetoric.

There are several examples of this approach to developing indicators systems. Seattle, Washington's "Sustainable Seattle" is a nonprofit group that gauges the city's progress on a variety of indicators. In response to this nonprofit initiative, the City created the Office of Sustainability and Environment to encourage integration of the indicators system with overall city functions and planning activities (Sustainable Seattle 2014). Again, as with the Hampton and Jacksonville examples, many similar efforts at local levels are modeled after this program. There is a rich history with Sustainable Seattle—established in 1991, it was the first U.S. organization to develop local indicators of well-being as an alternative to GDP (Sustainable Seattle 2014).

Who's Happy?

The upsurge of interest in the happiness is both fun and informative to observe—it's serious research and yields insight into how well communities are faring on important measures of populace, both individually and collectively. It may be a backlash to modern society that sometimes impedes the ability to feel connected to others, the loss of local or regional culture, or changing needs of the populace—whatever the reason, many communities and countries are enticed by measuring and ranking themselves according to how happy they feel. One of the most notable efforts is that by the tiny country of Bhutan, when their leaders declared they would measure progress according to Gross National Happiness, as opposed to Gross National or Domestic Product. These efforts have been noticed by many, so much so that the France and the UK have developed major studies of happiness and well-being as part of efforts to assess conditions across the spectrum of domains (economic, social, and environmental). A major study of over 140 countries was conducted on happiness, *The World Happiness Report* (Helliwell et al. 2013) that shows the span and depth of well-being across the globe.

Another effort that grew out of Sustainable Seattle is The Happiness Initiative. On their 20th anniversary,

> Sustainable Seattle again took the lead, by collecting and publicizing the first set of happiness (or well-being) indicators for any city in the United States. The work in Seattle continues, with an emphasis on social justice and catalyzing citizen dialogue and action in pursuit of happiness—defined as sustainability and love.
>
> (Sustainable Seattle 2014)

All these efforts are important to community development because they lead to a greater understanding of the factors influencing well-being. Further, the happiness and related quality-of-life studies and approaches can be incorporated into visioning, planning, and evaluation of community-based processes and outcomes to gauge progress toward goals.

Best Practices and Benchmarking

"Best practices" and "benchmarking" are terms commonly used in business and industry. They came into vogue in the 1980s when a host of

■ ***Figure 21.1*** *Sustainable Seattle Happiness Initiative. (Source: http://sustainableseattle.org/sahi, used with permission)*

books on business competitiveness were published. Since that time, these terms have been widely applied to many disciplines including community and economic development.

How are best practices and benchmarking defined? Consider these definitions from various sources:

Best Practices

> The processes, practices and systems identified in public and private organizations that performed exceptionally well and are widely recognized as improving an organization's performance and efficiency in specific areas.
> (U.S. General Accounting Office)

Benchmarking

> The process of identifying, learning and adapting outstanding practices and processes from any organization, anywhere in the world, to help an organization improve its performance.
> (International Network for Small and Medium (Business) Enterprises)

> Measuring how well one country, business, industry, etc. is performing compared to other countries, businesses, industries and so on. The benchmark is the standard by which performance will be judged.
> (European Union)

As seen in these definitions, best practices and benchmarking are concepts with broad applicability in a number of disciplines. Community and economic development organizations at all levels have adapted these tools to gauge success and track progress toward their goals.

Applications in Community Development

As discussed throughout this book, community development is a broad discipline. Economic development is a part of community development and there is a broad literature on best practices and benchmarking in economic development. However, community development also entails leadership, infrastructure development, effective local government, health care, workforce development, etc. Therefore, when applying best practices and benchmarking to community development, one must specify a particular aspect of the field. If the focus is on strategic planning and visioning as applied to both public and private sectors, there is a wide body of best practices and benchmarking literature. Likewise, if the focus is on local transportation, there is a corresponding body of literature.

Utilizing case studies of particular situations in community and economic development is certainly one of the most useful ways to identify best practices and conduct benchmarking. However, by definition, case studies deal with a particular situation. One would need to gather many of them in a specific area in order to identify a trend or common best practices theme. Surveys of similar organizations can provide a broader set of best practices parameters. Both techniques can provide useful best practices and benchmark information. Case studies can provide depth of information while surveys can provide breadth.

Why Are Best Practices and Benchmarking Important?

As noted above, while every community is unique, they all face many of the same issues. How can more industry be recruited? How can transportation options for low-income workers be improved? How do can a strategic plan be developed when community factions oppose each other? These are just examples of the plethora of issues that communities may face.

Rather than try to address these issues in a vacuum, doesn't it make sense to learn from the ways that other communities have addressed them? So much can be learned from examining the innovative solutions that other communities have developed to address similar issues or problems.

Another reason best practices and benchmarking are important in community development is that they serve as evaluation tools for programs and policies. Take economic development as an example. Many communities set economic development goals such as "Create 100 net new jobs each year for the next five years by recruiting new companies." This is an admirable goal but, given the history of the community and resources devoted to economic development, is it realistic? What if the community doesn't meet the goal? Should they dismiss the economic development staff or lower their sights?

Before taking any action, the first question they should ask is "why didn't we meet the goal?" The answer could be that the economy has taken a downturn during the year and the number of new company locations and expansions is much lower as a result. When economic development goals are not met, it is usually for reasons outside the community's control. By the same token, when a goal is not met in a particular year, the mistake is often to start over by changing the economic development program.

Because there are reasons for not meeting community goals, whether in economic development or other areas outside the its control, program evaluation should consist of looking at the "outputs" of the program (i.e. did we meet our goals) and the "inputs" (i.e. resources, staff, structure, productivity). One way to measure inputs such as program resources inputs is to see if all the action steps and items agreed upon were executed in accordance with the strategic plan. Another way is to compare a community's programs to those of similar communities that have proved successful. Does a community allocate a comparable amount of budget and staff time to a particular program? Are the program elements similar to those of other successful communities? This exercise, according to the definitions above, is what best practices and benchmarking analysis is all about.

Finding Existing Best Practices/ Benchmarking Research

As discussed above, in applying best practices and benchmarking to community development, the first step is to define the specific field of interest. As an example of best practices and benchmarking, let's consider the field of economic development. Numerous and varied organizations have done considerable research into best practices in several areas of economic development including organizational structure, marketing, business retention and expansion, and business start-up.

Below are some organizations that can be valuable sources of information on best practices and benchmarking in economic development:

- National and international government organizations such as the U.S. Economic Development Administration, the Department of Commerce, the Small Business Administration, the Department of Housing and Urban Development, and the Department of Labor.
- National and international nongovernmental organizations (NGOs) such as the United Nations, the World Bank, and the Organization for Economic Cooperation and Development (OECD).
- National and international professional associations such as the International Economic Development Council, the Community Development Council, the Council of State and Community Economic Development Agencies, and the National Association of State Development Agencies.
- Regional professional associations such as the Southern Economic Development Council, the Mid-American Economic Development Council, and the Northeast Economic Developers Association. In addition, most states have economic development professional associations that can be good sources of best practices and benchmarking information.
- State departments of economic development and related state agencies such as labor, community development, transportation, etc.
- Utilities. Many utilities have active economic and community development programs and sponsor or conduct best practices and benchmarking studies.
- Universities. Many universities have economic development, business institutes, or related departments that pursue best practices and benchmarking or related research.
- Consultants. While studies by private consultants are usually proprietary to the client, economic and community development consultants are often willing to share information with other communities.

In addition to these general sites, an internet search also turns up thousands of references to case studies and/or best practices analyses by communities who have dealt with specific economic development areas. Since such studies are usually public domain for their communities' benefit, many of them are available to download. Alternately, one can call the relevant community contact who is usually willing to share and discuss the study.

A third way to identify information on best practices and benchmarking is to search research and newspaper data bases.

Doing a Best Practices/ Benchmarking Study

It can sometimes be difficult to find best practices and benchmarking studies, or information dealing with a community's specific topic of interest, that are based on communities of similar size and circumstances. For example, a community might want to know how regional economic development marketing programs have been set up in other areas of like population. What kind of budget and staff requirements have been involved? How has the regional marketing entity be organized? The community might also want to benchmark themselves against closely comparable areas in terms of employment and industry base.

Fortunately, it is not very difficult to develop a survey form and solicit other communities or regions to participate in a best practices/benchmarking study. Communities are usually happy to participate in such studies, especially if they receive the results in order to benefit from the time and effort they invested.

Best practices/benchmarking studies can be anonymous or not. The decision would be influenced by the questions in the survey and the preferences of the participating community. In the case of economic development information, much of that is usually public (e.g. marketing budget, staff size) because public monies are often involved.

Here are the steps in conducting a best practices/benchmarking study for a community:

- Identify the topic. The more specific and definable the better.
- Identify the communities or areas with which you want to be compared. Usually four to six communities are sufficient for a best practices comparison. As mentioned above, often the comparison communities are of similar size and situation. However, some prefer to benchmark themselves against exemplary communities they hope to emulate.
- Call the appropriate representative(s) from the comparison communities and explain the process. Determine whether the participants prefer to remain anonymous or not. Decide whether you will share the results with the participants in return for their cooperation.
- Develop the survey form; an example is given in Table 21.2. Decide whether the survey is going to be administered in written form—either hard copy or internet—by telephone, or in person. If possible, travel to the comparison communities to administer the survey in person in order to get the best results. Among these are the many opportunities for elaboration and "off-the-record" conversations.
- Administer the best practices/benchmarking survey and collate the results. Compare your community against the others on all questions and analyze how yours is different or alike. Based on the survey results, develop program recommendations for the future.

Conclusion

Evaluation is critical as communities and community development organizations cannot rely solely on intrinsic perceptions of success. Evaluations must be integrated into the continuous cycle of organizational management and operations. There are numerous approaches to evaluation. One of these, community indicators systems, holds much promise as a means to integrate ongoing evaluation into overall development efforts.

Best practices and benchmarking are useful tools to address common issues and problems and benefit from collective wisdom and experience. These tools can also be used to evaluate the structure and effectiveness of ongoing programs. Community development is an extremely broad field so, in order to be manageable and meaningful, a best practice topic must be as narrowly defined as possible.

Best practices/benchmarking studies and data can be obtained from internet searches which yield relevant organizations, associations, and literature. Probably the best way to obtain specific, applicable best practices results is to conduct a study, picking participant communities that provide "apples to apples" comparisons. Best practices/benchmarking studies can be a relatively low-cost way to ensure that a community's programs are on track to produce the desired results.

Table 21.2 *Sample Best Practices Survey Instrument*

Survey Questions on Regional Economic Development Marketing Programs

1 Name of regional marketing organization
2 Names of counties and municipalities covered by the marketing organization
3 Total population of area served by marketing organization
4 How many years has the regional marketing organization been in existence?
5 Please indicate if the organization engages in the following types of marketing activities and how effective you believe they are:

	Effectiveness			
Activity	*Very effective*	*Somewhat effective*	*Not effective*	*Don't do*
National advertising				
Local/regional advertising				
Direct mail				
Email				
Trade show attendance				
Consultant visits				
Prospect visits				
Networking				
Telemarketing				
Other (please list below)				

6 How many full-time staff members are there? Part-time?
7 Would you please provide the titles of each staff person and a brief description of their primary job responsibilities?
8 If possible, can you share with us what the total annual budget is for the regional marketing organization? Can you provide us with any breakdown of budget allocation across major areas (e.g. advertising and other marketing activities, staff, rent, etc.)?
9 Please describe the organization's funding. Is it public, private, or a mixture? How do you determine the financial contribution from each county or municipality?
10 What local in-kind contributions does the regional marketing organization receive?
11 Do you have a written marketing plan? How often do you update it?
12 Please provide any other information that you believe will help contribute to a successful regional marketing effort.

Thank you for your time and cooperation!

CASE STUDY: THE ART OF BEING HAPPY: THE STORY OF COSTA RICA'S APOT (ASOCIACIÓN DE PRODUCTORES ORGÁNICOS DE TURRIALBA)

Costa Rica, a country with no military, dominated the Happy Planet Index in 2012 (www.happyplanetindex.org/countries/costa-rica). They beat out countries with higher disposable income levels, better roads, and higher income tax rates. They have relatively low average income tax rates (in the bottom 20 of all countries in the world at 15%), continuously battle the ever-present potholes in the roads due to the tropical environment and difficult terrain, and have an income that is lower than more than 55 countries in the world. And yet, they are happy. Doesn't happiness come from having a large wallet, good infrastructure, and lots of social programs paid for through taxes? Obviously Costa Rica has somehow found the right recipe for their country to overcome any shortcomings. So what does it mean to be a happy country? If we think about it from the perspective of a small coffee-producing community in the middle of the country, Turrialba, we can gain a better understanding of what makes the people in this country so happy.

APOT (Association of Organic Producers of Turrialba, www.fincamonteclaro.com/apot.html#.UxO75EJdWIl), was started nearly 20 years ago when a group of coffee farmers felt that their livelihoods, family life, health, and finances had deteriorated by bending to the push for maximizing profits by using more chemicals and eliminating any other plant growing in their coffee farms other than coffee. They were not happy with the changes considering that their grandparents' plantations had held numerous edible plants. The saying goes that you could walk into a coffee plantation with an empty stomach and walk out of it on the other side with your stomach filled up. They were tired of the fluctuating prices in coffee sales and having to leave their families to work in banana plantations to ensure an income to provide for their families. They wanted to take care of the life giving soils and produce a crop that was not damaging to the environment and would bring money into their family farm.

Modestly they started out by converting their farms to certified organic coffee plantations. Because they were a legally formed group, they were able to secure group certification, reducing the overall costs to each individual farmer. Gathering the harvests from each farm ensured a higher supply, reducing transaction costs for companies in the USA and Europe to allow them to purchase directly from the farmers. Farmers received an increased price for their coffee by selling at the organic certified and Fair Trade prices, thus enabling them to increase their incomes and be less reliant on other outside jobs such as working on banana plantations far away from their families. As time passed, APOT grew larger with more farmers becoming members. Their association linked with the local research institutions such as CATIE (www.catie.ac.cr) and universities (Universidad de Costa Rica) to gain knowledge on how to produce and sell organically certified coffee and produce. They started the first organic farmers market in the country and found ways to trade with other parts of the country that held comparative advantages for producing food that could not be grown in the area. By all measures of a sustainable rural development model, APOT was exemplary.

APOT members receiving their certificates of participation from members of the U.S. Embassy after having completed their two-year farmer-to-farmer training with Purdue Extension Educators from Indiana.

However, as APOT grew they encountered problems with the management of their business. They struggled trying to deal with changes needed for their organizational structure that comes with a business when it has outgrown its old management style. They had little experience with all that is entailed in running a sound business, such as running a meeting effectively, developing a strategic plan, getting all members of the association to contribute, and maintaining transparency and accountability. It was at this point that Purdue University Extension happened upon their group through a PU research scientist stationed at CATIE in Turrialba, Tamara Benjamin. Over two years, 15 Extension Educators worked with APOT to develop a strategic plan for their organization and to help them work through some of the transparency and accountability problems that were plaguing the association. APOT members worked hard to develop a sound business plan that they could use to move forward and become and even more sustainable producer group.

So what does this have to do with being happy? Well, APOT, like many Costa Rican citizens, are blessed with a country that believes that they should spend their tax dollars on public education rather than investing in a military. They believe that the progress of their country and community is not just based on economic activity but on the well-being of their population and having healthy, educated children, now as well as in the future. By forming a group that was concerned about the ecological well-being of their farms, the health and well-being of their families, and the financial sustainability of their crops, APOT was able to support their members to stay on their farms. These farmers may not be the wealthiest people in the world, but they probably can maintain they are some of the happiest.

Tamara Benjamin, Ph.D.
Sustainable Agriculture and Natural Resources Scientist, Purdue University, USA

Keywords

Evaluation, performance evaluation, community indicators, happiness studies, best practices, community well-being, best practice, benchmarking, case studies, comparison communities.

Review Questions

1 What are community development indicators and why are they useful?
2 What are some types of community development indicators?
3 Identify some benefits of incorporating benchmarking and indicators into community development planning processes.
4 Explore the *World Happiness Report* to find out where your country ranks on the scale.
5 The concept of happiness is garnering more interest worldwide. Can you identify a case within your own community, state, or province where happiness was a topic of conversation, research or activity?

Bibliography

Brand, D. (2004) "The New Transit Town: Best Practices in Transit-Oriented Development," *Choice*, 42(1): 156.

Budrik, M., and Phillips, R. (eds.) (2011) *Quality-of-Life Community Indicators for Parks, Recreation and Tourism*. Dordrecht: Springer.

Canada, E. (1993) "TQM Benchmarking for Economic Development Programs," *Economic Development Review*, 11(3): 34–39.

Canadian Index of Well-Being. (2014). Accessed February 14, 2014, at: https: //uwaterloo.ca/canadian-index-wellbeing/

Council for Community and Economic Research (Formerly "American Chamber of Commerce Research Association"). December 14, 2007, at: www.c2er.org/

Davidson, E.J. (2005) *Evaluation Methodology Basics*, Thousand Oaks, CA: Sage.

European Union, Brussels. (2007). Accessed December 19 at: http: //ec.europa.eu/enterprise /library/lib-competitiveness/series_competitiveness.htm.

Federal Reserve Bank of Chicago, Consumer and Economic Development Research Information Center, LesLe (Lessons Learned). (2007). Accessed December 19 at: www.chicagofed.org/cedric/lesle.

Hampton, VA Healthy Families Partnership. (2007). Accessed December 4 at: www.hampton.va.us/healthyfamilies.

Helliwell, J., Sachs, J., and Layard, R. (2013). *World Happiness Report*. Accessed December 13 at: www.earth.columbia.edu/sitefiles/file/Sachs%20 Writing/2012/World%20Happiness%20Report.pdf

International Network for Small and Medium (Business) Enterprises. (2007). Accessed December 19 at: www.insme.org/t_fulltext_search.asp?target =benchmarking&x=35&y=12.

National Governors Association. (2007) Accessed December 19 at: www.nga.org

National Neighborhood Indicators Partnership. (2013) Accessed July 10 at: www.urban.org/nnip

Nowakowski, J.M. (2003) "Creating Regional Wealth in the Innovation Economy: Models, Perspectives and Best Practices," *Choice*, 40(10): 1795.

Phillips, R. (ed.) (2005) *Community Indicators Measuring Systems*, London, UK: Ashgate.

Phillips, R. (2003) *Community Indicators. PAS Report No. 517*, Chicago, IL: American Planning Association.

Phillips, R., Sung, H., and Whitsett, A. (2013) State Level Applications: Developing a Policy Support and Public Awareness Indicator Project. In J. Sirgy, R. Phillips and D. Rahtz, (eds.) *Community Quality-of-Life Indicators: Best Cases VI*. Dordrecht: Springer, pp. 99–118.

Redefining Progress. (2007) Accessed December 4 at: www.rprogress.org

Reese, L. and Fasenfest, D. (2004) *Critical Evaluations of Economic Development Policies*, Detroit, MI: Wayne State University Press.

Renzas, J.H. (1994) "Economic Development Best Practices: The Best and the Worst," *Economic Development Review*, 12(4): 83–88.

Santa Monica Sustainable City Program. (2014) Accessed February 1, 2014, at: www.santa-monica.org/environment/policy

Steller, T. (2004) "Tucson AZ Panel Evaluates Best Practices in Economic Development," *Knight Ridder Tribune Business News* (June 13): 1.

Sustainable Communities Network. (2014) Accessed February 1, 2014, at: www.sustainable.org

Sustainable Seattle. Online. (2014) Accessed January 10, 2014, at: www.sustainableseattle.org

Swain, D. (2002) *Measuring Progress: Community Indicators and the Quality of Life*. Accessed December 4 at: www.jcci.org/measuringprogress.pdf

Toledo Business Journal. (1996) "Best Practices Study Takes a Look at NW Ohio Development Effort," *Toledo Business Journal*, 12(7): 1–11.

Tornatzky, L.G. (2001) "Benchmarking University-Industry Technology Transfer," *Journal of Technology Transfer*, 26(3): 269.

U.S. General Accounting Office, Washington, DC. (2007). Accessed December 19 at: www.gao.gov

U.S. National Agricultural Library, U.S. Department of Agriculture (USDA), Rural Information Center, Rural Resource Center. (2014). Accessed January 4, 2014, at: www. nal.usda.gov/ric/ruralres

Walker, B. (2004) "European Cities Join up to Share Best Practices," *Regeneration and Renewal*, (March 19): 1.

Connections

Explore which countries are happy with the Happy Planet Index: www.happyplanetindex.org/about

Discover tools for initiating happiness and well-being in your community: www.happycounts.org

Check out amazing statistical presentation with Hans Rosling's "200 countries, 200 years, in 4 minutes" at: www.youtube.com/watch?v=jbkSRLYSojo

Examine the National Neighborhood Indicators Partnership at: www.neighborhoodindicators.org/

Learn about a multitude of communities developing indicator projects at: www.communityindicators.net/projects

The Community Tool Box is provided by the University of Kansas Work Group for Community Health and Development, as a designated World Health Organization Collaborating Centre for Health and Development. Access at: http: //ctb.ku.edu/en/assessing-community-needs-and-resources

Watch a video, Bhutan: The Kingdom Where GDP Is Measured In Happiness, at: https: //www.youtube.com/watch?v=CXJwNSkdTH0

Various sources for information useful for benchmarking include:

U.S. National Agricultural Library, USDA, Rural Information Center, Rural Resource Center: www.nal.usda.gov/ric/ruralres

National Governors Association: www.nga.org

Federal Reserve Bank of Chicago, Consumer and Economic Development Research Information Center, LesLe (Lessons Learned): www.chicagofed.org/cedric/lesle

PART IV

Issues Impacting Community Development

22 Perspectives on Current Issues

Paul R. Lachapelle

Overview

Community development is increasingly described and defined as a dynamic, diverse, multi-faceted, and robust discipline. In an era characterized by change, complexity, uncertainty, and controversy, the theoretical and applied perspectives of both practitioners and academics of community development have evolved dramatically in recent years. Simultaneously, contemporary community development theory and practice is both trans-disciplinary crossing fields of study and practice, and interdisciplinary combining subjects and specialties that were once thought to be exclusive or unrelated. This chapter provides a scan of perspectives from around the globe on community development. It is presented in the form of responses on topics by each person interviewed. Note that referrals to cases are embedded within many of the responses, as opposed to providing one case at the end of the chapter.

Introduction

With many strange but fruitful community development bedfellows emerging, praxis continues to borrow from, collaborate with, and facilitate cooperative partnerships, affiliations, and relationships that increasingly prove to be synergistic and symbiotic. Academics and practitioners continue to explore theory and practice using perspectives that are both unique and unconventional in terms of how issues are defined, how opportunities are identified, and how organizations or communities demonstrate creativity or innovation to approach or address situations.

A growing body of literature is exploring the many novel and unique approaches and perspectives related to identifying and addressing community development's current issues (Brennan and Brown, 2008; Domahidy, 2003). Community development is increasingly collaborative with active partnerships between academics and practitioners (Mizrahi et al., 2008). The field is also described as a type of social movement with an implied focus on the construction of community, collective identity, and cultural rights (Green, 2008). While there is a growing body of literature exploring and revealing detailed case studies of specific community development issues, needs, and actions, there has been little empirical study of community development praxis from the perspective of community development academics and practitioners themselves. Moreover, there appears a blurring of these vocational lines leading to a new breed of community development "pracademics," both academic and active practitioner (Posner, 2009), who are discussing, designing, and implementing new and innovative praxis. Understanding these evolving perspectives can contribute to a growing body of literature, expand the current knowledge base, and provide further insights for community developers across varied contexts. The perspective of community developers in terms of how they perceive issues, pursue opportunities, and create innovations is essential to building more effective community development praxis.

This chapter provides perspectives on current issues from 13 community development leaders, academics, and practitioners, representing a variety of fields, experiences, and geographic locations. These community development leaders were asked to comment on and provide detailed narratives related to three questions and statements:

1 What are the most pressing issues in community development currently?
2 Share insights on potential opportunities or approaches to address these issues.
3 Provide an example of how organizations or communities have shown creativity or innovation in community development practices.

Following a case study approach (Yin, 1984) and using a qualitative method to provide rich description (Goetz, 1973), individuals were asked to provide detailed narratives based on their own interests, backgrounds, and experiences. This case study, while admittedly not representative in terms of summarizing the myriad perspectives, experiences, or spatial and temporal contexts that exist in the community development field, is an attempt to provide a brief, albeit descriptive set of diverse and unique perspectives on current issues. Below are summary statements by individuals appearing in alphabetical order. A main theme at the top of each box attempts to represent a unique or prominent perspective from the individual. The name of the individual, their title, department and university or organization, and city and country are provided after each response.

BOX 22.1 COMMUNITY DEVELOPMENT AND DEMOCRACY

From my vantage point, the most pressing issue lies at the intersection of community development and democracy. As we contemplate community development practice, it is imperative we consider how we uphold and strengthen essential democratic tenets. Additionally, we must examine how our work in community development may strengthen the system of democracy in our communities and countries. The role of dignity in development is a keen opportunity to reenvision community development practice. Showing high regard for people and the situations in which they exist and recognizing the inherent dignity of humanity has the potential to impact the ways we interface with communities and citizens. It is through this perspective, integrating it into our personal behaviors and professional practice, we can help to bridge the divide between expert and local citizen.

I see the intersection of democracy and community development as one of the greatest potential opportunities to simultaneously advance community development practice and meet the needs of our citizens and communities. We must re-conceptualize the roles of experts and citizens in a way that would allow for the kind of participatory democratic problem-solving envisioned by scholars like Boyte (2003, 2009) and Barber (1998). In practical terms, most models of engagement have maintained the technocratic, expert–citizen dichotomy (Bridger and Alter, 2010). Expert knowledge is still viewed and practiced in narrowly technical terms. There is little room for judgment, practical wisdom (or phronesis), and the democratic participation that is necessary to solve problems that have moral, political, and ethical dimensions (Fischer, 2009; Flyvbjerg, 2001, 2012). Citizens must be involved in all stages of exploring solutions, from identification of the issue to developing policies and practices to address the issue. Focusing on dignity in development is an approach that I believe is critical to ensuring mutual respect throughout community development practice and processes. By emphasizing dignity as a core principle of our work, we create the space for a more egalitarian approach to community development. This approach requires we suspend judgment and recognize the inherent value of each individual and their contribution to the process of community development.

Entrepreneurship development presents one promising option for cultivating local economies rooted in local leadership, ideas, talent, markets, and tastes. Often called economic gardening (Morgan et al. 2009) or entrepreneurship development systems (Lichtenstein and Lyons 2006), this style of development refocuses economic development as the cultivation of opportunity within a community rather than focusing efforts on attracting economic opportunities from elsewhere. These are generalized approaches requiring an existing core of entrepreneurial expertise. For those rural and declining regions where such expertise is either lacking or has low public visibility, a more specific—and in my view, very promising—approach, stems from the Intentional Innovation Communities (IIC) framework, which I have utilized in my work with rural communities in Australia. These IIC's are catalytic economic development programs that convene potential innovators from across the community to actively create a context for real and feasible community-wide entrepreneurial action. A major part of its intention is to surface new, collaborative opportunities for everything from innovative partnerships and supply chains to the development of novel local businesses, value-added agriculture, and solutions that build a more integrated, holistic local economy.

Theodore Alter, Professor of Agricultural, Environmental and Regional Economics, Co-Director, Center for Economic and Community Development, Pennsylvania State University, University Park, PA, USA

BOX 22.2 MARGINALIZED COMMUNITIES

Although community development ought to be a practice that makes a fundamental difference in the lives of people within our most marginalized communities, much of what we actually see takes the form of amelioration rather than real change. I identify three interconnected issues which are internal to our practice and which undermine effectiveness in dealing with any issue. (1) Dealing with peripheral issues: The central concern of community development is to work alongside people to release their personal and collective resources in order to effect change, challenge injustice and help create a better world. When that commitment is lost, community development workers organize events, run projects, fundraise, offer training etc. but none of that addresses the central concerns of people's lives. (2) Picking easy targets: Due to our increasingly managerialist culture and outcome-driven practice, community development projects target their interventions on people who can succeed, by externally defined criteria, rather than working with people in the greatest need or taking action which will have the greatest long-term impact. (3) Lack of critical reflection: This operates on two levels. First, it seems to me that practitioners are becoming less politicized and therefore have lost many tools needed to analyze their social world and make decisions about what to do about root causes of issues. Second, many examples of practice lack opportunities for local people to critically reflect on their lived experience and then develop visions of the world they want. Local people are therefore condemned to merely take on the ideas of others, further pacifying rather than empowering individuals. We must return to approaches which work with the have-nots of our world, engage them in dialogue about their experience and aspirations and help them organize to build their collective power to make those aspirations realities.

Slum Dwellers International is a grass-roots nongovernmental organization that organizes urban poor to negotiate with politicians, planners and policy makers thus taking an active role in shaping the decisions that affect their lives. The organization offers practical solutions to recognized and intractable problems with a moral legitimacy to its activities arising from high levels of local participation that in turn makes it hard for more powerful forces to organize against the urban poor (Muller and Mitlin 2007). Their development process links radical action with research and reflection, producing sustainable social change that for example, encourages households to join community-level savings and credit programs to build financial assets and local capacity; undertakes housing and infrastructure projects that provide needed improvements while demonstrating solutions and building the capacity of the poor; organizes community-led Surveys and maps to create a powerful informational base for strategizing and negotiations; facilitates peer exchanges among groups on local, regional and national levels so that communities can learn from each other; organizes housing exhibitions that showcase affordable solutions and act as a tool for mobilization and dialogue with officials; and advocates pro-poor policy changes on the basis of grass-roots experience and demonstration of good governance practices.

Dave Beck, Lecturer in Community Development, School of Education, University of Glasgow, UK

BOX 22.3 GOVERNANCE AND POWER

The most pressing issues in community development involve governance and civic engagement, social and economic innovation, and power. Governance and civic engagement involve a better understanding of how local residents can take on ownership of decision making. There is a strong need to identify best practices in these areas. Equally important, there is a need to bridge different theoretical perspectives. Comparative national and international work will be very helpful. Social and economic innovation involves exploring new ways for social and economic/entrepreneurial development. National and international research in these areas is needed to determine best practices and opportunities for advancement. Power in a community includes both an applied and theoretical understanding of the emergence of power by grass-roots and other stakeholders. We need applied research to understand why some communities can pull together residents and form strong power bases, while others fail to achieve power despite collective numbers.

Mark Brennan, Professor and UNESCO Chair in Rural Community,
Leadership, and Youth Development,
Pennsylvania State University, University Park, PA, USA

BOX 22.4 HOLISTIC APPROACH

A pressing issue is ensuring that there is a holistic approach to praxis that integrates the material, political, relational, and spiritual/values. I think that we are still caught in a situation where praxis, both theory and practice, are focused on any one of these aspects, and have a tough time putting them all together. I think that any attempt at community development that does not take all four of these into account runs the risk of missing things that are important both for the sustainability of the endeavor but also in making the most impact. In addition, there is a need for well-trained and skilled community development facilitators. The lack of facilitators who really understand the complexities of the craft and can adapt tools to context and create a ground-up approach for sustainability is a major problem and has been since the beginning of community development. There is also a need to ensure that metrics (planning, monitoring, and evaluation) are used in group, individual, and organizational settings in a holistic way. Metrics are crucial because of the power they hold to lead community development efforts, rather than serving as tools to advance community development efforts that are established based on local priorities and the important points of focus.

Examples of innovations include the following: (1) The Mondragón Cooperative Movement in the Basque region of Spain is a worker-owned movement that has had a huge impact on its community and beyond; (2) Manvoday is a center of learning in values-based development in India (Vidyarthi and Wilson, 2008; www.manavodaya.org.in); (3) Moucecore is based in Rwanda and is a church-based group focusing on community development projects, spirituality, and sustainability (moucecore.org/projects.html); (4) Christian Commission for Development based in Honduras focuses on health, agriculture, and small productive projects geared toward building knowledge and organizational capacity in the context of a "politicized" approach to development (Bronkema, 2005).

David Bronkema, Associate Professor, Director, International Development Programs,
School of Leadership and Development, Eastern University, St. Davids, PA, USA

BOX 22.5 CONTROL AND INEQUITIES

I see three core community development challenges. First is the long-term and continuing loss of community autonomy. Compared to the past, the ability of a local community or region to control its own economic, environmental, and social destiny is diminished. This undermines the essential democratic connection between the values people hold and their ability to realize those in everyday settings. Second is the degradation of political discourse, which I believe is exacerbated by the first trend. Lacking control, politics becomes a spectator sport for most and discourse becomes less anchored in everyday reality. Culturally, we are trained to be good consumers but learn little about how to conduct public life. Shrillness and demagoguery follow. Third is increasing segregation by income and class. Dramatic inequalities rooted in tax and spending policies are key drivers as well as increasing distrust of others. Lacking the ability to communicate past immediate differences, we withdraw into isolated cultural islands. Diversity exists across regions, but at the neighborhood and local community level, we are increasingly segregated by class and race. Because these challenges are rooted in macroeconomic structures backed by corporate and state political power, they are difficult to alter. Over the long term, however, we can envision approaches that (1) strengthen local control incrementally, (2) appeal to left and right critics of current political arrangements while building on trusted and typically more moderate local leadership, and (3) expand access to resources and opportunities for the lower class. Politically, the approach must include bottom-up and top-down strategies, as well as both insider and outsider strategies that combine incremental reform with a broader restructuring of institutions.

One example of creativity and innovation in meeting the challenges involves the movement to create local and regional food systems. Local food systems are collaborative networks that integrate sustainable food production, processing, distribution, consumption and waste management in order to enhance the environmental, economic and social health of a particular place. Local and regional food system networks engage a wide range of community partners in projects to promote more locally based, self-reliant food economies (see for example: www.sarep.ucdavis.edu/sfs/def).

David Campbell, Community Studies Specialist, UC Cooperative Extension, Department of Human Ecology, University of California Davis, Davis, CA, USA

BOX 22.6 RURAL COMMUNITY VITALITY

While many rural communities are small but relatively vibrant, surrounding rural areas struggle with fundamental social and economic challenges. Many rural communities have an agricultural economic base and many small family farms operate on narrow margins with increasing costs and debt levels. There are five main approaches to enhancing the vitality of rural communities. First, investment is needed in basic infrastructure and services. Governments increasingly see this as inefficient investment of public funds, and private investors see this as not profitable. Nevertheless, communities need basic services and infrastructure. Second, a set of diverse economic development strategies are needed to anticipate economic change and make gradual but purposeful changes in the economic base of communities such as increasing tourism, or value-adding to agriculture. Third, increases in the capacity the communities, particularly the leadership and ability to redefine assets are crucial. Fourth, changing the perception of rural communities might seem in effectual but it is important to their future. More positive examples and stories of life and success in rural areas are needed to gradually reposition the image of rural life. Finally, development is only improvement when it is in line with community values.

Considerable innovation was used in the development and implementation of the community plan for the Scenic Rim area in southeast Queensland in Australia. The Scenic Rim is a local government area of farmland, rural communities and rural residential living close to the major urban centers of Brisbane and Ipswich in eastern Australia. The Scenic Rim Regional Council facilitated the engagement of residents in the development of a community plan that described key directions and actions for the future of the area. The approach was innovative in that it used community-directed engagement, engagement based on natural communities and sectors, deliberation, arrangements of actions, and feedback forums.

Jim Cavaye, Associate Professor of Rural Development, School of Agriculture and Food Sciences, University of Queensland, Australia

BOX 22.7 COMMUNITY SUSTAINABILITY

From a rural and domestic perspective, the most pressing issues include community sustainability, mainly in terms of ecosystem, but other significant factors such as an aging population, institutions that suffer from scarce human and financial resources, and decay of physical infrastructure. Furthermore, rural places are fragmented by the geography of the workplace that creates "bedroom communities" and government and economic devolution that leads to reduced economic activity, technical assistance, and social services. Community development practice is both art and science based on understanding of what people want and need in terms of personal relationships. The potential opportunities in the face of resource scarcity and loss of community and personal empowerment involve linking past and present in places with willingness and ability to dream about positive choices and outcomes for the future. Models based on Schumacher's (1989) *Small is Beautiful* philosophy offer an opportunity to leverage negativity into positive action because it can be adapted to values of individual self-help and local associational activities. Community development's basic paradigm of democratic action is crucial to helping smaller, localized settings.

We need to think of community development techniques that are already part of our growing list of sustainable practices to build places with a higher quality of life and more in harmony with the environment; for example, building small, green businesses (earthtrepreneurship) that are based on local foods, environmental conservation and reclamation, and developing alternative energy sources for local use. At the same time, we need to work with communities to enhance local government and build voluntary community associations. In the long run, devolution of government and the economy could be a blessing in disguise for communities willing to tap into old values and use them to build new places. If people are opposed to big government—and many rural residents seem to be—then the task of community developers may well be to define our art and science more effectively so that it appeals to the needs of people in smaller places who want to have more control over their destinies.

Timothy Collins, Assistant Director, Illinois Institute for Rural Affairs,
Western Illinois University, Macomb, IL, USA

BOX 22.8 THE SPIRITUAL DIVIDE

There is a great spiritual divide. It's more than just between fundamentalist and liberal types; there is a growing disconnect with self and self. The World Health Organization (WHO) has gathered data which indicates that suicides in the world exceed the victims of violent crime and war. People are demonizing each other with political and ideological rhetoric. In spite of this, my colleagues in other fields report that something important is emerging just like the great cities that emerged almost simultaneously 3,000 years ago across the planet. We can't put our finger on the big change that's occurring but I've heard from many people that it's there. It's a different kind of Renaissance. Scharmer (1999), author of *Theory U: Leading from the Future as It Emerges*, suggests we can see the future. This future integrates conventional strategic planning with venues in which people can be silent, express gratitude and find their authentic self (rather be defined by convention). I believe this is a challenge for many of us who work in secular and Cartesian rationalist-type organizations. I believe the global presencing movement initiated by Peter Senge (1990), Scharmer (1999) and others is part of a change that is allowing a balance of intellect with heart. There are also growing disparities between humans and the natural world. We're consuming more than we can replace; it isn't sustainable. There are thousands of organizations that are doing alternative work such as those described in the film, *Awakening the Dreamer, Changing the Dream* (http: //vimeo.com/52048121). There is also a growing economic divide. The statistics indicate it's becoming more pronounced in the US—more so than all of the industrialized countries. It isn't a question of taking from the rich, but finding new ways for others to prosper.

The MIT Community Innovations Lab (http: //web.mit.edu/colab/) is one venue where people are involved in communities across the world to creatively address complex issues. The Grameen Bank (www.grameen-info.org) is a for-profit entity that provides modest loans for those with low incomes. This approach toward lending has been duplicated across the globe. It is one sign in which civic and social concerns are blended with the market system. We are seeing similar changes in the markets: faith-based groups are also accomplishing a lot with limited resources.

Ron Hustedde, Professor, Department of Community and Leadership Development
University of Kentucky, Lexington, KY, USA

BOX 22.9 DEFINING COMMUNITY DEVELOPMENT

From the standpoint of the field of community development, the most pressing issues may be agreement on a commonly held definition of community development. Over the 50-plus years of development of the community development body of knowledge, because of its multidisciplinary nature, there are many names, perspectives, and reinvented methodologies that, to a generalist, look to be the same. Community improvement, community betterment, community innovation, community engagement, for example, all fit within the community development frame. Without a well-defined frame, the Community Development field (especially as practiced by those who follow the Community Development Society's Principles of Good Practice: comm-dev.org/) has the tendency to become everything to everybody. This dynamic makes it difficult to organize, research, vet, and then diffuse the collection of knowledge fitting into that somewhat leaky frame. That may be the beauty of community development; it's a field that flexes for the changing people, places, and times of each era of citizen-participation society seeking to improve their communities.

As a practice, a pressing issue in community development is the tension in the field (and certainly at the community implementation level) to solve somewhat narrowly defined problems in an attempt to quickly show progress. I am concerned about the tendency toward specialization (housing experts, community health experts, achievement gap experts, environmental experts, for example) even as complexity demands multi-dimensional, multidisciplinary awareness, thinking, and action. There's not enough adherence to the understanding that community development works best when citizens (and practitioners) are able to cross sectors, geographies, and disciplines, and work collaboratively, inclusively and sustainably to solve problems and seek opportunities together. With the advent of usable, intuitive information and communication technologies, we have within our power the ability to better map, interpret, and apply social analysis network knowledge or network mapping.

One example of community innovation is the Minneapolis (MN) Youth Coordinating Board (www.ycb.org). This organization began in the mid-1980s as a cross-jurisdictional collaboration of community leaders and agencies coming together to address growing inequities in children and youth development opportunities, especially for kids of color. Today, it stands as a beacon of innovative collaboration and continues to serve as a catalyst for group action and citizen involvement in the ever-dynamic conditions of Minnesota's most populous city and county.

Jane Leonard, community development consultant and past president
of the Community Development Society, St. Paul, Minnesota, USA

BOX 22.10 EVALUATION AND MEASUREMENT

I think evaluation/measurement, volunteer leadership and funding are core issues that are linked together. Funders want to see measurement to indicate impact and volunteers want to be involved in something that makes a difference with forward momentum. I tend to think we have to bring communities back to thinking about their vision and developing an implementable plan of action to get there. We could benefit most from developing a few recognized, effective models of evaluation metrics (collection and analysis) that communities could compare and benchmark against. These metrics should help to tell "the story," track impact, and be accessible.

Our current obsession with measuring impact to justify value to funders, while making some sense, also creates challenges for promoting and encouraging collaboration and partnerships since so much pressure exists to claim credit for impact and thus competition escalates between organizations. It is becoming increasingly difficult to find those community leaders who have the time to apply to community development work. So often it is a community champion or small group of champions, who are dedicated to making things happen in a community. With more and more demands on people's time it becomes harder and harder to find people who will commit what it takes to be a catalyst and make a difference. The Community Progress Initiative is an example of creative and innovative practice: http: //ncrcrd.msu.edu/uploads/files/133/ncrcrd-rrd180-print.pdf

Connie Loden, Executive Director, Economic Development Corporation of Manitowoc County, Manitowoc WI, USA

BOX 22.11 FINANCING COMMUNITY DEVELOPMENT

A key issue is maintaining a vibrant and supported financial existence. For example, in the Irish case, there has been serious retrenchment in funding for the community and voluntary sector. Most of the funding for community development has come from the central state or from European institutions. With the downturn in the economy, estimates suggest that funding for the community and voluntary sector over the 2008–2012 period was cut by 35%. At the same time, various projects have been and are being subsumed into larger local structures. Employment and career pathways, therefore, remain quite a worrying and unclear aspect for existing and new entrants. A further problem concerns the independence and autonomy of the community and voluntary sector. The structures of governance have changed in Ireland in the last number of years. The central government announced in 2013 that local government (city and county levels), which was previously not a major stakeholder, will play an increasingly executive role in the future direction and oversight of programmes and initiatives at local community level. The increased involvement of local government within local community development could potentially provide a long-term future for community development, perhaps in the form of service contracts. While a lead agency therefore could be a potential strength and give continuity to the sector, it must recognize the existing expertise, knowledge and skill already in the community sector. If the role of community development is taken seriously, it may also require considerable training for local government staff and a renewed interest and concern for how well local communities are faring. Less direct funding from the state may also mobilize the community sector to seek more self-sustaining funding opportunities through social economy initiatives or potential growth areas of the economy such as "green energy."

There are many examples of creativity/innovation within the rural community sector, particularly around harnessing natural resources more effectively, or adding value to food and the environment. The green economy is heralded by government as a significant growth sector for the future and some local communities are in a strong position to capitalize on this in my view. As an example, some local communities have engaged in renewable energy development such as wind energy. A good case would be West Clare Renewable Energy Ltd, which is a community-driven cooperative, comprising of around 30 farm families in County Clare on the western coast of Ireland. It was recently granted planning permission to build 29 wind turbines and is described as "the largest community-owned wind farm development in Ireland and one of the biggest in Europe" (www.energyco-ops.ie/renewable-producers/our-community-partners-in-munster/west-clare-renewable-energy-2).

Brian McGrath, Lecturer/Programme Director, School of Political Science and Sociology, National University of Ireland, Galway, Ireland

BOX 22.12 LEADERSHIP

Community leadership has long been a necessity for community growth and development. Communities have seemingly struggled with two particular issues. First, is the issue of continued leadership with the potential need of new and innovative leaders to "join the fold." Second, is the issue of leadership turnover, with little exchange of information or passing down of community/institutional knowledge so that new leaders feel as though they must "reinvent the wheel." The challenge for the first issue is to avoid a coup over a long-standing leader, but find new ways to engage new leaders. In the second challenge, it is important to ensure that when turn-over is quick or unexpected, new leaders are not left struggling to pick up pieces of the organization with little guidance from institutional memory. Related to leadership is the need for diversified participation. Communities all over the country, and indeed the world, are becoming increasingly diversified. It is essential for us to tap into the talent that surrounds us. Our communities can continue to be and become even more vibrant through the recruitment of diverse populations to participate. As with the inclusion of new people in our community organizations, we must look for new ways building our communities. The key is to branch out and learn what other people are doing in other parts of the community or the world and see if applying those efforts to our communities might just make sense.

The neighborhood of Pleasant Ridge, in Cincinnati, OH, became the first community in Ohio to receive the Community Entertainment District designation. Previously reserved for big development, particularly in suburban areas, the Community Entertainment District designation provides discounted liquor licenses to the community for use by new or existing restaurants/bars/cafes, etc. In Pleasant Ridge, a community member brought the designation idea to the Pleasant Ridge Development Corporation and they worked together to complete the application and appeal to the city to lower the application fee from $15,000 to $1,500. This new fee, reserved for neighborhood organizations, opened up the potential of the CED designation not only to Pleasant Ridge, but to other Cincinnati neighborhoods as well.

Whitney McIntyre Miller, Assistant Professor of Leadership Studies,
College of Educational Studies, Chapman University, Orange, CA, USA

BOX 22.13 TOKENISTIC APPROACHES

Trying to make community development processes fit into external, imposed timelines can lead to tokenistic approaches. Often governments and organizations value a community development approach and want to incorporate it into work they are doing with communities; however, they fail to appreciate the time that genuine community development really takes. It is not a genuine community development approach if a process has to be 'completed' by a funding deadline or by fiscal year end, instead of when the process really is completed according to the community. Partly due to time pressures and lack of resources to sustain community development processes, and partly due to the relative ease with which mainstream groups can be reached, community development processes are not always genuine, and do not always include everyone. It is much more difficult to engage people who don't show up at a community meeting, so sometimes we simply work with the people who do show up—and they are not usually people from marginalized groups. In addition, arts and creative processes can truly engage communities.

The Municipality of the County of Kings (Nova Scotia, Canada) used a community-based process to develop the Action Plan for Ending Racism and Discrimination (the first municipality in Nova Scotia to do so), and then hired a community development team to implement the plan for five years. What's innovative is that while the Municipality is following the plan, it is also open to (and seeking) community ideas and innovations and engagement, and is keeping the process transparent by regularly and publicly reporting on progress—this encourages engagement and ownership of the process by the community and opens the Municipality to new ideas and strategies, which it is embracing.

Cari Patterson, Director, Horizons Community Development Associates, Inc.,
Wolfville, Nova Scotia, Canada

Conclusion

The responses offered here suggest both a diversity of perspectives as well as symmetry among the individuals. It should be noted that there was considerable overlap between some respondents in their full narratives; however, each response was edited for brevity with a focus on a prominent perspective. Nevertheless, several notable themes have emerged across all the narratives.

First, there appears to be a strong emphasis on people-to-people relationships. Many of the respondents noted the positive outcomes and creative energy that results from personal interaction. Constructive relationships lead to both a sense of optimism as well as productive collective action. A unique spiritual awakening is said to be emerging that is intimately linked to the health of our relationships and the corresponding positive influences on governance structures and democratic practices. As Ostrom (1997: 3) notes, "person-to-person, citizen-to-citizen relationships are what life in democratic societies is all about."

Second, beneficial relationships can produce powerful outcomes with positive unintended consequences. Indeed, there is a perceivable force that comes from collaboration and collective action resulting in positive change. This change can result when communities have ownership over a situation. Ownership of a community change process confronts issues of power and empowerment and ensures that all citizens can influence decisions of process and outcome (Lachapelle, 2008). The application of ownership influences how we define and measure a given situation, how we view opportunities, and how we take collective action. When this ownership is lacking, tokenism ensues with "citizens acting only in a consultative role but without some form of delegated power" (Arnstein, 1969: 217). The result is often marginalized populations experiencing gross inequities and lacking genuine opportunities for engagement. The increasing focus on technical solutions and the reliance on "expert advice" can lead to "little confidence ... in the skills, intelligence, and experience of ordinary people" (Scott, 1998: 346). Powerful alliances formed through community development efforts can bring about holistic approaches, build trust, promote civic engagement, enhance leadership, and lead to plans and actions that are more socially and politically acceptable. Moreover, positive community development outcomes and consequences can lead to a propensity to collaborate in the future. Seen as a type of social contract, this behavior can reinforce the notion of reciprocal altruism whereby an individual assists in a present effort with the expectation for support in the future (Trivers, 1971). Ostrom and others (1999) highlight many examples of successful long-term self-organized processes to manage public resources with residual reciprocal benefits. The power of collaborative community development efforts that seek out and use diverse perspectives can lead to many positive outcomes.

Last, there seems to be a genuine need for innovation and creativity in community development. Creative ways of financing community development, engaging stakeholders, and building community leadership can lead to opportunities for mutual learning and empathy. Dietz et al. (2003: 1907) report that the key components of successful collaborative processes include, "dialogue among interested parties, officials, and scientists; complex, redundant, and layered institutions; a mix of institutional types; and designs that facilitate experimentation, learning, and change." Moreover, innovation and creativity, when explicitly sought and applied can produce more vibrant, vital and resilient communities that are responsive to myriad needs and issues.

These narratives provide a concise but detailed examination of unique perspectives involving community development's current issues. Further analysis of a range of perspectives will only expand the knowledge base and, with it, bring more effective community development outcomes and impacts.

Keywords

Community innovation, democracy, devolution, empowerment, engagement, entrepreneurship, evaluation, governance, leadership, local government, ownership, praxis, reciprocal altruism, rural communities, sustainability, volunteerism.

Review Questions

1 What principal themes emerge from the respondents?
2 Identify commonalities between the different perspectives as well as those that are unique and original.
3 How does creativity relate to innovation within and between the different community development perspectives?
4 What is reciprocal altruism in a community development context? Provide examples.
5 What current issues does your own community face?

Bibliography

Arnstein, S.R. (1969) "A Ladder of Citizen Participation." *Journal of the American Institute of Planners*, 35(4), 216–224.

Barber, B.R. (1998) *A Place for Us: How to Make Society Civil and Democracy Strong*. New York: Hill and Wang.

Boyte, H.C. (2003) "A different kind of politics: John Dewey and the meaning of citizenship in the 21st century." *Good Society*, 12(2): 1–15.

Boyte, H.C. (2009) *Civic Agency and the Cult of the Expert*. Dayton, OH: Kettering Foundation.

Brennan, M.A. and R.B. Brown. (2008) "Community Theory: Current Perspectives and Future Directions." *Community Development*, (39)1: 1–4.

Bridger, J.C., and T.R. Alter. (2010) "Public Sociology, Public Scholarship, and Community Development." *Community Development*, 41(4): 405–416.

Bronkema, D. (2005) Development as a political gift: donor/recipient relationships, religion, knowledge, and praxis in a protestant development NGO in Honduras. Unpublished dissertation, Yale University, New Haven, CT.

Dietz, T., E. Ostrom, and P.C. Stern. (2003) "The Struggle to Govern the Commons." *Science*, 302, 1907–1912.

Domahidy, M. (2003) "Using Theory to Frame Community and Practice." *Community Development*. (34)1: 75–84.

Fischer, F. (2009) *Democracy and Expertise: Reorienting Policy Inquiry*. New York: Oxford University Press.

Flyvbjerg. B. (2001) *Making Social Science Matter: Why Social Inquiry Fails and How it can Succeed Again*. Cambridge, UK: Cambridge University Press.

Flyvbjerg. B., T. Landman, and S. Schram. (2012) *Real Social Science: Applied Phronesis*. Cambridge, UK: Cambridge University Press.

Geertz, C. (1973) Thick Description: Toward an Interpretive Theory of Culture. In C. Geertz (Ed.), *The Interpretation of Cultures*: Basic Books pp. 3–30.

Green, J.J. (2008) "Community Development as Social Movement: A Contribution to Models of Practice." *Community Development*, 39(1): 50–62.

Lachapelle, P.R. (2008) "A Sense of Ownership in Community Development: Understanding the Potential for Participation in Community Planning Efforts." *Community Development*, 39(2): 52–59.

Lichtenstein, G.A., and Lyons, T.S. (2006) "Managing the Community's Pipeline of Entrepreneurs and Enterprises: A New Way of Thinking about Business Assets." *Economic Development Quarterly*, 20(4), 377–386.

Mizrahi, T., M. Bayne-Smith, and M. Garcia. (2008) "Comparative Perspectives on Interdisciplinary Community Collaboration between Academics and Community Practitioners," *Community Development*, 39(3): 1–15.

Morgan, J.Q., W. Lambe, and A. Freyer (2009) "Homegrown Responses to Economic Uncertainty in Rural America." *Rural Realities*, 3(2), 1–15.

Muller A. and D. Mitlin (2007) "Securing Inclusion: Strategies for Community Empowerment and State Redistribution." *Environment and Urbanization* (19)2: 425–440.

Ostrom, V. (1997) *The Meaning of Democracy and the Vulnerability of Democracies: A Response to Tocqueville's Challenge*. Ann Arbor: University of Michigan Press.

Ostrom, E., J. Burger, C.B. Field, R.B. Norgaard, and D. Policansky. (1999) "Revisiting the Commons: Local Lessons, Global Challenges." *Science*, 284(5412), 278–281.

Posner, P.L. (2009) "The Pracademic: An Agenda for Re-Engaging Practitioners and Academics." *Public Budgeting and Finance*. 29(1): 12–26.

Scharmer, O. (1999) *Theory U: Leading from the Future as It Emerges*. San Francisco: Berrett-Koehler Publishers.

Scott, J. (1998) *Seeing Like a State: How Certain Schemes to Improve the Human Condition Have Failed*. New Haven: Yale University Press.

Senge, P. (1990) *The Fifth Discipline: The Art and Practice of the Learning Organization*, New York: Doubleday.

Trivers, R. (1971) "The Evolution of Reciprocal Altruism." *Quarterly Review of Biology*, 46(1), 35–56.

Vidyarthi, V. and P. Wilson. (2008) *Development from Within: Facilitating Collective Reflection for Sustainable Change*. Bloomington: Xlibris.

Yin, R. K. (1984) *Case Study Research: Design and Methods*. Beverly Hills: Sage.

Connections

Watch a series of podcasts on community planning and development tools at The Community Planning Toolkit sponsored by the UK's National Lottery Fund at: www.communityplanningtoolkit.org/podcasts/toolkit-how-to

Explore a variety of approaches and mini-cases at the following sites:

Orton Family Foundation's Heart & Soul community projects at: www.orton.org/projects/current

Community and Economic Development programs of the United States Department of Agriculture, Rural Development at: www.rurdev.usda.gov/LP_EconDevHome.html

23 Community-Based Energy

Rhonda Phillips

Overview

There are numerous ongoing and emerging issues in community development—one of the most pressing is that of "clean" energy. This chapter explores approaches to community-based energy systems, alternative energy sources, and ideas that communities are using to help become more resilient and sustainable. An overview of energy systems that are innovative and community-based is provided. The chapter concludes with cases on community-based energy cooperatives and organizations.

Introduction

Energy is a basic need in the sense that modern society cannot be supported or maintained without ready, affordable, and equitable sources. Traditional electric grid connected and centralized systems are the norm; at the same time, some communities and even entire countries are exploring and implementing alternatives that reduce vulnerability and increase sustainability (Kahn et al., 2007a). These traditional systems may be very efficient or there may be high costs involved, particularly on the environmental front, such as loss of biodiversity and emission of greenhouse gases (Kahn et al., 2007a). Interest continues to soar in "alternative" sources such as solar, biofuels, hydro, and wind powered. Part of the reason is that communities can help build sustainability into their planning and development by having affordable and reliable locally based energy systems.

In some cases, production of community-based energy may lead to creation of economic opportunity in local economies. For example, biofuel production can lead to improvements in sustainability with local wastes being used as a source for generating "zero-waste" approaches (Kahn et al., 2007b: 353). The authors also found that "community-based energy technologies, especially biodiesel and direct solar energy, are sustainable, considering their time-tested functionality and ecological, economic, and societal considerations" (403).

Some argue that there are not truly sustainable energy models, or clean energy, as all have drawbacks, challenges, and issues in design, implementation, and cost whether monetary, social, or environmental in nature. For example, wind energy seems like a natural choice yet there are impacts on the environment; the same holds true for wave-generated energy with noise impacting wildlife. While these concerns may not seem as large as obvious environmental impacts of coal-fired electrical plants, for example, they represent challenges nonetheless. Zatzman (2012: xvii) proposes that sustainable energy systems must be considered with the following caveat: Sustainable development of natural resources in general is actually premised on the sustainable engineering of sustainable energy resource development in particular. In other words, energy is essential to support community economic development.

Really Alternative Energy Sources

Much is happening on the frontiers of science in alternative energy sources. Bio-enzymatic research with algal biofuels, for example, is one such area. At Arizona State University's Sustainable Algal Biofuels Consortium, work toward biochemical conversion of aglal biomass is in full process. Test fields of algae grow in the desert, requiring little water and holding much promise.

The Consortium describes their work as follows:

> The primary objective is to evaluate biochemical (enzymatic) conversion as a potentially viable strategy for converting algal biomass into lipid-based and carbohydrate-based biofuels. Secondary objective is to test the acceptability of algal biofuels as replacements for petroleum-based fuels.
>
> (Sustainable Algal Biofuels Consortium 2014: 1)

A quote listed on their website captures the essence of the potential of biofuels research:

> Developing the next generation of biofuels is key to our effort to end our dependence on foreign oil and address the climate crisis—while creating millions of new jobs that can't be outsourced.
>
> (Dr. Steven Chu, U.S. Secretary of Energy)

There are other sources of alternative research, ranging from germs to photosynthesis. The options are not so far afield, considering the transition that biofuels have made from idea to reality in some communities. The U.S. Midwest, for example, produces large amounts of biofuel from corn crops, which has its own set of benefits and costs. The potential of alternative fuels productions holds much relevance for community and economic development, particularly in areas where natural resources are needed to aid in production (cropland, water basins, windfields, etc.).

A Carbon Neutral Challenge

The Scandinavian countries in particular have led the way in reducing carbon emissions. Denmark has achieved much in alternative energy production, as well as district heating systems (using the waste steam from electrical generation to heat homes). The island of Samsø is considered carbon neutral as it generates more energy from renewable resources than its residents use. While some sources such as wood and straw are burned to generate some of the energy, for the most part, the 4,000 residents are entirely energy self-sufficient. It serves as a model for small, community-based energy systems with lessons to learn from its experiences. While there are still carbon emissions (from vehicles), it is carbon neutral because of the excess energy production. (See the video link at the end of the chapter for more details.)

While both New Zealand and Costa Rica are not huge economies, nor do they have high levels of population, they are conducting experiments to see which one can become the first carbon neutral country. The lessons to be learned by their efforts will be valuable to observe for other countries and at the community level. Janet Sawin (2014), describes the efforts of these two countries in the following:

> Costa Rica's government announced (in 2007) it was drawing up plans to reduce net GHG emissions to zero before 2030. The country aims to reduce emissions from transport, farming, and industry, and to clean up its fossil fuel power plants, which account for 4% of the country's electricity (of the rest, 78% comes from hydropower and 18% from wind and geothermal power). In addition, through an innovative program begun in 1997 and funded by a gas tax, the government compen-

sates landowners for growing trees to absorb carbon while protecting watersheds and wildlife habitat. Costa Rica aims to be the first country to become carbon neutral. But Costa Rica could be in a race with New Zealand, which set the target of becoming "the first truly sustainable nation on earth." Prime Minister Helen Clark announced in that her country will adopt an economy-wide program to reduce all GHG emissions, with different economic sectors being gradually introduced into a national emissions trading program that should be in effect fully by 2013. Other commitments include an increase in renewable electricity to 90% by 2025 (up from 70% today), a major net increase in forest area, widespread introduction of electric vehicles, and a 50% reduction in transport-related emissions by 2040.

Other countries are reaching for results as well. Germany has set a goal to use 45% renewables by the year 2030. Norway and Iceland have declared intentions to become carbon neutral as well.

It is not just at the country level that being carbon neutral is popular. The David Suzuki Foundation lists the following ways that more people, organizations, and businesses are embracing the concepts with a few examples:

- Major sporting events such as the World Cup Soccer are going carbon neutral.
- Airlines and travel agents are starting to offer customers the option to offset their flights, and some airlines are offsetting all of their flights. Many hotels are also providing carbon neutral accommodations.
- Movie studios have offset the emissions from the production of feature films and documentaries, and media companies such as BSkyB, MTV, and News Corp are offsetting the emissions associated with their broadcasts.
- Major conferences (e.g. United Nations World Climate Research Programme) and conventions have offset their emissions.
- Organizations as diverse as Wells Fargo, Whole Foods, the EPA and the city of Vail, CO have purchased large quantities of renewable energy certificates to offset their electricity use.
- Businesses like HSBC, Swiss Re, Google, Nike, Dell, and Vancity have committed to making their entire operations carbon neutral. Other companies are offering carbon neutral products or services, such as carpeting, coffee, and deliveries.
- Some utilities are offsetting their emissions and allowing their customers to purchase carbon neutral energy.
- The World Bank has committed to being carbon neutral.
- Schools and churches are voluntarily offsetting their emissions.
- Rock bands like the Rolling Stones, Coldplay, and Dave Matthews Band have offset the emissions associated with their concerts and albums.
- Many people are now offsetting their weddings (including air travel by guests).

(David Suzuki Foundation 2014: 1)

BOX 23.1 FORT COLLINS, COLORADO'S NET ZERO ENERGY DISTRICT PLANNING

The Rocky Mountain Institute, a research-oriented nonprofit housed in Snowmass, Colorado has long advocated for renewable and sustainable energy sources. A recent plan for the City of Fort Collins, Colorado shows the approach and desire to reduce energy consumption and provide renewable sources. The following is excerpted from the plan.

Building on its strong history for delivering affordable, reliable power, Fort Collins Utilities, with strong support from the community, has developed aggressive environmental goals that are designed to make it a leader in clean energy. Fort Collins Utilities has been working to meet its clean energy goals including a flagship effort, called FortZED, to build a net zero energy district in downtown Fort Collins. Fort Collins Utilities and its partners worked with the Electricity Innovation Lab (e-Lab) to design and carry out. In 2009, the city and the utility set the following community-wide energy goals:

- Support the community greenhouse gas reduction goal of 20% reduction below 2005 levels by 2020 and 80% reduction by 2050.
- Achieve annual energy efficiency and conservation program savings of at least 1.5% of annual energy use (based on a three-year average history).
- Maintain a minimum fraction of renewable energy in compliance with Colorado's Renewable Energy Standard requirements for municipal utilities to meet 10% of energy demand with renewables by 2020.

In addition, the city is part of a collaborative effort, called FortZED, to transform the downtown area of Fort Collins and the main campus of Colorado State University into a net zero energy district through conservation, efficiency, renewable sources, and smart technologies.

Recognizing that individual efforts in efficiency and renewable energy might not be sufficient to achieve Fort Collins' goals, the Colorado Clean Energy Cluster and the nonprofit UniverCity Connections launched a new idea they called FortZED in 2007. FortZED has become a collaborative effort among the city, the utility, Colorado Clean Energy Cluster, and Colorado State University to "transform the downtown area of Fort Collins and the main campus of Colorado State University into a net zero energy district." The area encompasses two square miles and 45 MW of peak electricity demand or approximately 10–15% of Fort Collins Utilities' distribution system. The FortZED effort is governed through a steering committee made of members from the city, the utility, the Colorado Clean Energy Cluster, Colorado State University, and the community. Initial efforts in FortZED have centered around four projects:

- Renewable and distributed system integration—Represents the largest project and focuses on coordinating distributed resources to reduce peak electricity demand;
- New Energy Communities Grant—Reduces energy demand in city buildings and installs renewable energy technologies;
- Community Energy Challenge—Grass-roots outreach effort to reduce home energy use; and
- Green Restaurant Initiative—Encourages local restaurants to conserve energy.

Source

Building the electricity system of the future: Fort Collins & FortZED (ww.rmi.org). For more on FortZED, go to: www.fortzed.com

Conclusion

The range of options for renewable and alternative energy sources is increasing rapidly. As more technologies emerge, communities will have greater choices to pursue in planning and implementing community-based energy systems. The potential impacts are large, not only in terms of economic opportunities that may be created ("green" jobs in the renewable and alternative energy industry) but also by lessening impacts on environmental systems.

CASE STUDIES: COOPERATIVE ENERGY

As interest in cleaner energy sources and efficiency continues to grow and as technologies develop enabling smaller scale "home grown" projects to be more feasible, a variety of community energy projects are springing up. Such projects can allow energy consumers to be more connected to and responsible for their energy supply, and in many cases to actually own a part of it.

Community energy projects can be sponsored by a variety of local organizations including nonprofits and local governments. The following case studies highlight projects by different types of energy cooperatives, a business model well-suited to working at the community level.

Electric Co-ops and Community Solar Projects

As the cost of photovoltaic (PV) technology has dropped and as states have established laws and programs to encourage net metering, many individual homes and business have installed PV panels and are generating some or all of the electricity they use. In order for net metering to work, such systems need to be connected to the grid via the local electric utility, with the utility able to provide power at all hours regardless of the solar generation at any point in time.

Electric cooperatives were first formed in the US during the late 1930s as part of President Roosevelt's depression-era electrification program. Established to bring power to those no one else would serve, the 838 distribution co-ops today serve 12% of the meters on the electrical grid system. Still primarily serving rural America, they operate 42% of all local distribution lines covering 75% of the US's land mass. As consumer-owned utilities, their "community" is typically a larger rural area crossing town and county lines. Historically, electric cooperatives have relied primarily on large scale generation sources including coal and natural gas, nuclear and large hydro. More recently, some co-ops have become actively involved in commercial-scale renewable sources such as wind, and smaller landfill gas and farm methane facilities.

A number of electric co-ops in different regions have begun developing smaller scale solar projects following the "solar gardens" model, allowing co-op members to voluntarily participate in and obtain all or some of their power from the project. The cooperative is thus giving interested members the opportunity to "crowd fund" a project they support and want to directly benefit from. The solar garden model overcomes some obstacles that individual net metering presents. It allows all members to participate even if their property does not have good solar exposure, if they are renters, if there are aesthetic or other issues with location. It also avoids the need for individual permitting, installation and maintenance. By aggregating a larger number of members, it lowers the installed cost per unit considerably.

Wright-Hennepin Cooperative Electric Association (WH), headquartered in Rockford, Minnesota, serves approximately 46,000 member households and businesses in an area northwest of the Twin Cities.

In the summer of 2013, construction began on the WH Solar Community, a project located at the cooperative's headquarters. It is built and operated in partnership with the Clean Energy Collective, a Colorado-based group working with cooperatives and other organizations to make solar project ownership and output available to a broad population. Co-op members may purchase individual panels at U.S. $917/panel (U.S. $4.83/watt) up to but not exceeding their homes' electric consumption. An additional benefit derived from the aggregated scale of the project is the capacity to store some of the solar electricity generated during the day and to release it for use in the early evening when residential members typically use more electricity.

WH's website informs those interested that they can expect a 21-year payback on their investment, and a lifetime savings on electric bills of U.S. $3,128 per panel. The 171-panel project is fully subscribed and WH will consider building others, as a survey indicates that a third of the cooperative's members have some interest in participating.

WH's Chief Operating Officer for Power Supply Rod Nikula was asked to define the "community" of this project. "The Solar Community is a subset of the co-op's membership," he said. When asked whether the subscribers fit a particular profile, he said that they really did not. While there is a growing interest in renewable energy in the area, some of the subscribers are long-term members who are participating as much for a retirement hedge on their electric bills as they are for environmental reasons. "We are building the WH Solar Community so we can continue to be leaders for our members," Nikula explained. "They're asking us to help them get renewable energy; we're doing it in an economical and financially responsible way on their behalf" (Nikula 2013). It should be noted that the co-op is oriented toward the community, and has provided over U.S.$3 million in funding for scholarships and local projects to support the community.

Cherryland Electric Cooperative serves 34,000 members in northern Michigan and has developed a community solar project located at its facility in Grawn. Originally intending to construct a 48-panel array, member and community interest soon had them increasing the project's size to 224 panels. Cherryland's CEO Tony Anderson likens the Solar Up North (SUN) Alliance to a community garden. While he felt there was likely to be support for moving toward renewable energy, actually proceeding with the project was a tangible way of gauging support. "This is a very environmentally conscious community," Anderson said. "The project is actually a bricks and mortar survey."

The Solar Up North Alliance was formed after the co-op approached its neighboring municipally owned utility, Traverse City Light & Power, which agreed to add its 12,000 customers to the community of people able to participate. Shares in the project cost U.S. $470. Participants can expect a monthly credit on their bill of about U.S. $2, depending on actual output from the project, and payback is estimated at 20 years. The individual participation cost is reduced by funding from a variety of sources including the wholesale generation and transmission cooperative that Cherryland is a member of, and a state-sponsored energy optimization rebate. Some co-op members may also be able to apply a U.S. $75 rebate of capital credits (patronage refunds) toward the cost. A participant who takes advantage of all the rebates and discounted capital credits ultimately pays U.S. $320 which achieves a payback of fewer than 14 years. Project participants can monitor the project's output on Cherryland's website.

The community aspect of the project and its connection to the co-op's roots is emphasized in a promotional video which begins: "Cherryland Electric Cooperative began as a community of people coming together to meet the pressing need of building a system that could bring electricity to the far rural regions of northern Michigan. Now, over three quarters of a century later, we are still a community" (Anderson 2013). CEO Anderson notes that the people signing up for the Solar Up North Alliance represent a true cross section of the areas' residents. While some are motivated primarily by environmental concerns, many are drawn to the cooperative and community aspect of the project. "We have older members who want to support their co-op doing the right thing," he said, "traditional tried and true members."

A different type of energy cooperative

Co-op Power, based in West Hatfield, Massachusetts, is an energy co-op with 450 members that is not a utility and does not sell electricity or operate poles and wires like the rural electric cooperatives do. It takes an entrepreneurial approach, offering its members opportunities for starting local business in a variety of clean energy areas including solar photovoltaic and hot water systems, energy efficiency and fuels. As stated on their website:

> Co-op Power has created an innovative structure. We enable communities to create community-owned businesses and jobs. In addition to member equity, members can invest in the growth and development of businesses and projects that they would like to make reality.

Although Co-op Power has members and activities reaching across much of Massachusetts including the Boston metro area as well as in southern Vermont, it has a unique structure that keeps business activity and decision making local. Separate autonomous Local Organizing Councils are established in areas where there are members. "The Councils define their community—their geographic area," reports Lynne Benander, Co-op Power's CEO. "They decide what activities and businesses to develop in their own community as they define it." A major priority is that enterprises that are part of the Co-op Power network be community-owned and focused on service to the community.

"The Local Organizing Councils assure long-term stewardship," Benander says. "The business can be owned in large part by a nonprofit, a cooperative, or a local government entity. To support the development of local energy businesses, Co-op Power provides access to experts in engineering, finance, green energy construction, economic development and law" (Benander 2013).

The Co-op Power network offers a wide variety of goods and services. These include: energy efficiency products and services, solar photovoltaic systems, solar hot water and air heating systems, small community-owned generation projects, and heating fuel buying groups. One of the more ambitious projects still in the development stage is a 3.5 million gallon biodiesel facility that will convert used vegetable oil to fuel.

Avram Patt (ret.)
Washington Energy Coop, USA

Sources

Anderson, Tony, Cherryland CEO, telephone interview, 8/13/13.
Benander, Lynne, Co-op power CEO, telephone interview, 8/20/13.
Nikula, Rod, WH COO for Power Supply, telephone interview 8/9/13.

For additional information:

National Rural Electric Cooperative Association
www.nreca.coop

Clean Energy Collective
www.easycleanenergy.com

Solar Gardens Institute
www.solargardens.org

Wright-Hennepin Electric Cooperative Association
For general information and information about WH Solar Community
www.whe.org

Minneapolis Star Tribune
1/14/13: www.startribune.com/business/186565751.html?refer=y

Cherryland Electric Cooperative
For general information and information about Solar Up North Alliance
www.cecelec.com

Traverse City Record-Eagle
3/7/13:
www.record-eagle.com/local/x1503764051/Solar-panel-project-scheduled-for-GT-region
4/3/13
www.record-eagle.com/local/x1413935904/Cherrylands-solar-project-proves-popular

Traverse City Area Chamber of Commerce
www.tcchamber.org/cherryland-powers-up-solar-partnership

Co-op Power
www.cooppower.coop

Keywords

Clean energy, solar systems, community-based power systems, community energy.

Review Questions

1 What is the concept of community energy?
2 Explain the importance of reliable and sustainable energy systems in community recovery after disaster.
3 Can you identify a community that is becoming "carbon neutral" in terms of energy generation?
4 What is an approach that can help communities assess and gauge clean energy as a community economic development strategy?
5 Can you think of an example in your community where clean energy may be an appropriate pursuit? What would be the potential outcomes?

Bibliography

David Suzuki Foundation. (2014). Accessed March 5, 2014, at: http: //davidsuzuki.org/what-you-can-do/reduce-your-carbon-footprint/go-carbon-neutral

Khan, M.I., Chhetri, A.B., & Islam, M.R. (2007a). Analyzing Sustainability of Community-based Energy Technologies. *Energy Source, Part B*. 2(4): 403–419.

Khan, M.I., Chhetri, A.B., & Islam, M.R. (2007b). Community-based Energy Model: A Novel Approach to Developing Sustainable Energy. *Energy Sources, Part B*. 2(4): 353–370.

Sawin, J. (2014). Costa Rica and New Zealand on Path to Carbon Neutrality. Worldwatch Institute. Accessed March 5, 2014 at: www.worldwatch.org/node/5439

Sustainable Algal Biofuels Consortium. (2014). Accessed March 5, 2014, at: http: //asulightworks.com/solutions/sustainable-algal-biofuels-consortium

Zatzman, G.M. (2012). Sustainable Resource Development. Somerset, NJ: Wiley.

Connections

Watch a video on (really) alternative energy sources! Algae, germs, and a host of other sources at: https: //www.youtube.com/watch?v=BpWw7Cq2shA

Explore Denmark's Carbon Neutral Island; more details at: https: //www.youtube.com/watch?v=riYT4Oyh67w

LightWorks focuses on light-inspired research at Arizona State University, particularly in renewable energy fields including artificial photosynthesis, biofuels, and next-generation photovoltaics. Watch the video and learn more at: http: //asulightworks.com/solutions/sustainable-algal-biofuels-consortium #sthash.wwhkuQrQ.dpuf

24 Community and Economic Development Finance

Janet R. Hamer and Jessica LeVeen Farr

Overview

Success in community and economic development projects often hinges on financing. Since many projects may not qualify for conventional financing, it is important to be aware of federal, state and private sources of capital for community and economic development. This chapter provides an overview and brief explanation of different types of community and economic development financing. It also discusses how to put a financing deal together.

Introduction

Economic development is a comprehensive strategy that integrates a wide array of activities that help sustain and grow a local economy, and it is a critical component of community development programs at the neighborhood, city, and state level. Creative financing tools are usually required to accomplish economic development objectives. Most community and economic development finance programs are public–private partnerships designed to fill the funding gaps not covered by the private market alone (Seidman 2005). Programs are available to help finance economic development activities such as business recruitment and retention, job creation, small business assistance, and real estate development. The purpose of this chapter is to examine the broad community development finance industry focusing primarily on the tools that are provided through this industry for economic development project financing.

Basic Community Development Finance Vocabulary

The first step toward understanding community development finance is to become familiar with some basic terms.

Community Development Finance: A lending process designed to stimulate community and/or economic development. With the use of community partners and/or lending enhancements, financial institutions are able to lend to borrowers who cannot meet conventional credit underwriting standards.

Subsidy: Any financial assistance granted to an individual or organization.

Debt: Any money, goods or services owed to someone else. Debt can take the form of mortgages, other kinds of loans, notes or bonds.

Equity: The ownership interest in a project after debt and other liabilities are deducted.

Grants: A gift usually given by a foundation, a government agency or the philanthropic community that may take the form of money, land, or in-kind services. Grants may provide equity to a project and can reduce the amount of debt required.

Credit enhancements: Special arrangements and programs that mitigate the credit risk associated with the borrower or the project, thereby affecting the evaluation of a potential borrower's creditworthiness. The enhancements may include mortgage insurance, tax credits, rent supplements, interest rate subsidies, loan guarantees, favorable structure, terms, conditions and pricing of credit products, underwriting flexibility, loan-to-value ratios, and tax abatements.

Interest subsidy: A grant to reduce the interest a borrower is required to pay on a loan. Subsidies may take the form of a direct cash grant to a lending institution to lower (or buy down) the bank's interest rate; a government sponsored, low-interest loan subordinated to a participating lender; or a lower than market rate loan to a qualified borrower as a result of an advance from a public entity.

Loan guarantee: Repayment of loans may be guaranteed through private or public sector sources. Loan guarantees are used to reduce risk of loss to a lender and are usually considered a secondary source of repayment for a portion of the debt in the event of default. A loan guarantee can also improve a project's ability to secure private financing, or qualify loans for sale on the secondary market.

Doing a Deal: Who Are the Players in Community Development Finance?

Public–private partnerships play a key role in the success of most economic development projects. Frequently there are multiple partners involved in a project so it is very important that their roles be defined early in the planning process. Another key step is identifying the "visionary leader" for the project, who is often the person or organization that develops the first concept of the project and acts as motivator and spokesperson. This section looks at the various players most often involved in economic development projects.

Government: Local, state, and federal governments will typically be involved in an economic development project in three ways: (1) providing funding, (2) approving permits, and (3) acting as landowner or developer. Local government, in particular, is always a critical player and needs to be brought into the planning process as early as possible.

Financial institutions: The Community Reinvestment Act (CRA) specifically encourages banks to help meet the credit needs of the communities in which they operate including low- and moderate-income neighborhoods. The CRA creates an incentive for banks to participate in the financing of community and economic development activities. Financial institutions generally provide the market rate financing on projects, invest in loan funds, and purchase tax credits. Larger banks may involve their bank-owned community development corporation. This type of corporation, either for-profit or nonprofit, is capitalized by one or more banks for the purpose of making debt and/or equity investments in projects that promote community and economic development.

Community Development Financial Institution (CDFI): CDFIs are private sector financial intermediaries with community development as their primary mission. Some are chartered as banks or credit unions and others are nonregulated nonprofit institutions that gather private capital from a range of investors for community development lending and investing. CDFIs provide lending and equity financing for projects that typically could not be financed solely by a conventional financial institution.

Community Development Corporations (CDCs): CDCs are community-based organizations owned and controlled by community residents engaged in community and economic development activities, the majority of which are nonprofit 501(c)(3) organizations. The role of nonprofit CDCs is critical to the success of many economic development projects. CDCs have access to funding sources not available to the private sector or government partners, and they can also serve as the critical link with the surrounding neighborhood and community leaders. CDCs may take many forms including faith-based organizations. They also differ greatly in capacity and areas of focus. For instance, larger CDCs may have the capacity and experience to act as the developer and builder of a project, but many are required to outsource these activities to a third party.

National intermediary: National intermediaries are organizations that mediate between community-based organizations and large-scale sources of capital. These intermediaries function at the national level aggregating capital from sources such as foundations, corporations, and the government and disbursing it to local organizations for capital projects, operating support, and predevelopment financing along with technical assistance and capacity building. Local Initiative Support Corporation (LISC), Neighborhood Reinvestment Corporation (NRC), and Enterprise are the three largest national community development intermediaries. National intermediaries may serve limited geographic areas.

Economic Development Commissions/ Chambers of Commerce: These entities are an important link with the business community and potential investors and developers of the economic development project. They may be structured as member-supported private organizations, government-supported entities, or divisions of local government. These organizations are essential players in the recruitment of new businesses to the project and can also mobilize support for the project with local government officials and potential private investors.

Investor: An investor is any organization, corporation, individual, or other entity that acquires an ownership position in a project, thus assuming some risk of loss in exchange for anticipated returns. In the case of economic development projects, the investor may be any combination of entrepreneur, small business owner, developer, purchaser of tax credits or other large corporation.

BOX 24.1 A LITTLE BIT ABOUT MICROFINANCE

Over three billion people (half the world's population) live on less than a few dollars a day. Facing limited income and poverty, many do not have access to conventional financing sources. Microfinance refers to financial services to serve low-income people, and while not the only tool for poverty alleviation, it can help provide an opportunity. As Muhammed Yunis, founder of the Grameen Bank in Bangladesh and the innovator of microfinance, pointed out, "All people are entrepreneurs, but many don't have the opportunity to find that out." Yunis developed a model of small loans, with peer oversight, that has been replicated throughout the world, receiving a Nobel Peace Prize for his work. Now the World Bank estimates that about 160 million people have access to microfinance services in developing countries. Microfinance services can make a dramatic difference in lives, yielding positive outcomes not only for the recipient but also their families and communities. Other organizations have become involved, using online tools to help foster services. Nonprofit organizations such as Kiva connect people with as little as $25 to loan. Repayment rates are typically very high, as peers are highly involved with the process. For more information see www.grameenfoundation.org (an allied organization to the Grameen Bank), www.globalenvision.org (for information on issues ranging from poverty to financial resources for the developing world), and www.kiva.org

The Editors

Doing a Deal: Finding Sources of Money

Putting together the financing for an economic development project requires convening the relevant players and identifying the resources that they have available. Government has always played a significant role in economic development financing, but funding from some traditional federal government programs has been declining, making state and local government funding increasingly more important. There is a growing emphasis on public–private partnerships that bring multiple players together and use limited public resources to leverage private sector investment.

Financing for economic development activities is divided into several broad categories including low-cost loans and loan guarantees, subsidies or grants, and tax credits or abatements. This section reviews the financial resources offered by all levels of government and other players in the community development finance industry.

Federal Government Resources

The federal agencies that provide financing for community and economic development are the U.S. Department of Housing and Urban Development, the U.S. Department of Agriculture, the U.S. Department of Commerce, the Economic Development Administration, and the U.S. Environmental Protection Agency. In some cases, funding is passed through state or local agencies that use the funds for a variety of community development purposes. In other instances, funding is provided directly from the federal agency to a project. Actual grantees may be local or state government, nonprofit organizations, or for-profit entities depending on the source of funds and regulations.

U.S. Department of Housing and Urban Development (HUD)

HUD has several programs that provide grants, loans, and loan guarantees to support economic development.

- *Community Development Block Grant Program (CDBG):* CDBG is one of the most flexible federal programs intended for use by cities and counties to promote neighborhood revitalization, economic development, and improved community facilities and services principally to benefit low- and moderate-income persons and communities. Specific uses of the funds are left to the discretion of the local governments. Examples of eligible economic development activities include infrastructure improvements, small business loan programs, capitalization of revolving loan funds, purchase of land, and commercial rehabilitation activities. Local governments may provide funding for projects in the form of loans or grants. Historically, 8–15% of CDBG funds have been used for economic development purposes.

  Through the CDBG entitlement program, HUD allocates annual CDBG grants to entitlement communities. Entitlement communities are the principal cities in large Metropolitan Statistical Areas (MSAs), other metropolitan cities with a population greater than 50,000 and urban counties with populations greater than 200,000. These communities are allowed to develop their own community development programs and priorities. States administer CDBG funds for all non-entitlement communities; i.e. those communities that are not part of the CDBG entitlement program. Non-entitlement communities are cities with populations of fewer than 50,000 or counties with populations under 200,000. Annually, each state develops funding priorities and criteria for selecting projects and awards grants to local governments that carry out community development activities.

- *Section 108 Loan Guarantee Program (Section 108):* Section 108 is the loan guarantee provision of the CDBG program that allows jurisdictions to borrow against their future CDBG allocations. This program provides communities with a source of financing for economic development, housing rehabilitation, public facilities, and large-scale physical development projects.
- *Brownfields Economic Development Initiative (BEDI):* HUD also provides financial support for the redevelopment of "brownfields" which are environmentally contaminated industrial and commercial sites. The BEDI provides grants on a competitive basis to entitlement communities as defined above. Non-entitlement communities are eligible as supported by their state governments. BEDIs must be used in conjunction with loans guaranteed under the Section 108 Program. Communities fund projects with the BEDI grants and the Section 108 guaranteed loan financing to clean up and redevelop brownfields.

U.S. Department of Agriculture Rural Development (USDA)

The USDA provides financing to support economic development in rural communities through several programs that offer grants and zero-interest loans.

- *Rural Business Enterprise Grants:* The Rural Development, Business and Cooperative Programs (BCP) make grants under the Rural Business Enterprise Grants (RBEG) Program to local governments, private nonprofit corporations, and federally recognized Indian tribal groups to finance and facilitate development of small and emerging private business enterprises located in any area other than a city or town that has a population of greater than 50,000 and the urbanized area contiguous and adjacent to such a city or town. Local government, private nonprofit corporations, and federally recognized Indian tribes receive the grant to assist a business.
- *Rural Business Opportunity Grants:* This program provides grants to pay the costs of providing economic planning for rural communities, technical assistance for rural businesses, or training for rural entrepreneurs or economic development officials.
- *Rural Economic Development Loans and Grants:* This program provides zero-interest loans to electric and telephone utilities financed by the Utilities Program, an agency of the United States Department of Agriculture, to promote sustainable rural economic development and job creation projects.

U.S. Department of Commerce, Economic Development Administration (EDA), Public Works and Economic Development Program

The EDA is another source of financing for economic development projects. The Public Works and Economic Development Program provides matching grants for public works and economic development investments to help support the construction or rehabilitation of essential public infrastructure and facilities necessary to generate or retain private sector jobs and investments, attract private sector capital, and promote regional competitiveness including investments that expand and upgrade infrastructure to attract new industry, support technology-led development, redevelop brownfield sites, and provide eco-industrial development.

Small Business Administration (SBA)

The SBA provides a number of financial and technical assistance programs for small businesses including direct and guaranteed business loans, venture capital investments, and disaster loans. The Small Business Administration is the largest single financial backer and facilitator of technical assistance and contracting opportunities for the nation's small businesses.

U.S. Environmental Protection Agency (EPA), Brownfield Program

The EPA provides financial and technical assistance to support the redevelopment of brownfields for economic development purposes. The Brownfield Program provides funding to state, local, and tribal governments to make low-interest loans and grants to be used to assess environmental conditions of brownfield sites and fund cleanup activities.

Federal Tax Credit Programs

Federal tax credits have become some of the most important tools for financing economic development projects because they create an incentive for more private sector involvement, which allows greater leveraging of resources and encourages more public–private partnerships. Each federal tax credit program is structured differently, but all programs are designed to provide equity for projects by allowing the owner or developer of the project to sell or transfer the tax credits in return for up-front equity investments. Individuals or for-profit corporations purchase the tax credits for use against their federal income tax liability.

Tax credit programs help finance economic development activities ranging from business start-ups to real estate development. The major federal tax credit programs are:

- *New Markets Tax Credits:* The New Markets Tax Credit (NMTC) was created in 2000 to spur private investment in low-income urban and rural communities and to provide capital to businesses and economic development projects in these communities. Individuals or corporations can invest in Community Development Entities (CDEs) that provide loans, equity and other financing to qualified low-income businesses. Those investing in the CDE will receive a credit against their federal income taxes totaling 39% of the cost of the investment over a seven-year period. This program is administered by the Treasury Department's Community Development Financial Institutions (CDFI) Fund and the Internal Revenue Service (IRS).
- *Historic Preservation Tax Incentives:* Historic Preservation Tax Credits support the rehabilitation of historic and older buildings and encourage the preservation and revitalization of older cities, towns, and rural communities. The tax incentive rewards private investment in rehabilitating historic properties such as offices, rental housing, and retail stores. Currently, the tax credit for certified historic structures is equal to 20% of the renovation or construction costs, and a 10% tax credit is available for buildings built prior to 1936 that are nonresidential. This program is administered by the National Park Service, the IRS, and State Historic Preservation Offices.
- *Federal Brownfields Expensing Tax Incentive:* This cost deduction provides a business incentive to clean up sites contaminated with hazardous substances and is intended to offset the costs of cleanup. This program is administered by the EPA and the IRS.

Tax credit programs bring together many of the community development finance players including government entities, businesses, investors, financial institutions, and nonprofits. While the popularity of these programs has been increasing, they are considered some of the more complex forms of financing, and smaller developers and projects still tend to look for simpler and more traditional sources of financing.

State Resources

State governments play an important role in economic development financing, particularly in rural areas. Much of the federal funding is passed down through state agencies to reach the non-urban or non-entitlement communities. In addition, many of the tools that local governments rely on to finance economic development (e.g. tax increment financing, property tax abatements) are enabled by state legislation.

Every state has tax credit programs to encourage economic and community development. These programs are frequently tied to the state's individual economic development strategy and may be tailored to meet the specific goals of that strategy. Most state tax credit programs are tied to new business investment and job creation and are not typically as broad and flexible as the federal tax credit programs. Many states have created state enterprise zone programs. Enterprise zones are economically depressed areas targeted for revitalization through tax credits and other incentives given to companies that locate or expand their operations within the zone.

States also have the ability to offer favorable financing for economic development activities by issuing bonds, particularly Industrial Revenue Bonds (IRBs). Issued by a public authority, IRBs qualify for federal tax exemption. The proceeds of the bond issue can be used to offer lower-cost financing for new industrial development.

Municipal Funding Sources

Local governments also fund economic development projects and have worked to develop creative financing tools that do not generate higher tax burdens to local taxpayers. Tax Increment Financing is one of the most widely used of these tools along with tax abatements, tax credits, and providing access to lower-cost financing.

Tax Increment Financing (TIF)

TIF is a mechanism that allows the future tax benefits associated with new real estate development to pay for the present cost of the improvements. Generally, TIF is used to finance costs such as land acquisition, infrastructure improvements, utilities, parking structures, debt service, and other related development costs.

Almost every state has passed legislation that gives local governments the authority to designate TIF districts for redevelopment. These districts are typically physically or economically distressed areas where private investment is not likely to occur without some additional public subsidy. Based on the expectation that property values in the district will rise as a result of that redevelopment, the city splits the property tax revenues from the district into two streams: the first consisting of revenues based on the current assessed value; the second based on the increase in property values—the "tax increment." The tax increment is diverted away from normal property tax uses (i.e. schools, public safety) and can be used to repay the cost of new development in the TIF district.

TIF is popular with local officials because it is a flexible tool and, as a result, it is one of the most frequently used sources of funding for community economic development projects. However, because local officials have made TIF so widely available, there is some concern that developers have come to expect it for all projects. Thus, some critics contend that it no longer serves its purpose as a targeted development incentive.

In addition, the use of TIF has raised concerns because it diverts property tax revenue away from traditional uses. As a result, many states have tightened TIF legislation to control the amount of money that can be shifted from these uses to finance TIF debt. Some states have also established performance requirements or set-asides, such as the provision of affordable housing units, for projects financed with TIF.

Property Tax Abatement

Property tax abatements are another tool used by local governments to promote economic development in areas designated as blighted, distressed, or in state enterprise zones. Tax abatements are defined as either tax forgiveness or tax deferral until a certain date. Most states limit the maximum time that taxes can be forgiven or deferred to 10–12 years. Tax abatements are usually offered in targeted areas where local communities are hoping to encourage new investment and redevelopment or to create new job opportunities by attracting new businesses.

Tax abatements are one of the more frequently used tools because property taxes are one of the items that falls under state and local government control. Unlike TIF, tax abatements do not directly divert spending from other programs, so abatements may be more politically feasible.

Municipal Bonds

Local governments can provide favorable financing to encourage economic development by using their bonding authority to issue private activity bonds including industrial revenue bonds. These bonds are backed by the revenue generated by a project and not by the government. Because these bonds are tax exempt, local government can use the bond proceeds to offer a lower-cost source of project financing. Some local governments attach performance standards and accountability requirements to municipal bonds.

Land Acquisition

Local governments also help with the acquisition of land, either by providing land that the local government already owns or by acquiring property and transferring it to the private developer at a lower price. Eminent domain is one tool that local government can use to acquire land if it can show a public benefit from the project or if the property is blighted. Acquiring contiguous parcels for redevelopment can be costly and time-consuming for developers, so the ability of local government to assist with this process can be very valuable.

Fee Waivers or Deferrals

Fee waivers or deferrals are other tools used by local governments. Local governments may reduce fees for building permits, water and sewer connections, or impact fees[1] for projects that have a public benefit, such as redeveloping a blighted area, creating new commercial space in a distressed community, or developing affordable housing.

Other Financing Resources

In addition to government programs, there are several other important tools and programs used to finance economic development projects including revolving loans funds, the Main Street Preservation Program, and programs offered through the Federal Home Loan Bank System.

Revolving Loan Funds (RLFs)

A revolving loan fund is a self-replenishing pool of funds structured so that loan payments on old loans are used to make new loans. RLFs provide a lower-cost and more flexible source of capital that can be used in conjunction with conventional sources to reduce the total cost of debt to the borrower and to lower the risk to participating lenders. Initial funding for RLFs often comes from a combination of public sources (federal, state, or local government) and private sources (financial institutions and philanthropic organizations). State and local government funding may come from tax set-asides, bonds, or a direct appropriation. Federal funding to capitalize loan funds can come from HUD, USDA, and the Department of Commerce programs. The source of the public funds used for capitalization can impact the eligible uses for the loan fund proceeds. Revolving loan funds are most often used for small business assistance but can also be used to fund real estate development projects, land acquisition, and construction financing.

The Main Street Preservation Program

The Main Street Preservation Program is offered through the National Trust for Historic Preservation. In the 1970s, the National Trust developed its pioneering Main Street approach to commercial district revitalization, an innovative methodology that combines historic preservation with economic development to restore prosperity and vitality to downtowns and neighborhood business districts. It has been proven effective in communities of all sizes, from small towns to large cities. Main Street programs are locally operated by independent nonprofits or city agencies. There is typically a statewide (or citywide in larger cities) organization that coordinates a network of affiliated Main Street programs. The coordinating Main Street organization generally has an application process through which a community can be designated as a Main Street program and provide direct technical services, networking, and training opportunities to their affiliates.

The Federal Home Loan Bank System

The Federal Home Loan Bank System is a government-sponsored enterprise and another important partner for economic development finance. The 12 district banks provide stable, low-cost funds to financial institutions for home mortgage, small business, rural, and agricultural loans. Through grants, subsidized rate advances, reduced rate advances, and technical expertise, the district banks assist member financial institutions in their community economic development efforts. The Community Investment Program is a lending program that provides below market loans enabling member financial institutions to extend long-term financing for housing and economic development that benefits low- and moderate-income families and neighborhoods. This program is designed to be a catalyst for economic development because it supports projects that create and preserve jobs and help build infrastructure to support growth.

Foundations

Privately funded and community foundations have historically been a source of grant funds for nonprofit organizations and may be utilized by nonprofit partners in an economic development project. Additionally, to increase the impact of their resources, some foundations have added financial instruments known as program-related investments (PRIs) to their traditional grant portfolios.

BOX 24.2 "CROWDFUNDING" FOR THE COMMUNITY

Crowdfunding has dramatically increased in popularity, and no wonder—using the internet to garner funding for individuals, it makes funding more accessible for many. Private groups such as Indiegogo or Kickstarter have enabled numerous creative projects to obtain funding. Note there is also peer-to-peer lending too, which is often seen at the individual level. Zopa, a private firm in the UK, was one of the first to pioneer internet-based peer-to-peer lending without traditional financial organizations serving as intermediaries. What about at the community level? In some ways, crowdfunding for communities has always existed with special fundraisers, and other mechanisms for communities investing directly in its members. Sometimes it emerges out of pure necessity—there's no other source of funding and the community must meet a need. Take for example community-owned businesses (see Chapter 9 on establishing community-based organizations for more details). Little Muddy Dry Goods in Plentywood, Montana originated the idea of a community-owned department store when the chain store left town. Residents bought 18 shares at $10,000 each to raise the needed funds to open the department store, serving locals who otherwise would have a long commute to purchase goods. In addition to fostering community ownership, crowdfunding for these types of efforts helps local economies and creates a sense of place. Another first in this ilk is that of the Green Bay Packers, the professional football team housed permanently in Green Bay, Wisconsin. Organized as a nonprofit in 1923, with shares sold to community members, this "community-owned" team definitely contributes to the area.

The Editors

PRIs include loans, loan guarantees, and equity investments. They may finance a variety of community economic development projects including neighborhood shopping centers, revitalization projects, small businesses, microenterprise funds, and nonprofit intermediaries. Like grants, PRIs have as their primary purpose the achievement of the foundation's mission and philosophy. Unlike grants, PRIs are considered financial instruments which should produce financial returns to the foundation.

Conclusion: Putting It All Together

Community and economic development projects require many partners and sources of funding, so the real issue is how to bring it all together. How should these projects be structured, what role should each partner play, and how should potential risks be mitigated to successfully complete these projects?

All of the players fulfill an important role in financing an economic development project, and each player must understand their respective role and the role of the other partners. However, identifying a point person/organization to manage the project is critical to its success. This person should act as liaison between all of the different entities involved and provide oversight and management. It is the role of this point person to coordinate all project activity and define the roles of each participant. In addition, there is typically a visionary leader of the project who develops the first concept and serves as the "cheerleader" and spokesperson for the project.

It is also critical to the success of the project that an evaluation of the financial strengths and capacity of all partners be undertaken early in the process. This is particularly important when evaluating nonprofits and community development corporations. Many of these organizations are dependent on grant funds from public and private funding sources for their long-term administrative and operating expenses and organizational stability. Additionally, the mission and goals of many nonprofits are built on the strengths of their leaders (executive director and board leadership), and evaluation of their management structure and succession plan may also be necessary.

Most community and economic development projects require multiple financing sources, commonly known as "layered financing," which may include both traditional and nontraditional loans, investments, and grants from all levels of government and the private sector. Most financing programs are designed to bridge the gap between what the developer needs to finance a project and what the private lender is willing to lend. Gap financing can also help mitigate risk and may give a private lender or investor more confidence in the deal. A case study illustrating the financing of a community economic development project (Tangerine Plaza) is included at the end of the chapter.

There are several challenges to successfully completing an economic development project, largely due to the inherent complexity and the number of different players typically involved. First, each partner has a different tolerance for risk. Government and nonprofit partners are typically more willing to take risks than the private, for-profit players such as financial institutions and investors. This risk may be mitigated through tax abatements, tax credits, loan guarantees, interest rate subsidies, and other credit enhancements.

Second, funding programs may have unique guidelines and underwriting requirements. Some programs require a higher income ratio to service the debt (debt service ratio), while others may focus on loan-to-value ratios or the strength of its guarantees. Matching the different sources of funding to ensure that all requirements are met can be challenging but is an essential step to determine the viability of a project.

Third, the partners may not be aware of the available resources that other partners can bring to the project. Because these projects typically involve multiple sources of funding, it is critical that the lead player in the project be knowledgeable of all available resources and communicate levels of commitment to all parties.

Fourth, getting all of the partners together at the right time can be challenging. Most sources of funding for community and economic development projects have different funding cycles with unique eligibility requirements (such as requiring site control and land use approval prior to applying for funding). Coordinating the timing of the funding and approval processes is critical to the success of the project.

Fifth, it can be difficult to allocate the transaction costs for the project between the players. Who does what and when needs to be agreed upon at the beginning of the process and monitored by the lead person or organization.

Sixth, each partner may be looking for a different return on the project. The public sector partners evaluate the return on economic development projects in terms of the impact on the community including the amount of new investment, the increase in the tax base, the number of new jobs created, and the overall improvements in the quality of life. The private sector partners will be focusing on the financial return from the project. Thus, it is important to consider the "double bottom line" for these deals, and to weigh both the financial and social return to the community associated with each project.

Finally, it is important that the use of public financing and subsidies is evaluated to insure that the incentives are used responsibly to meet the needs of both the project and the local community. The use of public subsidies is critical for many community economic development projects, but, because these resources are limited, it is important that they are used for projects that provide a measurable benefit to the public.

CASE STUDY: TANGERINE PLAZA

Note: Please refer to the list below for an explanation of the abbreviations used in the following case study.

AMI: Area Median Income
CDE: Community Development Entity
CDFI: Community Development Financial Institution
EDGE: Economic Development and Growth Enhancement Program from the Federal Home Loan Bank of Atlanta
FHLB Atlanta: Federal Home Loan Bank of Atlanta
NLPWF: Neighborhood Lending Partners of West Florida
NMTC: New Markets Tax Credits
OCS: Office of Community Services

Project History and Overview

Tangerine Plaza was the first new significant commercial development to be done in the Midtown area of St. Petersburg, Florida, in many years. The initial concept for the shopping center was developed in the year 2000 by two individuals who were committed to redeveloping this area of St. Petersburg. Midtown is a low-income (average income is 47% of Area Median Income, 33% are below poverty level), predominantly African-American community that has been designated by the local government as a Community Redevelopment Area.

Tangerine Plaza is a 47,000-square-foot neighborhood shopping center that is anchored by a Sweetbay Supermarket, the first full-service grocery store in the neighborhood. The remaining retail space at the center is occupied by smaller local retail tenants.

Complex community economic development projects like Tangerine Plaza often involve a variety of partners and multiple sources of financing. The lead developer for Tangerine Plaza was Urban Development Solutions, a nonprofit organization that had previously been involved with affordable housing projects in the Midtown area. Urban Development Solutions had purchased several parcels of land that were abandoned and condemned, with the goal of building more affordable housing. As they continued to assemble land, however, it became clear that the real need in the community was commercial development and, specifically, a full-service grocery store. To develop the retail center, Urban Development Solutions formed a for-profit limited liability company.

Urban Development Solutions approached the City of St. Petersburg with this concept, and the city agreed to assist in the assembly of the remainder of the 32 parcels needed for the development. As is typical of urban redevelopment, obstacles such as nonconforming lots, inappropriate zoning classifications, and numerous lien and title problems had to be addressed. The city agreed to buy the lots that had been purchased by Urban Development Solutions, as well as the remaining lots that were not already under the developer's control. The city then cleared all of the liens and other encumbrances against the properties, rezoned all of the property for neighborhood commercial development, and replatted the lots into one parcel. Once the land was ready for construction, the city agreed to a 99-year lease with Urban Development Solutions with an annual payment of $5. The entire land assembly process took two and a half years.

Project Financing

A complex financing package that involved numerous partners was required for the development of Tangerine Plaza. Construction financing for the project was provided by Neighborhood Lending Partners of West Florida (NLPWF), a statewide CDFI. As described earlier, the City of St. Petersburg provided the land (valued at $1,150,000) to be leased for 99 years at $5 per year. Additional funding for the project came from a $700,000 federal grant from the Office of Community Services and a $10,000 donation from the principals of the developer.

During the time it took to assemble the property, construction costs escalated, almost doubling from the original estimates. In order to cover the increasing construction costs, New Markets Tax Credit (NMTC) funding was secured to provide the gap financing. The New Markets Tax Credit syndicator was LISC (Local Initiatives Support Corporation), a national intermediary, and the two investors in the transaction were large financial institutions. Finally, the project was awarded an EDGE (Economic Development and Growth Enhancement Program) Loan from the Federal Home Loan Bank of Atlanta for its permanent first mortgage financing at a below market rate of 3.4%.

The New Markets Tax Credit allocation was a very important component of the project financing. Due to the higher construction costs, there was insufficient funding to pay the developer's fee, so the New Market Tax Credit proceeds helped with the construction financing and payment to the developer. Second, the developers of the center had a strong desire to lease space in the shopping center to local small business owners, so the New Market Tax Credit proceeds were also used to assist in the tenant build outs, equipment purchases, and rent abatement for the small businesses locating in the center. The complete financing package for the project is shown in Tables 24.1a and 24.1b.

Table 24.1a *Summary of Sources and Uses of Funds*

Line	*Sources*	*Construction*	*NMTC*	*Total*
A	NLPWF edge loan (From FHLB Atlanta)	$3,500,000	$300,000	$3,200,000
B	City of St. Petersburg	$1,336,500		$1,336,500
C	NLPWF CDFI funds	$661,500		$661,500
D	OCS grant	$700,000		$700,000
E	Sweetbay/Kash-N-Karry	$394,960		$394,960
F	Sweetbay/Kash-N-Karry	$374,252		$374,252
G	City of St. Petersburg	$75,000		$75,000
H	Private donation	$10,000		$10,000
I	NMTC equity		$2,784,149	$2,784,149
J	Total sources	$7,052,212	$2,484,149	$9,536,361

Table 24.1b *Summary of Sources and Uses of Funds*

Line	*Uses*	*Construction*	*NMTC*	*Total*
1	Land lease (prepaid in full)	$495		$495
2	Site development	$750,270		$750,270
3	Building shell/local tenant improvements	$4,498,758		$4,498,758
4	Anchor tenant improvements	$1,000,000		$1,000,000
5	Permits and government fees	$39,466		$39,466
6	Professional fees	$377,310		$377,310
7	Pre-development soft costs	$133,000		$133,000
8	Construction financing costs	$209,301		$209,401
9	Insurance	$26,353		$26,353
10	Developer overhead	$17,159	$116,195	$133,354
11	Working capital		$504,859	$504,859
12	Developer profit		$429,943	$429,943
13	Community outreach program		$200,000	$200,000
14	Tenant assistance		$500,000	$500,000
15	NMTC financing costs		$264,003	$264,003
16	NMTC organization and placement fees		$456,609	$456,609
17	CDE/investment fund escrows		$12,540	$12,540
18	Total uses	$7,052,212	$2,484,149	$9,536,361

Project Impact and Results

Construction of Tangerine Plaza was completed in November 2005, and since opening, the Midtown Sweetbay store has set the record for the highest increase in sales for all Sweetbay Supermarkets. The property tax revenue on this property has increased from $6,000 in the year 2000 to over $110,000 in 2006. Since this is a designated Community Redevelopment Area, a portion of the tax revenue may now be utilized to fund additional redevelopment projects in the surrounding neighborhood. Additionally, Sweetbay received state tax credits for job creation and for hiring neighborhood residents.

Perhaps the most important achievement of this project is the "social equity" generated in the community, and the track record needed to continue redevelopment of the Midtown area. The success of Tangerine Plaza is widely recognized and, in 2006, the project was awarded the Florida Redevelopment Association's Roy F. Kenzie Award for "outstanding new building project" and also honored as "Best of the Best," the top redevelopment project in the state.

Lessons Learned and Best Practices

There are a number of lessons to be learned from the development of Tangerine Plaza and suggested best practices for the financing of similar community economic development projects.

Lessons Learned

The Tangerine case study illustrates some of the typical challenges of redevelopment projects, particularly those that are located in a low-income or distressed areas.

1 It can be extremely difficult to assemble the numerous parcels of property required for the project if they are owned by different individuals or have title problems and liens.
2 To secure the financing needed to develop a commercial center it is often necessary to secure the commitment of a major retail anchor (in the Tangerine Plaza case study it was the grocery store chain). Finding the right anchor tenant that is willing to commit up front can be a time-consuming process.
3 If the project is located in a depressed real estate market area, it will likely yield below market rental rates. Since rental rates are used to determine property values for purposes of obtaining construction and/or permanent financing, it can be difficult to obtain sufficient financing to develop projects in depressed real estate markets.

4 The complexity and length of time required to put together a deal can lead to significant increases in construction costs, which may require identifying additional sources of gap financing to cover the increased cost of the project.
5 It can be difficult to get the timing of the financing package aligned. In the case of Tangerine Plaza, the commitment for the permanent financing was not completed when the construction loan closed, as some lenders would have required.
6 There may be issues among the financial partners over control of the collateral. In the case of Tangerine Plaza, the city and the lenders had issues over control of the leased land.

Best Practices

This case study also highlights several examples of best practices for financing community economic development projects:

1 A nonprofit can be a key player to involve in a project because they can help with community creditability, and they often have better access to gap financing in the form of low-interest loans and grants.
2 If a nonprofit organization is involved, it must have the capacity and sophistication to take on the project or the financial ability to obtain this capacity through additional staff or outside consultants.
3 All players involved in the process must be focused, consistent, and patient in working through the problems, which may take an extended period of time.
4 Finally, a strong and committed local government that has the ability to attract funding and is willing to do what it takes is critical to turning a project idea into a reality.

Keywords

Community development, economic development, finance, Community Development Block Grant programs, new market tax credits, tax increment financing, revolving loan funds.

Review Questions

1 Who are some of the typical "players" in a community finance deal?
2 What are some of the major sources of Federal Government money for community development?
3 What are some state and local resources for community development financing?
4 What are some challenges often faced when putting a community development finance deal together?
5 Identify a few alternative financing sources for community projects.

Note

1 Impact fees are typically one-time charges imposed against new development to pay for the cost of development. Impact fees may be used to cover the costs of new roads, infrastructure, schools, or public services.

Bibliography

Federal Home Loan Banks. Online. Available at: <www.fhlbanks.com/>
Federal Reserve Board of Governors. Online. Available at: <www.federalreserve.gov>
International Economic Development Council. Online. Available at: <www.iedconline.org>
National Park Service Historic Preservation Tax Incentives. Online. Available at: <www.cr.nps.gov/tax.htm>
National Trust for Historic Preservation Main Street Program. Online. Available at: <www.mainstreet.org/>
Office of the Comptroller of the Currency. Online. Available at: <www.occ.treas.gov>

Opportunity Finance Network. Online. Available at: <www.opportunityfinance.org>

Seidman, K.F. (2005). *Economic Development Finance*. Thousand Oaks, California: Sage.

U.S. Department of Agriculture Rural Development. Online. Available at: <www.rurdev.usda.gov/>

U.S. Department of Commerce Economic Development Administration. Online. Available at: <www.eda.gov>

U.S. Department of Housing and Urban Development. Online. Available at: <www.hud.gov/economicdevelopment/>

U.S. Department of the Treasury CDFI and New Markets Tax Credit. Online. Available at: <www.cdfifund.gov/>

U.S. Environmental Protection Agency-Brownfields Program. Online. Available at: <www.epa.gov/brownfields>

U.S. Small Business Administration. Online. Available at: <http: //sba.gov/>

Connections

Family Income and Wealth Building program, Local Initiatives Support Corporation (LISC), www.lisc.org/section/ourwork/national/family

For additional information on grants and foundations, see The Foundation Center's site at: http: //foundationcenter.org

One source of training programs for economic and community development professionals is the National Development Council, a nonprofit organization, that offers certification (the Economic Development Finance Professional (EDFP). See www.nationaldevelopmentcouncil.org

25 Conclusions and Observations on the Future of Community Development

Rhonda Phillips and Robert H. Pittman

Overview

As indicated by the variety of material covered in this book, community development is a broad discipline, and it has grown considerably from its more narrowly focused genesis. Today's pressing local and global challenges call for the application of community development principles in even more places and ways, and the discipline must continue to evolve and draw from many different fields of study.

Introduction

As stated at the very beginning of this book, community development is a boundary-spanning field of study addressing not only the physical realm of community, but also the social, cultural, economic, political, and environmental aspects as well. Because of the interconnection and complexity of these dimensions of community, the field is constantly evolving to face new challenges to maintaining and improving quality of life.

We have taken quite a journey in this book. We started with the foundations of community development, studying frameworks and theoretical constructs, and then core principles such as asset-based community development, social capital, and capacity building. We moved on to preparation and planning, focusing on community visioning, community-based organizations, leadership skills and community assessments. Next we considered some key programming techniques and strategies such as community marketing, business development, cultural-based development, neighborhood renewal, and progress measurement. Finally, we focused on several topical issues including community development finance, and perspectives on a variety of current issues including clean energy. Along the way we learned that community and economic development are inextricably linked—success in one leads to success in the other. And, we learned that community and economic development are both processes and outcomes.

As the wide and varied territory covered by this book attests, community development is a broad field of study and practice. It has evolved from its roots in social advocacy and justice, neighborhood renewal, housing and related topics to a multidisciplinary field with applications in all sectors of society, all across the globe. Today, community development is being used to gain new insights into some of the most important contemporary issues such as climate change, global migration, and international diplomacy. As awareness and knowledge of the discipline of community development grows, so will its applications.

Continuing and Future Challenges

It has often been said that the only constant is change. While this may be trite, it remains true. We are witnessing social, economic, technological, and political change that is arguably of historic proportions. In this ever-changing environment, communities must literally reinvent themselves to survive and prosper. There is no better tool to help meet this challenge than community development. Communities and regions need to embrace the process of community development to adapt to rapidly changing environments. To add to the challenge, community and economic transformation must be undertaken with sustainability in mind in light of increasing demands for limited resources. It will also require communities to build more resilience for long-term durability. Resilience literally means to spring back—to recover and adapt. The Stockholm Resilience Center provides the following view: "resilience is the capacity of a system to continually change and adapt yet remain within critical thresholds" (Phillips et al. 2013: 1). Building resilience in a community's local economy, environment, and society "is a dynamic process, and requires continuing effort, as well as adaptation to changing conditions" (Phillips et al. 2013: 2). As mentioned previously, community development started out as a way to address a more narrowly defined set of issues, both socially and geographically. It is apparent now that the stage on of community development is much broader in scope.

Ongoing and future challenges will likely be even more intense as climate, economic, and social/political change impacts migration, health and well-being, and ability of systems to thrive, among other dimensions impacting community well-being. Let's take a moment to look at these three areas, as they are what some describe as a "perfect storm" threatening many of our communities. Within the climate and environmental fronts, there are myriad issues facing local to global communities. One of these is preparation for and response to disasters (both human and natural in origin). This is a rapidly growing area of interest, particularly as urban populations increase rapidly along with potential risks to larger numbers of people. Predictions are that incredibly large percentages of global population will be impacted by disaster, with the need to migrate to safer areas. More interest is emerging in community-based and -oriented approaches to disaster planning and recovery as opposed to typical response and control type responses (LaLone 2012). A major way to incorporate community-based approaches is by including resident or citizen participation in local and regional disaster planning efforts. It is believed that by incorporating citizen participation approaches greater political stability and citizen satisfaction can be fostered (Phillips 2014). Does this sound familiar? The basic foundation of community development is that having strong citizen participation (which also is an instrumental component of social capital and social community capacity building) makes a measurable difference in terms of process and outcomes. In this case, it makes a difference in impact and recovery efforts.

The economic challenges are large, with variations in income levels between the bottom and top earners the largest ever in recent history. This in turn widens the gap with a shrinking middle class in many areas, increasing poverty, and suffering. Poverty is a major factor aligned with suffering, as in daily quality of life (and amplified when disaster strikes) (Phillips 2014). The factor of suffering has long been witnessed throughout history—humans against humans with violence and discrimination based on disability, or low social or economic status, to name a few (Anderson 2012). Further, racism as a form of suffering (both at the individual and community levels) impacts quality of life, and in some societies, it "extends to apartheid, genocide, or other institutionalized forms of oppression and elimination of one group by and for the privilege of another" (Karraker, 2014). With the highest rate of incarceration of minority residents in the world, U.S. communities have much to question about racism, suffering, and poverty factors. Here again is the need for community because surviving as a collective (a community) is a story replayed through time, with the need for strong community more important than ever.

A social issue that demands attention is that of health in many communities, with poor health at unprecedented levels including a majority in many populations being obese or diabetic. In the US, as many as one in three children are overweight, and in many underrepresented populations, the intensity of these epidemics is amplified. Antidepressants are the second most prescribed medication in the US, while the percentage of children receiving them has soared; at the same time, three-quarters of high school students cannot pass basic fitness exams (Jackson 2012). The rate at which these epidemics has increased is truly astounding (see the Center for Disease Control data animated map link at the end of this chapter) and is speculated to be a result of many factors. The bottom line is that communities must deal with declining health of their residents. The impacts are severe and the need to address issues of community walkability and design features to encourage physical activity, access to health care, and access to quality food sources are just a few of the domains in this arena.

It is important to remember that community development is not only about responding to problems, it is also about the beauty of people coming together to improve their lot in life. (And it is equally important to remain hopeful!) This emphasis on quality of life is pervasive, and can lead to empowering positive directions and outcomes in our communities. The ability of people working together and moving toward common goals translates into the power to change for the better. The idea of caring in communities is reemerging: "Our caring—for ourselves and our community and for future generations—must shape what we build" (Jackson 2012: xii). In other words, our communities' built environments can help strengthen a sense of connection, leading to better quality of life across a spectrum of outcomes. With effective and equitable designs, for example, that help us care for ourselves physically, we can in turn create community environments that support well-being and sense of place with excellent recreational, leisure, and social interaction spaces. Caring capital is a concept that holds much relevance for community development:

> Just as the essence of social capital is valued, networks of interpersonal and institutional relationships, caring capital should be defined as those networked social relationships that consist of actions intended to improve the welfare of the other(s). Actions that are seen as based only on expectations of present rewards or future reciprocation would not contribute to caring capital. Caring capital typically involves diverse types of giving of care not dependent upon formal exchanges of goods or services. For this reason, caring capital tends to be described in words like compassionate, caregiving, generous, kind, altruistic, charitable, and humanitarian.
>
> (Anderson 2012: 1)

The idea that community development processes and outcomes emanate from a basis of caring is not too far afield. It is a way of thinking about community development that can help improve conditions and quality of life. A culture of caring could go far in fostering positive change in response to these and other challenges facing our communities currently and in the future.

The Ongoing Evolution of Community Development

To embrace its destiny and achieve its potential, community development must continue to evolve and improve both as a theoretical and applied discipline. As we have seen, community development is inherently a multidisciplinary field since it covers virtually all aspects of community. While community development has benefited greatly from its close association with sociology, there are many other disciplines that can continue to make important contributions to the field including urban and regional planning, political science, economics, business (marketing, organizational behavior, finance, etc.), and psychology. Community development offers a unique laboratory in which researchers from various backgrounds can collaborate to create new theories and tools based on multiple disciplines. It is hoped that scholars from these different fields will embrace community development not only as a promising area of research, but also as a way to help improve society.

As we have seen, many benefits from community development approaches and applications can be gained. The study and practice of community development is not just for students and professional community developers; it is also for elected officials, board members, community volunteers and residents from all walks of life who want to make their communities better places to live, work, and play. This is the broad audience that needs to learn about community development. This book was undertaken with the intention of helping this entire constituency understand and embrace the theory and practice of community development. It is the profound hope of the editors and chapter authors that we have made some small progress toward this goal.

Keywords

Change, process, outcomes, global, caring capital, recovery, and resilience.

Review Questions

1. What are some of the key areas covered by the field of community development?
2. How/why is it applicable to a wide variety of global issues today?
3. How do you think the field of community development will evolve in the future?
4. Can you identify at least three pressing issues facing your own community that could be addressed by an approach, technique, or application discussed in this book?
5. In your opinion, what is the major issue facing communities today that will influence community development outcomes?

Bibliography

Anderson, R.E. (2012). Caring Capital Websites. *Information, Communication & Society*, 15(4): 479–501. DOI: 10.1080/1369118X.2012.667817

Anderson, R.L. (2014). *World Suffering and Quality of Life*. Dordrecht: Springer.

Jackson, R.J. (2012). *Designing Healthy Communities*. San Francisco, California: Jossey-Bass.

Karraker, M.W. (2014, in press). Community action to alleviate suffering from racism: the role of religion and caring capital in small city USA. In R. Anderson, *World Suffering and Quality of Life*, Dordrecht, Netherlands: Springer.

LaLone, M.B. (2012). Neighbors helping neighbors: An examination of the social capital mobilization process for community resilience to environmental disasters. *Journal of Applied Social Science*, *6*(2), 209–37. DOI: 10.1177/1936724412458483

Phillips, R. (2014). Community Quality of Life Indicators to Avoid Tragedies. In R. Anderson, *World Suffering and Quality of Life*. Dordrecht: Springer.

Phillips, R., Seifer, B., & Antczak, E. (2013). *Sustainable Communities: Creating a Durable Local Economy*. London: Routledge.

Stockholm Resilience Center. (2013). Accessed December 14, 2013, from www.stockholmresilience.org/research/whatisresilience

Connections

Take a look at the rate of obesity increases over a 25-year period in the US. Data provided by the Center for Disease Control shows epidemic proportions.

www.theatlantic.com/health/archive/2013/04/look-how-quickly-the-us-got-fat-1985–2010-animated-map/274878

Find out about the status of children's well-being by exploring KIDS COUNT®, a project of the Annie E. Casey Foundation. This project tracks well-being of children in the US, state by state. www.aecf.org/MajorInitiatives/KIDSCOUNT.aspx

Index

Figures and Tables are indicated by *italic page numbers*; Boxes and Case Studies by **emboldened numbers**; Notes by suffix "n" (e.g. "311n1" means page 311, note 1)